OPTICAL FIBER COMMUNICATIONS

Principles and Practice

John M. Senior

Dept. of Electrical and Electronic Engineering
Manchester Polytechnic

Prentice/Hall International

Englewood Cliffs, NJ   London   New Delhi   Rio de Janeiro
Singapore   Sydney   Tokyo   Toronto   Wellington

Library of Congress Cataloging in Publication Data

Senior, John, M., 1951–
  Optical fiber communications.

  Bibliography: p.
  Includes index.
  1. Optical communications. 2. Fiber optics. I. Title.
  TK5103.59.S46   1984   621.38'0414   84–8315
  ISBN 0–13–638248–7 (case)
  ISBN 0–13–638222–3 (pbk.)

British Library Cataloging in Publication Data

Senior, John M.
  Optical fiber communications.
  1. Optical communications   2.   Fiber optics
  I. Title
  621.38'0414   TK5103.59

ISBN 0–13–638248–7
ISBN 0–13–638222–3 Pbk

ISBN 0-13-638248 7
ISBN 0-13-638222 3 {PBK}

Prentice-Hall International, Inc., *London*
Prentice-Hall of Australia Pty. Ltd., *Sydney*
Prentice-Hall Canada, Inc., *Toronto*
Prentice-Hall of India Private Ltd., *New Delhi*
Prentice-Hall of Southeast Asia Pte., Ltd., *Singapore*
Prentice-Hall Inc., *Englewood Cliffs, New Jersey*
Prentice-Hall do Brasil Ltda., *Rio de Janeiro*
Whitehall Books Ltd., *Wellington, New Zealand*

10 9 8 7 6 5

Printed in Great Britain by A. Wheaton and Co. Ltd., Exeter

Contents

Preface

The concept of guided lightwave communication along optical fibers has stimulated a major new technology which has come to maturity over the last fifteen years. During this period tremendous advances have been achieved with optical fibers and components as well as with the associated optoelectronics. As a result this new technology has now reached the threshold of large scale commercial exploitation. Installation of optical fiber communication systems is progressing within both national telecommunication networks and more localized data communication and telemetry environments. Furthermore, optical fiber communication has become synonymous with the current worldwide revolution in information technology. The relentless onslaught will undoubtedly continue over the next decade and the further predicted developments will ensure even wider application of optical fiber communication technology in this 'information age'.

The practical realization of wide-scale optical fiber communications requires suitable education and training for engineers and scientists within the technology. In this context the book has been developed from both teaching the subject to final year undergraduates and from a successful series of short courses on optical fiber communications conducted for professional engineers at Manchester Polytechnic. This book has therefore been written as a comprehensive introductory textbook for use by undergraduate and postgraduate engineers and scientists to provide them with a firm grounding in the major aspects of this new technology whilst giving an insight into the possible future developments within the field. The reader should therefore be in a position to appreciate developments as they occur. With these aims in mind the book has been produced in the form of a teaching text enabling the reader to progress onto the growing number of specialist texts concerned with optical fiber waveguides, optoelectronics, integrated optics, etc.

In keeping with the status of an introductory text the fundamentals are included where necessary and there has been no attempt to cover the entire field in full mathematical rigor. However, selected proofs are developed in important areas throughout the text. It is assumed that the reader is conversant with differential and integral calculus and differential equations. In addition, the reader will find it useful to have a grounding in optics as well as a reasonable familiarity with the fundamentals of solid state physics.

Chapter 1 gives a short introduction to optical fiber communications by considering the historical development, the general system and the major advantages provided by this new technology. In Chapter 2 the concept of the optical fiber as a transmission medium is introduced using a simple ray theory approach. This is followed by discussion of electromagnetic wave theory applied to optical fibers prior to consideration of lightwave transmission within the various fiber types. The major transmission characteristics of optical fibers are then discussed in some detail in Chapter 3.

Chapters 4 and 5 deal with the more practical aspects of optical fiber communications and therefore could be omitted from an initial teaching program. In Chapter 4 the manufacture and cabling of the various fiber types are described, together with fiber to fiber connection or jointing. Chapter 5 gives a general treatment of the major measurements which may be undertaken on optical fibers in both the laboratory and the field. This chapter is intended to provide sufficient background for the reader to pursue useful laboratory work with optical fibers.

Chapters 6 and 7 discuss the light sources employed in optical fiber communications. In Chapter 6 the fundamental physical principles of photoemission and laser action are covered prior to consideration of the various types of semiconductor and nonsemiconductor laser currently in use, or under investigation, for optical fiber communications. The other important semiconductor optical source, namely the light emitting diode, is dealt with in Chapter 7.

The next two chapters are devoted to the detection of the optical signal and the amplification of the electrical signal obtained. Chapter 8 discusses the basic principles of optical detection in semiconductors; this is followed by a description of the various types of photodetector currently utilized. The optical fiber receiver is considered in Chapter 9 with particular emphasis on its performance in noise.

Chapter 10 draws together the preceding material in a detailed discussion of optical fiber communication systems, aiming to provide an insight into the design criteria and practices for all the main aspects of both digital and analog fiber systems. A brief account of coherent optical fiber systems is also included to give an appreciation of this area of future development. Finally, Chapter 11 describes the many current and predicted application areas for optical fiber communications by drawing on examples from research and development work which has already been undertaken. This discussion is expanded into consideration of other likely future developments with a brief account of the current technology involved in integrated optics and optoelectronic integration.

Worked examples are interspersed throughout the text to assist the learning process by illustrating the use of equations and by providing realistic values for the various parameters encountered. In addition, problems have been provided at the end of relevant chapters (Chapters 2 to 10 inclusive) to examine the reader's understanding of the text and to assist tutorial work. A Teacher's Manual containing the solutions to these problems may be obtained from the publisher. Extensive end-of-chapter references provide a guide for further reading and indicate a source for those equations which have been quoted without derivation. A complete glossary of symbols, together with a list of common abbreviations employed in the text, is provided. SI units are used throughout the text.

I am very grateful for the many useful comments and suggestions provided by reviewers which have resulted in significant improvements to this text. Thanks must also be given to the authors of numerous papers, articles and books which I have referenced whilst preparing the text, and especially to those authors, publishers and companies who have kindly granted permission for the reproduction of diagrams and photographs. Further, I would like to thank my colleagues in the Dept. of Electrical and Electronic Engineering at Manchester Polytechnic for their many helpful comments on the text; in particular Dr. Norman Burrow, Dr. John Edwards and Stewart Cusworth for the time spent checking the manuscript. I am also grateful to my family and friends for tolerating my infrequent appearances over the period of the writing of this book. Finally, words cannot express my thanks to my wife, Marion, for her patience and encouragement with this project and for her skilful typing of the manuscript.

J. M. Senior
Manchester Polytechnic

Glossary of Symbols and Abbreviations

A constant, area (cross-section, emission), far field pattern size, mode amplitude, wave amplitude (A_o)

A_{21} Einstein coefficient of spontaneous emission

A_c peak amplitude of the subcarrier waveform (analog transmission)

a fiber core radius, parameter defining the asymmetry of a planar guide (eqn. 11.6), baseband message signal ($a(t)$)

a_k integer 1 or 0

B constant, electrical bandwidth (post detection), magnetic flux density, mode amplitude, wave amplitude (B_0)

B_{12}, B_{21} Einstein coefficients of absorption, stimulated emission

B_F modal birefringence

B_m bandwidth of an intensity modulated optical signal $m(t)$

B_{opt} optical bandwidth

B_r recombination coefficient for electrons and holes

B_T bit rate, when the system becomes dispersion limited (B_T (DL))

b normalized propagation constant for a fiber, ratio of luminance to composite video

C constant, capacitance, crack depth (fiber), wave coupling coefficient per unit length

C_a effective input capacitance of an optical fiber receiver amplifier

C_d optical detector capacitance

C_f capacitance associated with the feedback resistor of a transimpedance optical fiber receiver amplifier

C_L total optical fiber channel loss in decibels, including the dispersion–equalization penalty (C_{LD})

C_0 wave amplitude

C_T total capacitance

c velocity of light in a vacuum, constant (c_1, c_2)

c_i tap coefficients for a transversal equalizer

D amplitude coefficient, electric flux density, distance, corrugation period, decision threshold in digital optical fiber transmission

D_f frequency deviation ratio (subcarrier FM)

D_L dispersion–equalization penalty in decibels

D_p frequency deviation ratio (subcarrier PM)

d fiber core diameter, distance, width of the absorption region (photodetector), pin diameter (mode scrambler)

d_o fiber outer (cladding) diameter

E electric field, energy, Youngs modulus, expected value of a random variable

E_a	activation energy of homogeneous degradation for an LED
E_F	Fermi level (energy), quasi-Fermi level located in the conduction band (E_{Fc}), valence band (E_{Fv}) of a semiconductor
E_g	separation energy between the valence and conduction bands in a semiconductor (bandgap energy)
$E_m(t)$	subcarrier electric field (analog transmission)
E_o	optical energy
E_q	separation energy of the quasi-Fermi levels
e	electronic charge, base for natural logarithms
F	probability of failure, transmission factor of a semiconductor–external interface, excess avalanche noise factor $(F(M))$
$\mathcal{F}$	Fourier transformation
F_n	noise figure (amplifier)
f	frequency
f_D	peak to peak frequency deviation (PFM–IM)
f_d	peak frequency deviation (subcarrier FM and PM)
f_o	pulse rate (PFM–IM)
G	open loop gain of an optical fiber receiver amplifier
$G_i(r)$	amplitude function in the WKB method
G_o	optical gain (phototransistor)
Gsn	Gaussian (distribution)
g	degeneracy parameter
$\bar{g}$	gain coefficient per unit length (laser cavity)
g_m	transconductance of a field effect transistor
$\bar{g}_{th}$	threshold gain per unit length (laser cavity)
H	magnetic field
$H(\omega)$	optical power transfer function (fiber), circuit transfer function
$H_A(\omega)$	optical fiber receiver amplifier frequency response (including any equalization)
$H_{CL}(\omega)$	closed loop current to voltage transfer function (receiver amplifier)
$H_{eq}(\omega)$	equalizer transfer function (frequency response)
$H_{OL}(\omega)$	open loop current to voltage transfer function (receiver amplifier)
$H_{out}(\omega)$	output pulse spectrum from an optical fiber receiver
h	Planck's constant, thickness of a planar waveguide, power impulse response for an optical fiber $(h(t))$
$h_A(t)$	optical fiber receiver amplifier impulse response (including any equalization)
h_{eff}	effective thickness of a planar waveguide
h_{FE}	common emitter current gain for a bipolar transistor
$h_f(t)$	optical fiber impulse response
$h_{out}(t)$	output pulse shape from an optical fiber receiver
$h_p(t)$	input pulse shape to an optical fiber receiver
$h_t(t)$	transmitted pulse shape on an optical fiber link
I	electrical current, optical intensity
I_b	background radiation induced photocurrent (optical receiver)
I_{bias}	bias current for an optical detector
I_c	collector current (phototransistor)
I_d	dark current (optical detector)
I_o	maximum optical intensity
I_p	photocurrent generated in an optical detector
I_{th}	threshold current (injection laser)
i	electrical current

i_a	optical receiver preamplifier shunt noise current
i_{amp}	optical receiver preamplifier total noise current
i_D	decision threshold current (digital transmission)
i_d	photodiode dark noise current
i_{det}	output current from an optical detector
i_f	noise current generated in the feedback resistor of an optical fiber receiver transimpedance preamplifier
i_N	total noise current at a digital optical fiber receiver
i_n	multiplied shot noise current at the output of an APD excluding dark noise current
i_s	shot noise current on the photocurrent for a photodiode
i_{SA}	multiplied shot noise current at the output of an APD including the dark noise current
i_{sig}	signal current obtained in an optical fiber receiver
i_t	thermal noise current generated in a resistor
i_{TS}	total shot noise current for a photodiode without internal gain
J	Bessel function, current density
J_{th}	threshold current density (injection laser)
j	$\sqrt{-1}$
K	Boltzmann's constant, constant dependent on the optical fiber properties, modified Bessel function
K_1	stress intensity factor, for an elliptical crack (K_{IC})
k	wave propagation constant in a vacuum (free space wave number), wave vector for an electron in a crystal, ratio of ionization rates for holes and electrons, integer
k_f	angular frequency deviation (subcarrier FM)
k_p	phase deviation constant (subcarrier PM)
L	length (fiber), distance between mirrors (laser)
L_B	beat length in a single mode optical fiber
L_{bc}	coherence length in a single mode optical fiber
L_c	characteristic length (fiber)
L_o	constant with dimensions of length
L_t	lateral misalignment loss at an optical fiber joint
$\mathcal{L}$	transmission loss factor (transmissivity) of an optical fiber
l	azimuthal mode number, distance, length
l_a	atomic spacing (bond distance)
l_o	wave coupling length
M	avalanche multiplication factor, material dispersion parameter, total number of guided modes or mode volume; for a multimode step index fiber (M_s); for multimode graded index fiber (M_g), mean value (M_1) and mean square value (M_2) of a random variable
M_a	safety margin in an optical power budget
M_{op}	optimum avalanche multiplication factor
M^x	excess avalanche noise factor, (also denoted as $F(M)$)
m	radial mode number, Weibull distribution parameter, intensity modulated optical signal ($m(t)$), mean value of a random variable, integer
m_a	modulation index
N	integer, density of atoms in a particular energy level (e.g. N_1, N_2, N_3), minority carrier concentration in n type semiconductor material, group index of an optical waveguide (N_1)
NA	numerical aperture of an optical fiber
NEP	noise equivalent power

N_o	defined by equation 10.80
n	refractive index (e.g. n_1, n_2, n_3), stress corrosion susceptibility, negative type semiconductor material
n_e	effective refractive index of a planar waveguide
n_o	refractive index of air
P	electrical power, minority carrier concentration in p type semiconductor material, probability, of error ($P(e)$), of detecting a zero level ($P(0)$), of detecting a one level ($P(1)$), of detecting z photons in a particular time period ($P(z)$), conditional probability, of detecting a zero when a one is transmitted ($P(0/1)$), of detecting a one when a zero is transmitted ($P(1/0)$)
P_a	total power in a baseband message signal $a(t)$
P_B	threshold optical power for Brillouin scattering
P_c	optical power coupled into a step index fiber
P_D	optical power density
P_{dc}	d.c. optical output power
P_e	optical power emitted from an optical source
P_G	optical power in a guided mode
P_i	mean input (transmitted) optical power launched into a fiber
P_{int}	internally generated optical power (optical source)
P_m	total power in an intensity modulated optical signal $m(t)$
P_o	mean output (received) optical power from a fiber
P_{opt}	mean optical power travelling in a fiber
P_{out}	initial output optical power (prior to degradation) from an optical source
P_{po}	peak received optical power
P_r	reference optical power level
P_R	threshold optical power for Raman scattering
$P_{Ra}(t)$	backscattered optical power (Rayleigh) within a fiber
P_{sc}	optical power scattered from a fiber
$P_i(\omega)$	frequency spectrum of the mean input optical power launched into a fiber
$P_o(\omega)$	frequency spectrum of the mean output optical power received from a fiber
p	crystal momentum, average photoelastic coefficient, positive type semiconductor material, probability density function ($p(x)$)
q	integer, fringe shift
R	photodiode responsivity, radius of curvature of a fiber bend, electrical resistance (e.g. R_{in}, R_{out})
R_{12}	upward transition rate for electrons from energy level 1 to level 2
R_{21}	downward transition rate for electrons from energy level 2 to level 1
R_a	effective input resistance of an optical fiber receiver preamplifier
R_b	bias resistance, for optical fiber receiver preamplifier (R_{ba})
R_c	critical radius of an optical fiber
R_D	radiance of an optical source
RE_{dB}	ratio of electrical input power in decibels for an optical fiber system
R_f	feedback resistance in an optical fiber receiver transimpedance pre-amplifier
R_L	load resistance associated with an optical fiber detector
RO_{dB}	ratio of optical output power to optical input power in decibels for an optical fiber system
R_{TL}	total load resistance within an optical fiber receiver
r	radial distance from the fiber axis, Fresnel reflection coefficient, mirror reflectivity, electro-optic coefficient

r_e	generated electron rate in an optical detector
r_{ER}, r_{ET}	reflection and transmission coefficients respectively for the electric field at a planar, guide-cladding interface
r_{HR}, r_{HT}	reflection and transmission coefficients respectively for the magnetic field at a planar guide-cladding interface
r_p	incident photon rate at an optical detector
S	fraction of captured optical power, macroscopic stress
S_f	fracture stress
$S_i(r)$	phase function in the WKB method
$S_m(\omega)$	spectral density of the intensity modulated optical signal $m(t)$
S/N	peak signal power to rms noise power ratio, with peak to peak signal power $[(S/N)_{p-p}]$, with rms signal power $[(S/N)_{rms}]$
S_t	theoretical cohesive strength
s	pin spacing (mode scrambler)
T	temperature, time
T_a	insertion loss resulting from an angular offset between jointed optical fibers
T_c	10–90% rise time arising from intramodal dispersion on an optical fiber link
T_D	10–90% rise time for an optical detector
T_F	fictive temperature
T_l	insertion loss resulting from a lateral offset between jointed optical fibers
T_n	10–90% rise time arising from intermodal dispersion on an optical fiber link
T_o	threshold temperature (injection laser), nominal pulse period (PFM–IM)
T_R	10–90% rise time at the regenerator circuit input (PFM–IM)
T_S	10–90% rise time for an optical source
T_{syst}	total 10–90% rise time for an optical fiber system
T_T	total insertion loss at an optical fiber joint
T_t	temperature rise at time t
T_∞	maximum temperature rise
t	time
t_c	time constant
t_d	switch on delay (laser)
t_e	1/e pulse width from the center
t_r	10–90% rise time
U	eigenvalue of the fiber core
V	electrical voltage, nomalized frequency for an optical fiber or planar waveguide
V_{bias}	bias voltage for a photodiode
V_c	cutoff value of normalized frequency
V_{CC}	collector supply voltage
V_{CE}	collector–emitter voltage (bipolar transistor)
V_{EE}	emitter supply voltage
V_{opt}	voltage reading corresponding to the total optical power in a fiber
V_{sc}	voltage reading corresponding to the scattered optical power in a fiber
v	electrical voltage
v_a	amplifier series noise voltage
$v_A(t)$	receiver amplifier output voltage
v_c	crack velocity
v_g	group velocity
$v_{out}(t)$	output voltage from an RC filter circuit

v_p	phase velocity
W	eigenvalue of the fiber cladding, random variable
W_e	electric pulse width
W_o	optical pulse width
X	random variable
x	coordinate, distance, constant, evanescent field penetration depth, slab thickness
Y	constant, shunt admittance, random variable
y	coordinate, lateral offset at a fiber joint
Z	random variable
Z_0	electrical impedance
z	coordinate, number of photons
z_m	average or mean number of photons arriving at a detector in a time period τ
z_{md}	average number of photons detected in a time period τ
α	characteristic refractive index profile for fiber (profile parameter), optimum profile parameter (α_{op})
$\bar{\alpha}$	loss coefficient per unit length (laser cavity)
α_{cr}	connector loss at transmitter and receiver in decibels
α_{dB}	signal attenuation in decibels per unit length
α_{fc}	fiber cable loss in decibels per kilometer
α_j	fiber joint loss in decibels per kilometer
α_N	signal attenuation in nepers
α_o	absorption coefficient
α_r	radiation attenuation coefficient
β	wave propagation constant
$\bar{\beta}$	gain factor (injection laser cavity)
β_c	isothermal compressibility
β_o	proportionality constant
β_r	degradation rate
γ	angle, attenuation coefficient per unit length for a fiber
γ_p	surface energy of a material
γ_R	Rayleigh scattering coefficient for a fiber
Δ	relative refractive index difference between the fiber core and cladding
δ_E	phase shift associated with transverse electric waves
δ_f	uncorrected source frequency width
δ_H	phase shift associated with transverse magnetic waves
$\delta\lambda$	optical source spectral width (linewidth)
δT	intermodal dispersion time in an optical fiber
δT_g	delay difference between an extreme meridional ray and an axial ray for a graded index fiber
δT_s	delay difference between an extreme meridional ray and an axial ray for a step index fiber, with mode coupling (δT_{sc})
ε	electrical permittivity, of free space (ε_o), relative (ε_r)
ζ	solid acceptance angle
η	quantum efficiency (optical detector)
η_{ang}	angular coupling efficiency (fiber joint)
η_c	coupling efficiency (optical source to fiber)
η_D	differential external quantum efficiency (optical source)
η_{ep}	external power efficiency (optical source)
η_i	internal quantum efficiency (optical source)
η_{lat}	lateral coupling efficiency (fiber joint)
η_{pc}	overall power conversion efficiency (optical source)
η_T	total external quantum efficiency (optical source)

θ	angle, fiber acceptance angle (θ_a), Bragg diffraction angle (θ_B)
Λ	acoustic wavelength, period for perturbations in a fiber
Λ_c	cutoff period for perturbations in a fiber
λ	optical wavelength
λ_c	long wavelength cutoff (photodiode)
λ_o	wavelength at which first order dispersion is zero
μ	magnetic permeability, relative permeability (μ_r), permeability of free space (μ_o)
ν	optical source bandwidth in gigahertz
ρ	polarization rotation in a single mode optical fiber
ρ_f	spectral density of the radiation energy at a transition frequency f
σ	standard deviation, (rms pulse width), variance (σ^2)
σ_c	rms pulse broadening resulting from intramodal dispersion in a fiber
σ_m	rms pulse broadening resulting from material dispersion in a fiber
σ_n	rms pulse broadening resulting from intermodal disperion, in a graded index fiber (σ_g), in a step index fiber (σ_s)
σ_T	total rms pulse broadening in a fiber or fiber link
σ_λ	rms spectral width of emission from optical source
τ	time period, bit period, pulse duration, 3 dB pulse width ($\tau(3\ \text{dB})$)
τ_{21}	spontaneous transition lifetime between energy levels 2 and 1
τ_E	time delay in a transversal equalizer
τ_e	$1/e$ full width pulse broadening due to dispersion on an optical fiber link
τ_g	group delay
τ_i	injected (minority) carrier lifetime
τ_r	radiative minority carrier lifetime
Φ	linear retardation
ϕ	angle, critical angle (ϕ_c)
ψ	scalar quantity representing E or H field
ω	angular frequency, of the subcarrier waveform in analog transmission (ω_c), of the modulating signal in analog transmission (ω_m)
ω_o	spot size of the fundamental mode
∇	vector operator, Laplacian operator (∇^2)

A–D	analog to digital	CMOS	complementary metal oxide silicon
a.c.	alternating current		
AGC	automatic gain control	CNR	carrier to noise ratio
		CPU	central processing unit
AM	amplitude modulation	CSP	channelled substrate planar (injection laser)
APD	avalanche photodiode		
ASK	amplitude shift keying	CW	continuous wave or operation
BER	bit error rate		
BH	buried heterostructure (injection laser)	D–A	digital to analog
		dB	decibel
BOD	bistable optical device	D–IM	direct intensity modulation
CAM	computer aided manufacture		
		DFB	distributed feedback (injection laser)
CATV	common antenna television		
		DBR	distributed Bragg reflector (injection laser)
CCTV	close circuit television		
CDH	constricted double heterojunction (injection laser)	d.c.	direct current
		DH	double heterostructure or heterojunction (injection laser or LED)
CMI	coded mark inversion		

DSB	double sideband (amplitude modulation)
EH	traditional mode designation
EMI	electromagnetic interference
EMP	electromagnetic pulse
erf	error function
erfc	complementary error function
FDM	frequency division multiplexing
FET	field effect transistor
FM	frequency modulation
FSK	frequency shift keying
FWHP	full width half power
HDB	high density bipolar
HE	traditional mode designation
He–Ne	helium–neon (laser)
HF	high frequency
HV	high voltage
IF	intermediate frequency
ILD	injection laser diode
IM	intensity modulation
IO	integrated optics
I/O	input/output
ISI	intersymbol interference
LAN	local area network
LED	light emitting diode
LOC	large optical cavity (injection laser)
LP	linearly polarized (mode notation)
LPE	liquid phase epitaxy
MCVD	modified chemical vapor deposition
MESFET	metal Schottky field effect transistor
MISFET	metal integrated-semiconductor field effect transistor
Nd:YAG	neodymium-doped yttrium–aluminum–garnet (laser)
NRZ	nonreturn to zero
OTDR	optical time domain reflectometry
OVPO	outside vapor phase oxidation
PAM	pulse amplitude modulation
PCM	pulse code modulation
PCS	plastic-clad silica (fiber)

PCVD	plasma-activated chemical vapor deposition
PCW	plano-convex waveguide (injection laser)
PDF	probability density function
PFM	pulse frequency modulation
PIN–FET	p–i–n photodiode followed by a field effect transistor
PM	phase modulation
PPM	pulse position modulation
PSK	phase shift keying
PTT	Post, Telegraph and Telecommunications
PWM	pulse width modulation
RAPD	reach-through avalanche photodiode
RFI	radio frequency interference
rms	root mean square
RO	relaxation oscillation
RZ	return to zero
SAW	surface acoustic wave
SDM	space division multiplexing
SHF	super high frequency
SML	separated multiclad layer (injection laser)
SNR	signal to noise ratio
TDM	time division multiplexing
TE	transverse electric
TEM	transverse electromagnetic
TJS	transverse junction stripe (injection laser)
TM	transverse magnetic
TTL	transistor–transistor logic
UHF	ultra high frequency
VAD	vapor axial deposition
VCO	voltage controlled oscillator
VHF	very high frequency
VPE	vapor phase epitaxy
WDM	wavelength division multiplexing
WKB	Wentzel, Kramers, Brillouin (analysis technique) for graded fiber
WPS	wideband switch point
ZD	zenner diode

1

Introduction

Communication may be broadly defined as the transfer of information from one point to another. When the information is to be conveyed over any distance a communication system is usually required. Within a communication system the information transfer is frequently achieved by superimposing or modulating the information onto an electromagnetic wave which acts as a carrier for the information signal. This modulated carrier is then transmitted to the required destination where it is received and the original information signal is obtained by demodulation. Sophisticated techniques have been developed for this process using electromagnetic carrier waves operating at radio frequencies as well as microwave and millimeter wave frequencies. However, 'communication' may also be achieved using an electromagnetic carrier which is selected from the optical range of frequencies.

1.1 HISTORICAL DEVELOPMENT

The use of visible optical carrier waves or light for communication has been common for many years. Simple systems such as signal fires, reflecting mirrors and, more recently, signalling lamps have provided successful, if limited, information transfer. Moreover, as early as 1880 Alexander Graham Bell reported the transmission of speech using a light beam |Ref. 1|. The photophone proposed by Bell just four years after the invention of the telephone modulated sunlight with a diaphragm giving speech transmission over a distance of 200 m. However, although some investigation of optical communication continued in the early part of the 20th Century |Refs. 2 and 3| its use was limited to mobile, low capacity communication links. This was due to both the lack of suitable light sources and the problem that light transmission in the atmosphere is restricted to line of sight and severely affected by disturbances such as rain, snow, fog, dust and atmospheric turbulence. Nevertheless lower frequency and hence longer wavelength electromagnetic waves* (i.e. radio and microwave) proved suitable carriers for information transfer in the

---

* For the propagation of electromagnetic waves in free space, the wavelength λ equals the velocity of light in a vacuum c times the reciprocal of the frequency f in hertz or $\lambda = c/f$.

atmosphere, being far less affected by these atmospheric conditions. Depending on their wavelengths these electromagnetic carriers can be transmitted over considerable distances but are limited in the amount of information they can convey by their frequencies (i.e. the information-carrying capacity is directly related to the bandwidth or frequency extent of the modulated carrier, which is generally limited to a fixed fraction of the carrier frequency). In theory, the greater the carrier frequency, the larger the available transmission bandwidth and thus the information-carrying capacity of the communication system. For this reason radio communication was developed to higher frequencies (i.e. VHF and UHF) leading to the introduction of the even higher frequency microwave and, latterly, millimeter wave transmission. The relative frequencies and wavelengths of these types of electromagnetic wave can be observed from the electromagnetic spectrum shown in Fig. 1.1. In this context it may also be noted that communication at optical frequencies offers an increase in the potential usable bandwidth by a factor of around 10^4 over high frequency microwave transmission. An additional benefit of the use of high carrier frequencies is the general ability of the communication system to concentrate the available power within the transmitted electromagnetic wave, thus giving an improved system performance [Ref. 4].

A renewed interest in optical communication was stimulated in the early 1960s with the invention of the laser [Ref. 5]. This device provided a powerful coherent light source together with the possibility of modulation at high frequency. In addition the low beam divergence of the laser made enhanced free space optical transmission a practical possibility. However, the previously mentioned constraints of light transmission in the atmosphere tended to restrict these systems to short distance applications. Nevertheless, despite the problems some modest free space optical communication links have been implemented for applications such as the linking of a television camera to a base vehicle and for data links of a few hundred meters between buildings. There is also some interest in optical communication between satellites in outer space using similar techniques [Ref. 6].

Although the use of laser for free space optical communication proved somewhat limited, the invention of the laser instigated a tremendous research effort in the study of optical components to achieve reliable information transfer using a lightwave carrier. The proposals for optical communication via dielectric waveguides or optical fibers fabricated from glass were made almost simultaneously in 1966 by Kao and Hockham [Ref. 7] and Werts [Ref. 8] to avoid degradation of the optical signal by the atmosphere. Such systems were viewed as a replacement for coaxial cable or carrier transmission systems. Initially the optical fibers exhibited very high attenuation (i.e. 1000 dB km^{-1}) and were therefore not comparable with the coaxial cables they were to replace (i.e. 5–10 dB km^{-1}). There were also serious problems involved with jointing the fiber cables in a satisfactory manner to achieve low loss and to enable the process to be performed relatively easily and repeatedly in the

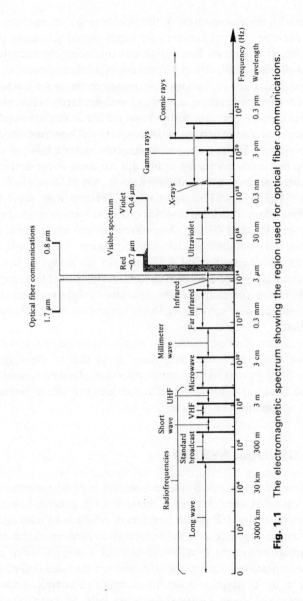

Fig. 1.1 The electromagnetic spectrum showing the region used for optical fiber communications.

field. Nevertheless, within the space of ten years optical fiber losses were reduced to below 5 dB km^{-1} and suitable low loss jointing techniques were perfected.

In parallel with the development of the fiber waveguide, attention was also focused on the other optical components which would constitute the optical fiber communication system. Since optical frequencies are accompanied by extremely small wavelengths the development of all these optical components essentially required a new technology. Thus semiconductor optical sources (i.e. injection lasers and light emitting diodes), as well as detectors (i.e. photodiodes and to a certain extent phototransistors) compatible in size with optical fibers were designed and fabricated to enable successful implementation of the optical fiber system. Initially the semiconductor lasers exhibited very short lifetimes of at best a few hours, but significant advances in the device structure enabled lifetimes greater than 1000 hr |Ref. 9| and 7000 hr |Ref. 10| to be obtained by 1973 and 1977 respectively.* These devices were originally fabricated from alloys of gallium arsenide (AlGaAs) which emitted in the near infrared between 0.8 and 0.9 μm. More recently this wavelength range has been extended to include the 1.1–1.6 μm region by the use of other semiconductor alloys (see Section 6.3.6) to take advantage of the enhanced performance characteristics displayed by optical fibers over this range. Similar developments in the generally simpler structure of light emitting diodes and detector photodiodes also contributed to the realization of reliable optical fiber communication.

The achievement of these impressive results has stemmed from the enormous amount of work directed into these areas due to the major distinct advantages offered by optical fiber communications. However, prior to discussion of these advantages we will briefly consider the salient features of the optical fiber communication system.

1.2 THE GENERAL SYSTEM

An optical fiber communication system is similar in basic concept to any type of communication system. A block schematic of a general communication system is shown in Fig. 1.2(a), the function of which is to convey the signal from the information source over the transmission medium to the destination. The communication system therefore consists of a transmitter or modulator linked to the information source, the transmission medium, and a receiver or demodulator at the destination point. In electrical communications the information source provides an electrical signal, usually derived from a message

---

* Projected semiconductor laser lifetimes are currently in the region of 10^5 to 10^6 h (see Section 6.9.6) indicating a substantial improvement since 1977.

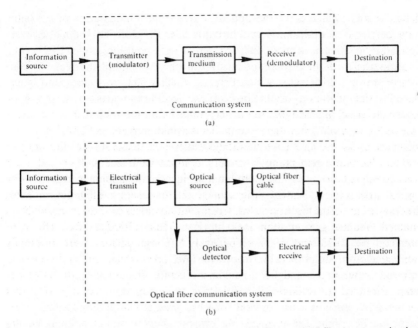

Fig. 1.2 (a) The general communication system. (b) The optical fiber communication system.

signal which is not electrical (e.g. sound), to a transmitter comprising electrical and electronic components which converts the signal into a suitable form for propagation over the transmission medium. This is often achieved by modulating a carrier which, as mentioned previously, may be an electromagnetic wave. The transmission medium can consist of a pair of wires, a coaxial cable or a radio link through free space down which the signal is transmitted to the receiver, where it is transformed into the original electrical information signal (demodulated) before being passed to the destination. However it must be noted that in any transmission medium the signal is attenuated, or suffers loss, and is subject to degradations due to contamination by random signals and noise as well as possible distortions imposed by mechanisms within the medium itself. Therefore, in any communication system there is a maximum permitted distance between the transmitter and the receiver beyond which the system effectively ceases to give intelligible communication. For long haul applications these factors necessitate the installation of repeaters or line amplifiers (see Section 10.4) at intervals, both to remove signal distortion and to increase signal level before transmission is continued down the link.

For optical fiber communications the system shown in Fig. 1.2(a) may be considered in slightly greater detail, as in Fig. 1.2(b). In this case the information source provides an electrical signal to a transmitter comprising an electrical stage which drives an optical source to give modulation of the lightwave carrier. The optical source which provides the electrical–optical conversion may be either a semiconductor laser or light emitting diode (LED). The transmission medium consists of an optical fiber cable and the receiver consists of an optical detector which drives a further electrical stage and hence provides demodulation of the optical carrier. Photodiodes (p–n, p–i–n or avalanche) and, in some instances, phototransistors are utilized for the detection of the optical signal or the optical–electrical conversion. Thus there is a requirement for electrical interfacing at either end of the optical link and at present the signal processing is usually performed electrically.*

The optical carrier may be modulated using either an analog or digital information signal. In the system shown in Fig. 1.2(b) analog modulation involves the variation of the light emitted from the optical source in a continuous manner. With digital modulation, however, discrete changes in the light intensity are obtained (i.e. on–off pulses). Although often simpler to implement, analog modulation with an optical fiber communication system is less efficient, requiring a far higher signal to noise ratio at the receiver than digital modulation. Also the linearity needed for analog modulation is not always provided by semiconductor optical sources, especially at high modulation frequencies. For these reasons, analog optical fiber communication links are generally limited to shorter distances and lower bandwidths than digital links.

Figure 1.3 shows a block schematic of a typical digital optical fiber link. Initially the input digital signal from the information source is suitably encoded for optical transmission. The laser drive circuit directly modulates the intensity of the semiconductor laser with the encoded digital signal. Hence a digital optical signal is launched into the optical fiber cable. The avalanche photodiode (APD) detector is followed by a front-end amplifier and equalizer or filter to provide gain as well as linear signal processing and noise bandwidth reduction. Finally, the signal obtained is decoded to give the original digital information. The various elements of this and alternative optical fiber system configurations are discussed in detail in the following chapters. However, at this stage it is instructive to consider the advantages provided by lightwave

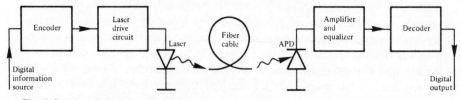

Fig. 1.3   A digital optical fiber link using a semiconductor laser source and an avalanche photodiode (APD) detector.

---

* Significant developments are taking place in optical signal processing which may alter this situation in the future (see Sections 11.7 and 11.8).

communication via optical fibers in comparison with other forms of line and radio communication which have brought about the introduction of such systems in many areas throughout the world.

1.3 ADVANTAGES OF OPTICAL FIBER COMMUNICATION

Communication using an optical carrier wave guided along a glass fiber has a number of extremely attractive features, several of which were apparent when the technique was originally conceived. Furthermore, the advances in the technology to date have surpassed even the most optimistic predictions creating additional advantages. Hence it is useful to consider the merits and special features offered by optical fiber communications over more conventional electrical communications. In this context we commence with the originally foreseen advantages and then consider additional features which have become apparent as the technology developed.

(a) Enormous potential bandwidth
The optical carrier frequency in the range 10^{13} to 10^{16} Hz (generally in the near infrared around 10^{14} Hz or 10^5 GHz) yields a far greater potential transmission bandwidth than metallic cable systems (i.e. coaxial cable bandwidth up to around 500 MHz) or even millimeter wave radio systems (i.e. systems currently operating with modulation bandwidths of 700 MHz). At present, the bandwidth available to fiber systems is not fully utilized but modulation at several gigahertz over a few kilometers and hundreds of megahertz over tens of kilometers without intervening electronics (repeaters) is possible. Therefore, the information-carrying capacity of optical fiber systems is already proving far superior to the best copper cable systems. By comparison the losses in wideband coaxial cable systems restrict the transmission distance to only a few kilometers at bandwidths over a hundred megahertz. Moreover, it is certain that the usable fiber system bandwidth will be extended further towards the optical carrier frequency in the future to provide an information-carrying capacity far in excess of that obtained using copper cables or a wideband radio system.

(b) Small size and weight
Optical fibers have very small diameters which are often no greater than the diameter of a human hair. Hence, even when such fibers are covered with protective coatings they are far smaller and much lighter than corresponding copper cables. This is a tremendous boon towards the alleviation of duct congestion in cities, as well as allowing for an expansion of signal transmission within mobiles such as aircraft, satellites and even ships.

(c) Electrical isolation
Optical fibers which are fabricated from glass or sometimes a plastic polymer

are electrical insulators and therefore, unlike their metallic counterparts, they do not exhibit earth loop and interface problems. Furthermore, this property makes optical fiber transmission ideally suited for communication in electrically hazardous environments as the fibers create no arcing or spark hazard at abrasions or short circuits.

(d) Immunity to interference and crosstalk
Optical fibers form a dielectric waveguide and are therefore free from electromagnetic interference (EMI), radiofrequency interference (RFI), or switching transients giving electromagnetic pulses (EMP). Hence the operation of an optical fiber communication system is unaffected by transmission through an electrically noisy environment and the fiber cable requires no shielding from EMI. The fiber cable is also not susceptible to lightning strikes if used overhead rather than underground. Moreover, it is fairly easy to ensure that there is no optical interference between fibers and hence, unlike communication using electrical conductors, crosstalk is negligible, even when many fibers are cabled together.

(e) Signal security
The light from optical fibers does not radiate significantly and therefore they provide a high degree of signal security. Unlike the situation with copper cables, a transmitted optical signal cannot be obtained from a fiber in a non-invasive manner (i.e. without drawing optical power from the fiber). Therefore, in theory, any attempt to acquire a message signal transmitted optically may be detected. This feature is obviously attractive for military, banking and general data transmission (i.e. computer network) applications.

(f) Low transmission loss
The development of optical fibers over the last 15 years has resulted in the production of optical fiber cables which exhibit very low attenuation or transmission loss in comparison with the best copper conductors. Fibers have been fabricated with losses as low as 0.2 dB km^{-1} (see Section 3.3.2) and this feature has become a major advantage of optical fiber communications. It facilitates the implementation of communication links with extremely wide repeater spacing (long transmission distances without intermediate electronics), thus reducing both system cost and complexity. Together with the already proven modulation bandwidth capability of fiber cable this property provides a totally compelling case for the adoption of optical fiber communication in the majority of long-haul telecommunication applications.

(g) Ruggedness and flexibility
Although protective coatings are essential, optical fibers may be manufactured with very high tensile strengths (see Section 4.6.1). Perhaps surprisingly for a glassy substance, the fibers may also be bent to quite small radii or twisted

without damage. Furthermore, cable structures have been developed (see Section 4.7.4) which have proved flexible, compact and extremely rugged. Taking the size and weight advantage into account, these optical fiber cables are generally superior in terms of storage, transportation, handling and installation than corresponding copper cables whilst exhibiting at least comparable strength and durability.

(h) System reliability and ease of maintenance
These features primarily stem from the low loss property of optical fiber cables which reduces the requirement for intermediate repeaters or line amplifiers to boost the transmitted signal strength. Hence with fewer repeaters, system reliability is generally enhanced in comparison with conventional electrical conductor systems. Furthermore, the reliability of the optical components is no longer a problem with predicted lifetimes of 20–30 years now quite common. Both these factors also tend to reduce maintenance time and costs.

(i) Potential low cost
The glass which generally provides the optical fiber transmission medium is made from sand—not a scarce resource. So, in comparison with copper conductors, optical fibers offer the potential for low cost line communication. As yet this potential has not been fully realized because of the sophisticated, and therefore expensive, processes required to obtain ultra-pure glass, and the lack of production volume. At present, optical fiber cable is reasonably competitive with coaxial cable, but not with simple copper wires (e.g. twisted pairs). However, it is likely that in the future it will become as cheap to use optical fibers with their superior performance than almost any type of electrical conductor.

Moreover, overall system costs when utilizing optical fiber communication on long-haul links are generally reduced to those for equivalent electrical line systems because of the low loss and wideband properties of the optical transmission medium. As indicated in (f), the requirement for intermediate repeaters and the associated electronics is reduced, giving a significant cost advantage. However, although this cost benefit gives a net gain for long-haul links this is not usually the case in short-haul applications where the additional cost incurred, due to the electrical–optical conversion (and vice versa), may be a deciding factor. Nevertheless, there are other possible cost advantages in relation to shipping, handling, installation and maintenance, as well as the features indicated in (c) and (d) which may prove significant in the system choice.

The low cost potential of optical fiber communications not only provides strong competition with electrical line transmission systems, but also with microwave and millimeter wave radio transmission systems. Although these systems are reasonably wideband the relatively short span 'line of sight' transmission necessitates expensive aerial towers at intervals no greater than a few tens of kilometers.

Many advantages are therefore provided by the use of a lightwave carrier within a transmission medium consisting of an optical fiber. The fundamental principles giving rise to these enhanced performance characteristics, together with their practical realization, are described in the following chapters. However, a general understanding of the basic nature and properties of light is assumed. If this is lacking, the reader is directed to the many excellent texts encompassing the topic, a few of which are indicated in Refs. 16–22.

REFERENCES

1   A. G. Bell, 'Selenium and the photophone', *The Electrician*, pp. 214, 215, 220, 221, 1880.
2   W. S. Huxford and J. R. Platt, 'Survey of near infra-red communication systems', *J. Opt. Soc. Am.*, **38**, pp. 253–268, 1948.
3   N. C. Beese, 'Light sources for optical communication', *Infrared Phys.*, **1**, pp. 5–16, 1961.
4   R. M. Gagliardi and S. Karp, *Optical Communications*, John Wiley, 1976.
5   T. H. Maiman, 'Stimulated optical radiation in ruby', *Nature, Lond.*, **187**, pp. 493–494, 1960.
6   A. R. Kraemer, 'Free-space optical communications', *Signal*, pp. 26–32, 1977.
7   K. C. Kao and G. A. Hockham, 'Dielectric-fiber surface waveguides for optical frequencies', *Proc. IEE*, **113**(7), pp. 1151–1158, 1966.
8   A. Werts, 'Propagation de la lumière cohérente dans les fibres optiques', *L'Onde Electrique*, **46**, pp. 967–980, 1966.
9   R. L. Hartman, J. C. Dyment, C. J. Hwang and H. Kuhn, 'Continuous operation of GaAs–Ga$_x$Al$_{1-x}$As, double heterostructure lasers with 30°C half lives exceeding 1000 h', *Appl. Phys. Lett.*, **23**(4), pp. 181–183, 1973.
10   A. R. Goodwin, J. F. Peters, M. Pion and W. O. Bourne, 'GaAs lasers with consistently low degradation rates at room temperature', *Appl. Phys. Lett.*, **30**(2), pp. 110–113, 1977.
11   P. Russer, 'Introduction to optical communication', M. J. Howes and D. V. Morgan (Eds.), *Optical Fibre Communications*, pp. 1–26, John Wiley, 1980.
12   J. E. Midwinter, *Optical Fibres for Transmission*, John Wiley, 1979.
13   B. Costa, 'Historical remarks', in *Optical Fibre Communication* by the Technical Staff of CSELT, McGraw-Hill, 1981.
14   C. P. Sandbank (Ed.), *Optical Fibre Communication Systems*, John Wiley, 1980.
15   H. F. Wolf (Ed.), *Handbook of Fiber Optics, Theory and Applications*, Granada, 1981.
16   F. A. Jenkins and H. E. White, *Fundamentals of Optics* (4th edn.), McGraw-Hill, 1976.
17   E. Hecht and A. Zajac, *Optics*, Addison-Wesley, 1974.
18   G. R. Fowles, *Introduction to Modern Optics* (2nd edn.), Holt, Rinehart & Winston, 1975.
19   R. S. Longhurst, *Geometrical and Physical Optics*, (3rd edn.), Longman, 1973.
20   F. G. Smith and J. H. Thomson, *Optics*, John Wiley, 1980.
21   S. G. Lipson and H. Lipson, *Optical Physics*, (2nd edn.), Cambridge University Press, 1981.
22   M. Born and E. Wolf, *Principles of Optics*, (6th edn.), Pergamon Press, 1980.

2

Optical Fiber Waveguides

2.1 INTRODUCTION

The transmission of light via a dielectric waveguide structure was first proposed and investigated at the beginning of the 20th Century. In 1910 Hondros and Debye |Ref. 1| conducted a theoretical study and experimental work was reported by Schriever in 1920 |Ref. 2|. However, a transparent dielectric rod, typically of silica glass with a refractive index of around 1.5, surrounded by air, proved to be an impractical waveguide due to its unsupported structure (especially when very thin waveguides were considered in order to limit the number of optical modes propagated) and the excessive losses at any discontinuities of the glass–air interface. Nevertheless, interest in the application of dielectric optical waveguides in such areas as optical imaging and medical diagnosis (e.g. endoscopes) led to proposals |Refs. 3 and 4| for a clad dielectric rod in the mid 1950s in order to overcome these problems. This structure is illustrated in Fig. 2.1 which shows a transparent core with a refractive index n_1 surrounded by a transparent cladding of slightly lower refractive index n_2. The cladding supports the waveguide structure whilst also, when sufficiently thick, substantially reducing the radiation loss into the surrounding air. In essence, the light energy travels in both the core and the cladding allowing the associated fields to decay to a negligible value at the cladding–air interface.

The invention of the clad waveguide structure led to the first serious proposals by Kao and Hockham |Ref. 5|, and Werts |Ref. 6| in 1966 to utilize optical fibers as a communications medium even though they had losses in excess of 1000 dB km^{-1}. These proposals stimulated tremendous efforts to

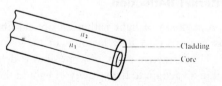

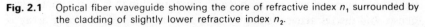

Fig. 2.1 Optical fiber waveguide showing the core of refractive index n_1 surrounded by the cladding of slightly lower refractive index n_2.

reduce the attenuation by purification of the materials. This has resulted in improved conventional glass refining techniques giving fibers with losses of around 4.2 dB km^{-1} [Ref. 7]. Also progress in glass refining processes such as depositing vapor-phase reagents to form silica [Ref. 8] has allowed fibers with losses below 1 dB km^{-1} to be fabricated.

Most of this work was focused on the 0.8–0.9 μm wavelength band because the first generation optical sources fabricated from gallium aluminum arsenide alloys operated in this region. However, as silica fibers were studied in further detail it became apparent that transmission at longer wavelengths (1.1–1.6 μm) would result in lower losses and reduced signal dispersion. This produced a shift in optical fiber source and detector technology in order to provide operation at these longer wavelengths. Hence at longer wavelengths, especially around 1.55 μm, fibers with losses as low as 0.2 dB km^{-1} have been reported [Ref. 9].

In order to appreciate the transmission mechanism of optical fibers with dimensions approximating to those of a human hair, it is necessary to consider the optical waveguiding of cylindrical glass fibers. Such a fiber acts as an open optical waveguide, which may be analyzed utilizing simple ray theory. However, the concepts of geometric optics are not sufficient when considering all types of optical fiber and electromagnetic mode theory must be used to give a complete picture. The following sections will therefore outline the transmission of light in optical fibers prior to a more detailed discussion of the various types of fiber.

In Section 2.2 we continue the discussion of light propagation in optical fibers using the ray theory approach in order to develop some of the fundamental parameters associated with optical fiber transmission (acceptance angle, numerical aperture, etc.). Furthermore, this provides a basis for the discussion of electromagnetic wave propagation presented in Section 2.3. In this section the electromagnetic mode theory is developed for the planar (rectangular) waveguide prior to consideration of the cylindrical fiber. Following, in Section 2.4, we discuss optical propagation in step index fibers (both multimode and single mode). Finally, Section 2.5 gives a brief account of the waveguiding mechanism within graded index fibers.

2.2   RAY THEORY TRANSMISSION

2.2.1   Total Internal Reflection

To consider the propagation of light within an optical fiber utilizing the ray theory model it is necessary to take account of the refractive index of the dielectric medium. The refractive index of a medium is defined as the ratio of the velocity of light in a vacuum to the velocity of light in the medium. A ray of light travels more slowly in an optically dense medium than in one that is less dense, and the refractive index gives a measure of this effect. When a ray is incident on the interface between two dielectrics of differing refractive indices (e.g. glass–air), refraction occurs as illustrated in Fig. 2.2(a). It may be observed that the ray approaching the interface is propagating in a dielectric of refractive index n_1 and is at an angle ϕ_1 to the normal at the surface of the interface. If the dielectric on the other side of the interface has a refractive index n_2 which is less than n_1 then the refraction is such that the ray path in this lower index medium is at an angle ϕ_2 to the normal, where ϕ_2 is greater than ϕ_1. The angles of incidence ϕ_1 and refraction ϕ_2 are related to each other and to the refractive indices of the dielectrics by Snell's law of refraction | Ref. 10|, which states that:

$$n_1 \sin \phi_1 = n_2 \sin \phi_2$$

or

$$\frac{\sin \phi_1}{\sin \phi_2} = \frac{n_2}{n_1} \qquad (2.1)$$

It may also be observed in Fig. 2.2(a) that a small amount of light is reflected back into the originating dielectric medium (partial internal reflection). As n_1 is greater than n_2, the angle of refraction is always greater than the angle of incidence. Thus when the angle of refraction is 90° and the refracted ray emerges parallel to the interface between the dielectrics the angle of incidence must be less than 90°. This is the limiting case of refraction and the angle of incidence is now known as the critical angle ϕ_c as shown in Fig.

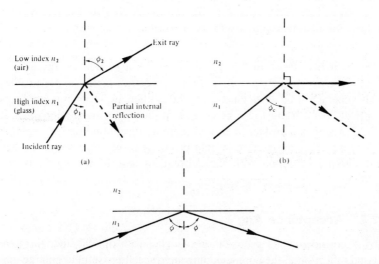

Fig. 2.2 Light rays incident on high to low refractive index interface (e.g. glass–air): (a) refraction; (b) the limiting case of refraction showing the critical ray at an angle ϕ_c; (c) total internal reflection where $\phi > \phi_c$.

2.2(b). From Eq. (2.1) the value of the critical angle is given by:

$$\sin \phi_c = \frac{n_2}{n_1} \tag{2.2}$$

At angles of incidence greater than the critical angle the light is reflected back into the originating dielectric medium (total internal reflection) with high efficiency (around 99.9%). Hence it may be observed in Fig. 2.2(c) that total internal reflection occurs at the interface between two dielectrics of differing refractive indices when light is incident on the dielectric of lower index from the dielectric of higher index, and the angle of incidence of the ray exceeds the critical value. This is the mechanism by which light at a sufficiently shallow angle (less than $90° - \phi_c$) may be considered to propagate down an optical fiber with low loss. Figure 2.3 illustrates the transmission of a light ray in an optical fiber via a series of total internal reflections at the interface of the silica core and the slightly lower refractive index silica cladding. The ray has an angle of incidence ϕ at the interface which is greater than the critical angle and is reflected at the same angle to the normal.

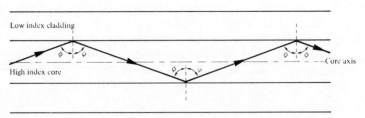

Fig. 2.3   The transmission of a light ray in a perfect optical fiber.

The light ray shown in Fig. 2.3 is known as a meridional ray as it passes through the axis of the fiber core. This type of ray is the simplest to describe and is generally used when illustrating the fundamental transmission properties of optical fibers. It must also be noted that the light transmission illustrated in Fig. 2.3 assumes a perfect fiber, and that any discontinuities or imperfections at the core–cladding interface would probably result in refraction rather than total internal reflection with the subsequent loss of the light ray into the cladding.

2.2.2   Acceptance Angle

Having considered the propagation of light in an optical fiber through total internal reflection at the core–cladding interface, it is useful to enlarge upon the geometric optics approach with reference to light rays entering the fiber. Since only rays with a sufficiently shallow grazing angle (i.e. with an angle to the normal greater than ϕ_c) at the core–cladding interface are transmitted by total

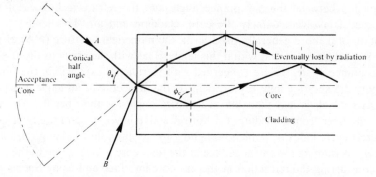

Fig. 2.4   The acceptance angle θ_a when launching light into an optical fiber.

internal reflection, it is clear that not all rays entering the fiber core will continue to be propagated down its length.

The geometry concerned with launching a light ray into an optical fiber is shown in Fig. 2.4 which illustrates a meridional ray A at the critical angle ϕ_c within the fiber at the core–cladding interface. It may be observed that this ray enters the fiber core at an angle θ_a to the fiber axis and is refracted at the air–core interface before transmission to the core–cladding interface at the critical angle. Hence, any rays which are incident into the fiber core at an angle greater than θ_a will be transmitted to the core–cladding interface at an angle less than ϕ_c, and will not be totally internally reflected. This situation is also illustrated in Fig. 2.4 where the incident ray B at an angle greater than θ_a is refracted into the cladding and eventually lost by radiation. Thus for rays to be transmitted by total internal reflection within the fiber core they must be incident on the fiber core within an acceptance cone defined by the conical half angle θ_a. Hence θ_a is the maximum angle to the axis that light may enter the fiber in order to be propagated and is often referred to as the acceptance angle* for the fiber.

If the fiber has a regular cross section (i.e. the core–cladding interfaces are parallel and there are no discontinuities) an incident meridional ray at greater than the critical angle will continue to be reflected and will be transmitted through the fiber. From symmetry considerations it may be noted that the output angle to the axis will be equal to the input angle for the ray, assuming the ray emerges into a medium of the same refractive index from which it was input.

2.2.3   Numerical Aperture

The acceptance angle for an optical fiber was defined in the previous section. However, it is possible to continue the ray theory analysis to obtain a

---

* θ_a is sometimes referred to as the maximum or total acceptance angle.

relationship between the acceptance angle and the refractive indices of the three media involved, namely the core, cladding and air. This leads to the definition of a more generally used term, the numerical aperture (NA) of the fiber. It must be noted that within this analysis, as with the previous discussion of acceptance angle, we are concerned with meridional rays within the fiber.

Figure 2.5 shows a light ray incident on the fiber core at an angle θ_1 to the fiber axis which is less than the acceptance angle for the fiber θ_a. The ray enters the fiber from a medium (air) of refractive index n_0, and the fiber core has a refractive index n_1, which is slightly greater than the cladding refractive index n_2. Assuming the entrance face at the fiber core to be normal to the axis, then considering the refraction at the air–core interface and using Snell's law given by Eq. (2.1):

$$n_0 \sin \theta_1 = n_1 \sin \theta_2 \qquad (2.3)$$

Considering the right-angled triangle ABC indicated in Fig. 2.5, then:

$$\phi = \frac{\pi}{2} - \theta_2 \qquad (2.4)$$

where ϕ is greater than the critical angle at the core–cladding interface. Hence Eq. (2.3) becomes

$$n_0 \sin \theta_1 = n_1 \cos \phi \qquad (2.5)$$

Using the trigonometrical relationship $\sin^2 \phi + \cos^2 \phi = 1$, Eq. (2.5) may be written in the form:

$$n_0 \sin \theta_1 = n_1 (1 - \sin^2 \phi)^{\frac{1}{2}} \qquad (2.6)$$

When the limiting case for total internal reflection is considered ϕ becomes equal to the critical angle ϕ_c for the core–cladding interface and is given by Eq. (2.2). Also in this limiting case θ_1 becomes the acceptance angle for the fiber θ_a. Combining these limiting cases into Eq. (2.6) gives:

$$n_0 \sin \theta_a = (n_1^2 - n_2^2)^{\frac{1}{2}} \qquad (2.7)$$

Equation (2.7), apart from relating the acceptance angle to the refractive

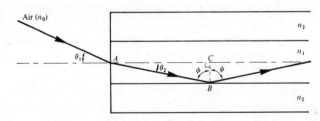

Fig. 2.5   The ray path for a meridional ray launched into an optical fiber in air at an input angle less than the acceptance angle for the fiber.

indices, serves as the basis for the definition of the important optical fiber parameter, the numerical aperture (NA). Hence the NA is defined as:

$$NA = n_0 \sin \theta_a = (n_1^2 - n_2^2)^{\frac{1}{2}} \qquad (2.8)$$

Since the NA is often used with the fiber in air where n_0 is unity, it is simply equal to $\sin \theta_a$. It may also be noted that incident meridional rays over the range $0 \leqslant \theta_1 \leqslant \theta_a$ will be propagated within the fiber.

The numerical aperture may also be given in terms of the relative refractive index difference Δ between the core and the cladding which is defined as:

$$\Delta = \frac{n_1^2 - n_2^2}{2n_1^2}$$

$$\simeq \frac{n_1 - n_2}{n_1} \qquad \text{for } \Delta \ll 1 \qquad (2.9)$$

Hence combining Eq. (2.8) with Eq. (2.9) we can write:

$$NA \simeq n_1 (2\Delta)^{\frac{1}{2}} \qquad (2.10)$$

The relationships given in Eqs. (2.8) and (2.10) for the numerical aperture are a very useful measure of the light-collecting ability of a fiber. They are independent of the fiber core diameter and will hold for diameters as small as 8 μm. However, for smaller diameters they break down as the geometric optics approach is invalid. This is because the ray theory model is only a partial description of the character of light. It describes the direction a plane wave component takes in the fiber but does not take into account interference between such components. When interference phenomena are considered it is found that only rays with certain discrete characteristics propagate in the fiber core. Thus the fiber will only support a discrete number of guided modes. This becomes critical in small core diameter fibers which only support one or a few modes. Hence electromagnetic mode theory must be applied in these cases |Ref. 12|.

Example 2.1

A silica optical fiber with a core diameter large enough to be considered by ray theory analysis has a core refractive index of 1.50 and a cladding refractive index of 1.47.

Determine: (a) the critical angle at the core–cladding interface; (b) the NA for the fiber; (c) the acceptance angle in air for the fiber.

Solution: (a) The critical angle ϕ_c at the core–cladding interface is given by Eq. (2.2) where:

$$\phi_c = \sin^{-1} \frac{n_2}{n_1} = \sin^{-1} \frac{1.47}{1.50}$$

$$= 78.5°$$

(b) From Eq. (2.8) the numerical aperture is:

$$NA = (n_1^2 - n_2^2)^{\frac{1}{2}} = (1.50^2 - 1.47^2)^{\frac{1}{2}}$$

$$= (2.25 - 2.16)^{\frac{1}{2}}$$

$$= 0.30$$

(c) Considering Eq. (2.8) the acceptance angle in air θ_a is given by:

$$\theta_a = \sin^{-1} NA = \sin^{-1} 0.30$$

$$= 17.4°$$

Example 2.2

A typical relative refractive index difference for an optical fiber designed for long distance transmission is 1%. Estimate the NA and the solid acceptance angle in air for the fiber when the core index is 1.46. Further calculate the critical angle at the core–cladding interface within the fiber. It may be assumed that the concepts of geometric optics hold for the fiber.

Solution: Using Eq. (2.10) with $\Delta = 0.01$ gives the numerical aperture as:

$$NA \simeq n_1 (2\Delta)^{\frac{1}{2}} = 1.46 (0.02)^{\frac{1}{2}}$$

$$= 0.21$$

For small angles the solid acceptance angle in air ζ is given by:

$$\zeta \simeq \pi\theta_a^2 = \pi \sin^2 \theta_a$$

Hence from Eq. (2.8):

$$\zeta \simeq \pi(NA)^2 = \pi 0.04$$

$$= 0.13 \text{ rad}$$

Using Eq. (2.9) for the relative refractive index difference Δ gives:

$$\Delta \simeq \frac{n_1 - n_2}{n_1} = 1 - \frac{n_2}{n_1}$$

Hence

$$\frac{n_2}{n_1} = 1 - \Delta = 1 - 0.01$$

$$= 0.99$$

From Eq. (2.2) the critical angle at the core–cladding interface is:

$$\phi_c = \sin^{-1} \frac{n_2}{n_1} = \sin^{-1} 0.99$$

$$= 81.9°$$

2.2.4 Skew Rays

In the previous sections we have considered the propagation of meridional rays in the optical waveguide. However, another category of ray exists which is transmitted without passing through the fiber axis. These rays, which greatly outnumber the meridional rays, follow a helical path through the fiber as illustrated in Fig. 2.6 and are called skew rays. It is not easy to visualize the skew ray paths in two dimensions but it may be observed from Fig. 2.6(b) that the helical path traced through the fiber gives a change in direction of 2γ at each reflection where γ is the angle between the projection of the ray in two dimensions and the radius of the fiber core at the point of reflection. Hence, unlike meridional rays, the point of emergence of skew rays from the fiber in air will depend upon the number of reflections they undergo rather than the input conditions to the fiber. When the light input to the fiber is nonuniform, skew rays will therefore tend to have a smoothing effect on the distribution of the light as it is transmitted, giving a more uniform output. The amount of smoothing is dependent on the number of reflections encountered by the skew rays.

A further possible advantage of the transmission of skew rays becomes apparent when their acceptance conditions are considered. In order to calculate the acceptance angle for a skew ray it is necessary to define the direction of the ray in two perpendicular planes. The geometry of the situation is illustrated in Fig. 2.7 where a skew ray is shown incident on the fiber core at the point A, at an angle θ_s to the normal at the fiber end face. The ray is refracted at the air–core interface before travelling to the point B in the same plane. The angles of incidence and reflection at the point B are ϕ which is greater than the critical angle for the core–cladding interface.

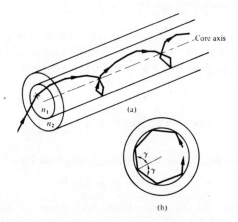

Fig. 2.6 The helical path taken by a skew ray in an optical fiber: (a) skew ray path down the fiber; (b) cross-sectional view of the fiber.

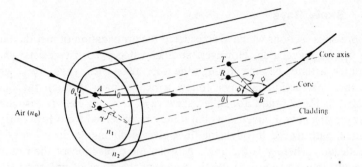

Fig. 2.7   The ray path within the fiber core for a skew ray incident at an angle θ_s to the normal at the air–core interface.

When considering the ray between A and B it is necessary to resolve the direction of the ray path AB to the core radius at the point B. As the incident and reflected rays at the point B are in the same plane, this is simply $\cos \phi$. However, if the two perpendicular planes through which the ray path AB traverses are considered, then γ is the angle between the core radius and the projection of the ray onto a plane BRS normal to the core axis, and θ is the angle between the ray and a line AT drawn parallel to the core axis. Thus to resolve the ray path AB relative to the radius BR in these two perpendicular planes, requires multiplication by $\cos \gamma$ and $\sin \theta$.

Hence, the reflection at point B at an angle ϕ may be given by:

$$\cos \gamma \sin \theta = \cos \phi \qquad (2.11)$$

Using the trigonometrical relationship $\sin^2 \phi + \cos^2 \phi = 1$, Eq. (2.11) becomes:

$$\cos \gamma \sin \theta = \cos \phi = (1 - \sin^2 \phi)^{\frac{1}{2}} \qquad (2.12)$$

If the limiting case for total internal reflection is now considered then ϕ becomes equal to the critical angle ϕ_c for the core–cladding interface and following Eq. (2.2) is given by $\sin \phi_c = n_2/n_1$. Hence Eq. (2.12) may be written as:

$$\cos \gamma \sin \theta \leqslant \cos \phi_c = \left(1 - \frac{n_2^2}{n_1^2} \right)^{\frac{1}{2}} \qquad (2.13)$$

Furthermore, using Snell's law at the point A following Eq. (2.1), we can write:

$$n_0 \sin \theta_a = n_1 \sin \theta \qquad (2.14)$$

where θ_a represents the maximum input axial angle for meridional rays as expressed in Section 2.2.2, and θ is the internal axial angle. Hence substituting for $\sin \theta$ from Eq. (2.13) into Eq. (2.14) gives:

$$\sin \theta_{as} = \frac{n_1}{n_0} \frac{\cos \phi_c}{\cos \gamma} = \frac{n_1}{n_0 \cos \gamma} \left(1 - \frac{n_2^2}{n_1^2} \right)^{\frac{1}{2}} \qquad (2.15)$$

where θ_{as} now represents the maximum input angle or acceptance angle for skew rays. It may be noted that the inequality shown in Eq. (2.13) is no longer necessary as all the terms in Eq. (2.15) are specified for the limiting case. Thus the acceptance conditions for skew rays are:

$$n_0 \sin \theta_{as} \cos \gamma = (n_1^2 - n_2^2)^{\frac{1}{2}} = NA \qquad (2.16)$$

and in the case of the fiber in air ($n_0 = 1$):

$$\sin \theta_{as} \cos \gamma = NA \qquad (2.17)$$

Therefore, by comparison with Eq. (2.8) derived for meridional rays, it may be noted that skew rays are accepted at larger axial angles in a given fiber than meridional rays, depending upon the value of cos γ. In fact for meridional rays cos γ is equal to unity and θ_{as} becomes equal to θ_a. Thus although θ_a is the maximum conical half angle for the acceptance of meridional rays, it defines the minimum input angle for skew rays. Hence as may be observed from Fig. 2.6, skew rays tend to propagate only in the annular region near the outer surface of the core, and do not fully utilize the core as a transmission medium. However, they are complementary to meridional rays and increase the light-gathering capacity of the fiber. This increased light-gathering ability may be significant for large NA fibers, but for most communication design purposes the expressions given in Eqs. (2.8) and (2.10) for meridional rays are considered adequate.

Example 2.3

An optical fiber in air has an NA of 0.4. Compare the acceptance angle for meridional rays with that for skew rays which change direction by 100° at each reflection.

Solution: The acceptance angle for meridional rays is given by Eq. (2.8) with $n_0 = 1$ as:

$$\theta_a = \sin^{-1} NA = \sin^{-1} 0.4$$

$$= 23.6°$$

The skew rays change direction by 100° at each reflection, therefore γ = 50° Hence using Eq. (2.17) the acceptance angle for skew rays is:

$$\theta_{as} = \sin^{-1} \left(\frac{NA}{\cos \gamma} \right) = \sin^{-1} \left(\frac{0.4}{\cos 50°} \right)$$

$$= 38.5°$$

In this example, the acceptance angle for the skew rays is about 15° greater than the corresponding angle for meridional rays. However, it must be noted that we have only compared the acceptance angle of one particular skew ray path. When the light input to the fiber is at an angle to the fiber axis, it is possible that γ will vary from zero for meridional rays to 90° for rays which enter the fiber at the core–cladding interface giving acceptance of skew rays over a conical half angle of $\pi/2$ radians.

2.3 ELECTROMAGNETIC MODE THEORY FOR OPTICAL PROPAGATION

2.3.1 Electromagnetic Waves

In order to obtain an improved model for the propagation of light in an optical fiber, electromagnetic wave theory must be considered. The basis for the study of electromagnetic wave propagation is provided by Maxwell's equations |Ref. 13|. For a medium with zero conductivity these vector relationships may be written in terms of the electric field E, magnetic field H, electric flux density D and magnetic flux density B as the curl equations:

$$\mathbf{V} \times E = -\frac{\partial B}{\partial t} \tag{2.18}$$

$$\mathbf{V} \times H = \frac{\partial D}{\partial t} \tag{2.19}$$

and the divergence conditions:

$$\mathbf{V} \cdot D = 0 \quad \text{(no free charges)} \tag{2.20}$$

$$\mathbf{V} \cdot B = 0 \quad \text{(no free poles)} \tag{2.21}$$

where $\mathbf{V}$ is a vector operator.

The four field vectors are related by the relations:

$$D = \varepsilon E$$
$$B = \mu H \tag{2.22}$$

where ε is the dielectric permittivity and μ is the magnetic permeability of the medium.

Substituting for D and B and taking the curl of Eqs. (2.18) and (2.19) gives

$$\mathbf{V} \times (\mathbf{V} \times E) = -\mu\varepsilon \frac{\partial^2 E}{\partial t^2} \tag{2.23}$$

$$\mathbf{V} \times (\mathbf{V} \times H) = -\mu\varepsilon \frac{\partial^2 H}{\partial t^2} \tag{2.24}$$

Then using the divergence conditions of Eqs. (2.20) and (2.21) with the vector identity

$$\mathbf{V} \times (\mathbf{V} \times Y) = \mathbf{V}(\mathbf{V} \cdot Y) - \mathbf{V}^2 (Y)$$

we obtain the nondispersive wave equations:

$$\mathbf{V}^2 E = \mu\varepsilon \frac{\partial^2 E}{\partial t^2} \qquad (2.25)$$

and

$$\mathbf{V}^2 H = \mu\varepsilon \frac{\partial^2 H}{\partial t^2} \qquad (2.26)$$

where $\mathbf{V}^2$ is the Laplacian operator. For rectangular Cartesian and cylindrical polar coordinates the above wave equations hold for each component of the field vector, every component satisfying the scalar wave equation:

$$\mathbf{V}^2 \psi = \frac{1}{v_p^2} \frac{\partial^2 \psi}{\partial t^2} \qquad (2.27)$$

where ψ may represent a component of the E or H field and v_p is the phase velocity (velocity of propagation of a point of constant phase in the wave) in the dielectric medium. It follows that

$$v_p = \frac{1}{(\mu\varepsilon)^{\frac{1}{2}}} = \frac{1}{(\mu_r \mu_0 \varepsilon_r \varepsilon_0)^{\frac{1}{2}}} \qquad (2.28)$$

where μ_r and ε_r are the relative permeability and permittivity for the dielectric medium and μ_0 and ε_0 are the permeability and permittivity of free space. The velocity of light in free space c is therefore

$$c = \frac{1}{(\mu_0 \varepsilon_0)^{\frac{1}{2}}} \qquad (2.29)$$

If planar waveguides, described by rectangular Cartesian coordinates (x, y, z), or circular fibers, described by cylindrical polar coordinates (r, ϕ, z) are considered, then the Laplacian operator takes the form:

$$\mathbf{V}^2 \psi = \frac{\partial^2 \psi}{\partial x^2} + \frac{\partial^2 \psi}{\partial y^2} + \frac{\partial^2 \psi}{\partial z^2} \qquad (2.30)$$

or

$$\mathbf{V}^2 \psi = \frac{\partial^2 \psi}{\partial r^2} + \frac{1}{r} \frac{\partial \psi}{\partial r} + \frac{1}{r^2} \frac{\partial^2 \psi}{\partial \phi^2} + \frac{\partial^2 \psi}{\partial z^2} \qquad (2.31)$$

respectively. It is necessary to consider both these forms for a complete treatment of optical propagation in the fiber although many of the properties of interest may be dealt with using Cartesian coordinates.

The basic solution of the wave equation is a sinusoidal wave, the most important form of which is a uniform plane wave given by:

$$\psi = \psi_0 \exp j(\omega t - k \cdot r) \tag{2.32}$$

where ω is the angular frequency of the field, t is the time, k is the propagation vector which gives the direction of propagation and the rate of change of phase with distance, whilst the components of r specify the coordinate point at which the field is observed. When λ is the optical wavelength in a vacuum, the magnitude of the propagation vector or the vacuum propagation constant k (where $k = |k|$) is given by:

$$k = \frac{2\pi}{\lambda} \tag{2.33}$$

It should be noted that in this case k is also referred to as the free space wave number.

2.3.2 Modes in a Planar Guide

The planar guide is the simplest form of optical waveguide. We may assume it consists of a slab of dielectric with refractive index n_1 sandwiched between two regions of lower refractive index n_2. In order to obtain an improved model for optical propagation it is useful to consider the interference of plane wave components within this dielectric waveguide.

The conceptual transition from ray to wave theory may be aided by consideration of a plane monochromatic wave propagating in the direction of the ray path within the guide (see Fig. 2.8(a)). As the refractive index within the guide is n_1, the optical wavelength in this region is reduced to λ/n_1 whilst the vacuum propagation constant is increased to $n_1 k$. When θ is the angle between the wave propagation vector or the equivalent ray and the guide axis, the plane wave can be resolved into two component plane waves propagating in the z and x directions as shown in Fig. 2.8(a). The component of the propagation constant in the z direction β_z is given by:

$$\beta_z = n_1 k \cos \theta \tag{2.34}$$

The component of the propagation constant in the x direction β_x is:

$$\beta_x = n_1 k \sin \theta \tag{2.35}$$

The component of the plane wave in the x direction is reflected at the interface between the higher and lower refractive index media. When the total

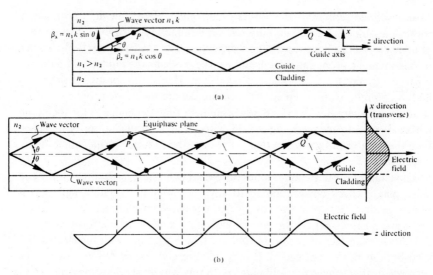

Fig. 2.8   The formation of a mode in a planar dielectric guide: (a) a plane wave propagating in the guide shown by its wave vector or equivalent ray—the wave vector is resolved into components in the z and x directions; (b) the interference of plane waves in the guide forming the lowest order mode ($m = 0$).

phase change* after two successive reflections at the upper and lower interfaces (between the points P and Q) is equal to 2 $m\pi$ radians, where m is an integer, then constructive interference occurs and a standing wave is obtained in the x direction. This situation is illustrated in Fig. 2.8(b) where the interference of two plane waves is shown. In this illustration it is assumed that the interference forms the lowest order (where $m = 0$) standing wave, where the electric field is a maximum at the center of the guide decaying towards zero at the boundary between the guide and cladding. However, it may be observed from Fig. 2.8(b) that the electric field penetrates some distance into the cladding, a phenomenon which is discussed in Section 2.3.4.

Nevertheless the optical wave is effectively confined within the guide and the electric field distribution in the x direction does not change as the wave propagates in the z direction. The sinusoidally varying electric field in the z direction is also shown in Fig. 2.8(b). The stable field distribution in the x direction with only a periodic z dependence is known as a mode. A specific mode is only obtained when the angle between the propagation vectors or the rays and the interface have a particular value as indicated in Fig. 2.8(b). In effect Eqs. (2.34) and (2.35) define a group or congruence of rays which in the

---

* It should be noted that there is a phase shift on reflection of the plane wave at the interface as well as a phase change with distance travelled. The phase shift on reflection at a dielectric interface is dealt with in Section 2.3.4.

case described represents the lowest order mode. Hence the light propagating within the guide is formed into discrete modes each typified by a distinct value of θ. These modes have a periodic z dependence of the form $\exp(-j\beta_z z)$ where β_z becomes the propagation constant for the mode as the modal field pattern is invariant except for a periodic z dependence. Hence for notational simplicity, and in common with accepted practice, we denote the mode propagation constant by β, where $\beta = \beta_z$. If we now assume a time dependence for the monochromatic electromagnetic light field with angular frequency ω of $\exp(j\omega t)$, then the combined factor $\exp j(\omega t - \beta z)$ describes a mode propagating in the z direction.

To visualize the dominant modes propagating in the z direction we may consider plane waves corresponding to rays at different specific angles in the planar guide. These plane waves give constructive interference to form standing wave patterns across the guide following a sine or cosine formula. Figure 2.9 shows examples of such rays for $m = 1, 2, 3$ together with the electric field distributions in the x direction. It may be observed that m denotes the number of zeros in this transverse field pattern. In this way m signifies the order of the mode and is known as the mode number.

When light is described as an electromagnetic wave it consists of a periodically varying electric field E and magnetic field H which are orientated

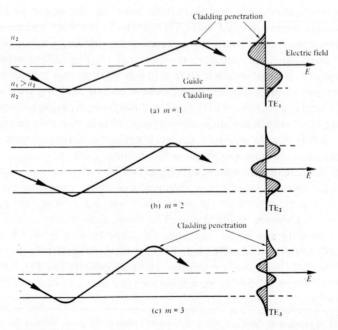

Fig. 2.9 Physical model showing the ray propagation and the corresponding transverse electric (TE) field patterns of three lower order modes ($m = 1, 2, 3$) in the planar dielectric guide.

at right angles to each other. The transverse modes shown in Fig. 2.9 illustrate the case when the electric field is perpendicular to the direction of propagation and hence $E_z = 0$, but a corresponding component of the magnetic field H is in the direction of propagation. In this instance the modes are said to be transverse electric (TE). Alternatively, when a component of the E field is in the direction of propagation, but $H_z = 0$, the modes formed are called transverse magnetic (TM). The mode numbers are incorporated into this nomenclature by referring to the TE_m and TM_m modes as illustrated for the transverse electric modes shown in Fig. 2.9. When the total field lies in the transverse plane, transverse electromagnetic (TEM) waves exist where both E_z and H_z are zero. However, although TEM waves occur in metallic conductors (e.g. coaxial cables) they are seldom found in optical waveguides.

2.3.3 Phase and Group Velocity

Within all electromagnetic waves, whether plane or otherwise, there are points of constant phase. For plane waves these constant phase points form a surface which is referred to as a wavefront. As a monochromatic light wave propagates along a waveguide in the z direction these points of constant phase travel at a phase velocity v_p given by:

$$v_p = \frac{\omega}{\beta} \tag{2.36}$$

where ω is the angular frequency of the wave. However, it is impossible in practice to produce perfectly monochromatic light waves, and light energy is generally composed of a sum of plane wave components of different frequencies. Often the situation exists where a group of waves with closely similar frequencies propagate so that their resultant forms a packet of waves. The formation of such a wave packet resulting from the combination of two waves of slightly different frequency propagating together is illustrated in Fig. 2.10. This wave packet does not travel at the phase velocity of the individual waves but is observed to move at a group velocity v_g given by

$$v_g = \frac{\delta\omega}{\delta\beta} \tag{2.37}$$

The group velocity is of greatest importance in the study of the transmission characteristics of optical fibers as it relates to the propagation characteristics of observable wave groups or packets of light.

If propagation in an infinite medium of refractive index n_1 is considered, then the propagation constant may be written as:

$$\beta = n_1 \frac{2\pi}{\lambda} = \frac{n_1 \omega}{c} \tag{2.38}$$

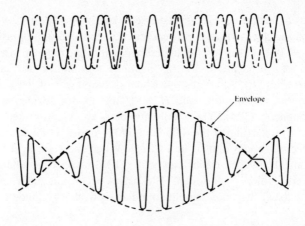

Fig. 2.10   The formation of a wave packet from the combination of two waves with nearly equal frequencies. The envelope of the wave packet or group of waves travels at a group velocity v_g.

where c is the velocity of light in free space. Equation (2.38) follows from Eqs. (2.33) and (2.34) where we assume propagation in the z direction only and hence $\cos \theta$ is equal to unity. Using Eq. (2.36) we obtain the following relationship for the phase velocity:

$$v_p = \frac{c}{n_1} \tag{2.39}$$

Similarly employing Eq. (2.37), where in the limit $\delta\omega/\delta\beta$ becomes $d\omega/d\beta$, the group velocity:

$$v_g = \frac{d\lambda}{d\beta} \cdot \frac{d\omega}{d\lambda} = \frac{d}{d\omega} \left(n_1 \frac{2\pi}{\lambda} \right)^{-1} \left(\frac{-\omega}{\lambda} \right)$$

$$= \frac{-\omega}{2\pi\lambda} \left(\frac{1}{\lambda} \frac{dn_1}{d\lambda} - \frac{n_1}{\lambda^2} \right)^{-1}$$

$$= \frac{c}{\left(n_1 - \lambda \frac{dn_1}{d\lambda} \right)} = \frac{c}{N_1} \tag{2.40}$$

The parameter N_1 is known as the group index of the guide.

2.3.4 Phase Shift with Total Internal Reflection and the Evanescent Field

The discussion of electromagnetic wave propagation in the planar waveguide given in Section 2.3.2 drew attention to certain phenomena that occur at the guide–cladding interface which are not apparent from ray theory considerations of optical propagation. In order to appreciate these phenomena it is necessary to use the wave theory model for total internal reflection at a planar interface. This is illustrated in Fig. 2.11, where the arrowed lines represent wave propagation vectors and a component of the wave energy is shown to be transmitted through the interface into the cladding. The wave equation in Cartesian coordinates for the electric field in a lossless medium is:

$$\mathbf{V}^2 E = \mu\varepsilon \frac{\partial^2 E}{\partial t^2} = \frac{\partial^2 E}{\partial x^2} + \frac{\partial^2 E}{\partial y^2} + \frac{\partial^2 E}{\partial z^2} \qquad (2.41)$$

As the guide–cladding interface lies in the y–z plane and the wave is incident in the x–z plane onto the interface, then $\partial/\partial y$ may be assumed to be zero. Since the phase fronts must match at all points along the interface in the z direction, the three waves shown in Fig. 2.11 will have the same propagation constant β in this direction. Therefore from the discussion of Section 2.3.2 the wave propagation in the z direction may be described by $\exp j(\omega t - \beta z)$. In addition, there will also be propagation in the x direction. When the components are resolved in this plane:

$$\beta_{x1} = n_1 k \cos \phi_1 \qquad (2.42)$$

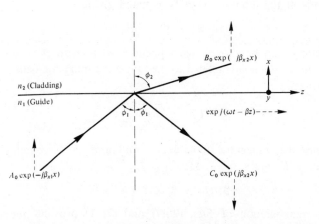

Fig. 2.11 A wave incident on the guide–cladding interface of a planar dielectric waveguide. The wave vectors of the incident, transmitted and reflected waves are indicated (solid arrowed lines) together with their components in the z and x directions (dashed arrowed lines).

$$\beta_{x2} = n_2 k \cos \phi_2 \tag{2.43}$$

where β_{x1} and β_{x2} are propagation constants in the x direction for the guide and cladding respectively. Thus the three waves in the waveguide indicated in Fig. 2.11, the incident, the transmitted and the reflected, with amplitudes A, B and C respectively will have the forms:

$$A = A_0 \exp -(j\beta_{x1} x) \exp j(\omega t - \beta z) \tag{2.44}$$

$$B = B_0 \exp -(j\beta_{x2} x) \exp j(\omega t - \beta z) \tag{2.45}$$

$$C = C_0 \exp (j\beta_{x1} x) \exp j(\omega t - \beta z) \tag{2.46}$$

Using the simple trigonometrical relationship $\cos^2 \phi + \sin^2 \phi = 1$:

$$\beta_{x1}^2 = (n_1^2 k^2 - \beta^2) = -\xi_1^2 \tag{2.47}$$

and

$$\beta_{x2}^2 = (n_2^2 k^2 - \beta^2) = -\xi_2^2 \tag{2.48}$$

When an electromagnetic wave is incident upon an interface between two dielectric media, Maxwell's equations require that both the tangential components of E and H and the normal components of D $(= \varepsilon E)$ and B $(= \mu H)$ are continuous across the boundary. If the boundary is defined at $x = 0$ we may consider the cases of the transverse electric (TE) and transverse magnetic (TM) modes.

Initially let us consider the TE field at the boundary. When Eqs. (2.44) and (2.46) are used to represent the electric field components in the y direction E_y and the boundary conditions are applied, then the normal components of the E and H fields at the interface may be equated giving:

$$A_0 + C_0 = B_0 \tag{2.49}$$

Furthermore it can be shown (see Appendix A) that an electric field component in the y direction is related to the tangential magnetic field component H_z following:

$$H_z = \frac{j}{\mu_r \mu_0 \omega} \frac{\partial E_y}{\partial x} \tag{2.50}$$

Applying the tangential boundary conditions and equating H_z by differentiating E_y gives:

$$-\beta_{x1} A_0 + \beta_{x2} C_0 = -\beta_{x2} B_0 \tag{2.51}$$

Algebraic manipulation of Eqs. (2.49) and (2.51) provides the following results:

$$C_0 = A_0 \left(\frac{\beta_{x1} - \beta_{x2}}{\beta_{x1} + \beta_{x2}} \right) = A_0 r_{ER} \tag{2.52}$$

$$B_0 = A_0 \left(\frac{2\beta_{x1}}{\beta_{x1} + \beta_{x2}} \right) = A_0 r_{ET} \tag{2.53}$$

where r_{ER} and r_{ET} are the reflection and transmission coefficients for the E field at the interface respectively. The expressions obtained in Eqs. (2.52) and (2.53) correspond to the Fresnel relationships |Ref. 10| for radiation polarized perpendicular to the interface (E polarization).

When both β_{x1} and β_{x2} are real it is clear that the reflected wave C is in phase with the incident wave A. This corresponds to partial reflection of the incident beam. However, as ϕ_1 is increased the component β_z (i.e. β) increases and following Eqs. (2.47) and (2.48), the components β_{x1} and β_{x2} decrease. Continuation of this process results in β_{x2} passing through zero, a point which is signified by ϕ_1 reaching the critical angle for total internal reflection. If ϕ_1 is further increased the component β_{x2} becomes imaginary and we may write it in the form $-j\xi_2$. During this process β_{x1} remains real because we have assumed that $n_1 > n_2$. Under the conditions of total internal reflection Eq. (2.52) may therefore be written as:

$$C_0 = A_0 \left(\frac{\beta_{x1} + j\xi_2}{\beta_{x2} - j\xi_2} \right) = A_0 \exp 2j\delta_E \tag{2.54}$$

where we observe there is a phase shift of the reflected wave relative to the incident wave. This is signified by δ_E which is given by:

$$\tan \delta_E = \frac{\xi_2^2}{\beta_{x1}} \tag{2.55}$$

Furthermore the modulus of the reflected wave is identical to the modulus of the incident wave ($|C_0| = |A_0|$). The curves of the amplitude reflection coefficient $|r_{ER}|$ and phase shift on reflection, against angle of incidence ϕ_1, for TE waves incident on a glass–air interface are displayed in Fig. 2.12 |Ref. 14|. These curves illustrate the above results, where under the conditions of total internal reflection the reflected wave has an equal amplitude to the incident wave, but undergoes a phase shift corresponding to δ_E degrees.

A similar analysis may be applied to the TM modes at the interface which leads to expressions for reflection and transmission of the form |Ref. 14|:

$$C_0 = A_0 \left(\frac{\beta_{x1} n_2^2 - \beta_{x2} n_1^2}{\beta_{x1} n_2^2 + \beta_{x2} n_1^2} \right) = A_0 r_{HR} \tag{2.56}$$

and

$$B_0 = A_0 \left(\frac{2\beta_{x1} n_2^2}{\beta_{x1} n_2^2 + \beta_{x2} n_1^2} \right) = A_0 r_{HT} \tag{2.57}$$

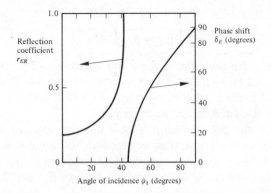

Fig. 2.12 Curves showing the reflection coefficient and phase shift on reflection for transverse electric waves against the angle of incidence for a glass–air interface ($n_1 = 1.5$, $n_2 = 1.0$). Reproduced with permission from J. E. Midwinter, *Optical Fibers for Transmission*, John Wiley & Sons Inc., 1979.

where r_{HR} and r_{HT} are respectively the reflection and transmission coefficients for the H field at the interface. Again the expressions given in Eqs. (2.56) and (2.57) correspond to Fresnel relationships |Ref. 10|, but in this case they apply to radiation polarized parallel to the interface (H polarization). Furthermore, considerations of an increasing angle of incidence ϕ_1, such that β_{x2} goes to zero and then becomes imaginary, again results in a phase shift when total internal reflection occurs. However, in this case a different phase shift is obtained corresponding to

$$C_0 = A_0 \exp(2j\delta_H) \tag{2.58}$$

where

$$\tan \delta_H = \left(\frac{n_1}{n_2}\right)^2 \tan \delta_E \tag{2.59}$$

Thus the phase shift obtained on total internal reflection is dependent upon both the angle of incidence and the polarization (either TE or TM) of the radiation.

The second phenomenon of interest under conditions of total internal reflection is the form of the electric field in the cladding of the guide. Before the critical angle for total internal reflection is reached and hence when there is only partial reflection, the field in the cladding is of the form given by Eq. (2.45). However, as indicated previously, when total internal reflection occurs, β_{x2} becomes imaginary and may be written as $-j\xi_2$. Substituting for β_{x2} in Eq. (2.45) gives the transmitted wave in the cladding as:

$$B = B_0 \exp(-\xi_2 x) \exp j(\omega t - \beta z) \tag{2.60}$$

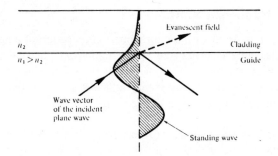

Fig. 2.13　The exponentially decaying evanescent field in the cladding of the optical waveguide.

Thus the amplitude of the field in the cladding is observed to decay exponentially* in the x direction. Such a field, exhibiting an exponentially decaying amplitude, is often referred to as an evanescent field. Figure 2.13 shows a diagrammatic representation of the evanescent field. A field of this type stores energy and transports it in the direction of propagation (z) but does not transport energy in the transverse direction (x). Nevertheless the existence of an evanescent field beyond the plane of reflection in the lower index medium indicates that optical energy is transmitted into the cladding.

The penetration of energy into the cladding underlines the importance of the choice of cladding material. It gives rise to the following requirements:

(a) The cladding should be transparent to light at the wavelengths over which the guide is to operate.

(b) Ideally the cladding should consist of a solid material in order to avoid both damage to the guide and the accumulation of foreign matter on the guide walls. These effects degrade the reflection process by interaction with the evanescent field. This in part explains the poor performance (high losses) of early optical waveguides with air cladding.

(c) The cladding thickness must be sufficient to allow the evanescent field to decay to a low value or losses from the penetrating energy may be encountered. In many cases, however, the magnitude of the field falls off rapidly with distance from the guide–cladding interface. This may occur within distances equivalent to a few wavelengths of the transmitted light.

Therefore the most widely used optical fibers consist of a core and cladding both made of glass. The cladding refractive index is thus higher than would be the case with liquid or gaseous cladding giving a lower numerical aperture for the fiber, but it provides a far more practical solution.

---

* It should be noted that we have chosen the sign of ξ_2 so that the exponential field decays rather than grows with distance into the cladding. In this case a growing exponential field is a physically improbable solution.

2.3.5   Goos–Haenchen Shift

The phase change incurred with the total internal reflection of a light beam on a planar dielectric interface may be understood from physical observation. Careful examination shows that the reflected beam is shifted laterally from the trajectory predicted by simple ray theory analysis as illustrated in Fig. 2.14. This lateral displacement is known as the Goos–Haenchen shift after its first observers.

The geometric reflection appears to take place at a virtual reflecting plane which is parallel to the dielectric interface in the lower index medium as indicated in Fig. 2.14. Utilizing wave theory it is possible to determine this lateral shift |Ref. 14| although it is very small ($d \simeq 0.06$–0.10 μm for a silvered glass interface at a wavelength of 0.55 μm) and difficult to observe. However, this concept provides an important insight into the guidance mechanism of dielectric optical waveguides.

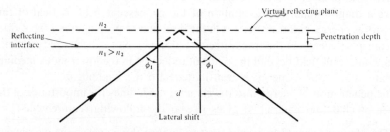

Fig. 2.14   The lateral displacement of a light beam on reflection at a dielectric interface (Goos–Haenchen shift).

2.3.6   Cylindrical Fiber

The exact solution of Maxwell's equations for a cylindrical homogeneous core dielectric waveguide* involves much algebra and yields a complex result |Ref. 15|. Although the presentation of this mathematics is beyond the scope of this text, it is useful to consider the resulting modal fields. In common with the planar guide (Section 2.3.2), TE (where $E_z = 0$) and TM (where $H_z = 0$) modes are obtained within the dielectric cylinder. The cylindrical waveguide, however, is bounded in two dimensions rather than one. Thus two integers, l and m, are necessary in order to specify the modes in contrast to the single integer (m) required for the planar guide. For the cylindrical waveguide we therefore refer to TE_{lm} and TM_{lm} modes. These modes correspond to meridional rays (see Section 2.2.1) travelling within the fiber. However, hybrid

---

*This type of optical waveguide with a constant refractive index core is known as a step index fiber (see Section 2.4).

modes where E_z and H_z are nonzero also occur within the cylindrical waveguide. These modes which result from skew ray propagation (see Section 2.2.4) within the fiber are designated HE_{lm} and EH_{lm} depending upon whether the components of H or E make the larger contribution to the transverse (to the fiber axis) field. Thus an exact description of the modal fields in a step index fiber proves somewhat complicated.

Fortunately the analysis may be simplified when considering optical fibers for communication purposes. These fibers satisfy the weakly guiding approximation [Ref. 16] where the relative index difference $\Delta \ll 1$. This corresponds to small grazing angles θ in Eq. (2.34). In fact Δ is usually less than 0.03 (3%) for optical communications fibers. For weakly guiding structures with dominant forward propagation, mode theory gives dominant transverse field components. Hence approximate solutions for the full set of HE, EH, TE and TM modes may be given by two linearly polarized components [Ref. 16]. These linearly polarized (LP) modes are not exact modes of the fiber except for the fundamental (lowest order) mode. However, as Δ in weakly guiding fibers is very small, then HE–EH mode pairs occur which have almost identical propagation constants. Such modes are said to be degenerate. The superpositions of these degenerating modes characterized by a common propagation constant correspond to particular LP modes regardless of their HE, EH, TE or TM field configurations. This linear combination of degenerate modes obtained from the exact solution produces a useful simplification in the analysis of weakly guiding fibers.

The relationship between the traditional HE, EH, TE and TM mode designations and the LP_{lm} mode designations are shown in Table 2.1. The mode subscripts l and m are related to the electric field intensity profile for a particular LP mode (see Fig. 2.15(d)). There are in general $2l$ field maxima around the circumference of the fiber core and m field maxima along a radius vector. Furthermore, it may be observed from Table 2.1 that the notation for labelling

Table 2.1 Correspondence between the lower order linearly polarized modes and the traditional exact modes from which they are formed

Linearly polarized	Exact
LP_{01}	HE_{11}
LP_{11}	$HE_{21}, TE_{01}, TM_{01}$
LP_{21}	HE_{31}, EH_{11}
LP_{02}	HE_{12}
LP_{31}	HE_{41}, EH_{21}
LP_{12}	$HE_{22}, TE_{02}, TM_{02}$
LP_{lm}	$HE_{2m}, TE_{0m}, TM_{0m}$
$LP_{lm} (l \neq 0 \text{ or } 1)$	$HE_{l+1,m}, EH_{l-1,m}$

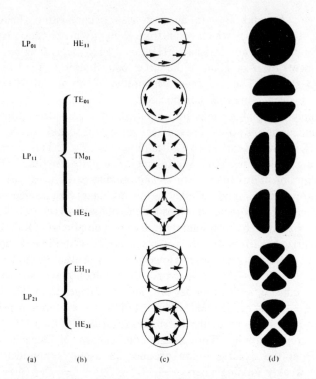

Fig. 2.15   The electric field configurations for the three lowest LP modes illustrated in terms of their constituent exact modes: (a) LP mode designations; (b) exact mode designations; (c) electric field distribution of the exact modes; (d) intensity distribution of E_x for the exact modes indicating the electric field intensity profile for the corresponding LP modes.

the HE and EH modes has changed from that specified for the exact solution in the cylindrical waveguide mentioned previously. The subscript l in the LP notation now corresponds to HE and EH modes with labels $l + 1$ and $l - 1$ respectively.

The electric field intensity profile for the lowest three LP modes, together with the electric field distribution of their constituent exact modes, are shown in Fig. 2.15. It may be observed from the field configurations of the exact modes that the field strength in the transverse direction (E_x or E_y) is identical for the modes which belong to the same LP mode. Hence the origin of the term 'linearly polarized'.

Using Eq. (2.31) for the cylindrical homogeneous core waveguide under the weak guidance conditions outlined above, the scalar wave equation can be written in the form [Ref. 17]:

$$\frac{d^2\psi}{dr^2} + \frac{1}{r}\frac{d\psi}{dr} + \frac{1}{r^2}\frac{d^2\psi}{d\phi^2} + (n_1^2 k^2 - \beta^2)\psi = 0 \qquad (2.61)$$

where ψ is the field (E or H), n_1 is the refractive index of the fiber core, k is the propagation constant for light in a vacuum, and r and ϕ are cylindrical coordinates. The propagation constants of the guided modes β lie in the range

$$n_2 k < \beta < n_1 k \qquad (2.62)$$

where n_2 is the refractive index of the fiber cladding. Solutions of the wave equation for the cylindrical fiber are separable, having the form:

$$\psi = E(r) \left\{ \frac{\cos l\phi}{\sin l\phi} \exp(\omega t - \beta z) \right\} \qquad (2.63)$$

where in this case ψ represents the dominant transverse electric field component. The periodic dependence on ϕ following $\cos l\phi$ or $\sin l\phi$ gives a mode of radial order l. Hence the fiber supports a finite number of guided modes of the form of Eq. (2.63).

Introducing the solutions given by Eq. (2.63) into Eq. (2.61) results in a

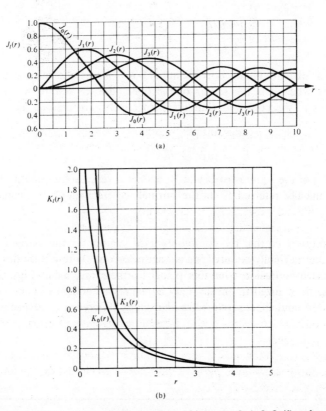

Fig. 2.16 (a) Variation of the Bessel function $J_l(r)$ for $l = 0, 1, 2, 3$ (first four orders), plotted against r. (b) Graph of the modified Bessel function $K_l(r)$ against r for $l = 0, 1$.

differential equation of the form:

$$\frac{d^2 E}{dr^2} + \frac{1}{r}\frac{dE}{dr} + \left[(n_1 k^2 - \beta^2) - \frac{l^2}{r^2}\right] E = 0 \qquad (2.64)$$

For a step index fiber with a constant refractive index core, Eq. (2.64) is a Bessel's differential equation and the solutions are cylinder functions. In the core region the solutions are Bessel functions denoted by J_l. A graph of these gradually damped oscillatory functions (with respect to r) is shown in Fig. 2.16(a). It may be noted that the field is finite at $r = 0$ and may be represented by the zero order Bessel function J_0. However, the field vanishes as r goes to infinity and the solutions in the cladding are therefore modified Bessel functions denoted by K_l. These modified functions decay exponentially with respect to r as illustrated in Fig. 2.16(b). The electric field may therefore be given by:

$$E(r) = GJ_l(UR) \qquad \text{for} \quad R < 1 \quad \text{(core)}$$

$$\qquad (2.65)$$

$$= GJ_l(U)\frac{K_l(WR)}{K_l(W)} \qquad \text{for} \quad R > 1 \quad \text{(cladding)}$$

where G is the amplitude coefficient and $R = r/a$ is the normalized radial coordinate when a is the radius of the fiber core. U and W which are the eigenvalues in the core and cladding respectively, are defined as:

$$U = a(n_1^2 k^2 - \beta^2)^{\frac{1}{2}} \qquad (2.66)$$

$$W = a(\beta^2 - n_2^2 k^2)^{\frac{1}{2}} \qquad (2.67)$$

The sum of the squares of U and W defines a very useful quantity |Ref. 18| which is usually referred to as the normalized frequency* V where

$$V = (U^2 + W^2)^{\frac{1}{2}} = ka(n_1^2 - n_2^2)^{\frac{1}{2}} \qquad (2.68)$$

It may be observed that the commonly used symbol for this parameter is the same as that normally adopted for voltage. However, within this chapter there should be no confusion over this point. Furthermore, using Eqs. (2.8) and (2.10), the normalized frequency may be expressed in terms of the numerical aperture NA and the relative refractive index difference Δ respectively as:

$$V = \frac{2\pi}{\lambda} a(\text{NA}) \qquad (2.69)$$

---

* When used in the context of the planar waveguide, V is sometimes known as the normalized film thickness as it relates to the thickness of the guide layer (see Section 11.7.1).

$$= \frac{2\pi}{\lambda} a n_1 (2\Delta)^{\frac{1}{2}} \qquad (2.70)$$

The normalized frequency is a dimensionless parameter and hence is also sometimes simply called the V number or value of the fiber.

It is also possible to define the normalized propagation constant b for a fiber in terms of the parameters of Eq. (2.68) so that:

$$b = 1 - \frac{U^2}{V^2} = \frac{(\beta/k)^2 - n_2^2}{n_1^2 - n_2^2}$$

$$= \frac{(\beta/k)^2 - n_2^2}{2n_1^2\Delta} \qquad (2.71)$$

Referring to the expression for the guided modes given in Eq. (2.62), the limits of β are $n_2 k$ and $n_1 k$, hence b must lie between 0 and 1.

In the weak guidance approximation the field matching conditions at the boundary require continuity of the transverse and tangential electrical field components at the core–cladding interface (at $r = a$). Therefore, using the Bessel function relations outlined previously, an eigenvalue equation for the LP modes may be written in the following form [Ref. 18]:

$$U \frac{J_{l\pm1}(U)}{J_l(U)} = \pm W \frac{K_{l\pm1}(W)}{K_l(W)} \qquad (2.72)$$

Solving Eq. (2.72) with Eqs. (2.66) and (2.67) allows the eigenvalue U and hence β to be calculated as a function of the normalized frequency. In this way the propagation characteristics of the various modes and their dependence on the optical wavelength and the fiber parameters may be determined.

Considering the limit of mode propagation when $\beta = n_2 k$, then the mode phase velocity is equal to the velocity of light in the cladding and the mode is no longer properly guided. In this case the mode is said to be cut off and the eigenvalue $W = 0$ (Eq. 2.67). Unguided or radiation modes have frequencies below cutoff where $\beta < kn_2$, and hence W is imaginary. Nevertheless, wave propagation does not cease abruptly below cutoff. Modes exist where $\beta < kn_2$ but the difference is very small, such that some of the energy loss due to radiation is prevented by an angular momentum barrier [Ref. 20] formed near the core–cladding interface. Solutions of the wave equation giving these states are called leaky modes, and often behave as very lossy guided modes rather than radiation modes. Alternatively as β is increased above $n_2 k$, less power is propagated in the cladding until at $\beta = n_1 k$ all the power is confined to the fiber core. As indicated previously, this range of values for β signifies the guided modes of the fiber.

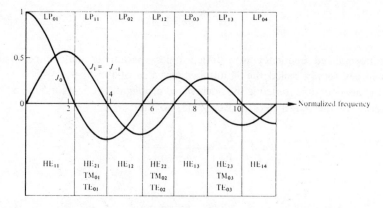

Fig. 2.17   The allowed regions for the LP modes of order $l = 0$, 1 against normalized frequency (V) for a circular optical waveguide with a constant refractive index core (step index fiber). Reproduced with permission from D. Gloge, *Appl. Opt.*, **10**, p. 2552, 1971.

The lower order modes obtained in a cylindrical homogeneous core waveguide are shown in Fig. 2.17 [Ref. 16]. Both the LP notation and the corresponding traditional HE, EH, TE and TM mode notations are indicated. In addition, the Bessel functions J_0 and J_1 are plotted against the normalized frequency and where they cross the zero gives the cutoff point for the various modes. Hence the cutoff point for a particular mode corresponds to a distinctive value of the normalized frequency (where $V = V_c$) for the fiber. It may be observed from Fig. 2.17 that the value of V_c is different for different modes. For example the first zero crossing J_1 occurs when the normalized frequency is 0 and this corresponds to the cutoff for the LP_{01} mode. However, the first zero crossing for J_0 is when the normalized frequency is 2.405, giving a cutoff value V_c of 2.405 for the LP_{11} mode. Similarly, the second zero of J_1 corresponds to a normalized frequency of 3.83, giving a cutoff value V_c for the LP_{02} mode of 3.83. It is therefore apparent that fibers may be produced with particular values of normalized frequency which allows only certain modes to propagate. This is further illustrated in Fig. 2.18 [Ref. 16] which shows the normalized propagation constant b for a number of LP modes as a function of V. It may be observed that the cutoff value of normalized frequency V_c which occurs when $\beta = n_2 k$ corresponds to $b = 0$.

The propagation of particular modes within a fiber may also be confirmed through visual analysis. The electric field distribution of different modes gives similar distributions of light intensity within the fiber core. These waveguide patterns (often called mode patterns) may give an indication of the predominant modes propagating in the fiber. The field intensity distributions for the three lower order LP modes were shown in Fig. 2.15. In Fig. 2.19 we illustrate the mode patterns for two higher order LP modes. However, unless

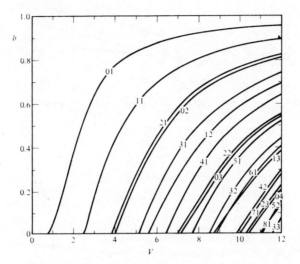

Fig. 2.18   The normalized propagation constant b as a function of normalized frequency V for a number of LP modes. Reproduced with permission from D. Gloge, *Appl. Opt.*, **10**, p. 2552, 1971.

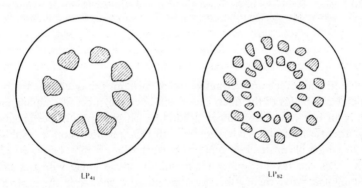

LP_{41} LP_{82}

Fig. 2.19   Sketches of fiber cross sections illustrating the distinctive light intensity distributions (mode patterns) generated by propagation of individual linearly polarized modes.

the fiber is designed for the propagation of a particular mode it is likely that the superposition of many modes will result in no distinctive pattern.

2.3.7   Mode Coupling

We have thus far considered the propagation aspects of perfect dielectric waveguides. However, waveguide perturbations such as deviations of the fiber axis from straightness, variations in the core diameter, irregularities at the

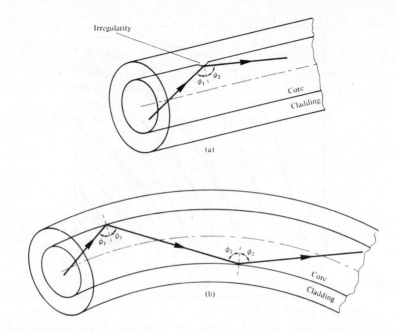

Fig. 2.20   Ray theory illustrations showing two of the possible fiber perturbations which give mode coupling: (a) irregularity at the core–cladding interface; (b) fiber bend.

core–cladding interface and refractive index variations may change the propagation characteristics of the fiber. These will have the effect of coupling energy travelling in one mode to another depending on the specific perturbation.

Ray theory aids the understanding of this phenomenon as shown in Fig. 2.20 which illustrates two types of perturbation. It may be observed that in both cases the ray no longer maintains the same angle with the axis. In electromagnetic wave theory this corresponds to a change in the propagating mode for the light. Thus individual modes do not normally propagate throughout the length of the fiber without large energy transfers to adjacent modes even when the fiber is exceptionally good quality and not strained or bent by its surroundings. This mode conversion is known as mode coupling or mixing. It is usually analyzed using coupled mode equations which can be obtained directly from Maxwell's equations. However, the theory is beyond the scope of this text and the reader is directed to Ref. 17 for a comprehensive treatment. Mode coupling affects the transmission properties of fibers in several important ways; a major one being in relation to the dispersive properties of fibers over long distances. This is pursued further in Sections 3.7–3.10.

2.4 STEP INDEX FIBERS

The optical fiber considered in the previous sections with a core of constant refractive index n_1 and a cladding of a slightly lower refractive index n_2 is known as step index fiber. This is because the refractive index profile for this type of fiber makes a step change at the core–cladding interface as indicated in Fig. 2.21 which illustrates the two major types of step index fiber. The refractive index profile may be defined as:

$$n(r) = \begin{cases} n_1 & r < a \quad \text{(core)} \\ n_2 & r \geqslant a \quad \text{(cladding)} \end{cases} \qquad (2.73)$$

in both cases.

Figure 2.21(a) shows a multimode step index fiber with a core diameter of around 50 μm or greater, which is large enough to allow the propagation of many modes within the fiber core. This is illustrated in Fig. 2.21(a) by the many different possible ray paths through the fiber. Figure 2.21(b) shows a single mode or monomode step index fiber which allows the propagation of only one transverse electromagnetic mode (typically HE_{11}), and hence the core diameter must be of the order of 2–10 μm. The propagation of a single mode is illustrated in Fig. 2.21(b) as corresponding to a single ray path only (usually shown as the axial ray) through the fiber.

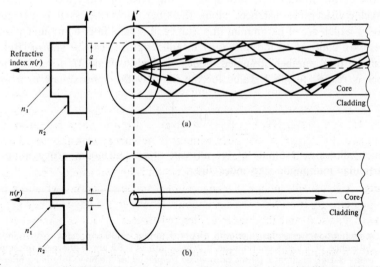

Fig. 2.21   The refractive index profile and ray transmission in step index fibers: (a) multimode step index fiber; (b) single mode step index fiber.

The single mode step index fiber has the distinct advantage of low intermodal dispersion (broadening of transmitted light pulses), as only one mode is transmitted, whereas with multimode step index fiber considerable dispersion may occur due to the differing group velocities of the propagating modes (see Section 3.9). This in turn restricts the maximum bandwidth attainable with multimode step index fibers, especially when compared with single mode fibers. However, for lower bandwidth applications multimode fibers have several advantages over single mode fibers. These are:

(a) The use of spatially incoherent optical sources (e.g. most light emitting diodes) which cannot be efficiently coupled to single mode fibers;
(b) Larger numerical apertures, as well as core diameters, facilitating easier coupling to optical sources;
(c) Lower tolerance requirements on fiber connectors.

2.4.1 Multimode Step Index Fibers

Multimode step index fibers allow the propagation of a finite number of guided modes along the channel. The number of guided modes is dependent upon the physical parameters (i.e. relative refractive index difference, core radius) of the fiber and the wavelengths of the transmitted light which are included in the normalized frequency V for the fiber. It was indicated in Section 2.3.6 that there is a cutoff value of normalized frequency V_c for guided modes below which they cannot exist. However, mode propagation does not entirely cease below cutoff. Modes may propagate as unguided or leaky modes which can travel considerable distances along the fiber. Nevertheless it is the guided modes which are of paramount importance in optical fiber communications as these are confined to the fiber over its full length. It can be shown |Ref. 16| that the total number of guided modes or mode volume M_s for a step index fiber is related to the V value for the fiber by the approximate expression:

$$M_s \simeq \frac{V^2}{2} \tag{2.74}$$

which allows an estimate of the number of guided modes propagating in a particular multimode step index fiber.

Example 2.4

A multimode step index fiber with a core diameter of 80 µm and a relative index difference of 1.5% is operating at a wavelength of 0.85 µm. If the core refractive index is 1.48, estimate: (a) the normalized frequency for the fiber; (b) the number of guided modes.

Solution: (a) The normalized frequency may be obtained from Eq. (2.70) where:

$$V \simeq \frac{2\pi}{\lambda} an_1 (2\Delta)^{\frac{1}{2}} = \frac{2\pi \times 40 \times 10^{-6} \times 1.48}{0.85 \times 10^{-6}} (2 \times 0.015)^{\frac{1}{2}}$$

$$= 75.8$$

(b) The total number of guided modes is given by Eq. (2.74) as:

$$M_s \simeq \frac{V^2}{2} = \frac{5745.6}{2}$$

$$= 2873$$

Hence this fiber has a V number of approximately 76 giving nearly 3000 guided modes.

Therefore as illustrated in example 2.4, the optical power is launched into a large number of guided modes each having different spatial field distributions, propagation constants, etc. In an ideal multimode step index fiber with properties (i.e. relative index difference, core diameter) which are independent of distance, there is no mode coupling, and the optical power launched into a particular mode remains in that mode and travels independently of the power launched into the other guided modes. Also the majority of these guided modes operate far from cutoff, and are well confined to the fiber core [Ref. 16]. Thus most of the optical power is carried in the core region and not in the cladding. The properties of the cladding (e.g. thickness) therefore do not significantly affect the propagation of these modes.

2.4.2 Single Mode Step Index Fibers

The advantage of the propagation of a single mode within an optical fiber is that the signal dispersion caused by the delay differences between different modes in a multimode fiber may be avoided (see Section 3.9). Multimode step index fibers do not lend themselves to the propagation of a single mode due to the difficulties of maintaining single mode operation within the fiber when mode conversion (i.e. coupling) to other guided modes takes place at both input mismatches and fiber imperfections. Hence for the transmission of a single mode the fiber must be designed to allow propagation of only one mode, whilst all other modes are attenuated by leakage or absorption.

Following the previous discussion of multimode fibers this may be achieved through choice of a suitable normalized frequency for the fiber. For single mode operation, only the fundamental LP_{01} mode can exist. Hence the limit of single mode operation depends on the lower limit of guided propagation for the LP_{11} mode. The cutoff normalized frequency for the LP_{11} mode occurs at $V_c = 2.405$ (see Section 2.3.6). Thus single mode propagation of the LP_{01}

mode is possible over the range:

$$0 \leqslant V < 2.405 \qquad (2.75)$$

as there is no cutoff for the fundamental mode. It must be noted that there are in fact two modes with orthogonal polarization over this range, and the term single mode applies to propagation of light of a particular polarization. Also, it is apparent that the normalized frequency for the fiber may be adjusted to within the range given in Eq. (2.75) by reduction of the core radius, and possibly the relative refractive index difference following Eq. (2.70).

Example 2.5

Estimate the maximum core diameter for an optical fiber with the same relative refractive index difference (1.5%) and core refractive index (1.48) as the fiber given in example 2.4 in order that it may be suitable for single mode operation. It may be assumed that the fiber is operating at the same wavelength (0.85 μm). Further, estimate the new maximum core diameter for single mode operation when the relative refractive index difference is reduced by a factor of ten.

Solution: Considering the relationship given in Eq. (2.75), the maximum V value for a fiber which gives single mode operation is 2.4. Hence from Eq. (2.70) the core radius a is:

$$a = \frac{V\lambda}{2\pi n_1 (2\Delta)^{\frac{1}{2}}} = \frac{2.4 \times 0.85 \times 10^{-6}}{2\pi \times 1.48 \times (0.03)^{\frac{1}{2}}}$$

$$= 1.3 \ \mu m$$

Therefore the maximum core diameter for single mode operation is approximately 2.6 μm.

Reducing the relative refractive index difference by a factor of 10 and again using Eq. (2.70) gives:

$$a = \frac{2.4 \times 0.85 \times 10^{-6}}{2\pi \times 1.48 \times (0.003)^{\frac{1}{2}}} = 4.0 \ \mu m.$$

Hence the maximum core diameter for single mode operation is now approximately 8 μm.

It is clear from example 2.5 that in order to obtain single mode operation with a maximum V number of 2.4 the single mode fiber must have a much smaller core diameter than the equivalent multimode step index fiber (in this case by a factor of 32). However, it is possible to achieve single mode operation with a slightly larger core diameter, albeit still much less than the diameter of multimode step index fiber, by reducing the relative refractive index difference of the fiber. Both these factors create difficulties with single mode fibers. The small core diameters pose problems with launching light into the fiber and with field jointing, and the reduced relative refractive index difference presents difficulties in the fiber fabrication process.

A further problem with single mode fibers with low relative refractive index differences and low V values is that the electromagnetic field associated with the LP_{10} mode extends appreciably into the cladding. For instance, with V values less than 1.4, over half the modal power propagates in the cladding [Ref. 20]. Thus the exponentially decaying evanescent field may extend significant distances into the cladding. It is therefore essential that the cladding is of a suitable thickness, and has low absorption and scattering losses in order to reduce attenuation of the mode. Estimates [Ref. 21] show that the necessary cladding thickness is of the order of 50 μm to avoid prohibitive losses (greater than 1 dB km^{-1}) in single mode fibers, especially when additional losses resulting from microbending (see Section 4.6.2) are taken into account. Therefore the total fiber cross section for single mode fibers is of a comparable size to multimode fibers.

Another approach to single mode fiber design which allows the V value to be increased above 2.405 is the W fiber [Ref. 23]. The refractive index profile for this fiber is illustrated in Fig. 2.22 where two cladding regions may be observed. Use of such two step cladding allows the loss threshold between the desirable and undesirable modes to be substantially increased. The fundamental mode will be fully supported with small cladding loss when its propagation constant lies in the range $kn_3 < \beta < kn_1$.

If the undesirable higher order modes which are excited or converted to have values of propagation constant $\beta < kn_3$, they will leak through the barrier layer between a_1 and a_2 (Fig. 2.22) into the outer cladding region n_3. Consequently these modes will lose power by radiation into the lossy surroundings. This design can provide single mode fibers with larger core diameters than the conventional single cladding approach which proves useful for easing jointing difficulties. W fibers also tend to give reduced losses at bends in comparison with conventional single mode fibers.

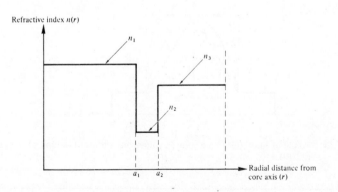

Fig. 2.22 The refractive index profile for the single mode W fiber.

2.5 GRADED INDEX FIBERS

Graded index fibers do not have a constant refractive index in the core* but a decreasing core index $n(r)$ with radial distance from a maximum value of n_1 at the axis to a constant value n_2 beyond the core radius a in the cladding. This index variation may be represented as:

$$n(r) = \begin{cases} n_1(1 - 2\Delta(r/a)^\alpha)^{\frac{1}{2}} & r < a \quad \text{(core)} \\ n_1(1 - 2\Delta)^{\frac{1}{2}} = n_2 & r \geqslant a \quad \text{(cladding)} \end{cases} \tag{2.76}$$

where Δ is the relative refractive index difference and α is the profile parameter which gives the characteristic refractive index profile of the fiber core. Equation (2.76) which is a convenient method of expressing the refractive index profile of the fiber core as a variation of α allows representation of the step index profile when $\alpha = \infty$, a parabolic profile when $\alpha = 2$ and a triangular profile when $\alpha = 1$. This range of refractive index profiles is illustrated in Fig. 2.23.

The graded index profiles which at present produce the best results for multimode optical propagation have a near parabolic refractive index profile core with $\alpha \approx 2$. Fibers with such core index profiles are well established and consequently when the term 'graded index' is used without qualification it usually refers to a fiber with this profile. For this reason in this section we consider the waveguiding properties of graded index fiber with a parabolic refractive index profile core.

A multimode graded index fiber with a parabolic index profile core is illustrated in Fig. 2.24. It may be observed that the meridional rays shown appear to follow curved paths through the fiber core. Using the concepts of

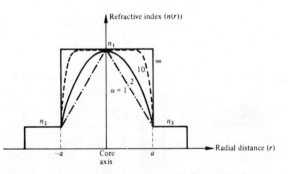

Fig. 2.23 Possible fiber refractive index profiles for different values of α (given in Eq. (2.76)).

---

* Graded index fibers are therefore sometimes referred to as inhomogeneous core fibers.

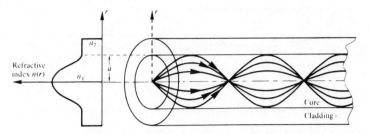

Fig. 2.24   The refractive index profile and ray transmission in a multimode graded index fiber.

geometric optics, the gradual decrease in refractive index from the center of the core creates many refractions of the rays as they are effectively incident on a large number of high to low index interfaces. This mechanism is illustrated in Fig. 2.25 where a ray is shown to be gradually curved, with an ever-increasing angle of incidence, until the conditions for total internal reflection are met, and the ray travels back towards the core axis, again being continuously refracted.

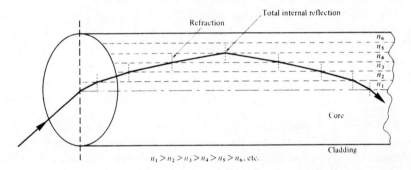

Fig. 2.25   An expanded ray diagram showing refraction at the various high to low index interfaces within a graded index fiber giving an overall curved ray path.

Multimode graded index fibers exhibit far less intermodal dispersion (see Section 3.9.2) than multimode step index fibers due to their refractive index profile. Although many different modes are excited in the graded index fiber, the different group velocities of the modes tend to be normalized by the index grading. Again considering ray theory, the rays travelling close to the fiber axis have shorter paths when compared with rays which travel into the outer regions of the core. However, the near axial rays are transmitted through a region of higher refractive index and therefore travel with a lower velocity than the more extreme rays. This compensates for the shorter path lengths and reduces dispersion in the fiber. A similar situation exists for skew rays which follow longer helical paths as illustrated in Fig. 2.26. These travel for the most

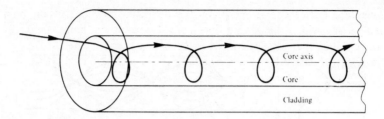

Fig. 2.26   A helical skew ray path within a graded index fiber.

part in the lower index region at greater speeds thus giving the same mechanism of mode transit time equalization. Hence multimode graded index fibers with parabolic or near parabolic index profile cores have transmission bandwidths which may be orders of magnitude greater than multimode step index fiber bandwidths. Consequently, although they are not capable of the bandwidths attainable with single mode fibers, such multimode graded index fibers have the advantage of large core diameters (greater than 30 μm) coupled with bandwidths suitable for long distance communication.

The parameters defined for step index fibers (i.e. NA, Δ, V) may be applied to graded index fibers and give a comparison between the two fiber types. However, it must be noted that for graded index fibers the situation is more complicated since the numerical aperture is a function of the radial distance from the fiber axis. Graded index fibers, therefore, accept less light than corresponding step index fibers with the same relative refractive index difference.

Electromagnetic mode theory may also be utilized with the graded profiles. Approximate field solutions of the same order as geometric optics are often obtained employing the WKB method from quantum mechanics after Wentzel, Kramers and Brillouin |Ref. 24|. Using the WKB method modal solutions of the guided wave are achieved by expressing the field in the form:

$$E_x = \tfrac{1}{2}|G_1(r)e^{jS(r)} + G_2(r)e^{-jS(r)}| \left(\frac{\cos l\phi}{\sin l\phi} \right) e^{j\beta z} \qquad (2.77)$$

where G and S are assumed to be real functions of the radial distance r.

Substitution of Eq. (2.77) into the scalar wave equation of the form given by Eq. (2.61) (in which the constant refractive index of the fiber core n_1 is replaced by $n(r)$) and neglecting the second derivative of $G_i(r)$ with respect to r provides approximate solutions for the amplitude function $G_i(r)$ and the phase function $S(r)$. It may be observed from the ray diagram shown in Fig. 2.24 that a light ray propagating in a graded index fiber does not necessarily reach every point within the fiber core. The ray is contained within two cylindrical caustic surfaces and for most rays a caustic does not coincide with the core–cladding interface. Hence the caustics define the classical turning points of the light ray within the graded fiber core. These turning points defined by the two caustics may be designated as occurring at $r = r_1$ and $r = r_2$.

The result of the WKB approximation yields an oscillatory field in the region $r_1 < r < r_2$ between the caustics where:

$$G_1(r) = G_2(r) = D/|(n^2(r)k^2 - \beta^2)r^2 - l^2|^{\frac{1}{4}} \tag{2.78}$$

(where D is an amplitude coefficient) and

$$S(r) = \int_{r_1}^{r_2} |(n^2(r)k^2 - \beta^2)r^2 - l^2|^{\frac{1}{2}} \frac{dr}{r} - \frac{\pi}{4} \tag{2.79}$$

Outside the interval $r_1 < r < r_2$ the field solution must have an evanescent form. In the region inside the inner caustic defined by $r < r_1$ and assuming r_1 is not too close to $r = 0$, the field decays towards the fiber axis giving:

$$G_1(r) = De^{jm\pi}/|l^2 - (n^2(r)k^2 - \beta^2)r^2|^{\frac{1}{4}} \tag{2.80}$$

$$G_2(r) = 0 \tag{2.81}$$

where the integer m is the radial mode number and

$$S(r) = j \int_r^{r_1} |l^2 - (n^2(r)k^2 - \beta^2)r^2|^{\frac{1}{2}} \frac{dr}{r} \tag{2.82}$$

Also outside the outer caustic in the region $r > r_2$, the field decays away from the fiber axis and is described by the equations:

$$G_1(r) = De^{jm\pi}/|l^2 - (n^2(r)k^2 - \beta^2)r^2|^{\frac{1}{4}} \tag{2.83}$$

$$G_2(r) = 0 \tag{2.84}$$

$$S(r) = j \int_{r_2}^r |l^2 - (n^2(r)k^2 - \beta^2)r^2|^{\frac{1}{2}} \frac{dr}{r} \tag{2.85}$$

The WKB method does not initially provide valid solutions of the wave equation in the vicinity of the turning points. Fortunately this may be amended by replacing the actual refractive index profile by a linear approximation at the location of the caustics. The solutions at the turning points can then be expressed in terms of Hankel functions of the first and second kind of order $\frac{1}{3}$ |Ref. 25|. This facilitates the joining together of the two separate solutions described previously for inside and outside the interval $r_1 < r < r_2$. Thus the WKB theory provides an approximate eigenvalue equation for the propagation constant β of the guided modes which cannot be determined using ray theory. The WKB eigenvalue equation of which β is a solution is given by |Ref. 25|:

$$\int_{r_1}^{r_2} |(n^2(r)k^2 - \beta^2)r^2 - l^2|^{\frac{1}{2}} \frac{dr}{r} = (2m - 1)\frac{\pi}{2} \tag{2.86}$$

where the radial mode number $m = 1, 2, 3 \ldots$ and determines the number of maxima of the oscillatory field in the radial direction. This eigenvalue equation can only be solved in a closed analytical form for a few simple refractive index profiles. Hence, in most cases it must be solved approximately or with the use of numerical techniques.

Finally the amplitude coefficient D may be expressed in terms of the total optical power P_G within the guided mode. Considering the power carried between the turning points r_1 and r_2 gives a geometric optics approximation of |Ref. 28|,

$$D = \frac{4(\mu_0/\varepsilon_0)^{\frac{1}{4}} P_G^{\frac{1}{2}}}{n_1 \pi a^2 I} \tag{2.87}$$

where

$$I = \int_{r_1/a}^{r_2/a} \frac{x\,dx}{|(n^2(ax)k^2 - \beta^2)a^2 x^2 - l^2|^{\frac{1}{2}}} \tag{2.88}$$

The properties of the WKB solution may be observed from a graphical representation of the integrand given in Eq. (2.79). This is shown in Fig. 2.27 together with the corresponding WKB solution. Figure 2.27 illustrates the

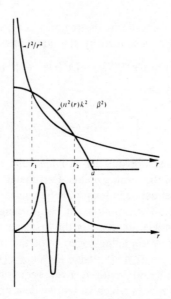

Fig. 2.27 Graphical representation of the functions $(n^2(r)k^2 - \beta^2)$ and (l^2/r^2) that are important in the WKB solution and which define the turning points r_1 and r_2. Also shown is an example of the corresponding WKB solution for a guided mode where an oscillatory wave exists in the region between the turning points.

functions $(n^2(r)k^2 - \beta^2)$ and (l^2/r^2). The two curves intersect at the turning points $r = r_1$ and $r = r_2$. The oscillatory nature of the WKB solution between the turning points (i.e. when $l^2/r^2 < n^2(r)k^2 - \beta^2$) which changes into a decaying exponential (evanescent) form outside the interval $r_1 < r < r_2$ (i.e. when $l^2/r^2 > n^2(r)k^2 - \beta^2$) can also be clearly seen.

It may be noted that as the azimuthal mode number l increases, the curve l^2/r^2 moves higher and the region between the two turning points becomes narrower. In addition, even when l is fixed the curve $(n^2(r)k^2 - \beta^2)$ is shifted up and down with alterations in the value of the propagation constant β. Therefore modes far from cutoff which have large values of β exhibit more closely spaced turning points. As the value of β decreases below $n_2 k$, $(n^2(r)k^2 - \beta^2)$ is no longer negative for large values of r and the guided mode situation depicted in Fig. 2.27 changes to one corresponding to Fig. 2.28. In this case a third turning point $r = r_3$ is created when at $r = a$ the curve $(n^2(r)k^2 - \beta^2)$ becomes constant, thus allowing the curve (l^2/r^2) to drop below it. Now the field displays an evanescent, exponentially decaying form in the region $r_2 < r < r_3$ as shown in Fig. 2.28. Moreover, for $r > r_3$ the field resumes an oscillatory behavior and therefore carries power away from the fiber core. Unless mode cutoff occurs at $\beta = n_2 k$ the guided mode is no longer fully contained within the fiber core but loses power through leakage or tunnelling into the cladding. This situation corresponds to the leaky modes mentioned previously in Section 2.3.6.

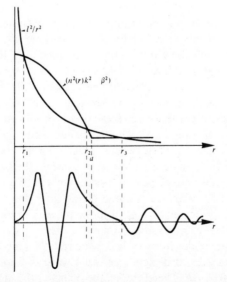

Fig. 2.28 Similar graphical representation as that illustrated in Fig. 2.27. Here the curve $(n^2(r)k^2 - \beta^2)$ no longer goes negative and a third turning point r_3 occurs. This corresponds to leaky mode solutions in the WKB method.

The WBK method may be used to calculate the propagation constants for the modes in a parabolic refractive index profile core fiber where following Eq. (2.76):

$$n^2(r) = n_1^2 \left(1 - 2 \left(\frac{r}{a} \right)^2 \Delta \right) \quad \text{for} \quad r < a \qquad (2.89)$$

Substitution of Eq. (2.89) into Eq. (2.86) gives:

$$\int_{r_1}^{r_2} \left[n_1^2 k^2 - \beta^2 - 2n_1^2 k^2 \left(\frac{r}{a} \right)^2 \Delta - \frac{l^2}{r^2} \right]^{\frac{1}{2}} dr = (m + \tfrac{1}{2})\pi \qquad (2.90)$$

The integral shown in Eq. (2.90) can be evaluated using a change of variable from r to $u = r^2$. The integral obtained may be found in a standard table of indefinite integrals |Ref. 29|. As the square root term in the resulting expression goes to zero at the turning points (i.e. $r = r_1$ and $r = r_2$), then we can write

$$\left[\frac{a(n_1 k^2 - \beta^2)}{4n_1 k \sqrt{(2\Delta)}} - \frac{l}{2} \right] \pi = (m + \tfrac{1}{2})\pi \qquad (2.91)$$

Solving Eq. (2.91) for β^2 gives:

$$\beta^2 = n_1^2 k^2 \left[\frac{1 - 2\sqrt{(2\Delta)}}{n_1 k a} (2m + l + 1) \right] \qquad (2.92)$$

It is interesting to note that the solution for the propagation constant for the various modes in a parabolic refractive index core fiber given in Eq. (2.92) is exact even though it was derived from the approximate WKB eigenvalue equation (Eq. 2.86). However, although Eq. (2.92) is an exact solution of the scalar wave equation for an infinitely extended parabolic profile medium, the wave equation is only an approximate representation of Maxwell's equation. Furthermore, practical parabolic refractive index profile core fibers exhibit a truncated parabolic distribution which merges into a constant refractive index at the cladding. Hence Eq. (2.92) is not exact for real fibers.

Equation (2.92) does, however, allow us to consider the mode number plane spanned by the radial and azimuthal mode numbers m and l. This plane is displayed in Fig. 2.29 where each mode of the fiber described by a pair of mode numbers is represented as a point in the plane. The mode number plane contains guided, leaky and radiation modes. The mode boundary which separates the guided modes from the leaky and radiation modes is indicated by the solid line in Fig. 2.29. It depicts a constant value of β following Eq. (2.92) and occurs when $\beta = n_2 k$. Therefore, all the points in the mode number plane lying below the line $\beta = n_2 k$ are associated with guided modes whereas the region above this line is occupied by leaky and radiation modes. The concept of the mode plane allows us to count the total number of guided modes within the fiber. For each pair of mode numbers m and l the corresponding mode field can have azimuthal mode dependence $\cos l\phi$ or $\sin l\phi$ and can exist in two possible polarizations (see Section 3.12). Hence the modes are said to be fourfold degenerate.* If we define the mode boundary as the function $m = f(l)$ then the total number of guided modes M is given by:

$$M = 4 \int_0^{l_{max}} f(l)\, dl \qquad (2.93)$$

as each representation point corresponding to four modes occupies an element of unit area in the mode plane. Equation (2.93) allows the derivation of the total number of guided modes or mode volume M_g supported by the graded index fiber. It can be shown |Ref. 25| that:

$$M_g = \left(\frac{\alpha}{\alpha + 2} \right) (n_1 ka)^2 \Delta \qquad (2.94)$$

Furthermore, utilizing Eq. (2.70), the normalized frequency V for the fiber when $\Delta \ll 1$ is approximately given by:

$$V = n_1 ka(2\Delta)^{\frac{1}{2}} \qquad (2.95)$$

Substituting Eq. (2.95) into Eq. (2.94), we have:

$$M_g \simeq \left(\frac{\alpha}{\alpha + 2} \right) \left(\frac{V^2}{2} \right) \qquad (2.96)$$

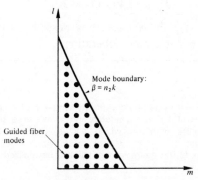

Fig. 2.29 The mode number plane illustrating the mode boundary and the guided fiber modes.

---

* An exception to this are the modes that occur when $l = 0$ which are only doubly degenerate as $\cos l\phi$ becomes unity and $\sin l\phi$ vanishes. However, these modes represent only a small minority and therefore may be neglected.

Hence for a parabolic refractive index profile core fiber ($\alpha = 2$), $M_g \approx V^2/4$ which is half the number supported by a step index fiber ($\alpha = \infty$) with the same V value.

Example 2.6

A graded index fiber has a core with a parabolic refractive index profile which has a diameter of 50 μm. The fiber has a numerical aperture of 0.2. Estimate the total number of guided modes propagating in the fiber when it is operating at a wavelength of 1 μm.
Solution: Using Eq. (2.69), the normalized frequency for the fiber is:

$$V = \frac{2\pi}{\lambda} a(NA) = \frac{2\pi \times 25 \times 10^{-6} \times 0.2}{1 \times 10^{-6}}$$

$$= 31.4$$

The mode volume may be obtained from Eq. (2.96) where for a parabolic profile:

$$M_g \simeq \frac{V^2}{4} = \frac{986}{4} = 247$$

Hence the fiber supports approximately 247 guided modes.

Graded index fibers may also be designed for single mode operation although there is no obvious advantage to this as in the step index case. However, it may be shown [Ref. 30] that the cutoff value of normalized frequency V_c to support a single mode in a graded index fiber is given by:

$$V_c = 2.405 \left(1 + 2/\alpha\right)^{\frac{1}{2}} \tag{2.97}$$

Therefore, as in the step index case, it is possible to determine the fiber parameters which give single mode operation.

Example 2.7

A graded index fiber with a parabolic refractive index profile core has a refractive index at the core axis of 1.5 and a relative index difference of 1%. Estimate the maximum possible core diameter which allows single mode operation at a wavelength of 1.3 μm.
Solution: Using Eq. (2.97) the maximum value of normalized frequency for single mode operation is

$$V = 2.4(1 + 2/\alpha)^{\frac{1}{2}} = 2.4(1 + 2/2)^{\frac{1}{2}}$$

$$= 2.4\sqrt{2}$$

The maximum core radius may be obtained from Eq. (2.95) where:

$$a = \frac{V\lambda}{2\pi n_1 (2\Delta)^{\frac{1}{2}}} = \frac{2.4\sqrt{2} \times 1.3 \times 10^{-6}}{2\pi \times 1.5 \times (0.02)^{\frac{1}{2}}}$$

$$= 3.3 \ \mu m$$

Hence the maximum core diameter which allows single mode operation is approximately 6.6 µm.

It may be noted that the critical value of normalized frequency for the parabolic profile graded index fiber is increased by a factor of $\sqrt{2}$ on the step index case. This gives a core diameter increased by a similar factor for the graded index fiber over a step index fiber with the equivalent core refractive index (equivalent to the core axis index), and the same relative refractive index difference.

The maximum V number which permits single mode operation can be increased still further when a graded index fiber with a triangular profile is employed. It is apparent from Eq. (2.97) that the increase in this case is by a factor of $\sqrt{3}$ over comparable step index fiber. Hence significantly larger core diameter single mode fibers may be produced utilizing this index profile. Such fibers have recently generated some interest |Ref. 38| for use in single mode transmission at wavelengths of 1.55 µm.

PROBLEMS

2.1 Using simple ray theory, describe the mechanism for the transmission of light within an optical fiber. Briefly discuss with the aid of a suitable diagram what is meant by the acceptance angle for an optical fiber. Show how this is related to the fiber numerical aperture and the refractive indices for the fiber core and cladding.

An optical fiber has a numerical aperture of 0.20 and a cladding refractive index of 1.59. Determine:
(a) the acceptance angle for the fiber in water which has a refractive index of 1.33;
(b) the critical angle at the core–cladding interface.
Comment on any assumptions made about the fiber.

2.2 The velocity of light in the core of a step index fiber is 2.01×10^8 m s^{-1}, and the critical angle at the core–cladding interface is 80°. Determine the numerical aperture and the acceptance angle for the fiber in air, assuming it has a core diameter suitable for consideration by ray analysis. The velocity of light in a vacuum is 2.998×10^3 m s^{-1}.

2.3 Define the relative refractive index difference for an optical fiber and show how it may be related to the numerical aperture.

A step index fiber with a large core diameter compared with the wavelength of the transmitted light has an acceptance angle in air of 22° and a relative refractive index difference of 3%. Estimate the numerical aperture and the critical angle at the core–cladding interface for the fiber.

2.4 A step index fiber has a solid acceptance angle in air of 0.115 radians and a relative refractive index difference of 0.9%. Estimate the speed of light in the fiber core.

2.5 Briefly indicate with the aid of suitable diagrams the difference between meridional and skew ray paths in step index fibers.

Derive an expression for the acceptance angle for a skew ray which changes direction by an angle 2γ at each reflection in a step index fiber in terms of the fiber NA and γ. It may be assumed that ray theory holds for the fiber.

A step index fiber with a suitably large core diameter for ray theory considerations has core and cladding refractive indices of 1.44 and 1.42 respectively. Calculate the acceptance angle in air for skew rays which change direction by 150° at each reflection.

2.6 Skew rays are accepted into a large core diameter (compared to the wavelength of the transmitted light) step index fiber in air at a maximum axial angle of 42°. Within the fiber they change direction by 90° at each reflection. Determine the acceptance angle for meridional rays for the fiber in air.

2.7 Explain the concept of electromagnetic modes in relation to a planar optical waveguide.

Discuss the modifications that may be made to electromagnetic mode theory in a planar waveguide in order to describe optical propagation in a cylindrical fiber.

2.8 Briefly discuss, with the aid of suitable diagrams, the following concepts in optical fiber transmission:
   (a) the evanescent field;
   (b) Goos–Haenchen shift;
   (c) mode coupling.
Describe the effects of these phenomena on the propagation of light in optical fibers.

2.9 Define the normalized frequency for an optical fiber and explain its use in the determination of the number of guided modes propagating within a step index fiber.

A step index fiber in air has a numerical aperture of 0.16, a core refractive index of 1.45 and a core diameter of 60 μm. Determine the normalized frequency for the fiber when light at a wavelength of 0.9 μm is transmitted. Further, estimate the number of guided modes propagating in the fiber.

2.10 Describe with the aid of simple ray diagrams:
   (a) the multimode step index fiber;
   (b) the single mode step index fiber.
Compare the advantages and disadvantages of these two types of fiber for use as an optical channel.

2.11 A multimode step index fiber has a relative refractive index difference of 1% and a core refractive index of 1.5. The number of modes propagating at a wavelength of 1.3 μm is 1100. Estimate the diameter of the fiber core.

2.12 A single mode step index fiber has a core diameter of 4 μm and a core refractive index of 1.49. Estimate the shortest wavelength of light which allows single mode operation when the relative refractive index difference for the fiber is 2%.

2.13 In problem 2.12, it is required to increase the fiber core diameter to 10 μm whilst maintaining single mode operation at the same wavelength. Estimate the maximum possible relative refractive index difference for the fiber.

2.14 Explain what is meant by a graded index optical fiber, giving an expression for the possible refractive index profile. Using simple ray theory concepts, discuss the transmission of light through the fiber. Indicate the major advantage of this type of fiber with regard to multimode propagation.

2.15 The relative refractive index difference between the core axis and the cladding of a graded index fiber is 0.7% when the refractive index at the core axis is 1.45. Estimate values for the numerical aperature of the fiber when:
(a) the index profile is not taken into account; and
(b) the index profile is assumed to be triangular.
Comment on the results.

2.16 A multimode graded index fiber has an acceptance angle in air of 8°. Estimate the relative refractive index difference between the core axis and the cladding when the refractive index at the core axis is 1.52.

2.17 A graded index fiber with a parabolic index profile supports the propagation of 742 guided modes. The fiber has a numerical aperture in air of 0.3 and a core diameter of 70 μm. Determine the wavelength of the light propagating in the fiber.

Further estimate the maximum diameter of the fiber which gives single mode operation at the same wavelength.

2.18 A graded index fiber with a core axis refractive index of 1.5 has a characteristic index profile (α) of 1.90, a relative refractive index difference of 1.3% and a core diameter of 40 μm. Estimate the number of guided modes propagating in the fiber when the transmitted light has a wavelength of 1.55 μm, and determine the cutoff value of the normalized frequency for single mode transmission in the fiber.

Answers to Numerical Problems

2.1	(a) 8.6°; (b) 83.6°		**2.11**	92 μm
2.2	0.26, 15.2°		**2.12**	1.56 μm
2.3	0.37, 75.9°		**2.13**	0.32%
2.4	2.11×10^8 m s^{-1}		**2.15**	(a) 0.172; (b) 0.171
2.5	34.6°		**2.16**	0.42%
2.6	28.2°		**2.17**	1.2 μm, 4.4 μm
2.9	33.5, 561		**2.18**	94, 3.45

REFERENCES

1 D. Hondros and P. Debye 'Electromagnetic waves along long cylinders of dielectric', *Annal. Physik*, **32**(3), pp. 465–476, 1910.

2 O. Schriever, 'Electromagnetic waves in dielectric wires', *Annal. Physik*, **63**(7), pp. 645–673, 1920.

3 A. C. S. van Heel, 'A new method of transporting optical images without aberrations', *Nature, Lond.*, **173**, p. 39, 1954.

4   H. H. Hopkins and N. S. Kapany, 'A flexible fibrescope, using static scanning', *Nature, Lond.*, **173**, pp. 39–41, 1954.

5   K. C. Kao and G. A. Hockham, 'Dielectric-fibre surface waveguides for optical frequencies', *Proc IEE*, **113**, pp. 1151–1158, 1966.

6   A. Werts, 'Propagation de la lumière cohérente dans les fibres optiques', *L'Onde Electrique*, **46**, pp. 967–980, 1966.

7   S. Takahashi and T. Kawashima, 'Preparation of low loss multi-component glass fiber', *Tech. Dig. Int. Conf. Integr. Opt. and Opt. Fiber Commun.*, p. 621, 1977.

8   J. B. MacChesney, P. B. O'Connor, F. W. DiMarcello, J. R. Simpson and P. D. Lazay, 'Preparation of low-loss optical fibres using simultaneous vapour phase deposition and fusion', *Proc. 10th Int. Conf. on Glass*, paper 6–40, 1974.

9   T. Miya, Y. Terunuma, T. Hosaka and T. Miyashita, 'Ultimate low-loss single-mode fibre at 1.55 μm', *Electron Lett.*, **15**(4), pp. 106–108, 1979.

10 M. Born and E. Wolf, *Principles of Optics*, 6th edn., Pergamon Press, 1980.

11 W. B. Allan, *Fibre Optics*, Oxford University Press, 1980.

12 D. C. Agarwal, 'Ray concepts in optical fibres', *Indian J. Theoret. Phys.*, **28**(1), pp. 41–54, 1980.

13 R. P. Feyman, *The Feyman Lectures on Physics, Vol. 2*, Addison-Wesley, 1969.

14 J. E. Midwinter, *Optical Fibers for Transmission*, John Wiley, 1979.

15 E. Snitzer, 'Cylindrical dielectric waveguide modes', *J. Opt. Soc. Am.*, **51**, pp. 491–498, 1961.

16 D. Gloge, 'Weakly guiding fibers', *Appl. Opt.*, **10**, pp. 2252–2258, 1971.

17 D. Marcuse, *Theory of Dielectric Optical Waveguides*, Academic Press, New York, 1974.

18 A. W. Snyder, 'Asymptotic expressions for eigenfunctions and eigenvalues of a dielectric or optical waveguide', *Trans IEEE Microwave Theory Tech.*, MTT–17, pp. 1130–1138, 1969.

19 D. Gloge, 'Optical power flow in multimode fibers', *Bell Syst. Tech. J.*, **51**, pp. 1767–1783, 1972.

20 R. Olshansky, 'Propagation in glass optical waveguides', *Rev. Mod. Phys.*, **51**(2), pp. 341–366, 1979.

21 D. Gloge, 'The optical fibre as a transmission medium', *Rep. Prog. Phys.*, **42**, pp. 1777–1824, 1979.

22 M. M. Ramsay and G. A. Hockham, 'Propagation in optical fibre waveguides' in C. P. Sandbank (ed.) *Optical Fibre Communication Systems*, pp. 25–41, John Wiley, 1980.

23 S. Kawakami and S. Nishida, 'Characteristics of a doubly clad optical fiber with a low index cladding;, *IEEE J. Quantum Electron*, QE–10, pp. 879–887, 1974.

24 P. M. Morse and H. Fesbach, Methods of Theoretical Physics, Vol. II, McGraw-Hill, 1953.

25 D. Marcuse, *Light Transmission Optics*, 2nd edn., Van Nostrand Reinhold, 1982.

26 A. Ghatak and K. Thyagarajan, 'Graded index optical waveguides', in E. Wolf (ed.), *Progress in Optics Vol XVIII*, pp. 3–128, North-Holland 1980.

27 D. B. Beck, 'Optical Fiber Waveguides', in M. K. Barnoski (ed.), *Fundamentals of Optical Fiber Communications*, pp. 1–58, Academic Press, 1976.

28 D. Marcuse, D. Gloge, E. A. J. Marcatili, 'Guiding properties of fibers', in *Optical Fiber Telecommunications*, S. E. Miller and A. G. Chynoweth (eds), Academic Press, pp. 37–100, 1979.

29 I. S. Gradshteyn and I. M. Ryzhik, *Tables of Integrals, Series and Products*, 4th edn., Academic Press, 1965.

30 K. Okamoto and T. Okoshi, 'Analysis of wave propagation in optical fibers having core with α-power refractive-index distribution and uniform cladding', *IEEE Trans. Microwave Theory Tech.*, MTT–24, pp. 416–421, 1976.

31 C. W. Yeh, 'Optical waveguide theory', *IEEE Trans. Circuits and Syst.*, CAS–26 (12), pp. 1011–1019, 1979.

32 C. Pask and R. A. Sammut, 'Developments in the theory of fibre optics', *Proc. IREE Aust.*, **40**(3), pp. 89–101, 1979.

33 W. A. Gambling, A. H. Hartog and C. M. Ragdale, 'Optical fibre transmission lines', *The Radio Electron. Eng.*, **51**(7/8), pp. 313–325, 1981.

34 H. G. Unger, *Planar Optical Waveguides and Fibres*, Clarendon Press, 1977.

35 M. J. Adams, *An Introduction to Optical Waveguides*, John Wiley, 1981.

36 Y. Suematsu and K.-I. Iga, *Introduction to Optical Fibre Communications*, John Wiley, 1982.

37 T. Okoshi, *Optical Fibers*, Academic Press, 1982.

38 M. A. Saifi, 'Triangular index monomode fibres' in *Proc. SPIE Int. Soc. Opt. Eng. (USA)*, **374**, pp. 13–15, 1983.

3

Transmission Characteristics of Optical Fibers

3.1 INTRODUCTION

The basic transmission mechanisms of the various types of optical fiber waveguide have been discussed in Chapter 2. However, the factors which affect the performance of optical fibers as a transmission medium were not dealt with in detail. These transmission characteristics are of utmost importance when the suitability of optical fibers for communication purposes is investigated. The transmission characteristics of most interest are those of attenuation (or loss) and bandwidth.

The huge potential bandwidth of optical communications helped stimulate the birth of the idea that a dielectric waveguide made of glass could be used to carry wideband telecommunication signals. This occurred, as indicated in Section 2.1 in the celebrated papers by Kao and Hockham, and Werts in 1966. However, at the time the idea may have seemed somewhat ludicrous as a typical block of glass could support optical transmission for at best a few tens of meters before it was attenuated to an unacceptable level. Nevertheless, careful investigation of the attenuation showed that it was largely due to absorption in the glass, caused by impurities such as iron, copper, manganese and other transition metals which occur in the third row of the periodic table. Hence, research was stimulated on a new generation of 'pure' glasses for use in optical fiber communications.

A major breakthrough came in 1970 when the first fiber with an attenuation below 20 dB km^{-1} was reported |Ref. 1|. This level of attenuation was seen as the absolute minimum that had to be achieved before an optical fiber system could in any way compete economically with existing communication systems. Since 1970 tremendous improvements have been made leading to fibers with losses of less than 1 dB km^{-1} in the laboratory. Hence, comparatively low loss fibers have been incorporated into optical communication systems throughout the world.

The other characteristic of primary importance is the bandwidth of the fiber. This is limited by the signal dispersion within the fiber, which determines the number of bits of information transmitted in a given time period. Therefore, once the attenuation was reduced to acceptable levels attention was directed towards the dispersive properties of fibers. Again this has led to substantial improvements giving wideband fiber bandwidths of tens of gigahertz over a number of kilometers [Ref. 2]. In order to appreciate these advances and possible future developments, the optical transmission characteristics of fibers must be considered in greater depth. Therefore in this chapter we discuss the mechanisms within optical fibers which give rise to the major transmission characteristics mentioned previously (attenuation and dispersion), whilst also considering other perhaps less obvious effects when light is propagating down an optical fiber (modal noise and polarization).

We begin the discussion of attenuation in Section 3.2 with calculation of the total losses incurred in optical fibers. The various attenuation mechanisms (material absorption, linear scattering, nonlinear scattering, fiber bends) are then considered in detail in Sections 3.3 to 3.6. Following this, in Section 3.7, dispersion in optical fibers is described, together with the associated limitations on fiber bandwidth. Sections 3.8 and 3.9 deal with intramodal and intermodal dispersion mechanisms, prior to a discussion of overall fiber dispersion (in both multimode and single mode fibers) in Section 3.10. Modal noise in multimode optical fibers is then considered in Section 3.11. Finally, Section 3.12 presents a brief account of polarization within single mode optical fibers.

3.2 ATTENUATION

The attenuation or transmission loss of optical fibers has proved to be one of the most important factors in bringing about their wide acceptance in telecommunications. As channel attenuation largely determined the maximum transmission distance prior to signal restoration, optical fiber communications became especially attractive when the transmission losses of fibers were reduced below those of the competing metallic conductors (less than $5 \ dB \ km^{-1}$).

Signal attenuation within optical fibers, as with metallic conductors, is usually expressed in the logarithmic unit of the decibel. The decibel which is used for comparing two power levels may be defined for a particular optical wavelength as the ratio of the input (transmitted) optical power P_i into a fiber to the output (received) optical power P_o from the fiber as:

$$\text{number of decibels (dB)} = 10 \log_{10} \frac{P_i}{P_o} \qquad (3.1)$$

This logarithmic unit has the advantage that the operations of multiplication and division reduce to addition and subtraction, whilst powers and roots reduce to multiplication and division. However, addition and subtraction require a conversion to numerical values which may be obtained using the relationship:

$$\frac{P_i}{P_o} = 10^{(dB/10)} \tag{3.2}$$

In optical fiber communications the attenuation is usually expressed in decibels per unit length (i.e. dB km^{-1}) following:

$$\alpha_{dB} L = 10 \log_{10} \frac{P_i}{P_o} \tag{3.3}$$

where α_{dB} is the signal attenuation per unit length in decibels and L is the fiber length.

Example 3.1

When the mean optical power launched into an 8 km length of fiber is 120 μW, the mean optical power at the fiber output is 3 μW.
Determine:
(a) the overall signal attenuation or loss in decibels through the fiber assuming there are no connectors or splices;
(b) the signal attenuation per kilometer for the fiber.
(c) the overall signal attenuation for a 10 km optical link using the same fiber with splices at 1 km intervals each giving an attenuation of 1 dB;
(d) the numerical input/output power ratio in (c).
Solution: (a) Using Eq. (3.1), the overall signal attenuation in decibels through the fiber is:

$$\text{signal attenuation} = 10 \log_{10} \frac{P_i}{P_o} = 10 \log_{10} \frac{120 \times 10^{-6}}{3 \times 10^{-6}}$$

$$= 10 \log_{10} 40 = 16.0 \text{ dB}$$

(b) The signal attenuation per kilometer for the fiber may be simply obtained by dividing the result in (a) by the fiber length which corresponds using Eq. (3.3) where,

$$\alpha_{dB} L = 16.0 \text{ dB}$$

hence,

$$\alpha_{dB} = \frac{16.0}{8}$$

$$= 2.0 \text{ dB km}^{-1}$$

(c) As $\alpha_{dB} = 2$ dB km^{-1}, the loss incurred along 10 km of the fiber is given by

$$\alpha_{dB} L = 2 \times 10 = 20 \text{ dB}$$

However, the link also has nine splices (at 1 km intervals) each with an attenuation of 1 dB. Therefore, the loss due to the splices is 9 dB.
Hence, the overall signal attenuation for the link is:

$$\text{signal attenuation} = 20 + 9$$

$$= 29 \text{ dB}$$

(d) To obtain a numerical value for the input/output power ratio, Eq. (3.2) may be used where:

$$\frac{P_i}{P_o} = 10^{29/10} = 794.3$$

A number of mechanisms are responsible for the signal attenuation within optical fibers. These mechanisms are influenced by the material composition, the preparation and purification technique, and the waveguide structure. They may be categorized within several major areas which include material absorption, material scattering (linear and nonlinear scattering), curve and microbending losses, mode coupling radiation losses and losses due to leaky modes. There are also losses at connectors and splices as illustrated in example 3.1. However, in this chapter we are interested solely in the characteristics of the fiber; connector and splice losses are dealt with in Section 4.8. It is instructive to consider in some detail the loss mechanisms within optical fibers in order to obtain an understanding of the problems associated with the design and fabrication of low loss waveguides.

3.3 MATERIAL ABSORPTION LOSSES

Material absorption is a loss mechanism related to the material composition and the fabrication process for the fiber, which results in the dissipation of some of the transmitted optical power as heat in the waveguide. The absorption of the light may be intrinsic (caused by the interaction with one or more of the major components of the glass) or extrinsic (caused by impurities within the glass).

3.3.1 Intrinsic Absorption

An absolutely pure glass has little intrinsic absorption due to its basic material structure in the near infrared region. However, it does have two major intrinsic absorption mechanisms at optical wavelengths which leave a low intrinsic absorption window over the 0.8–1.7 μm wavelength range as illustrated in Fig. 3.1, which shows a possible optical attenuation against wavelength characteristic for absolutely pure glass [Ref. 3]. It may be observed that there is a fundamental absorption edge, the peaks of which are centered in the ultraviolet wavelength region. This is due to the stimulation of electron transitions within the glass by higher energy excitations. The tail of this peak may extend into the window region at the shorter wavelengths as illustrated in Fig. 3.1. Also in the infrared and far infrared, normally at wavelengths above 7 μm, fundamentals of absorption bands from the interaction of photons with molecular vibrations within the glass occur. These give absorption peaks which

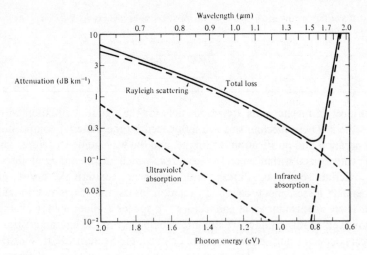

Fig. 3.1 The attenuation spectra for the intrinsic loss mechanisms in pure GeO_2–SiO_2 glass [Ref. 3].

again extend into the window region. The strong absorption bands occur due to oscillations of structural units such as Si—O (9.2 μm), P—O (8.1 μm), B—O (7.2 μm) and Ge—O (11.0 μm) within the glass. Hence, above 1.5 μm the tails of these largely far infrared absorption peaks tend to cause most of the pure glass losses.

However, the effects of both these processes may be minimized by suitable choice of both core and cladding compositions. For instance in some nonoxide glasses such as fluorides and chlorides, the infrared absorption peaks occur at much longer wavelengths which are well into the far infrared (up to 50 μm) giving less attenuation to longer wavelength transmission compared with oxide glasses.

3.3.2 Extrinsic Absorption

In practical optical fibers prepared by conventional melting techniques (see Section 4.3), a major source of signal attenuation is extrinsic absorption from transition metal element impurities. Some of the more common metallic impurities found in glasses are shown in the Table 3.1, together with the absorption losses caused by one part in 10^9 [Ref. 4]. It may be noted that certain of these impurities namely chromium and copper in their worst valence state can cause attenuation in excess of 1 dB km^{-1} in the near infrared region. Transition element contamination may be reduced to acceptable levels (i.e. one part in 10^{10}) by glass refining techniques such as vapor phase oxidation [Ref. 5] (see Section 4.4) which largely eliminates the effects of these metallic impurities.

Table 3.1   Absorption losses caused by some of the more common metallic ion impurities in glasses together with the absorption peak wavelength

	Peak wavelength (nm)	One part in 10^9 (dB km^{-1})
Cr^{3+}	625	1.6
C^{2+}	685	0.1
Cu^{2+}	850	1.1
Fe^{2+}	1100	0.68
Fe^{3+}	400	0.15
Ni^{2+}	650	0.1
Mn^{3+}	460	0.2
V^{4+}	725	2.7

However, another major extrinsic loss mechanism is caused by absorption due to water (as the hydroxyl or OH ion) dissolved in the glass. These hydroxyl groups are bonded into the glass structure and have fundamental stretching vibrations which occur at wavelengths between 2.7 and 4.2 μm depending on group position in the glass network. The fundamental vibrations give rise to overtones appearing almost harmonically at 1.38, 0.95 and 0.72 μm as illustrated in Fig. 3.2 [Ref. 6]. This shows the absorption spectrum for the hydroxyl group in silica. Furthermore, combinations between the overtones and the fundamental SiO_2 vibration occur at 1.24, 1.13 and 0.88 μm completing the absorption spectrum shown in Fig. 3.2.

It may also be observed in Fig. 3.2 that the only significant absorption band in the region below a wavelength of 1 μm is the second overtone at 0.95 μm which causes attenuation of about 1 dB km^{-1} for one part per million (ppm) of

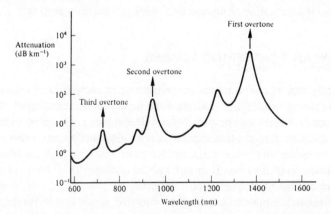

Fig. 3.2   The absorption spectrum for the hydroxyl (OH) group in silica. Reproduced with permission from D. B. Keck, K. D. Maurer and P. C. Schultz, *Appl. Phys. Lett.*, **22**, p. 307, 1973.

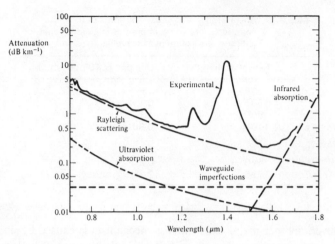

Fig. 3.3 The measured attenuation spectrum for an ultra low loss single mode fiber (solid line) with the calculated attenuation spectra for some of the loss mechanisms contributing to the overall fiber attenuation (dashed and dotted lines) [Ref. 3].

hydroxyl. At longer wavelengths the first overtone at 1.38 μm and its sideband at 1.24 μm are strong absorbers giving attenuation of about 2 dB km^{-1} ppm and 4 dB km^{-1} ppm respectively. Since most resonances are sharply peaked, narrow windows exist in the longer wavelength region around 1.3 and 1.55 μm which are essentially unaffected by OH absorption once the impurity level has been reduced below one part in 10^7. This situation is illustrated in Fig. 3.3 which shows the attenuation spectrum of an ultra low loss single mode fiber [Ref. 3]. It may be observed that the lowest attenuation for this fiber occurs at a wavelength of 1.55 μm and is 0.2 dB km^{-1}. This is approaching the minimum possible attenuation of around 0.18 dB km^{-1} at this wavelength [Ref. 8].

3.4 LINEAR SCATTERING LOSSES

Linear scattering mechanisms cause the transfer of some or all of the optical power contained within one propagating mode to be transferred linearly (proportionally to the mode power) into a different mode. This process tends to result in attenuation of the transmitted light as the transfer may be to a leaky or radiation mode which does not continue to propagate within the fiber core, but is radiated from the fiber. It must be noted that as with all linear processes there is no change of frequency on scattering.

Linear scattering may be categorized into two major types: Rayleigh and Mie scattering. Both result from the nonideal physical properties of the manufactured fiber which are difficult and in certain cases impossible to eradicate at present.

3.4.1 Rayleigh Scattering

Rayleigh scattering is the dominant intrinsic loss mechanism in the low absorption window between the ultraviolet and infrared absorption tails. It results from inhomogeneities of a random nature occurring on a small scale compared with the wavelength of the light. These inhomogeneities manifest themselves as refractive index fluctuations and arise from density and compositional variations which are frozen into the glass lattice on cooling. The compositional variations may be reduced by improved fabrication, but the index fluctuations caused by the freezing-in of density inhomogeneities are fundamental and cannot be avoided. The subsequent scattering due to the density fluctuations, which is in almost all directions, produces an attenuation proportional to $1/\lambda^4$ following the Rayleigh scattering formula [Ref. 9]. For a single component glass this is given by:

$$\gamma_R = \frac{8\pi^3}{3\lambda^4} n^8 p^2 \beta_c K T_F \tag{3.4}$$

where γ_R is the Rayleigh scattering coefficient, λ is the optical wavelength, n is the refractive index of the medium, p is the average photoelastic coefficient, β_c is the isothermal compressibility at a fictive temperature T_F, and K is Boltzmann's constant. The fictive temperature is defined as the temperature at which the glass can reach a state of thermal equilibrium and is closely related to the anneal temperature. Furthermore, the Rayleigh scattering coefficient is related to the transmission loss factor (transmissivity) of the fiber $\mathcal{L}$ following the relation [Ref. 10]:

$$\mathcal{L} = \exp(-\gamma_R L) \tag{3.5}$$

where L is the length of the fiber. It is apparent from Eq. (3.4) that the fundamental component of Rayleigh scattering is strongly reduced by operating at the longest possible wavelength. This point is illustrated in example 3.2.

Example 3.2

Silica has an estimated fictive temperature of 1400K with an isothermal compressibility of 7×10^{-11} m^2 N^{-1} [Ref. 11]. The refractive index and the photoelastic coefficient for silica are 1.46 and 0.286 respectively [Ref. 11]. Determine the theoretical attenuation in decibels per kilometer due to the fundamental Rayleigh scattering in silica at optical wavelengths of 0.63, 1.00 and 1.30 μm. Boltzmann's constant is 1.381×10^{-23} J K^{-1}.

Solution: The Rayleigh scattering coefficient may be obtained from Eq. (3.4) for each wavelength. However, the only variable in each case is the wavelength and therefore the constant of proportionality of Eq. (3.4) applies in all cases. Hence:

$$\gamma_R = \frac{8\pi^3 n^8 p^2 \beta_c K T_F}{3\lambda^4}$$

$$= \frac{248.15 \times 20.65 \times 0.082 \times 7 \times 10^{-11} \times 1.381 \times 10^{-23} \times 1400}{3 \times \lambda^4}$$

$$= \frac{1.895 \times 10^{-28}}{\lambda^4} \; m^{-1}$$

At a wavelength of 0.63 μm:

$$\gamma_R = \frac{1.895 \times 10^{-28}}{0.158 \times 10^{-24}} = 1.199 \times 10^{-3} \, m^{-1}$$

The transmission loss factor for one kilometer of fiber may be obtained using Eq. (3.5),

$$\mathcal{L}_{km} = \exp(-\gamma_R L) = \exp(-1.199 \times 10^{-3} \times 10^3)$$

$$= 0.301$$

The attenuation due to Rayleigh scattering in dB km^{-1} may be obtained from Eq. (3.1) where:

$$\text{Attenuation} = 10 \log_{10} (1/\mathcal{L}_{km}) = 10 \log_{10} 3.322$$

$$= 5.2 \; dB \; km^{-1}$$

At a wavelength of 1.00 μm:

$$\gamma_R = \frac{1.895 \times 10^{-28}}{10^{-24}} = 1.895 \times 10^{-4} \, m^{-1}$$

Using Eq. (3.5):

$$\mathcal{L}_{km} = \exp(-1.895 \times 10^{-4} \times 10^3) = \exp(-0.1895)$$

$$= 0.827$$

and Eq. (3.1):

$$\text{Attenuation} = 10 \log_{10} 1.209 = 0.8 \; dB \; km^{-1}$$

At a wavelength of 1.30 μm:

$$\gamma_R = \frac{1.895 \times 10^{-28}}{2.856 \times 10^{-24}} = 0.664 \times 10^{-4}$$

Using Eq. (3.5):

$$\mathcal{L}_{km} = \exp(-0.664 \times 10^{-4} \times 10^3) = 0.936$$

and Eq. (3.1):

$$\text{Attenuation} = 10 \log_{10} 1.069 = 0.3 \; dB \; km^{-1}$$

The theoretical attenuation due to Rayleigh scattering in silica at wavelengths of 0.63, 1.00 and 1.30 μm, from example 3.2, is 5.2, 0.8 and 0.3 dB km^{-1} respectively. These theoretical results are in reasonable agreement with experimental work. For instance the lowest reported value for Rayleigh scattering in silica at a wavelength of 0.6328 μm is 3.9 dB km^{-1} |Ref. 11|. However, values of 4.8 dB km^{-1} |Ref. 12| and 5.4 dB km^{-1} |Ref. 13| have also been reported. The predicted attenuation due to Rayleigh scattering against wavelength is indicated by a broken line on the attenuation characteristics shown in Figs. 3.1 and 3.3.

3.4.2 Mie Scattering

Linear scattering may also occur at inhomogeneities which are comparable in size to the guided wavelength. These result from the nonperfect cylindrical structure of the waveguide and may be caused by fiber imperfections such as irregularities in the core–cladding interface, core–cladding refractive index differences along the fiber length, diameter fluctuations, strains and bubbles. When the scattering inhomogeneity size is greater than $\lambda/10$, the scattered intensity which has an angular dependence can be very large.

The scattering created by such inhomogeneities is mainly in the forward direction and is called Mie scattering. Depending upon the fiber material, design and manufacture Mie scattering can cause significant losses. The inhomogeneities may be reduced by:

(a) removing imperfections due to the glass manufacturing process;
(b) carefully controlled extrusion and coating of the fiber;
(c) increasing the fiber guidance by increasing the relative refractive index difference.

By these means it is possible to reduce Mie scattering to insignificant levels.

3.5 NONLINEAR SCATTERING LOSSES

Optical waveguides do not always behave as completely linear channels whose increase in output optical power is directly proportional to the input optical power. Several nonlinear effects occur, which in the case of scattering cause disproportionate attenuation, usually at high optical power levels. This nonlinear scattering causes the optical power from one mode to be transferred in either the forward or backward direction to the same, or other modes, at a different frequency. It depends critically upon the optical power density within the fiber and hence only becomes significant above threshold power levels.

The most important types of nonlinear scattering within optical fibers are stimulated Brillouin and Raman scattering, both of which are usually only observed at high optical power densities in long single mode fibers. These scattering mechanisms in fact give optical gain but with a shift in frequency thus contributing to attenuation for light transmission at a specific wavelength.

However, it may be noted that such nonlinear phenomena can also be used to give optical amplification in the context of integrated optical techniques (see Section 11.8.4).

3.5.1 Stimulated Brillouin Scattering

Brillouin scattering may be regarded as the modulation of light through thermal molecular vibrations within the fiber. The scattered light appears as upper and lower sidebands which are separated from the incident light by the modulation frequency. The incident photon in this scattering process produces a phonon* of acoustic frequency as well as a scattered photon. This produces an optical frequency shift which varies with the scattering angle because the frequency of the sound wave varies with acoustic wavelength. The frequency shift is a maximum in the backward direction reducing to zero in the forward direction making Brillouin scattering a mainly backward process.

As indicated previously, Brillouin scattering is only significant above a threshold power density. Assuming the polarization state of the transmitted light is not maintained (see Section 3.12), it may be shown |Ref. 16| that the threshold power P_B is given by:

$$P_B = 4.4 \times 10^{-3} d^2 \lambda^2 \alpha_{dB} v \text{ watts} \qquad (3.6)$$

where d and λ are the fiber core diameter and the operating wavelength respectively, both measured in micrometers, α_{dB} is the fiber attenuation in decibels per kilometer and v is the source bandwidth (i.e. injection laser) in gigahertz. The expression given in Eq. (3.6) allows the determination of the threshold optical power which must be launched into a single mode optical fiber before Brillouin scattering occurs (see example 3.3).

3.5.2 Stimulated Raman Scattering

Stimulated Raman scattering is similar to stimulated Brillouin scattering except that a high frequency optical phonon rather than an acoustic phonon is generated in the scattering process. Also, Raman scattering occurs in the forward direction and may have an optical power threshold of up to three orders of magnitude higher than the Brillouin threshold in a particular fiber.

Using the same criteria as those specified for the Brillouin scattering threshold given in Eq. (3.6), it may be shown [Ref. 16] that the threshold optical power for stimulated Raman scattering P_R in a long single mode fiber is given by:

---

* The phonon is a quantum of an elastic wave in a crystal lattice. When the elastic wave has a frequency f, the quantized unit of the phonon has energy hf joules, where h is Planck's constant.

$$P_R = 5.9 \times 10^{-2}d^2\lambda\alpha_{dB} \text{ watts} \qquad (3.7)$$

where d, λ and α_{dB} are as specified for Eq. (3.6).

Example 3.3

A long single mode optical fiber has an attenuation of 0.5 db km^{-1} when operating at a wavelength of 1.3 µm. The fiber core diameter is 6 µm and the laser source bandwidth is 600 MHz. Compare the threshold optical powers for stimulated Brillouin and Raman scattering within the fiber at the wavelength specified.

Solution: The threshold optical power for stimulated Brillouin scattering is given by Eq. (3.6) as:

$$P_B = 4.4 \times 10^{-3}d^2\lambda^2\alpha_{dB}\nu$$

$$= 4.4 \times 10^{-3} \times 6^2 \times 1.3^2 \times 0.5 \times 0.6$$

$$= 80.3 \text{ mW}$$

The threshold optical power for stimulated Raman scattering may be obtained from Eq. (3.7), where:

$$P_R = 5.9 \times 10^{-2}d^2\lambda\alpha_{dB}$$

$$= 5.9 \times 10^{-2} \times 6^2 \times 1.3 \times 0.5$$

$$= 1.38 \text{ W}$$

In example 3.3, the Brillouin threshold occurs at an optical power level of around 80 mW whilst the Raman threshold is approximately 17 times larger. It is therefore apparent that the losses introduced by nonlinear scattering may be avoided by use of a suitable optical signal level (i.e. working below the threshold optical powers). However, it must be noted that the Brillouin threshold has been reported |Ref. 17| as occurring at optical powers as low as 10 mW in single mode fibers. Nevertheless, this is still a high power level for optical communications and may be easily avoided. Brillouin and Raman scattering are not usually observed in multimode fibers because their relatively large core diameters make the threshold optical power levels extremely high. Moreover it should be noted that the threshold optical powers for both these scattering mechanisms may be increased by suitable adjustment of the other parameters in Eqs. (3.6) and (3.7). In this context, operation at the longest possible wavelength is advantageous although this may be offset by the reduced fiber attenuation (from Rayleigh scattering and material absorption) normally obtained.

3.6 FIBER BEND LOSS

Optical fibers suffer radiation losses at bends or curves on their paths. This is due to the energy in the evanescent field at the bend exceeding the velocity of

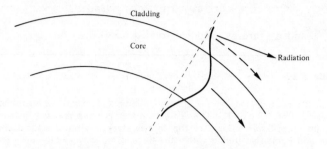

Fig. 3.4 An illustration of the radiation loss at a fiber bend. The part of the mode in the cladding outside the dashed arrowed line may be required to travel faster than the velocity of light in order to maintain a plane wavefront. Since it cannot do this, the energy contained in this part of the mode is radiated away.

light in the cladding and hence the guidance mechanism is inhibited, which causes light energy to be radiated from the fiber. An illustration of this situation is shown in Fig. 3.4. The part of the mode which is on the outside of the bend is required to travel faster than that on the inside so that a wavefront perpendicular to the direction of propagation is maintained. Hence part of the mode in the cladding needs to travel faster than the velocity of light in that medium. As this is not possible, the energy associated with this part of the mode is lost through radiation. The loss can generally be represented by a radiation attenuation coefficient which has the form [Ref. 19]:

$$\alpha_r = c_1 \exp\left(-c_2 R\right)$$

where R is the radius of curvature of the fiber bend and c_1, c_2 are constants which are independent of R. Furthermore, large bending losses tend to occur at a critical radius of curvature R_c which may be estimated from [Ref. 20]:

$$R_c \simeq \frac{3 n_1^2 \lambda}{4\pi (n_1^2 - n_2^2)^{3/2}} \tag{3.8}$$

It may be observed from the expression given in Eq. (3.8) that possible bending losses may be reduced by:

(a) designing fibers with large relative refractive index differences;
(b) operating at the shortest wavelength possible.

Both these factors therefore have the effect of reducing the critical bending radius as illustrated in the following example.

Example 3.4

Two step index fibers have the following characteristics:

(a) A core refractive index of 1.500 with a relative refractive index difference of 0.2% and an operating wavelength of 1.55 μm.
(b) A core refractive index the same as (a) but a relative refractive index difference of 3% and an operating wavelength of 0.82 μm.

Estimate the critical radius of curvature at which large bending losses occur in both cases.

Solution: (a) The relative refractive index difference Δ is given by Eq. (2.9) as:

$$\Delta = \frac{n_1^2 - n_2^2}{2n_1^2}$$

Hence

$$n_2^2 = n_1^2 - 2\Delta n_1^2 = 2.250 - 0.004 \times 2.250$$

$$= 2.241$$

Using Eq. (3.8) for the critical radius of curvature:

$$R_c \approx \frac{3n_1^2\lambda}{4\pi(n_1^2 - n_2^2)^{3/2}} = \frac{3 \times 2.250 \times 1.55 \times 10^{-6}}{4\pi(0.009)^{3/2}}$$

$$\simeq 975 \, \mu m$$

(b) Again, from Eq. (2.9):

$$n_2^2 = n_1^2 - 2\Delta n_1^2 = 2.250 - 0.06 \times 2.250$$

$$= 2.115$$

Substituting into Eq. (3.8):

$$R_c \simeq \frac{3 \times 2.250 \times 0.82 \times 10^{-6}}{4\pi \times (0.135)^{3/2}}$$

$$\simeq 9 \, \mu m$$

Example 3.4 shows that the critical radius of curvature for guided modes can be made extremely small (e.g. 9 μm), although this may be in conflict with the preferred design and operational characteristics. Nevertheless for most practical purposes, the critical radius of curvature is sufficiently small (even when considering case (a) which characterizes a long wavelength single mode fiber, it is approximately 1 mm) to avoid severe attenuation of the guided mode(s) at fiber bends. However, modes propagating close to cutoff, which are no longer fully guided within the fiber core, may radiate at substantially larger radii of curvature. Thus it is essential that sharp bends, with a radius of curvature approaching the critical radius, are avoided when optical fiber cables are installed. Finally, it is important that microscopic bends with radii of curvature approximating to the fiber radius are not produced in the fiber cabling process. These so-called microbends, which can cause significant losses from cabled fiber, are discussed further in Section 4.6.2.

3.7 DISPERSION

Dispersion of the transmitted optical signal causes distortion for both digital and analog transmission along optical fibers. When considering the major implementation of optical fiber transmission which involves some form of digital modulation, then dispersion mechanisms within the fiber cause broadening of the transmitted light pulses as they travel along the channel. The phenomenon is illustrated in Fig. 3.5 where it may be observed that each pulse broadens and overlaps with its neighbors, eventually becoming indistinguishable at the receiver input. The effect is known as intersymbol interference (ISI). Thus an increasing number of errors may be encountered on the digital optical channel as the ISI becomes more pronounced. The error rate is also a function of the signal attenuation on the link and the subsequent signal to noise ratio (SNR) at the receiver. This factor is not pursued further here but is considered in detail in Section 10.6.3. However, signal dispersion alone limits the maximum possible bandwidth attainable with a particular optical fiber to the point where individual symbols can no longer be distinguished.

For no overlapping of light pulses down on an optical fiber link the digital bit rate B_T must be less than the reciprocal of the broadened (through dispersion) pulse duration (2τ). Hence:

$$B_T \leqslant \frac{1}{2\tau} \qquad (3.9)$$

This assumes that the pulse broadening due to dispersion on the channel is τ which dictates the input pulse duration which is also τ. Hence Eq. (3.9) gives a conservative estimate of the maximum bit rate that may be obtained on an optical fiber link as $1/2\tau$.

Another more accurate estimate of the maximum bit rate for an optical channel with dispersion may be obtained by considering the light pulses at the output to have a Gaussian shape with an rms width of σ. Unlike the relationship given in Eq. (3.9), this analysis allows for the existence of a certain amount of signal overlap on the channel, whilst avoiding any SNR penalty which occurs when intersymbol interference becomes pronounced. The maximum bit rate is given approximately by (see Appendix F):

$$B_T \text{ (max)} \simeq \frac{0.2}{\sigma} \text{ bit s}^{-1} \qquad (3.10)$$

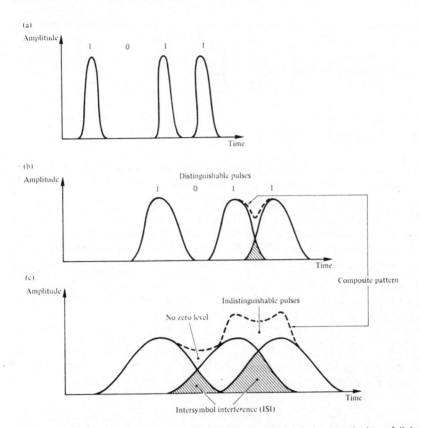

Fig. 3.5 An illustration using the digital bit pattern 1011 of the broadening of light pulses as they are transmitted along a fiber: (a) fiber input; (b) fiber output at a distance L_1; (c) fiber output at a distance $L_2 > L_1$.

It must be noted that certain sources |Refs. 25, 26| give the constant term in the numerator of Eq. (3.10) as 0.25. However, we take the slightly more conservative estimate given, following Olshansky |Ref. 9| and Gambling *et al.* |Ref. 27|. Equation (3.10) gives a reasonably good approximation for other pulse shapes which may occur on the channel resulting from the various dispersive mechanisms within the fiber. Also σ may be assumed to represent the rms impulse response for the channel as discussed further in Section 3.9.1.

The conversion of bit rate to bandwidth in hertz depends on the digital coding format used. For metallic conductors when a non return to zero code is employed, the binary one level is held for the whole bit period τ. In this case there are two bit periods in one wavelength (i.e. two bits per second per hertz), as illustrated in Fig. 3.6(a). Hence the maximum bandwidth B is one half the maximum data rate or

$$B_T \text{ (max)} = 2B \qquad (3.11)$$

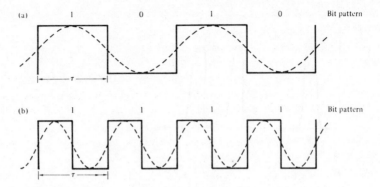

Fig. 3.6 Schematic illustration of the relationships of the bit rate to wavelength for digital codes: (a) non return to zero (NRZ); (b) return to zero (RZ).

However, when a return to zero code is considered as shown in Fig. 3.6(b), the binary one level is held for only part (usually half) the bit period. For this signalling scheme the data rate is equal to the bandwidth in hertz (i.e. one bit per second per hertz) and thus $B_T = B$.

The bandwidth B for metallic conductors is also usually defined by the electrical 3 dB points (i.e. the frequencies at which the electrical power has dropped to one half of its constant maximum value). However, when the 3 dB optical bandwidth of a fiber is considered it is significantly larger than the corresponding 3 dB electrical bandwidth for the reasons discussed in Section 7.4.3. Hence, when the limitations in the bandwidth of a fiber due to dispersion are stated (i.e. optical bandwidth B_{opt}), it is usually with regard to a return to zero code where the bandwidth in hertz is considered equal to the digital bit rate. Within the context of dispersion the bandwidths expressed in this chapter will follow this general criterion unless otherwise stated. However, as is made clear in Section 7.4.3, when electro-optical devices and optical fiber systems are considered it is more usual to state the electrical 3 dB bandwidth, this being the more useful measurement when interfacing an optical fiber link to electrical terminal equipment. Unfortunately the terms of bandwidth measurement are not always made clear and the reader must be warned that this omission may lead to some confusion when specifying components and materials for optical fiber communication systems.

Figure 3.7 shows the three common optical fiber structures, multimode step index, multimode graded index and single mode step index, whilst diagrammatically illustrating the respective pulse broadening associated with each fiber type. It may be observed that the multimode step index fiber exhibits the greatest dispersion of a transmitted light pulse and that the multimode graded index fiber gives a considerably improved performance. Finally, the single mode fiber gives the minimum pulse broadening and thus is capable of the greatest transmission bandwidths which are currently in the gigahertz range, whereas transmission via multimode step index fiber is usually limited to bandwidths of a few tens of megahertz. However, the amount of pulse broadening is dependent upon the distance the pulse travels within the fiber and hence for a given optical fiber link the restriction on usable bandwidth is dictated by the distance between regenerative repeaters (i.e. the distance the light pulse travels before it is reconstituted). Thus the measurement of the dispersive properties of a particular fiber is usually stated as the pulse broadening in time over a unit length of the fiber (i.e. ns km^{-1}).

Hence, the number of optical signal pulses which may be transmitted in a given period, and therefore the information-carrying capacity of the fiber is restricted by the amount of pulse dispersion per unit length. In the absence of mode coupling or filtering, the pulse broadening increases linearly with fiber length and thus the bandwidth is inversely proportional to distance. This leads to the adoption of a more useful parameter for the information-carrying

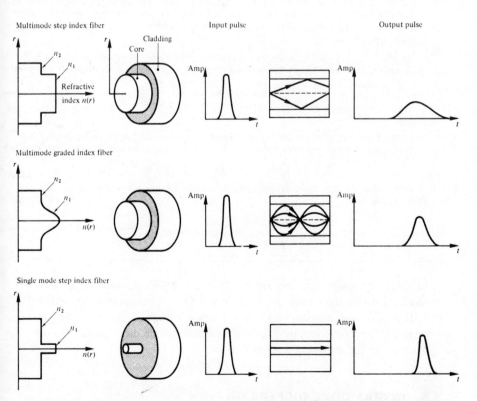

Fig. 3.7 Schematic diagram showing a multimode step index fiber, multimode graded index fiber and single mode step index fiber, and illustrating the pulse broadening due to intermodal dispersion in each fiber type.

capacity of an optical fiber which is known as the bandwidth–length product (i.e. $B_{opt} \times L$). The typical best bandwidth–length products for the three fibers shown in Fig. 3.7, are 20 MHz km, 1 Ghz km and 100 GHz km for multimode step index, multimode graded index and single mode step index fibers respectively.

Example 3.5

A multimode graded index fiber exhibits total pulse broadening of 0.1 μs over a distance of 15 km. Estimate:

(a) the maximum possible bandwidth on the link assuming no intersymbol interference;
(b) the pulse dispersion per unit length;
(c) the bandwidth–length product for the fiber.

Solution: (a) The maximum possible optical bandwidth which is equivalent to the maximum possible bit rate (for return to zero pulses) assuming no ISI may be obtained from Eq. (3.9), where:

$$B_{opt} = B_T = \frac{1}{2\tau} = \frac{1}{0.2 \times 10^{-6}} = 5 \text{ MHz}$$

(b) The dispersion per unit length may be acquired simply by dividing the total dispersion by the total length of the fiber.

$$\text{dispersion} = \frac{0.1 \times 10^{-6}}{15} = 6.67 \text{ ns km}^{-1}$$

(c) The bandwidth–length product may be obtained in two ways. Firstly by simply multiplying the maximum bandwidth for the fiber link by its length. Hence:

$$B_{opt}L = 5 \text{ MHz} \times 15 \text{ km} = 75 \text{ MHz km}$$

Alternatively it may be obtained from the dispersion per unit length using Eq. (3.9) where:

$$B_{opt}L = \frac{1}{2 \times 6.67 \times 10^{-9}} = 75 \text{ MHz km}$$

In order to appreciate the reasons for the different amounts of pulse broadening within the various types of optical fiber, it is necessary to consider the dispersive mechanisms involved. These include material dispersion, waveguide dispersion, intermodal dispersion and profile dispersion which are considered in the following sections.

3.8 INTRAMODAL DISPERSION

Intramodal or chromatic dispersion may occur in all types of optical fiber and results from the finite spectral linewidth of the optical source. Since optical sources do not emit just a single frequency but a band of frequencies (in the case of the injection laser corresponding to only a fraction of a per cent of the center frequency, whereas for the LED it is likely to be a significant percentage), then there may be propagation delay differences between the different spectral components of the transmitted signal. This causes broadening of each transmitted mode and hence intramodal dispersion. The delay differences may be caused by the dispersive properties of the waveguide material (material dispersion) and also guidance effects within the fiber structure (waveguide dispersion).

3.8.1 Material Dispersion

Pulse broadening due to material dispersion results from the different group velocities of the various spectral components launched into the fiber from the optical source. It occurs when the phase velocity of a plane wave propagating in the dielectric medium varies nonlinearly with wavelength, and a material is said to exhibit material dispersion when the second differential of the refractive index with respect to wavelength is not zero (i.e. $d^2 n/d\lambda^2 \neq 0$). The pulse spread due to material dispersion may be obtained by considering the group delay τ_g in the optical fiber which is the reciprocal of the group velocity v_g defined by Eqs. (2.37) and (2.40). Hence the group delay is given by:

$$\tau_g = \frac{d\beta}{d\omega} = \frac{1}{c} \left(n_1 - \lambda \frac{dn_1}{d\lambda} \right) \tag{3.12}$$

where n_1 is the refractive index of the core material. The pulse delay τ_m due to material dispersion in a fiber of length L is therefore:

$$\tau_m = \frac{L}{c} \left(n_1 - \lambda \frac{dn_1}{d\lambda} \right) \tag{3.13}$$

For a source with rms spectral width σ_λ and a mean wavelength λ, the rms pulse broadening due to material dispersion σ_m may be obtained from the expansion of Eq. (3.13) in a Taylor series about λ where:

$$\sigma_m = \sigma_\lambda \frac{d\tau_m}{d\lambda} + \sigma_\lambda \frac{2d^2\tau_m}{d\lambda^2} + \cdots \tag{3.14}$$

As the first term in Eq. (3.14) usually dominates, especially for sources operating over the 0.8–0.9 μm wavelength range, then:

$$\sigma_m \simeq \sigma_\lambda \frac{d\tau_m}{d\lambda} \tag{3.15}$$

Hence the pulse spread may be evaluated by considering the dependence of τ_m on λ, where from Eq. (3.13):

$$\frac{d\tau_m}{d\lambda} = \frac{L\lambda}{c} \left[\frac{dn_1}{d\lambda} - \frac{d^2 n_1}{d\lambda^2} - \frac{dn_1}{d\lambda} \right]$$

$$= \frac{-L\lambda}{c} \frac{d^2 n_1}{d\lambda^2} \tag{3.16}$$

Therefore substituting the expression obtained in Eq. (3.16) into Eq. (3.15), the rms pulse broadening due to material dispersion is given by:

$$\sigma_m \simeq \frac{\sigma_\lambda L}{c} \left| \lambda \frac{d^2 n_1}{d\lambda^2} \right| \tag{3.17}$$

The material dispersion for optical fibers is sometimes quoted as a value for $\left| \lambda^2 (d^2 n_1/d\lambda^2) \right|$ or simply $\left| d^2 n_1/d\lambda^2 \right|$.

However, it may be given in terms of a material dispersion parameter M which is defined as:

$$M = \frac{1}{L} \frac{d\tau_m}{d\lambda} = \frac{\lambda}{c} \left| \frac{d^2 n_1}{d\lambda^2} \right| \tag{3.18}$$

and which is often expressed in units of $ps\ nm^{-1}\ km^{-1}$.

Example 3.6

A glass fiber exhibits material dispersion given by $\left| \lambda^2 (d^2 n_1/d\lambda^2) \right|$ of 0.025. Determine the material dispersion parameter at a wavelength of 0.85 μm, and estimate the rms pulse broadening per kilometer for a good LED source with an rms spectral width of 20 nm at this wavelength.

Solution: The material dispersion parameter may be obtained from Eq. (3.18):

$$M = \frac{\lambda}{c} \left| \frac{d^2 n_1}{d\lambda^2} \right| = \frac{1}{c\lambda} \left| \lambda^2 \frac{d^2 n_1}{d\lambda^2} \right|$$

$$= \frac{0.025}{2.998 \times 10^5 \times 850}\ s\ nm^{-1}\ km^{-1}$$

$$= 98.1\ ps\ nm^{-1}\ km^{-1}$$

The rms pulse broadening is given by Eq. (3.17) as:

$$\sigma_m \simeq \frac{\sigma_\lambda L}{c} \left| \lambda \frac{d^2 n_1}{d\lambda^2} \right|$$

Therefore in terms of the material dispersion parameter M defined by Eq. (3.18):

$$\sigma_m \simeq \sigma_\lambda LM$$

Hence, the rms pulse broadening per kilometer due to material dispersion:

$$\sigma_m \, (1 \text{ km}) = 20 \times 1 \times 98.1 \times 10^{-12} = 1.96 \text{ ns km}^{-1}$$

Figure 3.8 shows the variation of the material dispersion parameter M with wavelength for pure silica |Ref. 28|. It may be observed that the material dispersion tends to zero in the longer wavelength region around 1.3 µm (for pure silica). This provides an additional incentive (other than low attenuation) for operation at longer wavelengths where the material dispersion may be minimized. Also the use of an injection laser with a narrow spectral width rather than an LED as the optical source leads to a substantial reduction in the pulse broadening due to material dispersion, even in the shorter wavelength region.

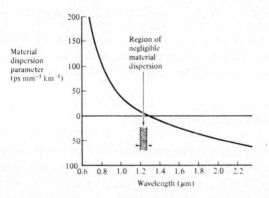

Fig. 3.8 The material dispersion parameter for silica as a function of wavelength. Reproduced with permission from D. N. Payne and W. A. Gambling, *Electron. Lett.*, **11**, p. 176, 1975.

Example 3.7

Estimate the rms pulse broadening per kilometer for the fiber in example 3.6 when the optical source used is an injection laser with a relative spectral width σ_λ/λ of 0.0012 at a wavelength of 0.85 µm.

Solution: The rms spectral width may be obtained from the relative spectral width by:

$$\sigma_\lambda = 0.0012\lambda = 0.0012 \times 0.85 \times 10^{-6}$$

$$= 1.02 \text{ nm}$$

The rms pulse broadening in terms of the material dispersion parameter following example 3.6 is given by:

$$\sigma_m \simeq \sigma_\lambda L M$$

Therefore the rms pulse broadening per kilometer due to material dispersion is:

$$\sigma_m \simeq 1.02 \times 1 \times 98.1 \times 10^{-12} = 0.10 \text{ ns km}^{-1}$$

Hence, in this example the rms pulse broadening is reduced by a factor of around 20 (i.e. equivalent to the reduced rms spectral width of the injection laser source) compared with that obtained with the LED source of example 3.6.

3.8.2 Waveguide Dispersion

The waveguiding of the fiber may also create intramodal dispersion. This results from the variation in group velocity with wavelength for a particular mode. Considering the ray theory approach it is equivalent to the angle between the ray and the fiber axis varying with wavelength which subsequently leads to a variation in the transmission times for the rays, and hence dispersion. For a single mode whose propagation constant is β, the fiber exhibits waveguide dispersion when $(d^2 \beta)/(d\lambda^2) \neq 0$. Multimode fibers, where the majority of modes propagate far from cutoff, are almost free of waveguide dispersion and it is generally negligible compared with material dispersion $(\approx 0.1-0.2 \text{ ns km}^{-1})$ [Ref. 28]. However, with single mode fibers where the effects of the different dispersion mechanisms are not easy to separate, waveguide dispersion may be significant (see Section 3.10.2).

3.9 INTERMODAL DISPERSION

Pulse broadening due to intermodal dispersion (sometimes referred to simply as modal or mode dispersion) results from the propagation delay differences between modes within a multimode fiber. As the different modes which constitute a pulse in a multimode fiber travel along the channel at different group velocities, the pulse width at the output is dependent upon the transmission times of the slowest and fastest modes. This dispersion mechanism creates the fundamental difference in the overall dispersion for the three types of fiber shown in Fig. 3.7. Thus multimode step index fibers exhibit a large amount of intermodal dispersion which gives the greatest pulse broadening. However, intermodal dispersion in multimode fibers may be reduced by adoption of an optimum refractive index profile which is provided by the near parabolic profile of most graded index fibers. Hence the overall pulse broadening in multimode graded index fibers is far less than that obtained in multimode step index fibers (typically by a factor of 100). Thus graded index fibers used with a multimode source give a tremendous bandwidth advantage over multimode step index fibers.

Under purely single mode operation there is no intermodal dispersion and therefore pulse broadening is solely due to the intramodal dispersion mechanisms. In theory this is the case with single mode step index fibers where only a single mode is allowed to propagate. Hence they exhibit the least pulse broadening and have the greatest possible bandwidths, but may only be usefully operated with single mode sources.

In order to obtain a simple comparison for intermodal pulse broadening between multimode step index and multimode graded index fibers it is useful to consider the geometric optics picture for the two types of fiber.

3.9.1 Multimode Step Index Fiber

Using the ray theory model, the fastest and slowest modes propagating in the step index fiber may be represented by the axial ray and the extreme meridional ray (which is incident at the core–cladding interface at the critical angle ϕ_c) respectively. The paths taken by these two rays in a perfectly structured step index fiber are shown in Fig. 3.9. The delay difference between these two rays when travelling in the fiber core allows estimation of the pulse broadening resulting from intermodal dispersion within the fiber. As both rays are travelling at the same velocity within the constant refractive index fiber core then the delay difference is directly related to their respective path lengths within the fiber. Hence the time taken for the axial ray to travel along a fiber of length L gives the minimum delay time T_{Min} and:

$$T_{Min} = \frac{distance}{velocity} = \frac{L}{(c/n_1)} = \frac{Ln_1}{c} \tag{3.19}$$

where n_1 is the refractive index of the core and c is the velocity of light in a vacuum.

The extreme meridional ray exhibits the maximum delay time T_{Max} where:

$$T_{Max} = \frac{L/\cos\theta}{c/n_1} = \frac{Ln_1}{c\cos\theta} \tag{3.20}$$

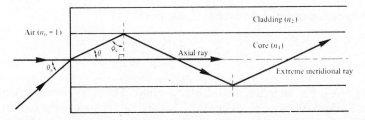

Fig. 3.9 The paths taken by the axial ray and an extreme meridional ray in a perfect multimode step index fiber.

Using Snell's law of refraction at the core–cladding interface following Eq. (2.2):

$$\sin \phi_c = \frac{n_2}{n_1} = \cos \theta \tag{3.21}$$

where n_2 is the refractive index of the cladding. Furthermore, substituting into Eq. (3.20) for $\cos \theta$ gives:

$$T_{Max} = \frac{Ln_1^2}{cn_2} \tag{3.22}$$

The delay difference δT_s between the extreme meridional ray and the axial ray may be obtained by subtracting Eq. (3.19) from Eq. (3.22). Hence:

$$\delta T_s = T_{Max} - T_{Min} = \frac{Ln_1^2}{cn_2} - \frac{Ln_1}{c}$$

$$= \frac{Ln_1^2}{cn_2} \left(\frac{n_1 - n_2}{n_1} \right) \tag{3.23}$$

$$\simeq \frac{Ln_1^2 \Delta}{cn_2} \quad \text{when } \Delta \ll 1 \tag{3.24}$$

where Δ is the relative refractive index difference. However, when $\Delta \ll 1$, then from the definition given by Eq. (2.9), the relative refractive index difference may also be given approximately by:

$$\Delta \simeq \frac{n_1 - n_2}{n_2} \tag{3.25}$$

Hence rearranging Eq. (3.23):

$$\delta T_s = \frac{Ln_1}{c} \left(\frac{n_1 - n_2}{n_2} \right) \simeq \frac{Ln_1 \Delta}{c} \tag{3.26}$$

Also substituting for Δ from Eq. (2.10) gives:

$$\delta T_s \simeq \frac{L(NA)^2}{2n_1 c} \tag{3.27}$$

where NA is the numerical aperture for the fiber. The approximate expressions for the delay difference given in Eq. (3.26) and (3.27) are usually employed to estimate the maximum pulse broadening in time due to intermodal dispersion in multimode step index fibers. It must be noted that this simple analysis only considers pulse broadening due to meridional rays and totally ignores skew rays with acceptance angles $\theta_{as} > \theta_a$ (see Section 2.2.4).

Again considering the perfect step index fiber, another useful quantity with regard to intermodal dispersion on an optical fiber link is the rms pulse broadening resulting from this dispersion mechanism along the fiber. When the optical input to the fiber is a pulse $p_i(t)$ of unit area, as illustrated in Fig. 3.10, then |Ref. 31|:

$$\int_{-\infty}^{\infty} p_i(t)\, dt = 1 \tag{3.28}$$

It may be noted that $p_i(t)$ has a constant amplitude of $1/\delta T_s$ over the range

$$\frac{-\delta T_s}{2} \leqslant p(t) \leqslant \frac{\delta T_s}{2}$$

The rms pulse broadening at the fiber output due to intermodal dispersion for the multimode step index fiber σ_s (i.e. the standard deviation) may be given in terms of the variance σ_s^2 as (see Appendix B):

$$\sigma_s^2 = M_2 - M_1^2 \tag{3.29}$$

where M_1 is the first temporal moment which is equivalent to the mean value of the pulse and M_2 the second temporal moment is equivalent to the mean square value of the pulse. Hence:

$$M_1 = \int_{-\infty}^{\infty} t p_i(t)\, dt \tag{3.30}$$

and

$$M_2 = \int_{-\infty}^{\infty} t^2 p_i(t)\, dt \tag{3.31}$$

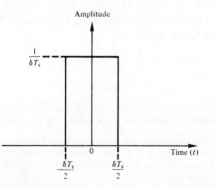

Fig. 3.10   An illustration of the light input to the multimode step index fiber consisting of an ideal pulse or rectangular function with unit area.

The mean value M_1 for the unit input pulse of Fig. 3.10 is zero, and assuming this is maintained for the output pulse, then from Eqs. (3.29) and (3.31):

$$\sigma_s^2 = M_2 = \int_{-\infty}^{\infty} t^2 p_i(t)\, dt \tag{3.32}$$

Integrating over the limits of the input pulse (Fig. 3.10) and substituting for $p_i(t)$ in Eq. (3.32) over this range gives:

$$\sigma_s^2 = \int_{-\delta T_s/2}^{\delta T_s/2} \frac{1}{\delta T_s} t^2 \, dt$$

$$= \frac{1}{\delta T_s} \left[\frac{t^3}{3} \right]_{-\delta T_s/2}^{\delta T_s/2} = \frac{1}{3} \left(\frac{\delta T_s}{2} \right)^2 \tag{3.33}$$

Hence substituting from Eq. (3.26) for δT_s gives:

$$\sigma_s \simeq \frac{L n_1 \Delta}{2\sqrt{3}\, c} \simeq \frac{L(NA)^2}{4\sqrt{3}\, n_1 c} \tag{3.34}$$

Equation (3.34) allows estimation of the rms impulse response of a multimode step index fiber if it is assumed that intermodal dispersion dominates and there is a uniform distribution of light rays over the range $0 \leqslant \theta \leqslant \theta_a$. The pulse broadening is directly proportional to the relative refractive index difference Δ and the length of the fiber L. The latter emphasizes the bandwidth–length trade-off that exists, especially with multimode step index fibers, and which inhibits their use for wideband long haul (between repeaters) systems. Furthermore, the pulse broadening is reduced by reduction of the relative refractive index difference Δ for the fiber. This suggests that weakly guiding fibers (see Section 2.3.6) with small Δ are best for low dispersion transmission. However, as may be seen from Eq. (3.34) this is also subject to a trade-off as a reduction in Δ reduces the acceptance angle θ_a and the NA, thus worsening the launch conditions.

Example 3.8

A 6 km optical link consists of multimode step index fiber with a core refractive index of 1.5 and a relative refractive index difference of 1%. Estimate:

(a) the delay difference between the slowest and fastest modes at the fiber output;
(b) the rms pulse broadening due to intermodal dispersion on the link;
(c) the maximum bit rate that may be obtained without substantial errors on the link assuming only intermodal dispersion;
(d) the bandwidth–length product corresponding to (c).

Solution: (a) The delay difference is given by Eq. (3.26) as:

$$\delta T_s \simeq \frac{Ln_1\Delta}{c} = \frac{6 \times 10^3 \times 1.5 \times 0.01}{2.998 \times 10^8}$$

$$= 300 \text{ ns}$$

(b) The rms pulse broadening due to intermodal dispersion may be obtained from Eq. (3.34) where:

$$\sigma_s \simeq \frac{Ln_1\Delta}{2\sqrt{3}\,c} = \frac{1}{2\sqrt{3}} \frac{6 \times 10^3 \times 1.5 \times 0.01}{2.998 \times 10^8}$$

$$= 86.7 \text{ ns}$$

(c) The maximum bit rate may be estimated in two ways. Firstly, to get an idea of the maximum bit rate when assuming no pulse overlap Eq. (3.9) may be used where:

$$B_T \text{ (max)} = \frac{1}{2\tau} = \frac{1}{2\delta T_s} = \frac{1}{600 \times 10^{-9}}$$

$$= 1.7 \text{ Mbit s}^{-1}$$

Alternatively an improved estimate may be obtained using the calculated rms pulse broadening in Eq. (3.10) where:

$$B_T \text{ (max)} = \frac{0.2}{\sigma_s} = \frac{0.2}{86.7 \times 10^{-9}}$$

$$= 2.3 \text{ Mbit s}^{-1}$$

(d) Using the most accurate estimate of the maximum bit rate from (c), and assuming return to zero pulses, the bandwidth–length product is:

$$B_{opt} \times L = 2.3 \text{ MHz} \times 6 \text{ km} = 13.8 \text{ MHz km}$$

Intermodal dispersion may be reduced by propagation mechanisms within practical fibers. For instance, there is differential attenuation of the various modes in a step index fiber. This is due to the greater field penetration of the higher order modes into the cladding of the waveguide. These slower modes therefore exhibit larger losses at any core–cladding irregularities which tends to concentrate the transmitted optical power into the faster lower order modes. Thus the differential attenuation of modes reduces intermodal pulse broadening on a multimode optical link.

Another mechanism which reduces intermodal pulse broadening in non-perfect (i.e. practical) multimode fibers is the mode coupling or mixing discussed in Section 2.3.7. The coupling between guided modes transfers optical power from the slower to the faster modes and vice versa. Hence with strong coupling the optical power tends to be transmitted at an average speed, which is a mean of the various propagating modes. This reduces the inter-modal dispersion on the link and makes it advantageous to encourage mode coupling within multimode fibers.

The expression for delay difference given in Eq. (3.26) for a perfect step index fiber may be modified for the fiber with mode coupling among all guided modes to |Ref. 32|:

$$\delta T_{sc} \simeq \frac{n_1 \Delta}{c} (LL_c)^{\frac{1}{2}} \qquad (3.35)$$

where L_c is a characteristic length for the fiber which is inversely proportional to the coupling strength. Hence the delay difference increases at a slower rate proportional to $(LL_c)^{\frac{1}{2}}$ instead of the direct proportionality to L given in Eq. (3.26). However, the most successful technique for reducing intermodal dispersion in multimode fibers is by grading the core refractive index to follow a near parabolic profile. This has the effect of equalizing the transmission times of the various modes as discussed in the following section.

3.9.2   Multimode Graded Index Fiber

Intermodal dispersion in multimode fibers is minimized with the use of graded index fibers. Hence multimode graded index fibers show substantial bandwidth improvement over multimode step index fibers. The reason for the improved performance of graded index fibers may be observed by considering the ray diagram for a graded index fiber shown in Fig. 3.11. The fiber shown has a parabolic index profile with a maximum at the core axis as illustrated in Fig. 3.11(a). Analytically, the index profile is given by Eq. (2.76) with $\alpha = 2$ as:

$$n(r) = n_1(1 - 2\Delta(r/a)^2)^{\frac{1}{2}} \qquad r < a \text{ (core)}$$
$$= n_1(1 - 2\Delta)^{\frac{1}{2}} = n_2 \qquad r \geqslant a \text{ (cladding)} \qquad (3.36)$$

Figure 3.11(b) shows several meridional ray paths within the fiber core. It may be observed that apart from the axial ray the meridional rays follow sinusoidal trajectories of different path lengths which result from the index

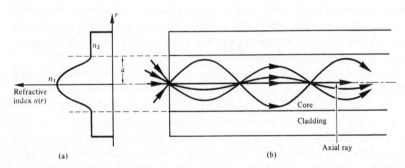

(a)   (b)

Fig. 3.11   A multimode graded index fiber: (a) parabolic refractive index profile; (b) meridional ray paths within the fiber core.

grading as was discussed in Section 2.5. However, following Eq. (2.40) the local group velocity is inversely proportional to the local refractive index and therefore the longer sinusoidal paths are compensated for by higher speeds in the lower index medium away from the axis. Hence there is an equalization of the transmission times of the various trajectories towards the transmission time of the axial ray which travels exclusively in the high index region at the core axis, and at the slowest speed. As these various ray paths may be considered to represent the different modes propagating in the fiber, then the graded profile reduces the disparity in the mode transit times.

The dramatic improvement in multimode fiber bandwidth achieved with a parabolic or near parabolic refractive index profile is highlighted by consideration of the reduced delay difference between the fastest and slowest modes for this graded index fiber δT_g. Using a ray theory approach the delay difference is given by |Ref. 33|:

$$\delta T_g \simeq \frac{Ln_1 \Delta^2}{2c} \simeq \frac{(NA)^4}{8n_1^3 c} \qquad (3.37)$$

As in the step index case Eq. (2.10) is used for conversion between the two expressions shown.

However, a more rigorous analysis using electromagnetic mode theory gives an absolute temporal width at the fiber output of |Ref. 34, 35|:

$$\delta T_g = \frac{Ln_1 \Delta^2}{8c} \qquad (3.38)$$

which corresponds to an increase in transmission time for the slowest mode of $\Delta^2/8$ over the fastest mode. The expression given in Eq. (3.38) does not restrict the bandwidth to pulses with time slots corresponding to δT_g as 70% of the optical power is concentrated in the first half of the interval. Hence the rms pulse broadening is a useful parameter for assessment of intermodal dispersion in multimode graded index fibers. It may be shown |Ref. 35| that the rms pulse broadening of a near parabolic index profile graded index fiber σ_g is reduced compared to the similar broadening for the corresponding step index fiber σ_s (i.e. with the same relative refractive index difference) following:

$$\sigma_g = \frac{\Delta}{D} \sigma_s \qquad (3.39)$$

where D is a constant between 4 and 10 depending on the precise evaluation and the exact optimum profile chosen.

The best minimum theoretical intermodal rms pulse broadening for a graded index fiber with an optimum characteristic refractive index profile for the core

α_{op} of |Refs. 35, 36|:

$$\alpha_{op} = 2 - \frac{12\Delta}{5} \tag{3.40}$$

is given by combining Eqs. (3.26) and (3.39) as |Refs. 27, 36|:

$$\sigma_g = \frac{Ln_1 \Delta^2}{20\sqrt{3}\, c} \tag{3.41}$$

Example 3.9

Compare the rms pulse broadening per kilometer due to intermodal dispersion for the multimode step index fiber of example 3.8 with the corresponding rms pulse broadening for an optimum near parabolic profile graded index fiber with the same core axis refractive index and relative refractive index difference.

Solution: In example 3.8, σ_s over 6 km of fiber is 86.7 ns. Hence the rms pulse broadening per kilometer for the multimode step index fiber is:

$$\frac{\sigma_s \ (1 \ km)}{L} = \frac{86.7}{6} = 14.4 \ ns \ km^{-1}$$

Using Eq. (3.41), the rms pulse broadening per kilometer for the corresponding graded index fiber is:

$$\sigma_g \ (1 \ km) = \frac{Ln_1\Delta^2}{20\sqrt{3}\, c} = \frac{10^3 \times 1.5 \times (0.01)^2}{20\sqrt{3} \times 2.998 \times 10^8}$$

$$= 14.4 \ ps \ km^{-1}$$

Hence, from example 3.9, the theoretical improvement factor of the graded index fiber in relation to intermodal rms pulse broadening is 1000. However, this level of improvement is not usually achieved in practice due to difficulties in controlling the refractive index profile radially over long lengths of fiber. Any deviation in the refractive index profile from the optimum results in increased intermodal pulse broadening. This may be observed from the curve shown in Fig. 3.12 which gives the variation in intermodal pulse broadening (δT_g) as a function of the characteristic refractive index profile α for typical graded index fibers (where $\Delta = 1\%$). The curve displays a sharp minimum at a characteristic refractive index profile slightly less than 2 ($\alpha = 1.98$). This corresponds to the optimum value of α in order to minimize intermodal dispersion. Furthermore, the extreme sensitivity of the intermodal pulse broadening to slight variations in α from this optimum value is evident. Thus at present improvement factors for practical graded index fibers over corresponding step index fibers with regard to intermodal dispersion are around 100 [Ref. 34].

Another important factor in the determination of the optimum refractive

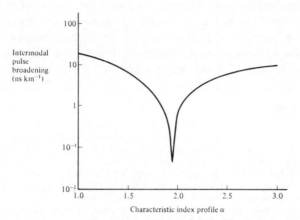

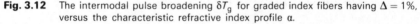

Fig. 3.12   The intermodal pulse broadening δT_g for graded index fibers having $\Delta = 1\%$, versus the characteristic refractive index profile α.

index profile for a graded index fiber is the dispersion incurred due to the difference in refractive index between the fiber core and cladding. It results from a variation in the refractive index profile with optical wavelength in the graded fiber and is often given by a profile dispersion parameter $d\Delta/d\lambda$. Thus the optimized profile at a given wavelength is not necessarily optimized at another wavelength. As all optical fiber sources (e.g. injection lasers and light emitting diodes) have a finite spectral width, the profile shape must be altered to compensate for this dispersion mechanism. Moreover the minimum overall dispersion for graded index fiber is also limited by the other intramodal dispersion mechanisms (i.e. material and waveguide dispersion). These give temporal pulse broadening of around 0.08 and 1 ns km^{-1} with injection lasers and light emitting diodes respectively. Therefore practical pulse broadening values for graded index fibers lie in the range 0.2–1 ns km^{-1}. This gives bandwidth–length products of between 0.5 and 2.5 GHz km when using lasers and optimum profile fiber.

3.10 OVERALL FIBER DISPERSION

3.10.1 Multimode Fibers

The overall dispersion in multimode fibers comprises both intramodal and intermodal terms. The total rms pulse broadening σ_T is given (see Appendix C) by:

$$\sigma_T = (\sigma_c^2 + \sigma_n^2)^{\frac{1}{2}} \tag{3.42}$$

where σ_c is the intramodal or chromatic broadening and σ_n is the intermodal broadening caused by delay differences between the modes (i.e. $\dot{\sigma}_s$ for multimode step index fiber and σ_g for multimode graded index fiber). The intramodal term σ_c consists of pulse broadening due to both material and waveguide dispersion. However, since waveguide dispersion is generally negligible compared with material dispersion in multimode fibers, then $\sigma_c \simeq \sigma_m$.

Example 3.10

A multimode step index fiber has a numerical aperture of 0.3 and a core refractive index of 1.45. The material dispersion parameter for the fiber is 250 ps nm^{-1} km^{-1} which makes material dispersion the totally dominating intramodal dispersion mechanism. Estimate (a) the total rms pulse broadening per kilometer when the fiber is used with an LED source of rms spectral width 50 nm and (b) the corresponding bandwidth–length product for the fiber.

Solution: (a) The rms pulse broadening per kilometer due to material dispersion may be obtained from Eq. (3.17), where

$$\sigma_m \ (1 \ \text{km}) \simeq \frac{\sigma_\lambda L \lambda}{c} \left| \frac{d^2 n_1}{d\lambda^2} \right| = \sigma_\lambda L M = 50 \times 1 \times 250 \ \text{ps km}^{-1}$$

$$= 12.5 \ \text{ns km}^{-1}$$

The rms pulse broadening per kilometer due to intermodal dispersion for the step index fiber is given by Eq. (3.34) as:

$$\sigma_s \ (1 \ \text{km}) \simeq \frac{L(\text{NA})^2}{4\sqrt{3} \, n_1 c} = \frac{10^3 \times 0.09}{4\sqrt{3} \times 1.45 \times 2.998 \times 10^8}$$

$$= 29.9 \ \text{ns km}^{-1}$$

The total rms pulse broadening per kilometer may be obtained using Eq. (3.42), where $\sigma_c \approx \sigma_m$ as the waveguide dispersion is negligible and $\sigma_n = \sigma_s$ for the multimode step index fiber. Hence:

$$\sigma_T = (\sigma_m^2 + \sigma_s^2)^{\frac{1}{2}} = (12.5^2 + 29.9^2)^{\frac{1}{2}}$$

$$= 32.4 \ \text{ns km}^{-1}$$

(b) The bandwidth–length product may be estimated from the relationship given in Eq. (3.10) where:

$$B_{\text{opt}} \times L = \frac{0.2}{\sigma_T} = \frac{0.2}{32.4 \times 10^{-9}}$$

$$= 6.2 \ \text{MHz km}$$

3.10.2 Single Mode Fibers

The pulse broadening in single mode fibers is solely due to intramodal dispersion as only a single mode is allowed to propagate. Hence the bandwidth is limited by the finite spectral width of the source. Unlike the situation in multimode fibers, the mechanisms giving intramodal dispersion in single mode fibers tend to be interrelated in a complex manner. The transit time or group delay τ_g

for a light pulse propagating along a unit length of fiber may be given as [Ref. 37]:

$$\tau_g = \frac{1}{c} \frac{d\beta}{dk} \qquad (3.43)$$

where c is the velocity of light in a vacuum, β is the propagation constant for a mode within the fiber core of refractive index n_1 and k is the propagation constant for the mode in a vacuum.

The fiber exhibits intramodal dispersion when β varies nonlinearly with wavelength. From Eq. (2.71) β may be expressed in terms of the relative refractive index difference Δ and the normalized propagation constant b as:

$$\beta = kn_1 [1 - 2\Delta(1 - b)]^{\frac{1}{2}} \qquad (3.44)$$

The rms pulse broadening caused by intramodal dispersion down a fiber of length L is given by the derivative of the group delay with respect to wavelength as [Ref. 27]:

$$\text{Total rms pulse broadening} = \sigma_\lambda L \left| \frac{d\tau_g}{d\lambda} \right|$$

$$= \frac{\sigma_\lambda L 2\pi}{c\lambda^2} \frac{d^2\beta}{dk^2} \qquad (3.45)$$

where σ_λ is the source rms spectral linewidth centered at a wavelength λ.

When Eq. (3.44) is substituted into Eq. (3.45), detailed calculation of the first and second derivatives with respect to k gives the dependence of the pulse broadening on the fiber material's properties and the normalized propagation constant b. This gives rise to three interrelated effects which involve complicated cross-product terms. However, the final expression may be separated into three composite dispersion components in such a way that one of the effects dominates each term [Ref. 38]. The dominating effects are:

(a) the material dispersion parameter defined by $\lambda/c \, |d^2 n/d\lambda^2|$ where $n = n_1$ or n_2 for the core or cladding respectively;
(b) the waveguide dispersion parameter defined as $V d^2(bV)/dV^2$ where V is the normalized frequency for the fiber;
(c) a profile dispersion parameter which is proportional to $d\Delta/d\lambda$.

This situation is different from multimode fibers where the majority of modes propagate far from cutoff and hence most of the power is transmitted in the fiber core. In the multimode case the composite dispersion components may be simplified and separated into two intramodal terms which depend on either material or waveguide dispersion as was discussed in Section 3.8. Also, especially when considering step index multimode fibers, the effect of profile dispersion is negligible. However, although material and waveguide dispersion tend to be dominant in single mode fibers, the composite profile dispersion term cannot be ignored.

Figure 3.8 shows the material dispersion parameter for pure silica plotted against wavelength. It may be observed that this characteristic goes through zero at a wavelength of 1.27 μm. This zero material dispersion (ZMD) point can be shifted anywhere in the wavelength range 1.2–1.4 μm by the addition of suitable dopants [Ref. 39]. For instance, the ZMD point shifts from 1.27 μm to approximately 1.37 μm as the GeO_2 dopant concentration is increased from 0 to 15%. However, the ZMD point (although of great interest in single mode fibers) does not represent a point of zero pulse broadening since the pulse dispersion is influenced by both waveguide and profile dispersion. With zero material dispersion the pulse spreading is dictated by the waveguide parameter $V d^2(bV)/dV^2$ which is illustrated in Fig. 3.13 as a function of normalized frequency for the LP_{01} mode. It may be seen that in the single mode region where the normalized frequency is less than 2.405 (see Section 2.4.2) the waveguide dispersion is always positive and has a maximum at $V = 1.15$. In this case the waveguide dispersion goes to zero outside the true single mode region at $V = 3.0$. However, a change in the fiber parameters (such as core radius) or in the operating wavelength alters the normalized frequency and therefore the waveguide dispersion.

The total fiber dispersion which depends on both the fiber material composition and dimensions may be minimized by trading off material and waveguide dispersion whilst limiting the profile dispersion (i.e. restricting the variation in refractive index with wavelength). The wavelength at which the first order dispersion is zero λ_0 may be selected in the range 1.3–2 μm by careful control of the core diameter and profile [Ref. 38]. This is illustrated in Fig. 3.14 which

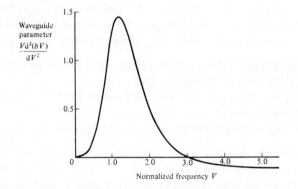

Fig. 3.13   The waveguide parameter $V d^2(bV)/dV^2$ as a function of the normalized frequency V for the LP_{01} mode, Reproduced with permission from W. A. Gambling, A. H. Hartog and C. M. Ragdale, *The Radio and Electron. Eng.*, **51**, p. 313, 1981.

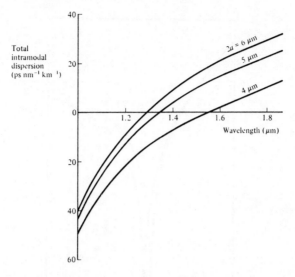

Fig. 3.14 The total first order intramodal dispersion as a function of wavelength for single mode fibers with core diameters of 4, 5 and 6 μm. Reproduced with permission from W. A. Gambling, A. H. Hartog and C. M. Ragdale, *The Radio and Electron. Eng.*, **51**, p. 313, 1981.

shows the total first order dispersion as a function of wavelength for three single mode fibers with core diameters of 4, 5 and 6 μm.

The effect of the interaction of material and waveguide dispersion on λ_0 is demonstrated in the dispersion against wavelength characteristics for a single mode silica core fiber shown in Fig. 3.15. It may be noted that the ZMD point occurs at a wavelength of 1.27 μm but that the influence of waveguide dispersion shifts the total dispersion minimum towards the longer wavelength giving a λ_0 of 1.32 μm.

The wavelength at which the first order dispersion is zero λ_0 may be extended to wavelengths beyond 1.55 μm by a combination of three techniques. These are:

(a) lowering the normalized frequency (V value) for the fiber;
(b) increasing the relative refractive index difference Δ for the fiber;
(c) suitable doping of the silica with germanium.

This should allow bandwidth–length products for single mode fibers in excess of 100 GHz km |Ref. 40| at the slight disadvantage of increased attenuation due to Rayleigh scattering within the doped silica. However, it must be noted that although there is zero first order dispersion at λ_0, higher order chromatic effects impose limitations on the possible bandwidth for single mode fibers. At present these give a fundamental lower limit to pulse spreading in silica-based fibers of, for example, around 2.50×10^{-2} ps nm^{-1} km^{-1} in fused silica at a

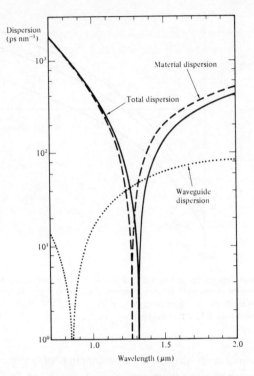

Fig. 3.15   The pulse dispersion as a function of wavelength in 11 km single mode fiber showing the major contributing dispersion mechanisms (dashed and dotted curves) and the overall total dispersion (solid curve). Reproduced with permission from J. I. Yamada, M. Saruwatari, K. Asatani, H. Tsuchiya, A. Kawana, K. Sugiyama and T. Kimura, 'High speed optical pulse transmission at 1.29 μm wavelength using low-loss single-mode fibers', *IEEE J. Quantum Electron.*, **QE-14**, p. 791, 1978. Copyright © 1980, IEEE.

wavelength of 1.273 μm [Ref. 41]. These secondary effects such as birefringence arising from ellipticity or mechanical stress in the fiber core are considered further in Section 3.12. However, they may cause dispersion, especially in the case of mechanical stress of between 2 and 40 ps km⁻¹. If mechanical stress is avoided pulse dispersion around the lower limit may be obtained in the longer wavelength region (i.e. 1.3–1.7 μm). By contrast the minimum pulse spread at a wavelength of 0.85 μm is around 100 ps nm⁻¹ km⁻¹ [Ref. 33].

3.11 MODAL NOISE

The intermodal dispersion properties of multimode optical fibers (see Section 3.9) create another phenomenon which affects the transmitted signals on the optical channel. It is exhibited within the speckle patterns observed in multimode fiber as fluctuations which have characteristic times longer than the resolution time of the detector, and is known as modal or speckle noise. The speckle patterns are formed by the interference of the modes from a coherent source when the coherence time of the source in greater than the intermodal dispersion time δT within the fiber. The coherence time for a source with uncorrelated source frequency width δf is simply $1/\delta f$. Hence, modal noise occurs when:

$$\delta f \gg \frac{1}{\delta T} \qquad (3.46)$$

Disturbances along the fiber such as vibrations, discontinuities, connectors, splices and source/detector coupling may cause fluctuations in the speckle patterns and hence modal noise. It is generated when the correlation between two or more modes which gives the original interference is differentially delayed by these disturbances. The conditions which give rise to modal noise are therefore specified as:

(a) a coherent source with a narrow spectral width and long coherence length (propagation velocity multiplied by the coherence time);
(b) disturbances along the fiber which give differential mode delay or modal and spatial filtering;
(c) phase correlation between the modes.

Measurements [Ref. 46] of rms signal to modal noise ratio using good narrow linewidth injection lasers show large signal to noise ratio penalties under the previously mentioned conditions. The measurements were carried out by misaligning connectors to create disturbances. They gave carrier to noise ratios reduced by around 10 dB when the attenuation at each connector was 20 dB due to substantial axial misalignment.

Modal noise may be avoided by removing one of the conditions (they must all be present) which give rise to this degradation. Hence modal noise free transmission may be obtained by:

(a) The use of a broad spectrum source in order to eliminate the modal interference effects. This may be achieved by either (1) increasing the width of the single longitudinal mode and hence decreasing its coherence time or (2) by increasing the number of longitudinal modes and averaging out of the interference patterns [Ref. 47].
(b) In conjunction with (a)(2) it is found that fibers with large numerical apertures support the transmission of a large number of modes giving a greater number of speckles, and hence reduce the modal noise generating effect of individual speckles [Ref. 48].
(c) The use of single mode fiber which does not support the transmission of different modes and thus there is no intermodal interference.
(d) The removal of disturbances along the fiber. This has been investigated

with regard to connector design [Ref. 49] in order to reduce the shift in speckle pattern induced by mechanical vibration and fiber misalignment.

Hence, modal noise may be prevented on an optical fiber link through suitable choice of the system components. However, this may not always be possible and then certain levels of modal noise must be tolerated. This tends to be the case on high quality analog optical fiber links where multimode injection lasers are frequently used. Analog transmission is also more susceptible to modal noise due to the higher optical power levels required at the receiver when quantum noise effects are considered (see Section 9.2.5). Therefore it is important that modal noise is taken into account within the design considerations for these systems.

3.12 POLARIZATION

Cylindrical optical fibers do not generally maintain the polarization state of the light input for more than a few meters, and hence for most applications involving optical fiber transmission some form of intensity modulation (see Section 7.5) of the optical source is utilized. The optical signal is thus detected by a photodiode which is insensitive to optical polarization or phase of the light wave within the fiber. Nevertheless, systems and applications have recently been investigated [Ref. 52] (see Sections 10.8 and 11.5.1) which could require the polarization states of the input light to be maintained over considerable distances, and fibers have been designed for this purpose. These fibers are single mode and the maintenance of the polarization state is described in terms of a phenomenon known as modal birefringence.

3.12.1 Modal Birefringence

Single mode fibers with nominal circular symmetry about the core axis allow the propagation of two nearly degenerate modes with orthogonal polarizations. They are therefore bimodal supporting HE_{11}^x and HE_{11}^y modes where the principal axes x and y are determined by the symmetry elements of the fiber cross section. Thus the fiber behaves as a birefringent medium due to the difference in the effective refractive indices and hence phase velocities for these two orthogonally polarized modes. The modes therefore have different propagation constants β_x and β_y which are dictated by the anisotropy of the fiber cross section. When the fiber cross section is independent of the fiber length L in the z direction then the modal birefringence B_F for the fiber is given by [Ref. 53],

$$B_F = \frac{(\beta_x - \beta_y)}{(2\pi/\lambda)} \qquad (3.47)$$

where λ is the optical wavelength. Light polarized along one of the principal axes will retain its polarization for all L.

The difference in phase velocities causes the fiber to exhibit a linear retardation $\Phi(z)$ which depends on the fiber length L in the z direction and is given by [Ref. 53]:

$$\Phi(z) = (\beta_x - \beta_y)L \tag{3.48}$$

assuming that the phase coherence of the two mode components is maintained. The phase coherence of the two mode components is achieved when the delay between the two transit times is less than the coherence time of the source. As indicated in Section 3.11 the coherence time for the source is equal to the reciprocal of the uncorrelated source frequency width $(1/\delta f)$.

It may be shown [Ref. 54] that birefringent coherence is maintained over a length of fiber L_{bc} (i.e. coherence length) when:

$$L_{bc} \simeq \frac{c}{B_F \delta f} = \frac{\lambda^2}{B_F \delta \lambda} \tag{3.49}$$

where c is the velocity of light in a vacuum and $\delta \lambda$ is the source linewidth.

However, when phase coherence is maintained (i.e. over the coherence length) Eq. 3.48 leads to a polarization state which is generally elliptical but which varies periodically along the fiber. This situation is illustrated in Fig. 3.16(a) [Ref. 53] where the incident linear polarization which is at 45° with respect to the x axis becomes circular polarization at $\Phi = \pi/2$, and linear again at $\Phi = \pi$. The process continues through another circular polarization at $\Phi = 3\pi/2$ before returning to the initial linear polarization at $\Phi = 2\pi$. The characteristic length L_B corresponding to this process is known as the beat length. It is given by:

$$L_B = \frac{\lambda}{B_F} \tag{3.50}$$

Substituting for B_F from Eq. (3.47) gives:

$$L_B = \frac{2\pi}{(\beta_x - \beta_y)} \tag{3.51}$$

It may be noted that Eq. (3.51) may be obtained directly from Eq. (3.48) where:

$$\Phi(L_B) = (\beta_x - \beta_y)L_B = 2\pi \tag{3.52}$$

Typical single mode fibers are found to have beat lengths of a few centimeters [Ref. 55], and the effect may be observed directly within a fiber via Rayleigh scattering with use of a suitable visible source (e.g. He–Ne laser) [Ref. 56]. It appears as a series of bright and dark bands with a period

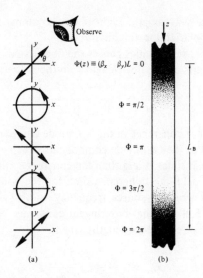

Fig. 3.16 An illustration of the beat length in a single mode optical fiber [Ref. 53]: (a) the polarization states against $\Phi(z)$; (b) the light intensity distribution over the beat length within the fiber.

corresponding to the beat length as shown in Fig. 3.16(b). The modal birefringence B_F may be determined from these observations of beat length.

Example 3.11

The beat length in a single mode optical fiber is 9 cm when light from an injection laser with a spectral linewidth of 1 nm and a peak wavelength of 0.9 μm is launched into it. Determine the modal birefringence and estimate the coherence length in this situation. In addition, calculate the difference between the propagation constants for the two orthogonal modes and check the result.

Solution: To find the modal birefringence Eq. (3.50) may be used where:

$$B_F = \frac{\lambda}{L_B} = \frac{0.9 \times 10^{-6}}{0.09} = 1 \times 10^{-5}$$

Knowing B_F, Eq. (3.49) may be used to obtain the coherence length:

$$L_{bc} \simeq \frac{\lambda^2}{B_F \delta\lambda} = \frac{0.81 \times 10^{-12}}{10^{-5} \times 10^{-9}} = 81 \text{ m}$$

The difference between the propagation constant for the two orthogonal modes may be obtained from Eq. (3.51) where:

$$\beta_x - \beta_y = \frac{2\pi}{L_B} = \frac{2\pi}{0.09} = 69.8$$

The result may be checked by using Eq. (3.47) where:

$$\beta_x - \beta_y = \frac{2\pi B_F}{\lambda} = \frac{2\pi \times 10^{-5}}{0.9 \times 10^{-6}}$$

$$= 69.8$$

In a nonperfect fiber various perturbations along the fiber length such as strain or variations in the fiber geometry and composition lead to coupling of energy from one polarization to the other. These perturbations are difficult to eradicate as they may easily occur in the fiber manufacture and cabling. The energy transfer is at a maximum when the perturbations have a period Λ, corresponding to the beat length, and defined by [Ref. 52]:

$$\Lambda = \frac{\lambda}{B_F} \tag{3.53}$$

However, the cross-polarizing effect may be minimized when the period of the perturbations is less than a cutoff period Λ_c (around 1 mm). Hence polarization-maintaining fibers may be designed by either:

(a) High (large) birefringence: the maximization of the modal birefringence, which following Eq. (3.50), may be achieved by reducing the beat length L_B to around 1 mm or less; or
(b) Low (small) birefringence: the minimization of the polarization coupling perturbations with a period of Λ. This may be achieved by increasing Λ_c giving a large beat length of around 50 m or more.

Example 3.12

Two polarization-maintaining fibers operating at a wavelength of 1.3 µm have beat lengths of 0.7 mm and 80 m. Determine the modal birefringence in each case and comment on the results.
Solution: Using Eq. (3.50), the modal birefringence is given by:

$$B_F = \frac{\lambda}{L_B}$$

Hence, for a beat length of 0.7 mm:

$$B_F = \frac{1.3 \times 10^{-6}}{0.7 \times 10^{-3}} = 1.86 \times 10^{-3}$$

This typifies a high birefringence fiber.
For a beat length of 80 m:

$$B_F = \frac{1.3 \times 10^{-6}}{80} = 1.63 \times 10^{-8}$$

which indicates a low birefringence fiber.

Techniques are being developed for the production of both high and low birefringence fibers to assist the implementation of coherent optical fiber communications. Fibers may be made highly birefringent by deliberately inducing large asymmetric radial stress. This may be achieved through thermal stress by using materials with widely different expansion coefficients coupled with an asymmetrical elliptical structure. A linear polarization state has been maintained over a kilometer of fiber with an extinction ratio of 30 dB using this technique [Ref. 57]. Further investigation of the optimal cross section geometry for high birefringence has suggested [Ref. 57] fiber core cross sections shaped as a bow tie (bow tie fiber).

To design low birefringence fibers it is necessary to reduce the possible perturbations within the fiber during manufacture. Therefore extreme care must be taken when jacketing and winding these fibers in order to reduce bends or twists that may contribute to birefringence. It is also necessary to use materials which minimize the thermal effects that may create birefringence. One technique used to minimize the temperature dependence of birefringence which has proved successful is to spin the fiber preform during manufacture |Ref. 58|. This method, which reduces the linear retardation within the fiber, has produced fibers with no birefringent properties and variations in output polarization result only from fiber packaging. However, even with these low birefringence spun fibers some form of polarization controller |Ref. 59| is necessary to stabilize the polarization state within the fiber.

PROBLEMS

3.1 The mean optical power launched into an optical fiber link is 1.5 mW and the fiber has an attenuation of 0.5 dB km^{-1}. Determine the maximum possible link length without repeaters (assuming lossless connectors) when the minimum mean optical power level required at the detector is 2 μW.

3.2 The numerical input/output mean optical power ratio in a 1 km length of optical fiber is found to be 2.5. Calculate the received mean optical power when a mean optical power of 1 mW is launched into a 5 km length of the fiber (assuming no joints or connectors).

3.3 A 15 km optical fiber link uses fiber with a loss of 1.5 dB km^{-1}. The fiber is jointed every kilometer with connectors which given an attenuation of 0.8 dB each. Determine the minimum mean optical power which must be launched into the fiber in order to maintain a mean optical power level of 0.3 μW at the detector.

3.4 Discuss absorption losses in optical fibers comparing and contrasting the intrinsic and extrinsic absorption mechanisms.

3.5 Briefly describe linear scattering losses in optical fibers with regard to:

(a) Rayleigh scattering;

(b) Mie scattering.

The photoelastic coefficient and the refractive index for silica are 0.286 and 1.46 respectively. Silica has an isothermal compressibility of 7×10^{-11} m^2 N^{-1} and an estimated fictive temperature of 1400 K. Determine the theoretical attenuation in decibels per kilometer due to the fundamental Rayleigh scattering in silica at optical wavelengths of 0.85 and 1.55 μm. Boltzmann's constant is 1.381×10^{-23} J K^{-1}.

3.6 A K$_2$O–SiO$_2$ glass core optical fiber has an attenuation resulting from Rayleigh scattering of 0.46 dB km^{-1} at a wavelength of 1 μm. The glass has an estimated fictive temperature of 758 K, isothermal compressibility of 8.4×10^{-11} m^2 N^{-1}, and a photoelastic coefficient of 0.245. Determine from theoretical considerations the refractive index of the glass.

3.7 Compare stimulated Brillouin and stimulated Raman scattering in optical fibers, and indicate the way in which they may be avoided in optical fiber communications.

The threshold optical powers for stimulated Brillouin and Raman scattering in a long 8 μm core diameter single mode fiber are found to be 190 mW and 1.70 W respectively when using an injection laser source with a bandwidth of 1 GHz. Calculate the operating wavelength of the laser and the attenuation in decibels per kilometer of the fiber at this wavelength.

3.8 The threshold optical power for stimulated Brillouin scattering at a wavelength of 0.85 μm in a long single mode fiber using an injection laser source with a bandwidth of 800 MHz is 127 mW. The fiber has an attenuation of 2 dB km^{-1} at this wavelength. Determine the threshold optical power for stimulated Raman scattering within the fiber at a wavelength of 0.9 μm assuming the fiber attenuation is reduced to 1.8 dB km^{-1} at this wavelength.

3.9 Explain what is meant by the critical bending radius for an optical fiber.

A single mode step index fiber has a critical bending radius of 2 mm when illuminated with light at a wavelength of 1.30 μm. Calculate the relative refractive index difference for the fiber.

3.10 A graded index fiber has a refractive index at the core axis of 1.46 with a cladding refractive index of 1.45. The critical radius of curvature which allows large bending losses to occur is 84 μm when the fiber is transmitting light of a particular wavelength. Determine the wavelength of the transmitted light.

3.11 (a) A multimode step index fiber gives a total pulse broadening of 95 ns over a 5 km length. Estimate the bandwidth–length product for the fiber when a non return to zero digital code is used.

(b) A single mode step index fiber has a bandwidth–length product of 10 GHz km. Estimate the rms pulse broadening over a 40 km digital optical link without repeaters consisting of the fiber, and using a return to zero code.

3.12 An 8 km optical fiber link without repeaters uses multimode graded index fiber which has a bandwidth–length product of 400 MHz km. Estimate:

(a) the total pulse broadening on the link;
(b) the rms pulse broadening on the link.

It may be assumed that a return to zero code is used.

3.13 Briefly explain the reasons for pulse broadening due to material dispersion in optical fibers.

The group delay T_g in an optical fiber is given by:

$$T_g = \frac{1}{c} \left(n_1 - \frac{\lambda dn_1}{d\lambda} \right)$$

where c is the velocity of light in a vacuum, n_1 is the core refractive index and λ is the wavelength of the transmitted light. Derive an expression for the rms pulse broadening due to material dispersion in an optical fiber and define the material dispersion parameter.

The material dispersion parameter for a glass fiber is 20 ps nm^{-1} km^{-1} at a wavelength of 1.5 μm. Estimate the pulse broadening due to material dispersion within the fiber when light is launched from an injection laser source with a peak wavelength of 1.5 μm and an rms spectral width of 2 nm into a 30 km length of the fiber.

3.14 The material dispersion in an optical fiber defined by $\left| d^2 n_1 / d\lambda^2 \right|$ is 4.0×10^{-2} μm^{-2}. Estimate the pulse broadening per kilometer due to material dispersion within the fiber when it is illuminated with an LED source with a peak wavelength of 0.9 μm and an rms spectral width of 45 nm.

3.15 Describe the mechanism of intermodal dispersion in a multimode step index fiber.

Show that the total broadening of a light pulse δT_s due to intermodal dispersion in a multimode step index fiber may be given by:

$$\delta T_s \simeq \frac{L(NA)^2}{2n_1 c}$$

where L is the fiber length, NA is the numerical aperture of the fiber, n_1 is the core refractive index and c is the velocity of light in a vacuum.

A multimode step index fiber has a numerical aperture of 0.2 and a core refractive index of 1.47. Estimate the bandwidth–length product for the fiber assuming only intermodal dispersion and a return to zero code when:

(a) there is no mode coupling between the guided modes;
(b) mode coupling between the guided modes gives a characteristic length equivalent to 0.6 of the actual fiber length.

3.16 Using the relation for δT_s given in problem 3.15, derive an expression for the rms pulse broadening due to intermodal dispersion in a multimode step index fiber. Compare this expression with a similar expression which may be obtained for an optimum near parabolic profile graded index fiber.

Estimate the bandwidth–length product for the step index fiber specified in problem 3.15 considering the rms pulse broadening due to intermodal dispersion within the fiber and comment on the result. Indicate the possible improvement in the bandwidth–length product when an optimum near parabolic profile graded index fiber with the same relative refractive index difference and core axis refractive index is used. In both cases assume only intermodal dispersion within the fiber and the use of a return to zero code.

3.17 An 11 km optical fiber link consisting of optimum near parabolic profile graded index fiber exhibits rms intermodal pulse broadening of 346 ps over its length. If the fiber has a relative refractive index difference of 1.5%, estimate the core axis refractive index. Hence determine the numerical aperture for the fiber.

3.18 A multimode, optimum near parabolic profile graded index fiber has a material dispersion parameter of 30 ps nm^{-1} km^{-1} when used with a good LED source of rms spectral width 25 nm. The fiber has a numerical aperture of 0.4 and a core axis refractive index of 1.48. Estimate the total rms pulse broadening per kilometer within the fiber assuming waveguide dispersion to be negligible. Hence estimate the bandwidth–length product for the fiber.

3.19 A multimode step index fiber has a relative refractive index difference of 1% and a core refractive index of 1.46. The maximum optical bandwidth that may be obtained with a particular source on a 4.5 km link is 3.1 MHz.

(a) Determine the rms pulse broadening per kilometer resulting from intramodal dispersion mechanisms.
(b) Assuming waveguide dispersion may be ignored, estimate the rms spectral width of the source used, if the material dispersion parameter for the fiber at the operating wavelength is 90 ps nm^{-1} km^{-1}.

3.20 Discuss dispersion mechanisms with regard to single mode fibers indicating the dominating effects. Hence describe how intramodal dispersion may be minimized within the single-mode region.

3.21 Describe the phenomenon of modal noise in optical fibers and suggest how it may be avoided.

3.22 Explain what is meant by:

(a) modal birefringence;
(b) the beat length;

in single mode fibers.
The difference between the propagation constants for the two orthogonal modes in a single mode fiber is 250. It is illuminated with light of peak wavelength 1.55 μm from an injection laser source with a spectral linewidth of 0.8 nm. Estimate the coherence length within the fiber.

3.23 A single mode fiber maintains birefringent coherence over a length of 100 km when it is illuminated with an injection laser source with a spectral linewidth of 1.5 nm and a peak wavelength of 1.32 μm. Estimate the beat length within the fiber and comment on the result.

Answers to Numerical Problems

3.1	57.5 km	**3.13**	1.2 ns
3.2	10.0 μW	**3.14**	5.4 ns km^{-1}
3.3	703 μW	**3.15**	(a) 11.0 MHz km; (b) 14.2 MHz
3.5	1.57 dB km^{-1}, 0.14 dB km^{-1}		km
3.6	1.49	**3.16**	15.3 MHz km; improvement to
3.7	1.50 μm, 0.30 dB km^{-1}		10.9 GHz km
3.8	2.4 W	**3.17**	1.45, 0.25
3.9	0.17%	**3.18**	774 ps km^{-1}, 258 MHz km
3.10	0.86 μm	**3.19**	(a) 2.82 ns km^{-1}; (b) 31 nm
3.11	(a) 13.2 MHz km; (b) 800 ps	**3.22**	48.6 m
3.12	(a) 10 ns; (b) 4 ns	**3.23**	113.6 m

REFERENCES

1   F. P. Kapron, D. B. Keck and R. D. Maurer, 'Radiation losses in optical waveguides', *Appl. Phys. Lett.*, **10**, pp. 423–425, 1970.

2   T. Kimura, 'Single-mode digital transmission technology', *Proc. IEEE*, **68**(10), pp. 1263–1268, 1980.

3   T. Miya, Y. Teramuna, Y. Hosaka and T. Miyashita, 'Ultimate low-loss single-mode fibre at 1.55 μm', *Electron. Lett.*, **15**(4), pp. 106–108, 1979.

4   P. C. Schultz, 'Preparation of very low loss optical waveguides', *J. Am. Ceram. Soc.*, **52**(4), pp. 383–385, 1973.

5   H. Osanai, T. Shioda, T. Morivama, S. Araki, M. Horiguchi, T. Izawa and H. Takata, 'Effect of dopants on transmission loss of low OH-content optical fibres', *Electron. Lett.*, **12**(21), pp. 549–550, 1976.

6   D. B. Keck, K. D. Maurer and P. C. Schultz, 'On the ultimate lower limit of attenuation in glass optical waveguides', *Appl. Phys. Lett.*, **22**(7), pp. 307–309, 1973.

7   A. R. Tynes, A. D. Pearson and D. L. Bisbee, 'Loss mechanisms and measurements in clad glass fibers and bulk glass', *J. Opt. Soc. Am.*, **61**, pp. 143–153, 1971.

8   K. J. Beales and C. R. Day, 'A review of glass fibres for optical communications', *Phys. Chem. Glasses*, **21**(1), pp. 5–21, 1980.

9   R. Olshansky, 'Propagation in glass optical waveguides', *Rev. Mod. Phys.*, **51**(2), pp. 341–367, 1979.

10   R. M. Gagliardi and S. Karp, *Optical Communications*, John Wiley, 1976.

11   J. Schroeder, R. Mohr, P. B. Macedo and C. J. Montrose, 'Rayleigh and Brillouin scattering in $K_2O–SiO_2$ glasses', *J. Am. Ceram. Soc.*, **56**, pp. 510–514, 1973.

12   R. D. Maurer, 'Glass fibers for optical communications', *Proc. IEEE*, **61**, pp. 452–462, 1973.

13   D. A. Pinnow, T. C. Rich, F. W. Ostermayer Jr and M. DiDomenico Jr, 'Fundamental optical attenuation limits in the liquid and glassy state with application to fiber optical waveguide materials', *App. Phys. Lett.*, **22**, pp. 527–29, 1973.

14   E. A. J. Marcatili, 'Objectives of early fibers: evolution of fiber types', in S. E. Miller and A. G. Chynoweth (Eds.), *Optical Fiber Telecommunications*, pp. 1–35, Academic Press, 1979.

15   D. Gloge, 'Propagation effects in optical fibers', *IEEE Trans. Microwave Theory Tech.*, **MTT-23**, pp. 106–120, 1975.

16   R. H. Stolen, 'Nonlinearity in fiber transmission', *Proc. IEEE*, **68**(10), pp. 1232–1236, 1980.

17 R. H. Stolen, 'Nonlinear properties of optical fibers', in S. E. Miller and A. G. Chynoweth (Eds.), *Optical Fiber Telecommunications*, pp. 125–150, Academic Press, 1979.

18 Y. Ohmori, Y. Sasaki and T. Edahiro, 'Fibre-length dependence of critical power for stimulated Raman scattering', *Electron. Lett.*, **17**(17), pp. 593–594, 1981.

19 M. M. Ramsay and G. A. Hockham, 'Propagation in optical fibre waveguides', in C. P. Sandbank (Ed.), *Optical Fibre Communication Systems*, pp. 25–41. John Wiley, 1980.

20 H. F. Wolf, 'Optical waveguides', in H. F. Wolf (Ed.), *Handbook of Fiber Optics Theory and Applications*, pp. 43–152, Granada, 1979.

21 A. W. Synder, 'Leaky-ray theory of optical waveguides of circular cross section', *Appl. Phys. (Germany)*, **4**(4), pp. 273–298, 1974.

22 T. Li, 'Structures, parameters and transmission properties of optical fibers', *Proc. IEEE*, **68**(10), pp. 1175–1180, 1980.

23 P. Baues, 'The anatomy of a fiber optic link', *Control Eng.*, pp. 46–49, August 1979.

24 S. D. Personick, 'Receiver design for digital fiber optic communication systems, Part I and II', *Bell Syst. Tech. J.*, **52**, pp. 843–886, 1973.

25 I. P. Kaminow, D. Marcuse and H. M. Presby, 'Multimode fiber bandwidth: theory and practice', *Proc. IEEE*, **68**(10), pp. 1209–1213, 1980.

26 M. J. Adams, D. N. Payne, F. M. Sladen and A. H. Hartog, 'Optimum operating wavelength for chromatic equalisation in multimode optical fibres', *Electron. Lett.*, **14**(3), pp. 64–66, 1978.

27 W. A. Gambling, A. H. Hartog and C. M. Ragdale, 'Optical fibre transmission lines', *Radio Electron. Eng. J. IERE*, **51**(7/8), pp. 313–325, 1981.

28 D. N. Payne and W. A. Gambling, 'Zero material dispersion in optical fibres', *Electron. Lett.*, **11**(8), pp. 176–178, 1975.

29 F. P. Kapron and D. B. Keck, 'Pulse transmission through a dielectric optical waveguide', *Appl. Opt.*, **10**(7), pp. 1519–1523, 1971.

30 M. DiDomenio Jr, 'Material dispersion in optical fiber waveguides', *Appl. Opt.*, **11**, pp. 652–654, 1972.

31 F. G. Stremler, *Introduction to Communication Systems*, 2nd Edn., Addison-Wesley, 1982.

32 D. Botez and G. J. Herkskowitz, 'Components for optical communication systems: a review', *Proc. IEEE*, **68**(6), pp. 689–730, 1980.

33 A. Ghatak and K. Thyagarajan, 'Graded index optical waveguides: a review', in E. Wolf (Ed.), *Progress in Optics*, pp. 1–109, North-Holland Publishing, 1980.

34 D. Gloge and E. A. Marcatili, 'Multimode theory of graded-core fibers', *Bell Syst. Tech. J.*, **52**, pp. 1563–1578, 1973.

35 J. E. Midwinter, *Optical Fibers for Transmission*, John Wiley, 1979.

36 R. Olshansky and D. B. Keck, 'Pulse broadening in graded-index optical fibers', *Appl. Opt.*, **15**(12), pp. 483–491, 1976.

37 D. Gloge, 'Dispersion in weakly guiding fibers', *Appl. Opt.*, **10**(11), pp. 2442–2445, 1971.

38 W. A. Gambling, H. Matsumura and C. M. Ragdale, 'Mode dispersion, material dispersion and profile dispersion in graded index single-mode fibers', *IEEJ. Microwaves, Optics and Acoustics (GB)*, **3** (6), pp. 239–246, 1979.

39 J. W. Fleming, 'Material dispersion in lightguide glasses', *Electron. Lett.*, **14**(11), pp. 326–328, 1978.

40 J. I. Yamada, M. Saruwatari, K. Asatani, H. Tsuchiya, A. Kawana, K. Sugiyama and T. Kimura, 'High speed optical pulse transmission at 1.29 μm wavelength using low-loss single-mode fibers', *IEEE J. Quantum Electron.*, **QE-14**, pp. 791–800, 1978.

41   F. P. Kapron, 'Maximum information capacity of fibre-optic waveguides', *Electron. Lett.*, **13**(4), pp. 96–97, 1977.

42   D. N. Payne and A. H. Hartog, 'Determination of the wavelength of zero material dispersion in optical fibres by pulse-delay measurements, *Electron. Lett.*, **13**(21), pp. 627–628, 1977.

43   L. G. Cohen, C. Lin and W. G. French, 'Tailoring zero chromatic dispersion into the 1.5–1.6 μm low-loss spectral region of single-mode fibres', *Electron. Lett.*, **15**(12), pp. 334–335, 1979.

44   A. W. Snyder and R. A. Sammut, 'Dispersion in graded single-mode fibres', *Electron. Lett.*, **15**(10), pp. 269–270, 1979.

45   W. A. Gambling, H. Matsumura and C. M. Ragdale, 'Zero mode dispersion in single mode fibres', *Electron. Lett.*, **14** (19), pp. 618–620, 1978.

46   R. E. Epworth, 'The phenomenon of modal noise in analogue and digital optical fibre systems', in *Proceedings of the 4th European Conference on Optical Communication*, Italy, pp. 492–501, 1978.

47   A. R. Godwin, A. W. Davis, P. A. Kirkby, R. E. Epworth and R. G. Plumb, 'Narrow stripe semiconductor laser for improved performance of optical communication systems', *Proceedings of the 5th European Conference on Optical Communications*, The Netherlands, paper 4-3, 1979.

48   K. Sato and K. Asatani, 'Analogue baseband TV transmission experiments using semiconductor laser diodes', *Electron. Lett.*, **15**(24), pp. 794–795, 1979.

49   B. Culshaw, 'Minimisation of modal noise in optical-fibre connectors', *Electron. Lett.*, **15**(17), pp. 529–531, 1979.

50   G. De Marchis, S. Piazzolla and B. Daino, 'Modal noise in optical fibres', in *Proceedings of the 6th European Conference on Optical Communication*, UK, pp. 76–79, 1980.

51   M. Monerie, D. Moutonnet and L. Jeunhomme, 'Polarisation studies in long length single mode fibres', in *Proceedings of the 6th European Conference on Optical Communication*, UK, pp. 107–111, 1980.

52   I. P. Kaminow, 'Polarization in fibers', *Laser Focus*, **16**(6), pp. 80–84, 1980.

53   I. P. Kaminow, 'Polarization in optical fibers', *IEEE J. Quantum Electron.*, **QE-17**(1), pp. 15–22, 1981.

54   S. C. Rashleigh and R. Ulrich, 'Polarization mode dispersion in single-mode fibers', *Opt. Lett.*, **3**, pp. 60–62, 1978.

55   V. Ramaswamy, R. D. Standley, D. Sze and W. G. French, 'Polarisation effects in short length, single mode fibres', *Bell Syst. Tech. J.*, **57**, pp. 635–651, 1978.

56   A. Papp and H. Harms, 'Polarization optics of index-gradient optical waveguide fibers', *Appl. Opt.*, **14**, pp. 2406–2411, 1975.

57   A. J. Barlow, D. N. Payne, M. P. Varnham and R. D. Birch, 'Polarisation characteristics of fibres for coherent detection systems', IEE Colloq. on Coherence in Opt. Fibre Syst., London, 25th May 1982.

58   D. N. Payne, A. J. Barlow and J. J. Ramskov Hansen, 'Development of low and high birefringence optical fibres', *IEEE J. Quantum Electron.*, **QE-18**(4), pp. 477–487, 1982.

59   R. Ulrich, 'Polarisation stabilisation on single-mode fibre', *Appl. Phys. Lett.*, **35**, pp. 840–842, 1979.

60   M. J. Adams, D. N. Payne and C. M. Ragdale, 'Birefringence in optical fibres with elliptical cross-section', *Electron. Lett.*, **15**(10), pp. 298–299, 1979.

61   T. Katsuyama, H. Matsumura and T. Suganuma, 'Low loss single-polarisation fibres', *Electron. Lett.*, **17**(13), pp. 473–474, 1981.

4

Optical Fibers, Cables and Connections

4.1 INTRODUCTION

Optical fiber waveguides and their transmission characteristics have been considered in some detail in Chapters 2 and 3. However, we have yet to discuss the practical considerations and problems associated with the production, application and installation of optical fibers within a line transmission system. These factors are of paramount importance if optical fiber communication systems are to be considered as viable replacements for conventional metallic line communication systems. Optical fiber communication is of little use if the many advantages of optical fiber transmission lines outlined in the previous chapters may not be applied in practice in the telecommunications network without severe degradation of their performance.

It is therefore essential that:

(a) Optical fibers may be produced with good stable transmission characteristics in long lengths at a minimum cost and with maximum reproducibility.

(b) A range of optical fiber types with regard to size, refractive indices and index profiles, operating wavelengths, materials, etc., be available in order to fulfill many different system applications.

(c) The fibers may be converted into practical cables which can be handled in a similar manner to conventional electrical transmission cables without problems associated with the degradation of their characteristics or damage.

(d) The fibers and fiber cables may be terminated and connected together (jointed) without excessive practical difficulties and in ways which limit the effect of this process on the fiber transmission characteristics to keep them within acceptable operating levels. It is important that these jointing techniques may be applied with ease in the field locations where cable connection takes place.

In this chapter we therefore pull together the various practical elements associated with optical fiber communications. Hence the various methods for preparing optical fibers (both liquid and vapor phase) with characteristics suitable for telecommunications applications are outlined in Sections 4.2 to 4.4.

This is followed in Section 4.5 with consideration of commercially available fibers describing in general terms both the types and their characteristics. The requirements for optical fiber cabling in relation to fiber protection are then discussed in Section 4.6 prior to consideration of cable design in Section 4.7. In Section 4.8 we deal with the losses incurred when optical fibers are connected together. This discussion provides a basis for consideration of the techniques employed for jointing optical fibers. Permanent fiber joints (or splices) are then dealt with in Section 4.9 prior to discussion of the various types of demountable fiber connector in Sections 4.10 to 4.12.

4.2 PREPARATION OF OPTICAL FIBERS

From the considerations of optical waveguiding of Chapter 2 it is clear that a variation of refractive index inside the optical fiber (i.e. between the core and the cladding) is a fundamental necessity in the fabrication of fibers for light transmission. Hence at least two different materials which are transparent to light over the operating wavelength range (0.8–1.6 µm) are required. In practice these materials must exhibit relatively low optical attenuation and they must therefore have low intrinsic absorption and scattering losses. A number of organic and inorganic insulating substances meet these conditions in the visible and near infrared regions of the spectrum.

However, in order to avoid scattering losses in excess of the fundamental intrinsic losses, scattering centers such as bubbles, strains and grain boundaries must be eradicated. This tends to limit the choice of suitable materials for the fabrication of optical fibers to either glasses (or glass-like materials) and monocrystalline structures (certain plastics).

It is also useful, and in the case of graded index fibers essential, that the refractive index of the material may be varied by suitable doping with another compatible material. Hence these two materials should have mutual solubility over a relatively wide range of concentrations. This is only achieved in glasses or glass-like materials, and therefore monocrystalline materials are unsuitable for the fabrication of graded index fibers, but may be used for step index fibers. However, it is apparent that glasses exhibit the best overall material characteristics for use in the fabrication of low loss optical fibers. They are therefore used almost exclusively in the preparation of fibers for telecommunications applications. Plastic clad [Ref. 1] and all plastic fibers find some use in short-haul, low bandwidth applications.

In this section the discussion will therefore be confined to the preparation of glass fibers. This is a two stage process in which initially the pure glass is produced and converted into a form (rod or preform) suitable for making the fiber. A drawing or pulling technique is then employed to acquire the end product. The methods of preparing the extremely pure optical glasses generally fall into two major categories which are:

(a) conventional glass refining techniques in which the glass is processed in the molten state (melting methods) producing a multicomponent glass structure;

(b) vapor phase deposition methods producing silica-rich glasses which have melting temperatures that are too high to allow the conventional melt process.

These processes, with their respective drawing techniques, are described in the following sections.

4.3 LIQUID PHASE (MELTING) TECHNIQUES

The first stage in this process is the preparation of ultra pure material powders which are usually oxides or carbonates of the required constituents. These include oxides such as SiO_2, GeO_2, B_2O_2 and A_2O_3, and carbonates such as

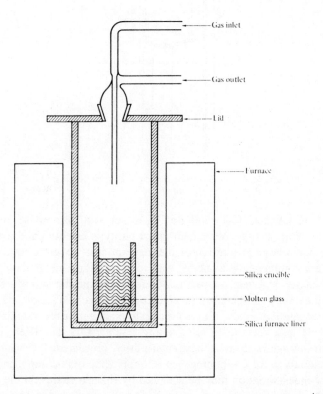

Fig. 4.1 Glassmaking furnace for the production of high purity glasses [Ref. 4].

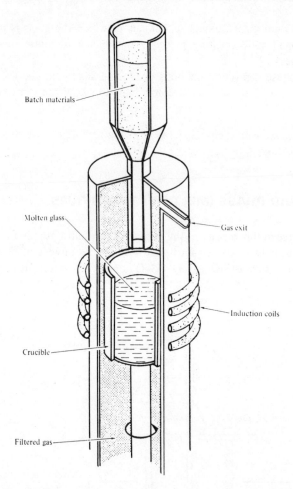

Fig. 4.2 High purity glass melting using a radiofrequency induction furnace [Refs. 6–8].

Na_2CO_3, K_2CO_3, $CaCO_3$ and $BaCO_3$ which will decompose into oxides during the glass melting. Very high initial purity is essential and purification accounts for a large proportion of the material cost; nevertheless these compounds are commercially available with total transition metal contents below 20 parts in 10^9 and below 1 part in 10^9 for some specific impurities [Ref. 2]. The purification may therefore involve combined techniques of fine filtration and coprecipitation, followed by solvent extraction before recrystallization and final drying in a vacuum to remove any residual OH ions [Ref. 3].

The next stage is to melt these high purity, powdered, low melting point glass materials to form a homogeneous, bubble-free multicomponent glass. A refractive index variation may be achieved by either a change in the composition of the various constituents or by ion exchange when the materials are in the molten phase. The melting of these multicomponent glass systems occurs at relatively low temperatures between 900 and 1300 °C and may take place in a silica crucible as shown in Fig. 4.1 [Ref. 4]. However, contamination can arise during melting from several sources including the furnace environment and the crucible. Both fused silica and platinum crucibles have been used with some success although an increase in impurity content was observed when the melt was held in a platinum crucible at high temperatures over long periods [Ref. 5].

Silica crucibles can give dissolution into the melt which may introduce inhomogeneities into the glass especially at high melting temperatures. A technique for avoiding this involves melting the glass directly into a radio-frequency (RF approximately 5 MHz) induction furnace while cooling the silica by gas or water flow as shown in Fig. 4.2 [Refs. 6–8]. The materials are preheated to around 1000 °C where they exhibit sufficient ionic conductivity to enable coupling between the melt and the RF field. The melt is also protected from any impurities in the crucible by a thin layer of solidified pure glass which forms due to the temperature difference between the melt and the cooled silica crucible.

In both techniques the glass is homogenized and dried by bubbling pure gases through the melt, whilst protecting against any airborne dust particles either originating in the melt furnace or present as atmospheric contamination. After the melt has been suitably processed, it is cooled and formed into long rods (cane) of multicomponent glass.

4.3.1 Fiber Drawing

The traditional technique for producing fine optical fiber waveguides is to make a preform using the rod in tube process. A rod of core glass is inserted into a tube of cladding glass and the preform is drawn in a vertical muffle furnace as illustrated in Fig. 4.3 [Ref. 9]. This technique is useful for the production of step index fibers with large core and cladding diameters where the achievement of low attenuation is not critical as there is a danger of including bubbles and particulate matter at the core–cladding interface.

Another technique which is also suitable for the production of large core diameter step index fibers, and reduces the core–cladding interface problems, is called the stratified melt process. This process, developed by Pilkington Laboratories [Ref. 10], involves pouring a layer of cladding glass over the core glass in a platinum crucible as shown in Fig. 4.4 [Ref. 11]. A bait glass rod is dipped into the molten combination and slowly withdrawn giving a composite core–clad preform which may be then drawn into a fiber.

Subsequent development in the drawing of optical fibers (especially graded index) produced by liquid phase techniques has concentrated on the double crucible method. In this method the core and cladding glass in the form of separate rods is fed into two concentric platinum crucibles as illustrated in

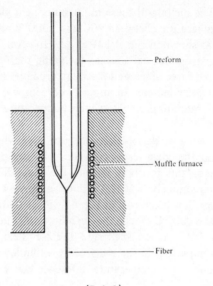

Fig. 4.3 Optical fiber from a preform [Ref. 9].

Fig. 4.5 [Ref. 4]. The assembly is usually located in a muffle furnace capable of heating the crucible contents to a temperature of between 800 and 1200 °C. The crucibles have nozzles in their bases from which the clad fiber is drawn directly from the melt as shown in Fig. 4.5. Index grading may be achieved through the diffusion of mobile ions across the core–cladding interface within the molten glass. It is possible to achieve a reasonable refractive index profile via this diffusion process, although due to lack of precise control it is not possible to obtain the optimum near parabolic profile which yields the

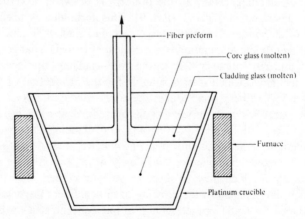

Fig. 4.4 The stratified melt process (glass on glass technique) for producing glass clad rods or preforms [Ref. 11].

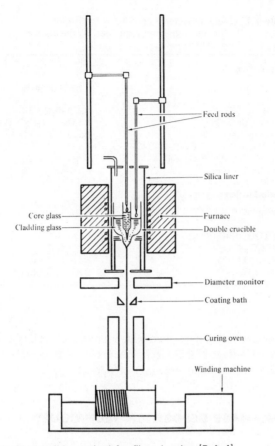

Fig. 4.5 The double crucible method for fiber drawing [Ref. 4].

minimum pulse dispersion (see Section 3.9.2). Hence graded index fibers produced by this technique are substantially less dispersive than step index fibers, but do not have the bandwidth–length products of optimum profile fibers. Pulse dispersion of 1–6 ns km^{-1} [Refs. 12, 13] is quite typical, depending on the material system used.

Some of the material systems used in the fabrication of multicomponent glass step index and graded index fibers are given in Table 4.1.

Using very high purity melting techniques and the double crucible drawing method, step index and graded index fibers with attenuations as low as 3.4 dB km^{-1} [Ref. 14] and 1.1 dB km^{-1} [Ref. 2] respectively have been produced. However, such low losses cannot be consistently obtained using liquid phase techniques and typical losses for multicomponent glass fibers prepared continuously by these methods are between 5 and 10 dB km^{-1}. Therefore, liquid phase techniques have the inherent disadvantage of obtaining

Table 4.1   Material systems used in the fabrication of multicomponent glass fibers by the double crucible technique

Step Index

Core glass	Cladding glass
$Na_2-B_2O_3-SiO_2$	$Na_2O-B_2O_3-SiO_2$
$Na_2-LiO-CaO-SiO_2$	$Na_2O-Li_2O-CaO-SiO_2$
$Na_2-CaO-GeO_2$	$Na_2O-CaO-SiO_2$
$Tl_2O-Na_2O-B_2O_3-GeO_2-BaO-CaO-SiO_2$	$Na_2O-B_2O_3-SiO_2$
$Na_2O-BaO-GeO_2-B_2O_3-SiO_2$	$Na_2O-B_2O_3-SiO_2$
$P_2O_5-Ga_2O_3-GeO_2$	$P_2O_5-Ga_2O_3-SiO_2$

Graded index

Base glass	Diffusion mechanism
$R_2O-GeO_2-CaO-SiO_2$	$Na^+ \rightleftharpoons K^+$
$R_2O-B_2O_3-SiO_2$	$Tl^+ \rightleftharpoons Na^+$
$Na_2O-B_2O_3-SiO_2$	Na_2O diffusion
$Na_2O-B_2O_3-SiO_2$	CaO, BaO diffusion

and maintaining extremely pure glass which limits their ability to produce low loss fibers. The advantage of these techniques is in the possibility of continuous production (both melting and drawing) of optical fibers.

4.4   VAPOR PHASE DEPOSITION TECHNIQUES

Vapor phase deposition techniques are used to produce silica-rich glasses of the highest transparency and with the optimal optical properties. The starting materials are volatile compounds such as $SiCl_4$, $GeCl_4$, SiF_4, BCl_3, O_2, BBr_3 and $POCl_3$ which may be distilled to reduce the concentration of most transition metal impurities to below one part in 10^9 giving negligible absorption losses from these elements. Refractive index modification is achieved through the formation of dopants from the nonsilica starting materials. These vapor phase dopants include TiO_2, GeO_2, P_2O_5, Al_2O_3, B_2O_3 and F, the effects of which on the refractive index of silica are shown in Fig. 4.6 [Ref. 2]. Gaseous mixtures of the silica-containing compound, the doping material and oxygen are combined in a vapor phase oxidation reaction where the deposition of oxides occurs. The deposition is usually onto a substrate or within a hollow tube and is built up as a stack of successive layers. Hence the dopant concentration may be varied gradually to produce a graded index profile or maintained to give a step index profile. In the case of the substrate this directly results in a solid rod or preform whereas the hollow tube must be collapsed to give a solid preform from which the fiber may be drawn.

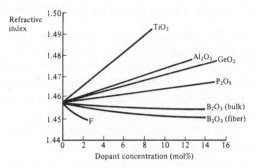

Fig. 4.6 The variation in the refractive index of silica using various dopants. Reproduced with permission from the publishers, Society of Glass Technology, *Phys. Chem. Glasses*, **21**, p. 5, 1980.

There are a number of variations of vapor phase deposition which have been successfully utilized to produce low loss fibers. The major techniques are illustrated in Fig. 4.7, which also indicates the plane (horizontal or vertical) in which the deposition takes place as well as the formation of the preform. These vapor phase deposition techniques fall into two broad categories: flame hydrolysis and chemical vapor deposition (CVD) methods. The individual techniques are considered in the following sections.

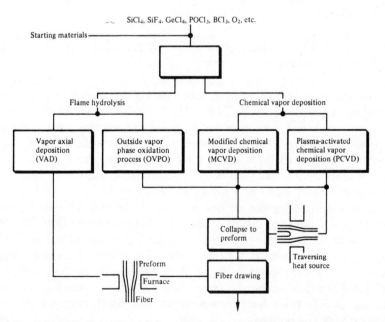

Fig. 4.7 Schematic illustration of the vapor phase deposition techniques used in the preparation of low loss optical fibers.

4.4.1   Outside Vapor Phase Oxidation (OVPO) Process

This process which uses flame hydrolysis stems from work on 'soot' processes originally developed by Hyde [Ref. 17] which were used to produce the first fiber with losses of less than 20 dB km^{-1} [Ref. 18]. The best known technique of this type is often referred to as the outside vapor phase oxidation process. In this process the required glass composition is deposited laterally from a 'soot' generated by hydrolyzing the halide vapors in an oxygen–hydrogen flame. Oxygen is passed through the appropriate silicon compound (i.e. $SiCl_4$) which is vaporized, removing any impurities. Dopants such as $GeCl_4$ or $TiCl_4$ are added and the mixture is blown through the oxygen–hydrogen flame giving the following reactions:

$$SiCl_4 + 2H_2O \xrightarrow{\text{heat}} SiO_2 + 4HCl \qquad (4.1)$$
$$\text{(vapor)} \quad \text{(vapor)} \qquad\qquad \text{(solid)} \quad \text{(gas)}$$

and

$$SiCl_4 + O_2 \xrightarrow{\text{heat}} SiO_2 + 2Cl_2 \qquad (4.2)$$
$$\text{(vapor)} \quad \text{(gas)} \qquad\qquad \text{(solid)} \quad \text{(gas)}$$

$$GeCl_4 + O_2 \xrightarrow{\text{heat}} GeO_2 + 2Cl_2 \qquad (4.3)$$
$$\text{(vapor)} \quad \text{(gas)} \qquad\qquad \text{(solid)} \quad \text{(gas)}$$

or

$$TiCl_4 + O_2 \xrightarrow{\text{heat}} TiO_2 + 2Cl_2 \qquad (4.4)$$
$$\text{(vapor)} \quad \text{(gas)} \qquad\qquad \text{(solid)} \quad \text{(gas)}$$

The silica is generated as a fine soot which is deposited on a cool rotating mandrel as illustrated in Fig. 4.8(a) [Ref. 19]. The flame of the burner is traversed back and forth over the length of the mandrel until a sufficient number of layers of silica (approximately 200) are deposited on it. When this process is completed the mandrel is removed and the porous mass of silica soot is sintered (to form a glass body) as illustrated in Fig. 4.8(b). The preform may contain both core and cladding glasses by properly varying the dopant concentrations during the deposition process. Several kilometers (around 10 km of 120 µm core diameter fiber have been produced [Ref. 2]) can be drawn from the preform by collapsing and closing the central hole as shown in Fig. 4.8(c). Fine control of the index gradient for graded index fibers may be achieved using this process as the gas flows can be adjusted at the completion of each traverse of the burner. Hence fibers with bandwidth–length products as high as 3 GHz km have been achieved [Ref. 20] through accurate index grading with this process.

The purity of the glass fiber depends on the purity of the feeding materials

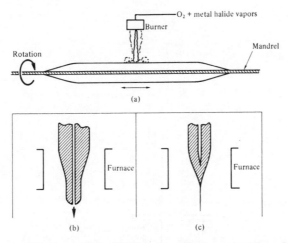

Fig. 4.8 Schematic diagram of the OVPO process for the preparation of optical fibers: (a) soot deposition; (b) preform sintering; (c) fiber drawing [Ref. 19].

and also upon the amount of OH impurity from the exposure of the silica to water vapor in the flame following the reactions given in Eqs. (4.1) to (4.4). Typically the OH content is between 50 and 200 parts per million and this contributes to the fiber attenuation. It is possible to reduce the OH impurity content by employing gaseous chlorine as a drying agent during sintering. This has given losses as low as 1 dB km^{-1} and 1.8 dB km^{-1} at wavelengths of 1.2 and 1.55 μm respectively [Ref. 21] in fibers prepared using the OVPO process.

Other problems stem from the use of the mandrel which can create some difficulties in the formation of the fiber preform. Cracks may form due to stress concentration on the surface of the inside wall when the mandrel is removed. Also the refractive index profile has a central depression due to the collapsed hole when the fiber is drawn. Therefore although the OVPO process is a useful fiber preparation technique, it has several drawbacks. Furthermore it is a batch process which limits its use for the volume production of optical fibers.

4.4.2 Vapor Axial Deposition (VAD)

This process was developed by Izawa *et al.* [Ref. 22] in the search for a continuous (rather than batch) technique for the production of low loss optical fibers. The VAD technique uses an end-on deposition onto a rotating fused silica target as illustrated in Fig. 4.9 [Ref. 23]. The vaporized constituents are injected from burners and react to form silica soot by flame hydrolysis. This is deposited on the end of the starting target in the axial direction forming a solid porous glass preform in the shape of a boule. The preform which is growing in the axial direction is pulled upwards at a rate which corresponds to the growth

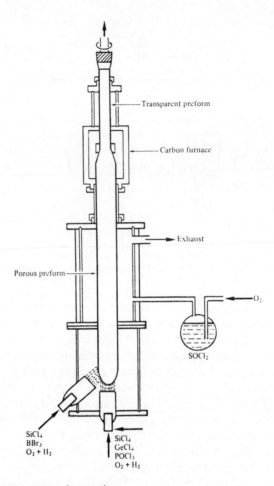

Fig. 4.9 The VAD process [Ref. 23].

rate. It is initially dehydrated by heating with $SOCl_2$ using the reaction:

$$H_2O + SOCl_2 \xrightarrow{\text{heat}} 2HCl + SO_2 \qquad (4.5)$$
$$\text{(vapor)} \quad \text{(vapor)} \qquad \qquad \text{(gas)} \quad \text{(gas)}$$

and is then sintered into a solid preform in a graphite resistance furnace at an elevated temperature of around 1500 °C. Therefore, in principle this process may be adapted to draw fiber continuously, although at present it tends to be operated as a batch process.

A spatial refractive index profile may be achieved using the deposition properties of SiO_2–GeO_2 particles within the oxygen–hydrogen flame. The concentration of these constituents deposited on the porous preform is controlled by the substrate temperature distribution which can be altered by changing the gas flow conditions. Fibers produced by the VAD process still suffer from some OH impurity content due to the flame hydrolysis and hence very low loss fibers have not been achieved using this method. Nevertheless, fibers with attenuation in the range 0.7–2.0 dB km^{-1} at a wavelength of 1.18 µm have been reported [Ref. 24].

4.4.3 Modified Chemical Vapor Deposition (MCVD)

Chemical vapor deposition techniques are commonly used at very low deposition rates in the semiconductor industry to produce protective SiO$_2$ films on silicon semiconductor devices. Usually an easily oxidized reagent such as SiH$_4$ diluted by inert gases and mixed with oxygen is brought into contact with a heated silicon surface where it forms a glassy transparent silica film. This heterogeneous reaction (i.e. requires a surface to take place) was pioneered for the fabrication of optical fibers using the inside surface of a fused quartz tube [Ref. 25]. However, these processes gave low deposition rates and were prone to OH contamination due to the use of hydride reactants. This led to the development of the modified chemical vapor deposition (MCVD) process by Bell Telephone Laboratories [Ref. 26] and Southampton University, UK [Ref. 27], which overcomes these problems and has found widespread application throughout the world.

The MCVD process is also an inside vapor phase oxidation (IVPO) technique taking place inside a silica tube as shown in Fig. 4.10. However, the vapor phase reactants (halide and oxygen) pass through a hot zone so that a substantial part of the reaction is homogeneous (i.e. involves only one phase; in

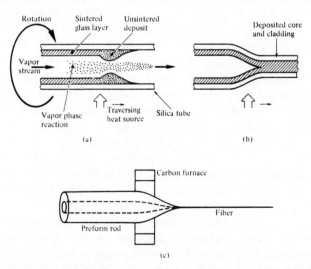

Fig. 4.10 Schematic diagram showing the MCVD method for the preparation of optical fibers: (a) deposition; (b) collapse to produce a preform; (c) fiber drawing.

this case the vapor phase). Glass particles formed during this reaction travel with the gas flow and are deposited on the walls of the silica tube. The tube may form the cladding material but usually it is merely a supporting structure which is heated on the outside by an oxygen–hydrogen flame to temperatures between 1400 °C and 1600 °C. Thus a hot zone is created which encourages high temperature oxidation reactions such as those given in Eqs. (4.2) and (4.3) or (4.4) (not Eq. (4.1)). These reactions reduce the OH impurity concentration to levels below those found in fibers prepared by hydride oxidation or flame hydrolysis.

The hot zone is moved back and forth along the tube allowing the particles to be deposited on a layer by layer basis giving a sintered transparent silica film on the walls of the tube. The film may be up to 10 μm in thickness and uniformity is maintained by rotating the tube. A graded refractive index profile can be created by changing the composition of the layers as the glass is deposited. Usually when sufficient thickness has been formed by successive traverses of the burner for the cladding, vaporized chlorides of germanium (GeCl$_4$) or phosphorus (POCl$_3$) are added to the gas flow. The core glass is then formed by the deposition of successive layers of germanosilicate or phosphosilicate glass. The cladding layer is important as it acts as a barrier which suppresses OH absorption losses due to the diffusion of OH ions from the silica tube into the core glass as it is deposited. After the deposition is completed the temperature is increased to between 1700 and 1900 °C. The tube is then collapsed to give a solid preform which may then be drawn into fiber at temperatures of 2000–2200 °C as illustrated in Fig. 4.10.

This technique is the most widely used at present as it allows the fabrication of fiber with the lowest losses. Apart from the reduced OH impurity contamination the MCVD process has the advantage that deposition occurs within an enclosed reactor which ensures a very clean environment. Hence gaseous and particulate impurities may be avoided during both the layer deposition and the preform collapse phases. The process also allows the use of a variety of materials and glass compositions. It has produced GeO$_2$ doped silica single mode fiber with minimum losses of only 0.2 dB km^{-1} at a wavelength of 1.55 μm [Ref. 28]. More generally the GeO$_2$–B$_2$O$_3$–SiO$_2$ system (B$_2$O$_3$ is added to reduce the viscosity and assist fining) has shown minimum losses of 0.34 dB km^{-1} with multimode fiber at a wavelength of 1.55 μm [Ref. 29]. Also graded index germanium phosphosilicate fibers have exhibited losses near the intrinsic level for their composition of 2.8, 0.45 and 0.35 dB km^{-1} at wavelengths of 0.82, 1.3 and 1.5 μm respectively [Ref. 30].

The MCVD process has also demonstrated the capability of producing fibers with very high bandwidths, although still well below the theoretical values which may be achieved. Multimode graded index fibers with measured bandwidth–length products of 4.3 GHz km and 4.7 GHz km at wavelengths of 1.25 and 1.29 μm have been reported [Ref. 31]. Large-scale batch production

(30,000 km) of 50 μm core graded index fiber has maintained bandwidth–length products of 825 MHz km and 735 MHz km at wavelengths of 0.825 and 1.3 μm respectively [Ref. 30]. The median attenuation obtained with this fiber was 3.4 dB km^{-1} at 0.825 μm and 1.20 dB km^{-1} at 1.3 μm. Hence, although it is not a continuous process, the MCVD technique has proved suitable for the mass production of high performance optical fiber.

4.4.4 Plasma-activated Chemical Vapor Deposition (PCVD)

A variation on the MCVD technique is the use of various types of plasma to supply energy for the vapor phase oxidation of halides. This method, first developed by Kuppers and Koenings [Ref. 32], involves plasma-induced chemical vapor deposition inside a silica tube as shown in Fig. 4.11. The essential difference between this technique and the MCVD process is the stimulation of oxide formation by means of a nonisothermal plasma maintained at low pressure in a microwave cavity (2.45 GHz) which surrounds the tube. Volatile reactants are introduced into the tube where they react heterogeneously within the microwave cavity and no particulate matter is formed in the vapor phase.

The reaction zone is moved backwards and forwards along the tube by control of the microwave cavity and a circularly symmetric layer growth is formed. Rotation of the tube is unnecessary and the deposition is virtually 100% efficient. Film deposition can occur at temperatures as low as 500 °C, but a high chlorine content may cause expansivity and cracking of the film. Hence the tube is heated to around 1000 °C during deposition using a stationary furnace.

The high deposition efficiency allows the composition of the layers to be accurately varied by control of the vapor phase reactants. Also when the plasma zone is moved rapidly backwards and forwards along the tube very thin layer deposition may be achieved giving the formation of up to 2000 individual layers. This enables very good graded index profiles to be realized which are a close approximation to the optimum near parabolic profile. Thus low pulse dispersion of less than 0.8 ns km^{-1}, for fibers with attenuations of between 3 and 4 dB km^{-1}, at a wavelength of 0.85 μm have been reported [Ref. 2].

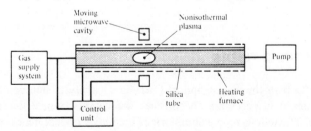

Fig. 4.11 The apparatus utilized in the PCVD process.

A further PCVD technique uses an inductively coupled radiofrequency argon plasma which operates at a frequency of 3.4 MHz |Ref. 33|. The deposition takes place at 1 atmosphere pressure and is predominantly a homogeneous vapor phase reaction which, via the high temperature discharge, causes the fusion of the deposited material into glass. This technique has proved to have a reaction rate five times faster than the conventional MCVD process. However, fiber attenuation is somewhat higher with losses of 6 dB km^{-1} at a wavelength of 1.06 μm. Variations on this theme operating at frequencies of 3–6 MHz and 27 MHz have produced $GeO_2-P_2O_5-SiO_2$ fibers with minimum losses of 4–5 dB km^{-1} at a wavelength of 0.85 μm [Ref. 34].

4.4.5 Summary of Vapor Phase Deposition Techniques

The salient features of the major vapor phase deposition techniques are summarized in Table 4.2 [Ref. 35].

Table 4.2   Summary of vapor phase deposition techniques used in the preparation of low loss optical fibers

Reaction type	
Flame hydrolysis	OVPO, VAD
High temperature oxidation	MCVD
Low temperature oxidation	PCVD
Depositional direction	
Outside layer deposition	OVPO
Inside layer deposition	MCVD, PCVD
Axial layer deposition	VAD
Refractive index profile formation	
Layer approximation	OVPO, MCVD, PCVD
Simultaneous formation	VAD
Process	
Batch	OVPO, MCVD, PCVD
Continuous	VAD

4.5 OPTICAL FIBERS

In order to plan the use of optical fibers in a variety of line communication applications it is necessary to consider the various optical fibers currently available. The following is a summary of the major optical fiber types with an indication of their general characteristics. The performance characteristics of the various fiber types discussed vary considerably depending upon the materials used in the fabrication process and the preparation technique involved. The values quoted are largely based upon manufacturers' and suppliers' data [Refs. 40–44] for commercially available fibers, presented in a general form rather than for specific fibers. Hence the fibers may appear to have somewhat poorer performance characteristics than those stated for the equivalent fiber types produced by the best possible techniques and in the best possible conditions which were indicated in Chapter 3. However, it must be remembered that the high performance values quoted in Chapter 3 were generally for fibers produced and tested in the laboratory. There the pursuit of enhanced performance was the predominant criterion, whereas the fibers considered in this section are those already manufactured in bulk for the commercial market.

This section, therefore, reflects the time delay between the achievement of fiber performance in the laboratory (and possibly achieved in a working environment by organizations which have a fiber production capability and are also in a position of servicing the telecommunications networks), and the general commercial availability of such fibers. Nevertheless, it is certain that as the momentum generated in this field increases, fibers with much improved performance characteristics will become more generally available, especially those in the longer wavelength region (1.1–1.6 μm). It must be noted that the performance values given throughout this section are for the shorter wavelength region (0.8–0.9 μm) unless otherwise stated, as commercially available fibers are more frequently specified at wavelengths in this region. However, although these fibers are not predominantly designed for use in the longer wavelength region it is generally the case that the silica glass fibers operate more efficiently over this wavelength range. Finally, the bandwidths quoted are specified over a 1 km length of fiber (i.e. $B_{opt} \times L$). These are generally obtained from manufacturers' data which does not always indicate whether the electrical or the optical bandwidth has been measured. It is likely that these are in fact optical bandwidths which are significantly greater than their electrical equivalents (see Section 7.4.3).

4.5.1 Multimode Step Index Fibers

Multimode step index fibers may be fabricated from either multicomponent glass compounds or doped silica. These fibers can have reasonably large core diameters and large numerical apertures to facilitate efficient coupling to incoherent light sources such as light emitting diodes (LEDs). The performance characteristics of this fiber type may vary considerably depending on the materials used and the method of preparation; the doped silica fibers exhibit the best performance. Multicomponent glass and doped silica fibers are often referred to as multicomponent glass/glass (glass-clad glass) and silica/silica (silica-clad silica) respectively, although the glass-clad glass terminology is sometimes used somewhat vaguely to denote both types. A typical structure for a glass multimode step index fiber is shown in Fig. 4.12.

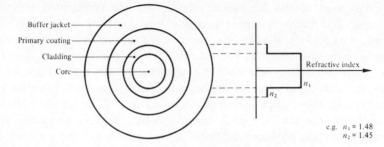

Buffer jacket

Primary coating

Cladding

Core

Refractive index

n_1

n_2

e.g. $n_1 = 1.48$
$n_2 = 1.45$

Fig. 4.12 Typical structure for a glass multimode step index fiber.

Structure

Core diameter:	50–400 µm.
Cladding diameter:	125–500 µm.
Buffer jacket diameter:	250–1000 µm.
Numerical aperture:	0.16–0.5.

Performance characteristics

Attenuation: 4–50 dB km^{-1} limited by absorption or scattering. The wide variation in attenuation is due to the large differences both within and between the two overall preparation methods

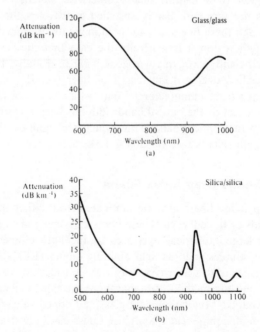

Fig. 4.13 Attenuation spectra for multimode step index fibers: (a) multicomponent glass fiber; (b) doped silica fiber. Reproduced with permission of Belling Lee Ltd.

(melting and deposition). To illustrate this point Fig. 4.13 shows the attenuation spectra from suppliers' data [Ref. 43] for a multicomponent glass fiber (glass-clad glass) and a doped silica fiber (silica-clad silica). It may be observed that the multicomponent glass fiber has an attenuation of around 40 dB km^{-1} at a wavelength of 0.85 μm whereas the doped silica fiber has an attenuation of less than 5 dB km^{-1} at a similar wavelength.

Bandwidth: 6–25 MHz km.

Applications: These fibers are best suited for short-haul, limited bandwidth and relatively low cost applications.

4.5.2 Multimode Graded Index Fibers

These multimode fibers which have a graded index profile may also be fabricated using multicomponent glasses or doped silica. However, they tend to be manufactured from materials with higher purity than the majority of multimode step index fibers in order to reduce fiber losses. The performance characteristics of multimode graded index fibers are therefore generally better than those for multimode step index fibers due to the index grading and lower attenuation. Multimode graded index fibers tend to have small core diameters than multimode step index fibers although the overall diameter including the buffer jacket is usually about the same. This gives the fiber greater rigidity to resist bending. A typical structure is illustrated in Fig. 4.14.

Structure

Core diameter: 30–100 μm, a standard of 50 μm has been established for telecommunications applications.

Cladding diameter: 100–150 μm, a standard of 125 μm has been established for telecommunications applications.

Buffer jacket diameter: 250–1000 μm.

Numerical aperture: 0.2–0.3.

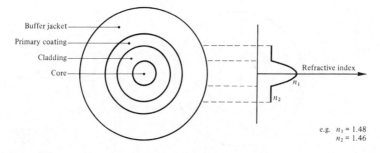

$$\text{e.g. } n_1 = 1.48$$
$$n_2 = 1.46$$

Fig. 4.14 Typical structure for a glass multimode graded index fiber.

Performance characteristics

Attenuation: 2–10 dB km^{-1}, generally a scattering limit.

Bandwidth: 150 MHz km to 2 GHz km.

Applications: These fibers are best suited for medium-haul, medium to high bandwidth applications using incoherent and coherent multimode sources (i.e. LEDs and injection lasers respectively).

It is useful to note that there are a number of partially graded index fibers commercially available. These fibers generally exhibit slightly better performance characteristics than corresponding multimode step index fibers but are somewhat inferior to the fully graded index fibers described above.

4.5.3   Single Mode Fibers

Single mode fibers can have either a step index or graded index profile. However, the benefits of using a graded index profile are by no means as significant as in the case of multimode fibers (see Section 2.5). Therefore at present commercially available single mode fibers are almost exclusively step index. They are high quality fibers for wideband, long-haul transmission and are generally fabricated from doped silica (silica-clad silica) in order to reduce attenuation.

Although single mode fibers have small core diameters to allow single mode propagation, the cladding diameter must be at least ten times the core diameter to avoid losses from the evanescent field. Hence with a buffer jacket to provide protection and strength, single mode fibers have similar overall diameters to multimode fibers. A typical example of a single mode step index fiber is shown in Fig. 4.15.

Structure

Core diameter: 3–10 µm.

Cladding diameter: 50–125 µm.

Buffer jacket diameter: 250–1000 µm.

Numerical aperture: 0.08–0.15, usually around 0.10.

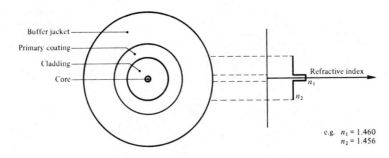

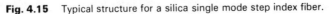

Fig. 4.15   Typical structure for a silica single mode step index fiber.

Performance characteristics

Attenuation: 2–5 dB km^{-1} with a scattering limit of around 1 dB km^{-1} at a wavelength of 0.85 µm. Significantly lower losses are possible in the longer wavelength region.

Bandwidth: Greater than 500 MHz km. In theory the bandwidth is limited by waveguide and material dispersion to approximately 40 GHz km at a wavelength of 0.85 µm.

Applications: These fibers are ideally suited for high bandwidth very long-haul applications using single mode injection laser sources.

4.5.4 Plastic-clad Fibers

Plastic-clad fibers are multimode and have either a step index or a graded index profile. They have a plastic cladding (often a silicone rubber) and a glass core which is frequently silica (i.e. plastic clad silica—PCS fibers). The PCS fibers exhibit lower radiation-induced losses than silica clad silica fibers and, therefore, have an improved performance in certain environments. Plastic-clad fibers are generally slightly cheaper than the corresponding glass fibers, but usually have more limited performance characteristics. A typical structure for a step index plastic-clad fiber (which is more common) is shown in Fig. 4.16.

Structure

Core diameter:	Step index	100–500 µm.
	Graded index	50–100 µm.
Cladding diameter:	Step index	300–800 µm.
	Graded index	125–150 µm.
Buffer jacket diameter:	Step index	500–1000 µm.
	Graded index	250–1000 µm.
Numerical aperture:	Step index	0.2–0.5.
	Graded index	0.2–0.3.

Performance characteristics

Attenuation:	Step index	5–50 dB/km.
	Graded index	4–15 dB/km.

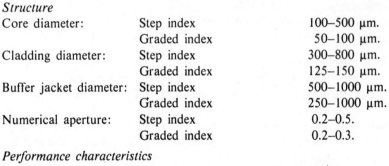
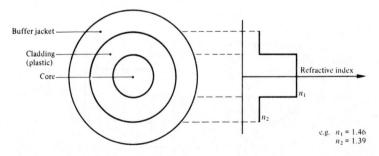

Fig. 4.16 Typical structure for a plastic-clad silica multimode step index fiber.

Bandwidth: Step index   5–25 MHz km.
               Graded index   200–400 MHz km.
Applications: These fibers are generally used on lower bandwidth, shorter-
               haul links where fiber costs need to be limited. They also have
               the advantage of easier termination over glass-clad multimode
               fibers.

4.5.5 All-plastic Fibers

All-plastic fibers are exclusively of the multimode step index type with large
core and cladding diameters. Hence there is a reduced requirement for a buffer
jacket for fiber protection and strengthening. These fibers are cheap to produce
and are easier to handle than the corresponding glass variety. However, their
performance (especially for optical transmission in the infrared) is severely
restricted, giving them very limited use in communication applications. All-plastic
fibers generally have large numerical apertures which allow easier coupling of
light into the fiber from a multimode source. A typical structure is illustrated in
Fig. 4.17.

Structure
Core diameter: 200–600 μm.
Cladding diameter: 450–1000 μm.
Numerical aperture: 0.5–0.6.

Performance characteristics
Attenuation: 150–1000 dB km^{-1} at a wavelength of 0.65 μm.
Bandwidth: This is not usually specified as transmission is generally
               limited to tens of meters.
Applications: These fibers can only be used for very short-haul (i.e. 'in-
               house') low cost links. However, fiber coupling and termina-
               tion are relatively easy and do not require sophisticated tech-
               niques.

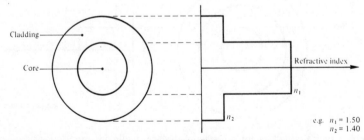

Fig. 4.17   Typical structure for an all-plastic fiber.

4.6   OPTICAL FIBER CABLES

It was indicated in Section 4.1 that if optical fibers are to be alternatives to electrical transmission lines it is imperative that they can be safely installed and maintained in all the environments (e.g. underground ducts) in which metallic conductors are normally placed. Therefore when optical fibers are to be installed in a working environment their mechanical properties are of prime importance. In this respect the unprotected optical fiber has several disadvantages with regard to its strength and durability. Bare glass fibers are brittle and have small cross-sectional areas which makes them very susceptible to damage when employing normal transmission line handling procedures. It is therefore necessary to cover the fibers to improve their tensile strength and to protect them against external influences. This is usually achieved by surrounding the fiber by a series of protective layers which are referred to as coating and cabling. The initial coating of plastic with high elastic modulus is applied directly to the fiber cladding as illustrated in Section 4.5. It is then necessary to incorporate the coated and buffered fiber into an optical cable to increase its resistance to mechanical strain and stress as well as adverse environmental conditions.

The functions of the optical cable may be summarized into four main areas. These are:

(a) Fiber protection. The major function of the optical cable is to protect against fiber damage and breakage both during installation and throughout the life of the fiber.
(b) Stability of the fiber transmission characteristics. The cabled fiber must have good stable transmission characteristics which are comparable with the uncabled fiber. Increases in optical attenuation due to cabling are quite usual and must be minimized within the cable design.
(c) Cable strength. Optical cables must have similar mechanical properties to electrical transmission cables in order that they may be handled in the same manner. These mechanical properties include tension, torsion, compression, bending, squeezing and vibration. Hence the cable strength may be improved by incorporating a suitable strength member and by giving the cable a properly designed thick outer sheath.
(d) Identification and jointing of the fibers within the cable. This is especially important for cables including a large number of optical fibers. If the fibers are arranged in a suitable geometry it may be possible to use multiple jointing techniques rather than jointing each fiber individually.

In order to consider the cabling requirements for fibers with regard to (a) and (b), it is necessary to discuss the fiber strength and durability as well as any possible sources of degradation of the fiber transmission characteristics which are likely to occur due to cabling.

4.6.1 Fiber Strength and Durability

Optical fibers for telecommunications usage are almost exclusively fabricated from silica or a compound of glass (multicomponent glass). These materials are brittle and exhibit almost perfect elasticity until their breaking point is reached. The bulk material strength of flawless glass is quite high and may be estimated for individual materials using the relationship [Ref. 44]:

$$S_t = \left(\frac{\gamma_p E}{4l_a}\right)^{\frac{1}{2}} \tag{4.6}$$

where S_t is the theoretical cohesive strength, γ_p is the surface energy of the material, E is the Young's modulus for the material (stress/strain), and l_a is the atomic spacing or bond distance. However, the bulk material strength may be drastically reduced by the presence of surface flaws within the material.

In order to treat surface flaws in glass analytically, the Griffith theory [Ref. 49] is normally used. This theory assumes that the surface flaws are narrow cracks with small radii of curvature at their tips as illustrated in Fig. 4.18. It postulates that the stress is concentrated at the tip of the crack which leads to crack growth and eventually catastrophic failure. Figure 4.18 shows the concentration of stress lines at the crack tip which indicates that deeper cracks have higher stress at their tips. The Griffith theory gives a stress intensity factor K_1 as:

$$K_1 = SYC^{\frac{1}{2}} \tag{4.7}$$

where S is the macroscopic stress on the fiber, Y is a constant dictated by the shape of the crack (e.g. $Y = \pi^{\frac{1}{2}}$ for an elliptical crack as illustrated in Fig. 4.18) and C is the depth of the crack (this is the semi-major axis length for an elliptical crack). Further, the Griffith theory gives an expression for the critical stress intensity factor K_{IC} where fracture occurs as:

$$K_{IC} = (2E\gamma_p)^{\frac{1}{2}} \tag{4.8}$$

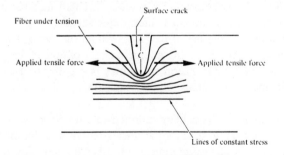

Fig. 4.18   An elliptical surface crack in a tensioned optical fiber.

Combining Eqs. (4.7) and (4.8) gives the Griffith equation for fracture stress of a crack S_f as:

$$S_f = \left(\frac{2E\gamma_p}{Y^2 C} \right)^{\frac{1}{2}}$$ (4.9)

It is interesting to note that S_f is proportional to $C^{-\frac{1}{2}}$. Therefore S_f decreases by a factor of 2 for a fourfold increase in the crack depth C.

Example 4.1

The Si—O bond has a theoretical cohesive strength of 2.6×10^6 psi which corresponds to a bond distance of 0.16 nm. A silica optical fiber has an elliptical crack of depth 10 nm at a point along its length. Estimate:

(a) the fracture stress in psi for the fiber if it is dependent upon this crack;
(b) the percentage strain at the break.

The Young's modulus for silica is approximately 9×10^{10} N m^{-2} and 1 psi $\equiv$ 6894.76 N m^{-2}.

Solution: (a) Using Eq. (4.6), the theoretical cohesive strength for the Si—O bond is:

$$S_t = \left(\frac{\gamma_p E}{4 l_a} \right)^{\frac{1}{2}}$$

Hence

$$\gamma_p = \frac{4 l_a S_t^2}{E} = \frac{4 \times 0.16 \times 10^{-9} (2.6 \times 10^6 \times 6894.76)^2}{9 \times 10^{10}}$$

$$= 2.29 \text{ J}$$

The fracture stress for the silica fiber may be obtained from Eq. (4.9) where:

$$S_f = \left(\frac{2E\gamma_p}{Y^2 C} \right)^{\frac{1}{2}}$$

For an elliptical crack:

$$S_f = \left(\frac{2E\gamma_p}{\pi C} \right)^{\frac{1}{2}} = \left(\frac{2 \times 9 \times 10^{10} \times 2.29}{\pi \times 10^{-8}} \right)^{\frac{1}{2}}$$

$$= 3.62 \times 10^9 \text{ N m}^{-2}$$

$$= 5.25 \times 10^5 \text{ psi}$$

It may be noted that the fracture stress is reduced from the theoretical value for flawless silica of 2.6×10^6 psi by a factor of approximately 5.

(b) Young's modulus is defined as:

$$E = \frac{\text{stress}}{\text{strain}}$$

Therefore

$$\text{strain} = \frac{\text{stress}}{E} = \frac{S_f}{E} = \frac{3.62 \times 10^9}{9 \times 10^{10}}$$

$$= 0.04$$

Hence the strain at the break is 4%, which corresponds to the change in length over the original length for the fiber.

In example 4.1 we considered only a single crack when predicting the fiber fracture. However, when a fiber surface is exposed to the environment and is handled, many flaws may develop. The fracture stress of a length of fiber is then dependent upon the dominant crack (i.e. the deepest) which will give a fiber fracture at the lowest strain. Hence the fiber surface must be protected from abrasion in order to ensure high fiber strength. A primary protective plastic coating is usually applied to the fiber at the end of the initial production process so that mechanically induced flaws may be minimized. Flaws also occur due to chemical and structural causes. These flaws are generally smaller than the mechanically induced flaws and may be minimized within the fiber fabrication process.

There is another effect which reduces the fiber fracture stress below that predicted by the Griffith equation. It is due to the slow growth of flaws under the action of stress and water and is known as stress corrosion. Stress corrosion occurs because the molecular bonds at the tip of the crack are attacked by water when they are under stress. This causes the flaw to grow until breakage eventually occurs. Hence stress corrosion must be taken into account when designing and testing optical fiber cables. It is usual for optical fiber cables to have some form of water-protective barrier as is the case for most electrical cable designs.

In order to predict the life of practical optical fibers under particular stresses it is necessary to use a technique which takes into account the many flaws a fiber may possess, rather than just the single surface flaw considered in example 4.1. This is approached using statistical methods due to the nature of the problem which involves many flaws of varying depths over different lengths of fiber.

Calculations of strengths of optical fibers are usually conducted using Weibull statistics [Ref. 50] which describe the strength behavior of a system that is dependent on the weakest link within the system. In the case of optical fibers this reflects fiber breakage due to the dominant or deepest crack. The empirical relationship established by Weibull and applied to optical fibers indicates that the probability of failure F at a stress S is given by:

$$F = 1 - \exp\left[-\left(\frac{S}{S_0}\right)^m \left(\frac{L}{L_0}\right)\right] \tag{4.10}$$

where m is the Weibull distribution parameter, S_0 is a scale parameter, L is the fiber length and L_0 is a constant with dimensions of length.

The expression given in Eq. (4.10) may be plotted for a fiber under test by breaking a large number of 10–20 m fiber lengths and measuring the strain at the break. The various strains are plotted against the cumulative probability of their occurrence to give the Weibull plot as illustrated in Fig. 4.19 [Ref. 51]. It may be observed from Fig. 4.19 that most of the fiber tested breaks at strain due to the prevalence of many shallow surface flaws. However, some of the fiber tested contains deeper flaws (possibly due to external damage) giving the failure at lower strain depicted by the tail of the plot. This reduced strength region is of greatest interest when determining the fiber's lifetime under stress.

Finally the additional problem of stress corrosion must be added to the information on the fiber under stress gained from the Weibull plot. The stress corrosion is usually predicted using an empirical relationship for the crack velocity v_c in terms of the applied stress intensity factor K_I, where [Ref. 51]:

$$v_c = AK_I^n \tag{4.11}$$

The constant n is called the stress corrosion susceptibility (typically in the range 15–50 for glass), and A is also a constant for the fiber material. Equation (4.11) allows estimation of the time to failure of a fiber under stress corrosion conditions. Therefore from a combination of fiber testing (Weibull plot) and stress corrosion information estimates of the maximum allowable fiber strain can be made available to the cable designer. These estimates may be confirmed by straining the fiber up to a specified level (proof testing) such as 1% strain. Fiber which survives this test can be accepted. However, proof testing presents further problems, as it may cause fiber damage. Also it is necessary to derate the maximum allowable fiber strain from the proof test value to increase confidence in fiber survival under stress conditions. It is suggested [Ref. 51] that a reasonable derating for use by the cable designer for fiber which has survived a 1% strain proof test is around 0.3% in order that the fiber has a reasonable chance of surviving with a continual strain for 20 years.

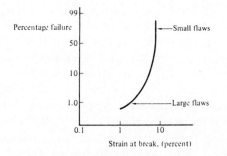

Fig. 4.19 A schematic representation of a Weibull plot. Reproduced with permission from M. H. Reeve, *The Radio and Electron. Eng.*, **51**, p. 327, 1981.

4.6.2 Stability of the Fiber Transmission Characteristics

Optical fiber cables must be designed so that the transmission characteristics of the fiber are maintained after the cabling process and cable installation. Therefore increases in the optical attenuation and reduction in the bandwidth of the cabled fiber must be avoided. A problem which occurs in the cabling of optical fiber is the meandering of the fiber core axis on a microscopic scale within the cable form. This phenomenon, known as microbending, results from small lateral forces exerted on the fiber during the cabling process. Such random bending of the fiber axis causes coupling of power between modes (see Section 2.3.7) which results in losses due to radiation in both multimode and single mode fibers. Thus excesssive microbending can easily create additional fiber losses to an unacceptable level. To avoid deterioration in the optical fiber transmission characteristics resulting from mode coupling induced by microbending, it is important that the fiber is free from irregular external pressure within the cable. Carefully controlled coating and cabling of the fiber is therefore essential in order to minimize the cabled fiber attenuation. Furthermore the fiber cabling must be capable of maintaining this situation under all the strain and environmental conditions envisaged within its lifetime.

4.7 CABLE DESIGN

The design of optical fiber cables must take account of the constraints discussed in Section 4.6. In practice these constraints may be overcome in various ways which are, to some extent, dependent upon the cable's application. Nevertheless, generally cable design may be separated into a number of major considerations. These can be summarized into the categories of fiber buffering, cable structural and strength members, and cable sheath and water barrier.

4.7.1 Fiber Buffering

It was indicated in Section 4.6. that the fiber is given a primary coating during production in order to prevent abrasion of the glass surface and subsequent flaws in the material. The primary coated fiber is then given a secondary or buffer coating (jacket) to provide protection against external mechanical and environmental influences. This buffer jacket is designed to protect the fiber from microbending losses and may take several different forms. These generally fall into one of three distinct types which are illustrated in Fig. 4.20 [Ref. 52]. A tight buffer jacket is shown in Fig. 4.20(a) which usually consists of a hard plastic (e.g. Nylon, Hytrel, Tefzel) and is in direct contact with the primary coated fiber. This thick buffer coating (0.25–1 mm in diameter) provides stiffening for the fiber against outside microbending influences, but it must be applied in such a manner as not to cause microbending losses itself.

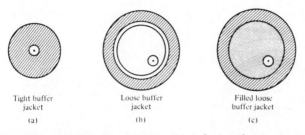

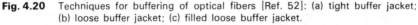

Fig. 4.20   Techniques for buffering of optical fibers [Ref. 52]: (a) tight buffer jacket;
(b) loose buffer jacket; (c) filled loose buffer jacket.

An alternative approach which is shown in Fig. 4.20(b) is the use of a loose buffer jacket. This produces an oversized cavity in which the fiber is placed and which mechanically isolates the fiber from external forces. Loose buffering is generally achieved by using a hard, smooth, flexible material in the form of an extruded tube, or sometimes a folded tape with a diameter between 1 and 2 mm.

Finally Fig. 4.20(c) shows a variation of the loose buffering in which the oversized cavity is filled with a moisture-resistant compound. This technique, which combines the advantages of the two previous methods, also provides a water barrier in the immediate vicinity of the fiber. The filling material must be soft, self-healing and stable over a wide range of temperatures and usually consists of specially blended petroleum or silicone-based compounds.

4.7.2   Cable Structural and Strength Members

One or more structural member is usually included in the optical fiber cable to serve as a core foundation around which the buffered fibers may be wrapped, or into which they may be slotted as illustrated in Fig. 4.21 [Refs. 51 and 52]. The structural member may also be a strength member if it consists of suitable material (i.e. solid or stranded steel wire or Kevlar (DuPont Ltd.) yarns). This situation is shown in Fig. 4.21(a) where the central steel wire acts as both a structural and strength member. In this case the steel wire is the primary load-bearing element. Figure 4.21(b) shows an extruded plastic structural member around a central steel strength member. The primary function of the structural member in this case is not load-bearing, but to provide suitable accommodation for the buffered fibers within the cable.

Structural members may be nonmetallic with plastics, fiberglass and Kevlar often being used. However, for strength members the preferred features include a high Young's modulus, high strain capability, flexibility and low weight per unit length. Therefore although similar materials are frequently utilized for both strength and structural members, the requirement for additional tensile strength of the strength member must be considered within the cable design.

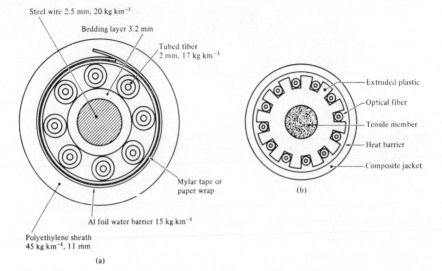

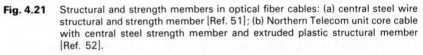

Fig. 4.21   Structural and strength members in optical fiber cables: (a) central steel wire structural and strength member [Ref. 51]; (b) Northern Telecom unit core cable with central steel strength member and extruded plastic structural member [Ref. 52].

Flexibility in strength members formed of materials with high Young's moduli may be improved by using a stranded or bunched assembly of smaller units as in the case of steel wire. Similar techniques are also employed with other materials used for strength members which include plastic monofilaments (i.e. specially processed polyester), textile fiber (Nylon, Terylene, Dacron and the widely used Kevlar) and carbon and glass fibers. These materials provide a variety of tensile strengths for different cable applications. However, it is worth noting that Kevlar, an aromatic polyester, has a very high Young's modulus (up to 13×10^{10} N m^{-2}) which gives it a strength to weight ratio advantage four times that of steel.

It is usual when utilizing a stranded strength member to cover it with a coating of extruded plastic, or helically applied tape. This is to provide the strength member with a smooth (cushioned) surface which is especially important for the prevention of microbending losses when the member is in contact with the buffered optical fibers.

4.7.3   Cable Sheath and Water Barrier

The cable is normally covered with a substantial outer plastic sheath in order to reduce abrasion and to provide the cable with extra protection against external mechanical effects such as crushing. The cable sheath is said to contain the cable core and may vary in complexity from a single extruded plastic jacket to a multilayer structure comprising two or more jackets with intermediate armoring. However, the plastic sheath material (e.g. polyethylene, polyurethane) tends to give very limited protection against the penetration of water into the cable. Hence an additional water barrier is usually incorporated. This may take the form of an axially laid aluminum foil/polyethylene laminated film immediately inside the sheath as used by British Telecom [Ref. 53] and illustrated in Fig. 4.21(a).

Alternatively the ingress of water may be prevented by filling the spaces in the cable with moisture-resistant compounds. Specially formulated silicone rubber or petroleum-based compounds are often used which do not cause difficulties in identification and handling of individual optical fibers within the cable form. These filling compounds are also easily removed from the cable and provide protection from corrosion for any metallic strength members within the fiber. Also the filling compounds must not cause degradation of the other materials within the cable and must remain stable under pressure and temperature variation.

4.7.4 Examples of Fiber Cables

Many different cable designs have been proposed and a large number have been adopted by different organizations throughout the world. At present there are no definite standards for optical fiber cables incorporating either a particular number of fibers or for specific applications. However, as discussed previously there is a general consensus on the overall design requirements and on the various materials that can be used for cable construction [Ref. 52]. In this section we therefore consider some further examples of optical fiber cable construction in order to give the reader a feel for the developments in this important field.

Figure 4.22 [Ref. 40] shows two examples of cable construction for single fibers. In Fig. 4.22(a) a tight buffer jacket of Hytrel is used surrounded by a layer of Kevlar for strengthening. In this construction the optical fiber itself acts as a central strength member.

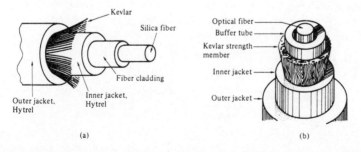

(a)   (b)

Fig. 4.22 Single fiber cables [Ref. 40]: (a) tight buffer jacket design; (b) loose buffer jacket design.

The cable construction illustrated in Fig. 4.22(b) uses a loose tube buffer around the central optical fiber. This is surrounded by a Kevlar strength member which is protected by an inner sheath or jacket before the outer sheath layer. The strength members of single optical fiber cables are not usually incorporated at the center of the cable (unless the fiber is acting as a strength member) but are placed in the surrounding cable form as illustrated in Fig. 4.22(b).

Cable designs for multifiber cables may also take this general form with the strength member surrounding the fibers at the center of the cable. Examples of this construction are illustrated in Fig. 4.23 [Ref. 52]. Figure 4.23(a) shows seven fibers at the cable center surrounded by a helically laid Kevlar strength member. Figure 4.23(b) shows a ribbon cable configuration with a strength member of polypropylene yarns in the surrounding cable form. It may also be noted that this design utilizes armoring of stainless steel wires placed in the outer sheath.

Two more cable designs which allow the incorporation of a larger number of fibers are shown in Fig. 4.24 [Ref. 52]. The configuration illustrated in Fig. 4.24(a) is a stranded design where the buffered fibers are arranged in one or more layers. Alternatively, Fig. 4.24(b) shows a multi-unit design where each unit contains seven buffered fibers. In this case the design allows 49 fibers to be included within the cable.

Finally a cable design which has proved successful in installations in the United States is shown in Fig. 4.25 [Ref. 54]. The cable has a central copper wire for strengthening and also to provide possible electrical condution surrounded by a plastic structural member. Up to 12 optical fibers are placed in a flat ribbon between plastic tapes and incorporated into a helical groove in the extruded plastic structural member. Another diametrically opposite groove is designed for the placement of up to seven plastic insulated metallic pairs or alternatively the incorporation of other ribbon of optical fibers. The principal strength member is a loose aluminum tube fitted over the cable core which also acts as a water barrier. This is surrounded by an inner polyethylene jacket or

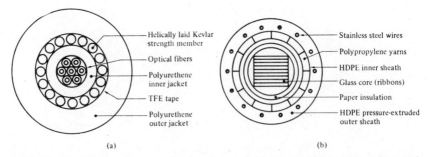

Fig. 4.23 Multifiber cables without central strength and structural member [Ref. 52]: (a) ITT seven fiber external strength member cable; (b) AT&T ribbon cable.

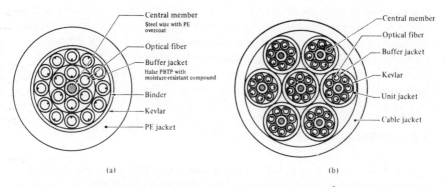

Fig. 4.24 Examples of multifiber cable design [Ref. 52]: (a) Siecor® 18 fiber duct cable; (b) Siecor® 49 fiber unit cable.

sheath followed by armoring consisting of corrugated steel tape with longitudinal overlap. A second polyethylene jacket acts as an outer cable sheath giving the cable an overall diameter of around 2.5 cm. The use of the aluminum tube also allows the cable to be operated under pressurized conditions which gives the additional advantages of:

(a) an alarm in the event of sheath perforation;
(b) sheath fault location;
(c) the exclusion or reduction of water ingress at a sheath fault.

Trials of various optical fiber cable designs have taken place throughout the world since 1977 with little indication of failure due to the possible degradation

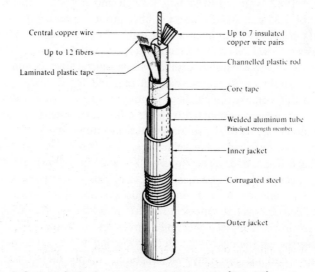

Fig. 4.25 A General Cable Corporation multifiber cable [Ref. 54].

mechanisms. It is therefore suggested [Ref. 51] that there is a possibility that current commercial optical fiber cables manufactured for telecommunications purposes have been 'over-engineered' and are thus working very successfully. Hence it is likely that the future development of optical fiber cables will concentrate on simpler designs which will bring both production and cost benefits as optical fiber systems are utilized more fully in telecommunication networks.

4.8 OPTICAL FIBER CONNECTION

Optical fiber links, in common with any line communication system, have a requirement for both jointing and termination of the transmission medium. The number of intermediate fiber connections or joints is dependent upon the link length (between repeaters), the continuous length of fiber cable that may be produced by the preparation methods outlined in Sections 4.2–4.4, and the length of fiber cable that may be practically or conveniently installed as a continuous section on the link. Current practice allows single lengths of fiber cable of around 1 km to be installed. However, it is anticipated [Ref. 56] that this will be increased to several kilometers, especially for submarine systems where continuous cable laying presents fewer problems.

Repeater spacing on optical fiber telecommunication links is a continuously increasing parameter with currently installed digital systems operating over spacings of up to 30 km together with the prospect of repeater spacings of many tens and even over 100 kilometers for the long wavelength single mode systems of the near future. (For example, 100 km operation without repeaters was achieved by British Telecom in the laboratory (uncabled fiber) at the beginning of 1982 with a 140 M bit s^{-1} single mode system operating at a wavelength of 1.55 μm. In this case the fiber produced by a MCVD process was jointed (spliced) at 6 km intervals.) It is therefore apparent that fiber–fiber connections with low loss and minimum distortion (e.g. modal noise with multimode fibers) is of increasing importance within optical fiber communications in order to sustain the repeater spacings required for developing systems. However, in this context optical fiber jointing has to a certain extent lagged behind the technologies associated with the other components of optical fiber communication systems (fiber, sources, detectors, etc.). Nevertheless in recent years there has been an increasing interest in this topic and significant advances have been made. Therefore in this and the following sections we review the theoretical and practical aspects of fiber–fiber connections with regard to both multimode and single mode systems. Fiber termination to sources and detectors is not considered as the important aspects of these topics are discussed in the chapters covering sources and detectors (Chapters 6, 7 and 8). Nevertheless the discussion on fiber jointing is relevant to both source and detector coupling, as many manufacturers supply these electro-optical devices already terminated to a fiber optic pigtail in order to facilitate direct fiber–fiber connection to an optical fiber link.

Before we consider fiber–fiber connection in further detail it is necessary to indicate the two major categories of fiber joint currently both in use and development. These are:

(a) Fiber splices: these are semipermanent or permanent joints which find major use in most optical fiber telecommunication systems (analogous to electrical soldered joints).
(b) Demountable fiber connectors or simply connectors: these are removable joints which allow easy fast manual coupling and uncoupling of fibers (analogous to electrical plugs and sockets).

A major consideration with all types of fiber–fiber connection is the optical loss encountered at the interface. Even when the two jointed fiber ends are smooth and perpendicular to the fiber axes, and the two fiber axes are perfectly aligned, a small proportion of the light may be reflected back into the transmitting fiber causing attenuation at the joint. This phenomenon, known as Fresnel reflection, is associated with the step changes in refractive index at the jointed interface (i.e. glass–air–glass). The magnitude of this partial reflection of the light transmitted through the interface may be estimated using the classical Fresnel formula for light of normal incidence and is given by [Ref. 57]:

$$r = \left(\frac{n_1 - n}{n_1 + n} \right)^2 \tag{4.12}$$

where r is the fraction of the light reflected at a single interface, n_1 is the refractive index of the fiber core and n is the refractive index of the medium between the two jointed fibers (i.e. for air $n = 1$). However in order to determine the amount of light reflected at a fiber joint, Fresnel reflection at both fiber interfaces must be taken into account. The loss in decibels due to Fresnel reflection at a single interface is given by:

$$\text{Loss}_{\text{Fres}} = -10 \log_{10} (1 - r) \tag{4.13}$$

Hence using the relationships given in Eqs. (4.12) and (4.13) it is possible to determine the optical attenuation due to Fresnel reflection at a fiber–fiber joint.

It is apparent that Fresnel reflection may give a significant loss at a fiber joint even when all other aspects of the connection are ideal. However, the effect of Fresnel reflection at a fiber–fiber connection can be reduced to a very low level through the use of an index matching fluid in the gap between the jointed fibers. When the index matching fluid has the same refractive index as the fiber core, losses due to Fresnel reflection are in theory eradicated.

Unfortunately Fresnel reflection is only one possible source of optical loss at a fiber joint. A potentially greater source of loss at a fiber–fiber connection is caused by misalignment of the two jointed fibers. In order to appreciate the development and relative success of various connection techniques it is useful to discuss fiber alignment in greater detail.

Example 4.2

An optical fiber has a core refractive index of 1.5. Two lengths of the fiber with smooth and perpendicular (to the core axes) end faces are butted together. Assuming the fiber axes are perfectly aligned, calculate the optical loss in decibels at the joint (due to Fresnel reflection) when there is a small air gap between the fiber end faces.

Solution: The magnitude of the Fresnel reflection at the fiber–air interface is given by Eq. (4.12) where:

$$r = \left(\frac{n_1 - n}{n_1 + n} \right)^2 = \left(\frac{1.5 - 1.0}{1.5 + 1.0} \right)^2$$

$$= \left(\frac{0.5}{2.5} \right)^2$$

$$= 0.04$$

The value obtained for r corresponds to a reflection of 4% of the transmitted light at the single interface. Further, the optical loss in decibels at the single interface may be obtained using Eq. (4.13) where:

$$\text{Loss}_{\text{Fres}} = -10 \log_{10} (1 - r) = -10 \log_{10} 0.96$$

$$= 0.18 \text{ dB}$$

A similar calculation may be performed for the other interface (air–fiber). However from considerations of symmetry it is clear that the optical loss at the second interface is also 0.18 dB.

Hence the total loss due to Fresnel reflection at the fiber joint is approximately 0.36 dB.

4.8.1 Fiber Alignment and Joint Loss

Any deviations in the geometrical and optical parameters of the two optical fibers which are jointed will affect the optical attenuation (insertion loss) through the connection. It is not possible within any particular connection technique to allow for all these variations. Hence there are inherent connection problems when jointing fibers with, for instance:

(a) different core and/or cladding diameters;
(b) different numerical apertures and/or relative refractive index differences;
(c) different refractive index profiles;
(d) fiber faults (core ellipticity, core concentricity, etc.).

The best results are therefore achieved with compatible (same) fibers which are manufactured to the lowest tolerance. In this case there is still the problem of the quality of the fiber alignment provided by the jointing mechanism. Examples of possible misalignment between coupled compatible optical fibers are illustrated in Fig. 4.26 [Ref. 58]. It is apparent that misalignment may

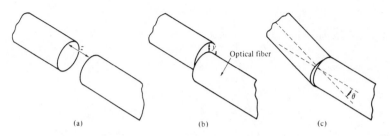

Fig. 4.26 The three possible types of misalignment which may occur when jointing compatible optical fibers [Ref. 58]: (a) longitudinal misalignment; (b) lateral misalignment; (c) angular misalignment.

occur in three dimensions, the separation between the fibers (longitudinal misalignment), the offset perpendicular to the fiber core axes (lateral/radial/axial misalignment) and the angle between the core axes (angular misalignment).

Optical losses resulting from these three types of misalignment depend upon the fiber type, core diameter and the distribution of the optical power between the propagating modes. Examples of the measured optical losses due to the various types of misalignment are shown in Fig. 4.27. Figure 4.27(a) [Ref. 58] shows the attenuation characteristic for both longitudinal and lateral misalignment of a 50 μm core diameter graded index fiber. It may be observed that the lateral misalignment gives significantly greater losses per unit displacement than the longitudinal misalignment. For instance in this case a lateral displacement of 10 μm gives about 1 dB insertion loss whereas a similar longitudinal displacement gives an insertion loss of around 0.1 dB. Figure 4.27(b) [Ref. 59] shows the attenuation characteristic for the angular misalignment of two multimode step index fibers with numerical apertures of 0.22 and 0.3. An insertion loss of around 1 dB is obtained with angular misalignment of 4° and 5° for the 0.22 NA and 0.3 NA fibers respectively. It may also be observed in Fig. 4.27(b) that the effect of an index matching fluid in the fiber gap causes increased losses with angular misalignment. Therefore it is clear that relatively small levels of lateral and/or angular misalignment can cause significant attenuation at a fiber joint. This is especially the case for small core diameter (less than 150 μm) fibers which are currently employed for most telecommunication purposes.

Theoretical and experimental studies of fiber misalignment in optical fiber connections [Refs. 60–72] allow approximate determination of the losses encountered with the various misalignments of different fiber types. We consider here some of the expressions used to calculate losses due to lateral and angular misalignment of optical fiber joints. Longitudinal misalignment is not discussed in detail as it tends to be the least important effect and may be largely avoided in fiber connection. Also there is some disagreement over the magnitude of the losses due to longitudinal misalignment when it is calculated

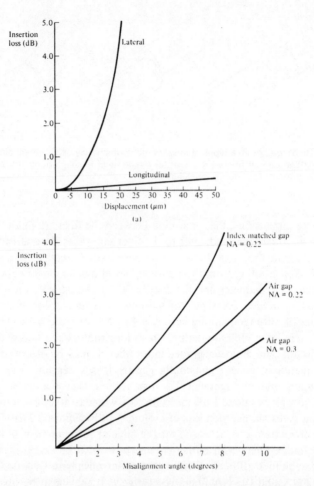

Fig. 4.27 Insertion loss characteristics for jointed optical fibers with various types of misalignment: (a) insertion loss due to lateral and longitudinal misalignment for a 50 μm core diameter graded index fiber, reproduced with permission from P. Mossman, *The Radio and Electron. Eng.*, **51**, p. 333, 1981; (b) insertion loss due to angular misalignment for joints in two multimode step index fibers with numerical apertures of 0.22 and 0.3. Reproduced with permission from C. P. Sandbank (ed), *Optical Fiber Communication Systems*, John Wiley & Sons, 1980.

theoretically between Miyazaki *et al.* [Ref. 61] and Tsuchiya *et al.* [Ref. 62]. Both groups of workers claim good agreement with experimental results which is perhaps understandable when considering the number of variables involved in the measurement. However, it is worth noting that the lower losses predicted by Tsuchiya *et al.* agree more closely with a third group of researchers [Ref. 63]. Also all groups predict higher losses for fibers with larger numerical apertures which is consistent with intuitive considerations (i.e. the larger the numerical aperture, the greater the spread of the output light and the higher the optical loss at a longitudinally misaligned joint).

Theoretical expressions for the determination of lateral and angular misalignment losses are by no means definitive although in all cases they claim reasonable agreement with experimental results. However, experimental results from different sources tend to vary (especially for angular misalignment losses) due to difficulties of measurement. It is therefore not implied that the expressions given in the text are necessarily the most accurate, as at present the choice appears somewhat arbitrary.

Lateral misalignment reduces the overlap region between the two fiber cores. Assuming uniform excitation of all the optical modes in a multimode step index fiber the overlapped area between both fiber cores approximately gives the lateral coupling efficiency η_{lat}. Hence the lateral coupling efficiency for two similar step index fibers may be written as [Ref. 62]:

$$\eta_{lat} \simeq \frac{16(n_1/n)^2}{(1 + (n_1/n))^4} \frac{1}{\pi} \left\{ 2 \cos^{-1}\left(\frac{y}{2a}\right) - \left(\frac{y}{a}\right) \left[1 - \left(\frac{y}{2a}\right)^2\right]^{\frac{1}{2}} \right\} \quad (4.14)$$

where n_1 is the core refractive index, n is the refractive index of the medium between the fibers, y is the lateral offset of the fiber core axes, and a is the fiber core radius. The lateral misalignment loss in decibels may be determined using:

$$\text{Loss}_{lat} = -10 \log_{10} \eta_{lat} \quad \text{dB} \quad (4.15)$$

The predicted losses obtained using the formulae given in Eqs. (4.14) and (4.15) are generally slightly higher than the measured values due to the assumption that all modes are equally excited. This assumption is only correct for certain cases of optical fiber transmission. Also certain authors [Refs. 61 and 71] assume index matching and hence no Fresnel reflection which makes the first term in Eq. (4.14) equal to unity (as $n_1/n = 1$). This may be valid if the two fiber ends are assumed to be in close contact (i.e. no air gap in between) and gives lower predicted losses. Nevertheless, bearing in mind these possible inconsistencies, useful estimates for the attenuation due to lateral misalignment of multimode step index fibers may be obtained.

Lateral misalignment loss in multimode graded index fibers assuming a uniform distribution of optical power throughout all guided modes was calculated by Gloge [Ref. 65]. He estimated that the lateral misalignment loss was dependent on the refractive index gradient α for small lateral offset and may be obtained from:

$$L_t = \frac{2}{\pi} \left(\frac{y}{a}\right) \left(\frac{\alpha + 2}{\alpha + 1}\right) \quad \text{for } 0 \leqslant y \leqslant 0.2a \quad (4.16)$$

where the lateral coupling efficiency was given by:

$$\eta_{lat} = 1 - L_t \tag{4.17}$$

Hence Eq. (4.17) may be utilized to obtain the lateral misalignment loss in decibels. With a parabolic refractive index profile where $\alpha = 2$, Eq. (4.16) gives:

$$L_t = \frac{8}{3\pi}\left(\frac{y}{a}\right) = 0.85\left(\frac{y}{a}\right) \tag{4.18}$$

A further estimate including the leaky modes, gave a revised expression for the lateral misalignment loss given in Eq. (4.17) of $0.75(y/a)$. This analysis was also extended to step index fibers (where $\alpha = \infty$) and gave lateral misalignment losses of $0.64(y/a)$ and $0.5(y/a)$ for the cases of guided modes only and both guided plus leaky modes respectively.

Example 4.3

A step index fiber has a core refractive index of 1.5 and a core diameter of 50 μm. The fiber is jointed with a lateral misalignment between the core axes of 5 μm. Estimate the insertion loss at the joint due to the lateral misalignment assuming a uniform distribution of power between all guided modes when:

(a) there is a small air gap at the joint;
(b) the joint is considered index matched.

Solution: (a) The coupling efficiency for a multimode step index fiber with uniform illumination of all propagating modes is given by Eq. (4.14) as:

$$\eta_{lat} \simeq \frac{16(n_1/n)^2}{(1 + (n_1/n))^4} \frac{1}{\pi}\left\{2\cos^{-1}\left(\frac{y}{2a}\right) - \left(\frac{y}{a}\right)\left[1 - \left(\frac{y}{2a}\right)^2\right]^{\frac{1}{2}}\right\}$$

$$= \frac{16(1.5)^2}{(1 + 1.5)^4}\frac{1}{\pi}\left\{2\cos^{-1}\left(\frac{5}{50}\right) - \left(\frac{5}{25}\right)\left[1 - \left(\frac{5}{50}\right)^2\right]^{\frac{1}{2}}\right\}$$

$$= 0.293\,\{2(1.471) - 0.2[0.99]^{\frac{1}{2}}\}$$

$$= 0.804$$

The insertion loss due to lateral misalignment is given by Eq. (4.15) where

$$Loss_{lat} = -10\log_{10}\eta_{lat} = -10\log_{10}0.804$$
$$= 0.95\text{ dB}$$

Hence assuming a small air gap at the joint the insertion loss is approximately 1 dB when the lateral offset is 10% of the fiber diameter.

(b) When the joint is considered index matched (i.e. no air gap) the coupling efficiency may be again obtained from Eq. 4.14 where:

$$\eta_{lat} \simeq \frac{1}{\pi}\left\{2\cos^{-1}\left(\frac{5}{50}\right) - \left(\frac{5}{25}\right)\left[1 - \left(\frac{5}{50}\right)^2\right]^{\frac{1}{2}}\right\}$$

$$= 0.318 \left\{2(1.471) - 0.2[0.99]^{\frac{1}{2}}\right\}$$

$$= 0.872$$

Therefore the insertion loss is:

$$\text{Loss}_{\text{lat}} = -10 \log_{10} 0.872 = 0.59 \text{ dB}$$

With index matching the insertion loss at the joint in example 4.3 is reduced to approximately 0.36 dB. It may be noted that the difference between the losses obtained in parts (a) and (b) corresponds to the optical loss due to Fresnel reflection at the similar fiber–air–fiber interface determined in example 4.2.

The result may be checked using the formulae derived by Gloge for a multimode step index fiber where the lateral misalignment loss assuming uniform illumination of all guided modes is obtained using:

$$L_t = 0.64 \left(\frac{y}{a}\right) = 0.64 \left(\frac{5}{25}\right) = 0.128$$

Hence the lateral coupling efficiency is given by Eq. (4.17) as:

$$\eta_{\text{lat}} = 1 - 0.128 = 0.872$$

Again using Eq. (4.15), the insertion loss due to the lateral misalignment assuming index matching is:

$$\text{Loss}_{\text{lat}} = -10 \log_{10} 0.872 = 0.59 \text{ dB}$$

Hence using the expression derived by Gloge we obtain the same value of approximately 0.6 dB for the insertion loss with the inherent assumption that there is no change in refractive index at the joint interface. Although this estimate of insertion loss may be shown to agree with certain experimental results [Ref. 61] a value of around 1 dB insertion loss for a 10% lateral displacement with regard to the core diameter (as estimated in example 4.3(a)) is more usually found to be the case with multimode step index fibers [Refs. 59, 72 and 73]. Further it is generally accepted that the lateral offset must be kept below 5% of the fiber core diameter in order to reduce insertion loss at a joint to below 0.5 dB [Ref. 72].

Example 4.4

A graded index fiber has a parabolic refractive index profile ($\alpha = 2$) and a core diameter of 50 μm. Estimate the insertion loss due to a 3 μm lateral misalignment at a fiber joint when there is index matching and assuming:

(a) there is uniform illumination of all guided modes only;
(b) there is uniform illumination of all guided and leaky modes.

Solution: (a) Assuming uniform illumination of guided modes only, the misalignment loss may be obtained using Eq. (4.18), where

$$L_t = 0.85 \left(\frac{y}{a} \right) = 0.85 \left(\frac{3}{25} \right) = 0.102$$

The coupling efficiency is given by Eq. (4.17) as:

$$\eta_{lat} = 1 - L_t = 1 - 0.102 = 0.898$$

Hence the insertion loss due to the lateral misalignment is given by Eq. (4.15), where:

$$\text{Loss}_{lat} = -10 \log_{10} 0.898 = 0.47 \text{ dB}$$

(b) When assuming the uniform illumination of both guided and leaky modes Gloge's formula becomes:

$$L_t = 0.75 \left(\frac{y}{a} \right) = 0.75 \left(\frac{3}{25} \right) = 0.090$$

Therefore the coupling efficiency is

$$\eta_{lat} = 1 - 0.090 = 0.910$$

and the insertion loss due to lateral misalignment is:

$$\text{Loss}_{lat} = -10 \log_{10} 0.910 = 0.41 \text{ dB}$$

It may be noted by observing Fig. 4.27(a) which shows the measured lateral misalignment loss for a 50 μm diameter graded index fiber that the losses predicted above are very pessimistic (the loss for 3 μm offset shown in Fig. 4.27(a) is less than 0.2 dB). A model which is found to predict insertion loss due to lateral misalignment in graded index fibers with greater accuracy was proposed by Miller and Mettler [Ref. 66]. In this model they assumed the power distribution at the fiber output to be of a Gaussian form. Unfortunately the analysis is too detailed for this text as it involves integration using numerical techniques. We therefore limit estimates of insertion losses due to lateral misalignment in multimode graded index fibers to the use of Gloge's formula.

Angular misalignment losses at joints in multimode step index fibers may be predicted with reasonable accuracy using an expression for the angular coupling efficiency η_{ang} given by [Ref. 62]:

$$\eta_{ang} \simeq \frac{16(n_1/n)^2}{(1 + (n_1/n))^4} \left[1 - \frac{n\theta}{\pi n_1 (2\Delta)^{\frac{1}{2}}} \right] \tag{4.19}$$

where θ is the angular displacement in radians and Δ is the relative refractive index difference for the fiber. The insertion loss due to angular misalignment may be obtained from the angular coupling efficiency in the same manner as the lateral misalignment loss following:

$$\text{Loss}_{ang} = -10 \log_{10} \eta_{ang} \tag{4.20}$$

The formulae given in Eqs. (4.19) and (4.20) predict that the smaller the values of Δ the larger the insertion loss due to angular misalignment. This appears intuitively correct as small values of Δ imply small numerical aperture fibers which will be more affected by angular misalignment. It is confirmed by the measurements shown in Fig. 4.27(b) and demonstrated in example 4.5.

Example 4.5

Two multimode step index fibers have numerical apertures of 0.2 and 0.4 respectively, and both have the same core refractive index which is 1.48. Estimate the insertion loss at a joint in each fiber caused by a 5° angular misalignment of the fiber core axes. It may be assumed that the medium between the fibers is air.

Solution: The angular coupling efficiency is given by Eq. (4.19) as

$$\eta_{ang} \simeq \frac{16(n_1/n)^2}{(1 + (n_1/n))^4} \left[1 - \frac{n\theta}{\pi n_1 (2\Delta)^{\frac{1}{2}}} \right]$$

The numerical aperture is related to the relative refractive index difference following Eq. (2.10) where:

$$NA \simeq n_1 (2\Delta)^{\frac{1}{2}}$$

Hence

$$\eta_{ang} \simeq \frac{16(n_1/n)^2}{(1 + (n_1/n))^4} \left[1 - \frac{n\theta}{\pi NA} \right]$$

For the 0.2 NA fiber:

$$\eta_{ang} \simeq \frac{16(1.48)^2}{(1 + 1.48)^4} \left[1 - \frac{5\pi/180}{\pi\,0.2} \right]$$

$$= 0.797$$

The insertion loss due to the angular misalignment may be obtained from Eq. (4.20), where:

$$Loss_{ang} = -10 \log_{10} \eta_{ang} = -10 \log_{10} 0.797$$

$$= 0.98 \text{ dB}$$

For the 0.4 NA fiber:

$$\eta_{ang} \simeq 0.926 \left[1 - \frac{5\pi/180}{\pi 0.4} \right]$$

$$\simeq 0.862$$

The insertion loss due to the angular misalignment is therefore:

$$Loss_{ang} = -10 \log_{10} 0.862$$

$$= 0.64 \text{ dB}$$

Hence it may be noted from example 4.5 that the insertion loss due to angular misalignment is reduced by using fibers with large numerical apertures. This is the opposite trend to the increasing insertion loss with numerical aperture for fiber longitudinal misalignment at a joint.

Misalignment losses at connections in single mode fibers have been theoretically considered by Marcuse [Ref. 68] and Gambling *et al.* [Refs. 69 and 70]. The theoretical analysis which was instigated by Marcuse is based upon the Gaussian or near Gaussian shape of the modes propagating in single mode fibers regardless of the fiber type (i.e. step index or graded index). Further development of this theory by Gambling *et al.* [Ref. 70] gave simplified formulae for both the lateral and angular misalignment losses at joints in single mode fibers. In the absence of angular misalignment Gambling *et al.* calculated that the loss T_l due to lateral offset y was given by:

$$T_l = 2.17 \left(\frac{y}{\omega_0} \right)^2 \text{dB} \qquad (4.21)$$

where ω_0 is the spot size of the fundamental mode. The spot size is usually defined as the width to $1/e$ intensity of the LP_{01} mode, or in terms of the spot size of an incident Gaussian beam which gives maximum launching efficiency [Ref. 14]. However, the spot size for the LP_{01} mode (corresponds to HE mode) may be obtained from the empirical formula [Refs. 68 and 69]:

$$\omega_0 = a \frac{(0.65 + 1.62V^{-1.5} + 2.88V^{-6})}{2^{\frac{1}{2}}} \qquad (4.22)$$

where ω_0 is the spot size in μm, a is the fiber core radius and V is the normalized frequency for the fiber. Alternatively the insertion loss T_a caused by an angular misalignment θ (in radians) at a joint in a single mode fiber may be given by:

$$T_a = 2.17 \left(\frac{\theta \omega_0 n_1 V}{a \, \text{NA}} \right)^2 \text{dB} \qquad (4.23)$$

where n_1 is the fiber core refractive index and NA is the numerical aperture of the fiber. It must be noted that the formulae given in Eqs. (4.21) and (4.23) assume that the spot sizes of the modes in the two coupled fibers are the same. Gambling *et al.* [Ref. 70] also derived a somewhat complicated formula which gave a good approximation for the combined losses due to both lateral and angular misalignment at a fiber joint. However they indicate that for small total losses (less than 0.75 dB) a reasonable approximation is obtained by simply combining Eqs. (4.21) and (4.23).

Example 4.6

A single mode fiber has the following parameters:
normalized frequency (V) = 2.40
core refractive index (n_1) = 1.46
core diameter $(2a)$ = 8 μm
numerical aperture (NA) = 0.1
Estimate the total insertion loss of a fiber joint with a lateral misalignment of 1 μm and an angular misalignment of 1°.
Solution: Initially it is necessary to determine the spot size in the fiber. This may be obtained from Eq. (4.22) where:

$$\omega_0 = a\frac{(0.65 + 1.62V^{-1.5} + 2.88V^{-6})}{2^{\frac{1}{2}}}$$

$$= 4\frac{(0.65 + 1.62(2.4)^{-1.5} + 2.88(2.4)^{-6})}{2^{\frac{1}{2}}}$$

$$= 3.12\ \mu m$$

The loss due to the lateral offset is given by Eq. (4.21) as:

$$T_l = 2.17\left(\frac{y}{\omega_0}\right)^2 = 2.17\left(\frac{1}{3.12}\right)^2$$

$$= 0.22\ dB$$

The loss due to angular misalignment may be obtained from Eq. (4.23) where:

$$T_a = 2.17\left(\frac{\theta\omega_0 n_1 V}{a\ NA}\right)^2$$

$$= 2.17\left(\frac{(\pi/180) \times 3.12 \times 1.46 \times 2.4}{4 \times 0.1}\right)$$

$$= 0.49\ dB$$

Hence the total insertion loss is

$$T_T \simeq T_l + T_a = 0.22 + 0.49$$

$$= 0.71\ dB$$

In this example the loss due to angular misalignment is significantly larger than that due to lateral misalignment. However, aside from the actual magnitudes of the respective misalignments, the insertion losses incurred are also strongly dependent upon the normalized frequency of the fiber. This is especially the case with angular misalignment at a single mode fiber joint where insertion losses of less than 0.3 dB may be obtained when the angular misalignment is $1°$ with fibers of appropriate V value. Nevertheless for low loss single mode fiber joints it is important that angular alignment is better than $1°$.

We have considered in some detail the optical attenuation at fiber–fiber connections. However we have not yet discussed the possible distortion of the transmitted signal at a fiber joint. Although work in this area is in its infancy, increased interest has been generated with the use of highly coherent sources (injection lasers) and very low dispersion fibers. It is apparent that fiber connections strongly affect the signal transmission causing modal noise (see Section 3.11) and nonlinear distortion [Ref. 76] when a coherent light source is utilized with a multimode fiber. Also it has been reported [Ref. 77] that the transmission loss of a connection in a coherent multimode system is extremely wavelength-dependent exhibiting a possible 10% change in the transmitted optical wavelength for a very small change (0.001 nm) in the laser emission wavelength. Nevertheless it has been found that these problems may be reduced by the use of single mode optical fiber [Ref. 76].

Furthermore the above modal effects become negligible when an incoherent source (light emitting diode) is used with multimode fiber. However, in this instance there is often some mode conversion at the fiber joint which can make the connection effectively act as a mode mixer or filter [Ref. 78]. Indications are that this phenomenon which has been investigated [Ref. 79] with regard to fiber splices, is more pronounced with fusion splices than with mechanical splices, both of which are described in Section 4.9.

4.9 FIBER SPLICES

A permanent joint formed between two individual optical fibers in the field or factory is known as a fiber splice. Fiber splicing is frequently used to establish long-haul optical fiber links where smaller fiber lengths need to be joined, and there is no requirement for repeated connection and disconnection. Splices may be divided into two broad categories depending upon the splicing technique utilized. These are fusion splicing or welding and mechanical splicing.

Fusion splicing is accomplished by applying localized heating (e.g. by a flame or an electric arc) at the interface between two butted, prealigned fiber ends causing them to soften and fuse. Mechanical splicing, in which the fibers are held in alignment by some mechanical means, may be achieved by various methods including the use of tubes around the fiber ends (tube splices) or V-grooves into which the butted fibers are placed (groove splices). All these techniques seek to optimize the splice performance (i.e. reduce the insertion loss at the joint) through both fiber end preparation and alignment of the two jointed fibers. Typical average splice insertion losses for multimode fibers are in the range 0.1–0.2 dB [Ref. 81] which is generally a better performance than that exhibited by demountable connections (see Sections 4.10–4.12). It may be

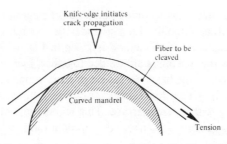

Fig. 4.28   Optical fiber end preparation: the principle of scribe and break cutting [Ref. 82].

noted that the insertion losses of fiber splices are generally much less than the possible Fresnel reflection loss at a butted fiber–fiber joint. This is because there is no large step change in refractive index with the fusion splice as it forms a continuous fiber connection, and some method of index matching (e.g. a fluid) tends to be utilized with mechanical splices. However, fiber splicing (especially fusion splicing) is at present a somewhat difficult process to perform in a field environment and suffers from practical problems in the development of field-usable tools.

A requirement with fibers intended for splicing is that they have smooth and square end faces. In general this end preparation may be achieved using a suitable tool which cleaves the fiber as illustrated in Fig. 4.28 [Ref. 82]. This process is often referred to as scribe and break or score and break as it involves the scoring of the fiber surface under tension with a cutting tool (e.g. sapphire, diamond, tungsten carbide blade). The surface scoring creates failure as the fiber is tensioned and a clean, reasonably square fiber end can be produced. Figure 4.28 illustrates this process with the fiber tensioned around a curved mandrel. However, straight pull, scribe and break tools are also utilized, which arguably give better results [Ref. 83].

4.9.1   Fusion Splices

The fusion splicing of single fibers involves the heating of the two prepared fiber ends to their fusing point with the application of sufficient axial pressure between the two optical fibers. It is therefore essential that the stripped (of cabling and buffer coating) fiber ends are adequately positioned and aligned in order to achieve good continuity of the transmission medium at the junction point. Hence the fibers are usually positioned and clamped with the aid of an inspection microscope.

Flame heating sources such as microplasma torches (argon and hydrogen) and oxhydric microburners (oxygen, hydrogen and alcohol vapor) have been utilized with some success [Ref. 84]. However, the most widely used heating source is an electric arc. This technique offers advantages of consistent, easily controlled heat with adaptability for use under field conditions. A schematic diagram of the basic arc fusion method is given in Fig. 4.29(a) [Refs. 81 and 85] illustrating how the two fibers are welded together. Figure 4.29(b) [Ref. 73] shows a development of the basic arc fusion process which involves the rounding of the fiber ends with a low energy discharge before pressing the fibers together and fusing with a stronger arc. This technique, known as prefusion, removes the requirement for fiber end preparation which has distinct

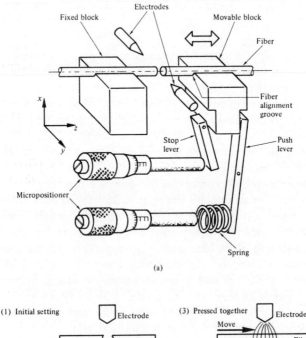

(a)

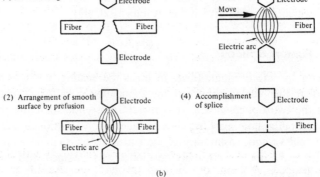

(b)

Fig. 4.29 Electric arc fusion splicing: (a) an example of fusion splicing apparatus [Refs. 81 and 85]; (b) schematic illustration of the prefusion method for accurately splicing optical fibers [Ref. 73].

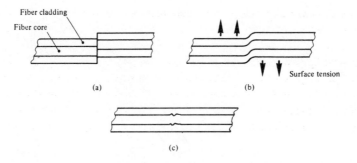

Fig. 4.30 Self-alignment phenomenon which takes place during fusion splicing: (a) before fusion; (b) during fusion; (c) after fusion [Refs. 85, 87 and 88].

advantage in the field environment. It has been utilized with multimode fibers giving average splice losses of 0.09 dB [Ref. 86].

Fusion splicing of single mode fibers with typical core diameters between 3 and 10 μm presents problems of more critical fiber alignment (i.e. lateral offsets of less than 1 μm are required for low loss joints). However, splice insertion losses below 0.3 dB may be achieved due to a self-alignment phenomenon which partially compensates for any lateral offset.

Self-alignment, illustrated in Fig. 4.30 [Refs. 85, 87 and 88], is caused by surface tension effects between the two fiber ends during fusing. A recently reported [Ref. 89] field trial of single mode fiber fusion splicing over a 31.6 km link gave mean splice insertion losses of 0.18 and 0.12 dB at wavelengths of 1.3 and 1.55 μm respectively.

A possible drawback with fusion splicing is that the heat necessary to fuse the fibers may weaken the fiber in the vicinity of the splice. It has been found that even with careful handling, the tensile strength of the fused fiber may be as low as 30% of that of the uncoated fiber before fusion [Ref. 91]. The fiber fracture generally occurs in the heat-affected zone adjacent to the fused joint. The reduced tensile strength is attributed [Refs. 91 and 92] to the combined effects of surface damage caused by handling, surface defect growth during heating and induced residential stresses due to changes in chemical composition. It is therefore necessary that the completed splice is packaged so as to reduce tensile loading upon the fiber in the vicinity of the splice.

4.9.2 Mechanical Splices

A number of mechanical techniques for splicing individual optical fibers have been developed. A common method involves the use of an accurately produced rigid alignment tube into which the prepared fiber ends are permanently bonded. This snug tube splice is illustrated in Fig. 4.31(a) [Ref. 94] and may utilize a glass or ceramic capillary with an inner diameter just large

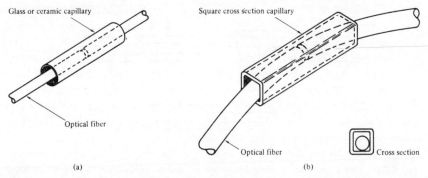

(a)   (b)

Fig. 4.31   Techniques for tube splicing of optical fibers: (a) snug tube splice [Ref. 94]; (b) loose tube splice utilizing square cross section capillary [Ref. 96].

enough to accept the optical fibers. Transparent adhesive, (e.g. epoxy resin) is injected through a transverse bore in the capillary to give mechanical sealing and index matching of the splice. Average insertion losses as low as 0.1 dB have been obtained [Ref. 95] with multimode graded index and single mode fibers using ceramic capillaries. However, in general, snug tube splices exhibit problems with capillary tolerance requirements.

A mechanical splicing technique which avoids the critical tolerance requirements of the snug tube splice is shown in Fig. 4.31(b) [Ref. 96]. This loose tube splice uses an oversized square section metal tube which easily accepts the prepared fiber ends. Transparent adhesive is first inserted into the tube followed by the fibers. The splice is self-aligning when the fibers are curved in the same plane, forcing the fiber ends simultaneously into the same corner of the tube, as indicated in Fig. 4.31(b). Mean splice insertion losses of 0.073 dB have been achieved [Refs. 88 and 97] using multimode graded index fibers with the loose tube approach.

An alternative method of obtaining a tight fitting splice is by use of the collapsed sleeve splicing technique which is illustrated in Fig. 4.32 [Ref. 98].

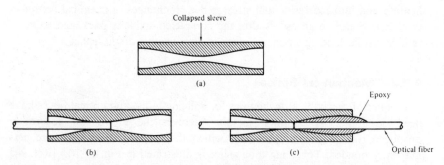

Fig. 4.32   The collapsed sleeve splicing technique [Ref. 98].

This method utilizes a Pyrex glass sleeve which has a lower melting point than the fibers to be jointed. When the sleeve is heated to its softening point it collapses due to surface tension, eventually forming a solid rod. Figure 4.32(a) shows a partially collapsed Pyrex sleeve formed by local heating of the sleeve. With the collapsed sleeve splicing technique the glass sleeve is collapsed around one of the prepared fiber ends to form a tight fitting socket as shown in Fig. 4.32(b). The second fiber is then inserted into the socket and the whole assembly is bonded with epoxy resin as illustrated in Fig. 4.32(c). Hence an index matched splice is created. This technique is useful in the splicing of two fibers with different diameters. In this case the sleeve is collapsed over the larger diameter fiber before the insertion of the second, smaller diameter fiber. The collapsing is then continued to form a socket of an appropriate size for a close fit to the smaller fiber. Collapsed sleeve splices are generally protected by enclosure in a metal ferrule. They have exhibited insertion losses in the range 0.2–0.3 dB [Ref. 59] when using multimode graded index fibers with average losses of 0.5 dB in the field.

Other common mechanical splicing techniques involve the use of grooves to secure the fibers to be jointed. A simple method utilizes a V-groove into which the two prepared fiber ends are pressed. The V-groove splice which is illustrated in Fig. 4.33(a) [Ref. 99] gives alignment of the prepared fiber ends through insertion in the groove. The splice is made permanent by securing the fibers in the V-groove with epoxy resin. Jigs for producing V-groove splices

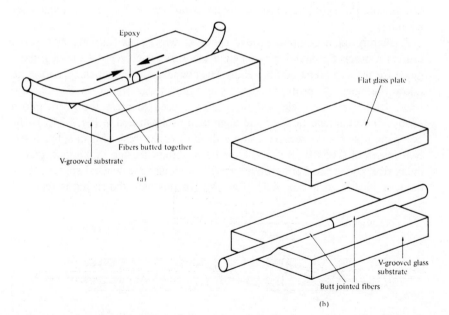

Fig. 4.33   V-groove splices [Ref. 99].

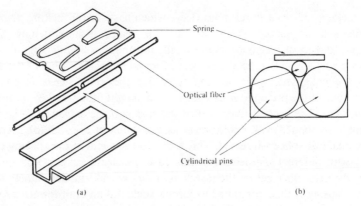

Fig. 4.34   The Springroove® splice [Ref. 101]: (a) expanded view of the splice; (b) schematic cross section of the splice.

have proved quite successful, giving joint insertion losses of around 0.1 dB [Ref. 82].

V-groove splices formed by sandwiching the butted fiber ends between a V-groove glass substrate and a flat glass plate as shown in Fig. 4.33(b) have also proved very successful in the laboratory. Splice insertion losses of less than 0.01 dB when coupling single mode fibers have been reported [Ref. 100] using this technique. However, reservations are expressed regarding the field implementation of these splices with respect to manufactured fiber geometry, and housing of the splice in order to avoid additional losses due to local fiber bending.

A slightly more complex groove splice known as the Springroove® splice utilizes a bracket containing two cylindrical pins which serve as an alignment guide for the two prepared fiber ends. The cylindrical pin diameter is chosen to allow the fibers to protrude above the cylinders as shown in Fig. 4.34(a) [Ref. 101]. An elastic element (a spring) is used to press the fibers into a groove and maintain the fiber end alignment as illustrated in Fig. 4.34(b). The complete assembly is secured using a drop of epoxy resin. Mean splice insertion losses of 0.05 dB [Ref. 88] have been obtained using multimode graded index fibers with the Springroove® splice. A similar mechanical splicing technique is illustrated in Fig. 4.35 [Ref. 94]. In this case the spring is replaced

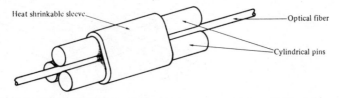

Fig. 4.35   The precision pin splice [Ref. 94].

by a third cylindrical pin and the whole assembly is held in place with a heat shrinkable sleeve. This precision pin splice has given mean insertion losses of around 0.2 dB [Ref. 88] with multimode fibers.

4.9.3 Multiple Splices

Multiple simultaneous splicing has mainly been attempted using mechanical splicing methods. Groove splicing techniques have been utilized for the simultaneous splicing of an array of fibers within a flat ribbon cable. Figure 4.36 [Ref. 102] shows a groove splice for a five fiber ribbon cable. It utilizes a grooved metal substrate with the groove spacing equal to the spacing of the fibers in the array. The plastic coating is removed from the fiber ends and they are prepared using a suitable scribe and break tool. Then the two ribbon ends are placed into the grooves, the fibers are pressed together and held in position with a rubber sheet and a cover plate. Finally epoxy resin is added to provide index matching as well as securing a permanent splice. A 12 fiber version of this splice using an injection moulded plastic substrate gave an average splice loss of 0.2 dB [Ref. 97].

A ribbon splice using etched silicon chips is shown in Fig. 4.37 [Ref. 103]. Similar chips may be utilized to form a 12×12 array of 12 ribbon cables each containing 12 fibers [Ref. 82]. The whole assembly is clamped together to form a single multiple splice.

Multiple fiber splicing of a circular cross section cable has also been achieved. A splice of this type is shown in Fig. 4.38 [Ref. 104] and consists of matched sets of precision moulded circular collars (one shown) which contain a series of grooves around the circumference. The fibers are inserted into the precision grooves and bonded with adhesive. They are then cut, polished and covered with index matching material at the jointing ends, before the two collars are brought together and fastened with semicylindrical shells and pins.

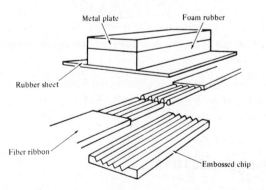

Fig. 4.36 Multiple fiber splicing of fiber ribbon cable using a groove alignment technique [Ref. 102].

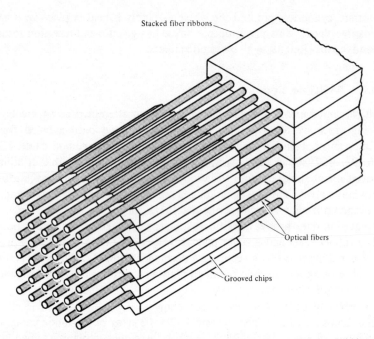

Fig. 4.37 Multiple fiber splicing of stacked ribbon cables using precision multi-V-groove silicon chips [Ref. 103].

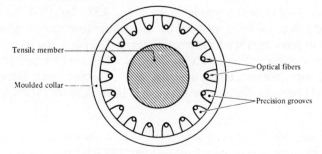

Fig. 4.38 Cross section of the star-core cable for multiple fiber mechanical splicing [Ref. 104].

When the two grooved collars are produced from a single piece of alumina, average splice insertion losses of 0.3 dB [Ref. 88] have been obtained with multimode step index fibers.

4.10 FIBER CONNECTORS

Demountable fiber connectors are more difficult to achieve than optical fiber splices. This is because they must maintain similar tolerance requirements to splices in order to couple light between fibers efficiently, but they must accomplish it in a removable fashion. Also the connector design must allow for repeated connection and disconnection without problems of fiber alignment which may lead to degradation in the performance of the transmission line at the joint. Hence to operate satisfactorily the demountable connector must provide reproducible accurate alignment of the optical fibers.

In order to maintain an optimum performance the connector must also protect the fiber ends from damage which may occur due to handling (connection and disconnection), must be insensitive to environmental factors (e.g. moisture and dust) and must cope with tensile load on the cable. Additionally, the connector should ideally be a low cost component which can be fitted with relative ease. Hence optical fiber connectors may be considered in three major areas, which are:

(a) the fiber termination which protects and locates the fiber ends;
(b) the fiber end alignment to provide optimum optical coupling;
(c) the outer shell which maintains the connection and the fiber alignment, protects the fiber ends from the environment and provides adequate strength at the joint.

The use of an index matching material in the connector between the two jointed fibers can assist the connector design in two ways. It increases the light transmission through the connection whilst keeping dust and dirt from between the fibers. However, this design aspect is not always practical with demountable connectors, especially where fluids are concerned. Apart from problems of sealing and replacement when the joint is disconnected and reconnected, liquids in this instance may have a detrimental affect, attracting dust and dirt to the connection.

There are a large number of demountable single fiber connectors, both commercially available and under development, which have insertion losses in the range 0.2–3 dB. Fiber connectors may be separated into two broad categories: butt jointed connectors and expanded beam connectors. Butt jointed connectors rely upon alignment of the two prepared fiber ends in close proximity (butted) to each other so that the fiber core axes coincide. Expanded beam connectors utilize interposed optics at the joint (i.e. lenses or tapers) in order to expand the beam from the transmitting fiber end before reducing it again to a size compatible with the receiving fiber end.

4.11 BUTT JOINTED CONNECTORS

Butt jointed connectors are the most widely used connector type and a substantial number have been reported. In this section we review some of the more common butt jointed connector designs which have been developed primarily for use with multimode fibers. Nevertheless in certain cases as indicated in the text, similar designs have been used successfully with single mode fibers.

4.11.1 Ferrule Connector

The basic ferrule connector (sometimes referred to as a concentric sleeve connector), which is perhaps the simplest optical fiber connector design, is illustrated in Fig. 4.39(a) [Ref. 58]. The two fibers to be connected are permanently bonded (with epoxy resin) in metal plugs known as ferrules which have an accurately drilled central hole in their end faces where the stripped (of buffer coating) fiber is located. Within the connector the two ferrules are placed in an alignment sleeve which, using accurately machined components, allows the fiber ends to be butt jointed. The ferrules are held in place via a retaining mechanism which in the example shown in Fig. 4.39(a) is a spring.

It is essential with this type of connector that the fiber end faces are smooth and square (i.e. perpendicular to the fiber axis). This may be achieved with varying success by either:

(a) cleaving the fiber before insertion into the ferrule;
(b) inserting and bonding before cleaving the fiber close to the ferrule end face;

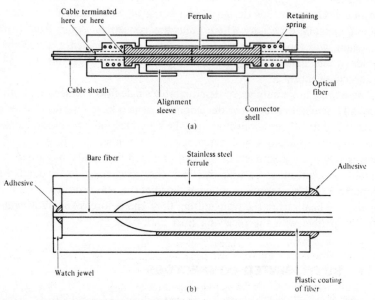

Fig. 4.39 Ferrule connectors: (a) structure of a basic ferrule connector [Ref. 58]; (b) structure of a watch jewel connector ferrule [Ref. 59].

(c) using either (a) or (b) and polishing the fiber end face until it is flush with the end of the ferrule.

Polishing the fiber end face after insertion and bonding provides the best results but it tends to be time-consuming and inconvenient especially in the field.

The fiber alignment accuracy of the basic ferrule connector is largely dependent upon the ferrule hole into which the fiber is inserted. Hence some ferrule connectors incorporate a watch jewel in the ferrule end face (jewelled ferrule connector) as illustrated in Fig. 4.39(b) [Ref. 59]. In this case the fiber is centered with respect to the ferrule through the watch jewel hole. The use of the watch jewel allows the close diameter and tolerance requirements of the ferrule end face hole to be obtained more easily than simply through drilling of the ferrule end face alone. Nevertheless, typical concentricity errors between the fiber core and the outside diameter of the jewelled ferrule are in the range 2–6 μm giving insertion losses in the range 1–2 dB with multimode step index fibers.

4.11.2 Biconical Connector

A ferrule type connector which is widely used as part of jumper cable in a variety of applications in the Bell system is the biconical plug connector [Refs. 81 and 105]. The plugs are either transfer moulded directly onto the fiber or cast around the fiber using a silica-loaded epoxy resin ensuring concentricity to within 5 μm. After plug attachment, the fiber end faces are polished before the plugs are inserted and aligned in the biconical moulded center sleeve as shown in Fig. 4.40 [Ref. 81]. Mean insertion losses as low as 0.21 dB have been reported [Ref. 105] when using this connector with 50 μm core diameter graded index fibers. In the original design transparent silicone resin pads were placed over the fiber end faces to provide index matching. However, currently the polished fiber end faces are butted directly, the gap and

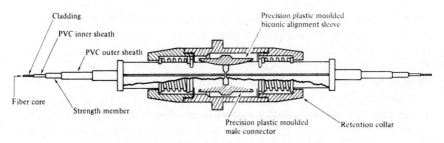

Fig. 4.40   Cross section of the biconical connector [Refs. 81 and 105].

parallelism of the end faces being controlled to a degree that gives insertion losses better than the level normally exhibited by Fresnel reflection.

4.11.3   Ceramic Capillary Connector

An approach giving accurate ferrule alignment is used in the ceramic capillary connector shown in Fig. 4.41 [Refs. 107 and 108]. The special feature of this concentric sleeve connector is the mounting of the optical fiber in a ceramic capillary which is set in the tip of the ferrule. This gives accurate positioning of the fiber core and allows the fiber end face to take a good polish. Average insertion losses of 0.4 dB have been reported [Ref. 107] when using multimode fibers. This connector has also been used for the coupling of single mode fibers giving average insertion losses of 0.5 dB [Ref. 108] when using a 10 μm core diameter step index fiber operating at a wavelength of 1.3 μm.

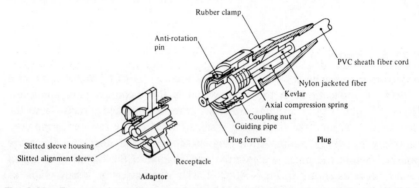

Fig. 4.41   The ceramic capillary connector showing the ferrule plug and the adaptor into which two plugs are located [Refs. 107 and 108].

4.11.4   Double Eccentric Connector

The double eccentric connector does not rely on a concentric fixed sleeve approach but is adjustable, allowing close alignment of the fiber axes. The mechanism, which is shown in Fig. 4.42 [Refs. 58 and 62], consists of two eccentric cylinders within the outer plug. It may be observed from Fig. 4.42 that the optical fiber is mounted eccentrically within the inner cylinder. Therefore when the two connector halves are mated it is always possible through rotation of the mechanism to make the fiber core axes coincide. This operation is performed on both plugs using either an inspection microscope or a peak optical adjustment. The mechanisms are then locked to give permanent alignment. This connector type has exhibited mean insertion losses of 0.48 dB with multimode graded index fibers: use of index matching fluid within the connector has reduced these losses to 0.2 dB. The double eccentric connector

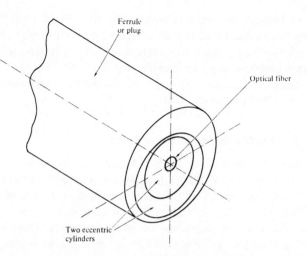

Fig. 4.42 Structure of the double eccentric connector plug [Refs. 58 and 62].

design has also been utilized with single mode fibers where its adjustable nature has proved advantageous for alignment of the small core diameter fibers.

4.11.5 Triple Ball Connector

There are a number of connectors which utilize kinematic design principles. A noteworthy example developed by British Telecom is the triple ball connector illustrated in Fig. 4.43 [Ref. 109]. Three accurately ground tungsten spheres

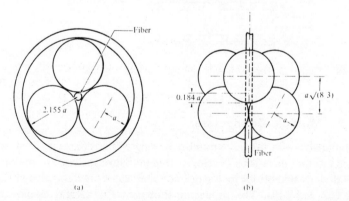

Fig. 4.43 The triple ball connector [Ref. 109]: (a) connector cross section showing the location of the fiber in a groove between three contacting balls; (b) plan view of the connector showing the two interlocking sets of three balls and the position of the fiber ends.

gripping the fiber are housed in a bush as shown in Fig. 4.43(a). Two sets of the spheres are nested together with a relative rotation of 60° (Fig. 4.43(b)) bringing the two butt jointed fibers into alignment. The reported [Ref. 109] average insertion loss using multimode step index fibers is 0.49 dB without index matching and 0.18 dB with matching. A recent improvement to this connector is described in Section 4.12.

4.11.6 Single Mode Fiber Connector

Although the ceramic capillary and the double eccentric connector have been utilized with single mode fibers they were not initially designed for this purpose. A connector which was designed specifically for use with single mode fibers is illustrated in Fig. 4.44 [Ref. 110]. It consists of a pair of fiber plugs and a sleeve with ball bearing arrays on its inside surface. These ball bearings provide plug alignment accuracy as well as smooth detachment. Each fiber plug contains two eccentric tubes which by means of rotation allow the single mode fiber to be accurately centered within the plug.

Average insertion losses of 0.46 dB [Ref. 110] were encountered with the connector when using 5.7 μm core diameter (150 μm cladding diameter) single mode fiber. A development [Ref. 110] of this connector design using a simpler cylindrical plug structure (not requiring eccentric tubes) with an accurately machined hole to locate the fiber exhibited similar average insertion losses (0.47 dB). However, the simpler plug design makes connector assembly possible without special equipment, making it far more convenient for practical use.

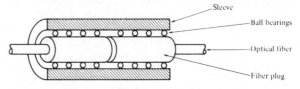

Fig. 4.44 Single mode optical fiber connector structure [Ref. 110].

4.11.7 Multiple Connectors

In comparison with the large number of single fiber connectors few multiple fiber connectors have been developed to date. Nevertheless two multiple connector designs suitable for jointing ribbon fiber cable are illustrated in Fig. 4.45 [Refs. 112 and 113]. The connector shown in Fig. 4.45(a) employs plastic moulded multiple terminations which are jointed in an alignment sleeve consisting of grooved silicon chips. This technique is very similar to the multiple groove splicing methods for ribbon cable described in Section 4.9.3. Using

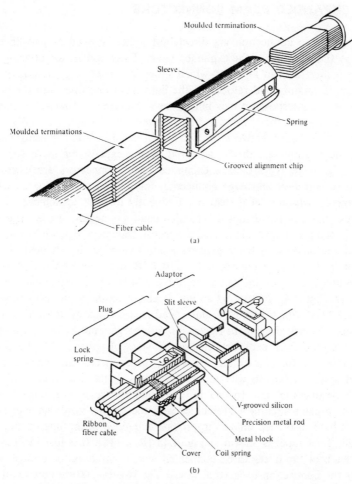

Fig. 4.45 Multiple fiber connectors: (a) connector with grooved alignment sleeve and moulded fiber ribbon end terminations [Ref. 112]; (b) fiber ribbon connector using V-grooved silicon chip [Ref. 113].

multimode graded index fibers, insertion losses in the range 0.2–0.32 dB were obtained [Ref. 112] with this device.

Figure 4.45(b) also shows a multiple connector design which utilizes V-grooved silicon chips. However, in this case ribbon fibers are mounted and bonded in the V-grooves in order to form a plug together with precision metal guiding rods and coil springs. The butt jointed fiber connections are accomplished by butt jointing the two pairs of guiding rods in slitted sleeves located in the adaptor also shown in Fig. 4.45(b). This connector exhibited average insertion losses of 0.8 dB which were reduced to 0.4 dB by the use of index matching fluid.

4.12 EXPANDED BEAM CONNECTORS

An alternative to connection via direct butt joints between optical fibers is offered by the principle of the expanded beam. Fiber connection utilizing this principle is illustrated in Fig. 4.46 which shows a connector consisting of two lenses for collimating and refocusing the light from one fiber into the other. The use of this interposed optics makes the achievement of lateral alignment much less critical than with a butt jointed fiber connector. Also the longitudinal separation between the two mated halves of the connector ceases to be critical. However, this is achieved at the expense of more stringent angular alignment. Nevertheless, expanded beam connectors are useful for multifiber connection and edge connection for printed circuit boards where lateral and longitudinal alignment are frequently difficult to achieve.

Two examples of lens coupled expanded beam connectors are illustrated in Fig. 4.47 [Refs. 114 and 115]. The connector shown in Fig. 4.47(a) utilizes spherical microlenses for beam expansion and reduction. It exhibited average losses of 1 dB which were reduced to 0.7 dB with the application of an antireflection coating on the lenses and the use of 50 μm core diameter graded index fiber. Figure 4.47(b) shows an improvement on the triple ball connector (Section 4.11.5) which utilizes a glass bead lens formed on the end of each fiber. This is achieved using an electric arc discharge, and the beads assist the centering of the fiber ends between the clusters of spheres. The bead lens reduces the need for index matching within the connector and gives average insertion losses of 0.8 dB with multimode fibers. It has also been used with single mode fibers [Refs. 116 and 117] exhibiting losses in the range 1–2 dB due to the greater alignment difficulties with the smaller core diameter fiber.

Tapers have also been utilized to provide expanded beam fiber connection. A cladded fiber taper is shown in Fig. 4.48 [Ref. 118]. It is fused to the end face of each of the fibers to be connected which increases the optical beam width at the connection by several times. The two tapers are butt jointed in order to complete the expanded beam connector. As with the lens connectors, the increased beam diameter reduces the effect of dust and dirt on connector loss. However, the difficulties involved in producing accurately dimensioned tapers currently preclude wide use of this connector type.

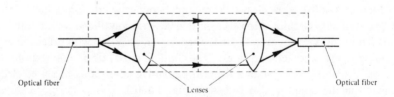

Fig. 4.46 Schematic illustration of an expanded beam connector showing the principle of operation.

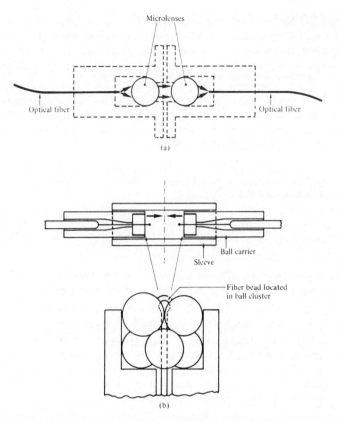

Fig. 4.47 Lens coupled expanded beam connectors: (a) schematic diagram of a connector with two microlenses making a 1:1 image of the emitting fiber upon the receiving one [Ref. 114]; (b) triple ball, fiber bead connector [Ref. 115].

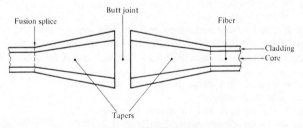

Fig. 4.48 Expanded beam connector using tapers [Ref. 118].

PROBLEMS

4.1 Describe in general terms liquid phase techniques for the preparation of multicomponent glasses for optical fibers. Discuss with the aid of a suitable diagram one melting method for the preparation of multicomponent glass.

4.2   Indicate the major advantages of vapor phase deposition in the preparation of glasses for optical fibers. Briefly describe the various vapor phase techniques currently in use.

4.3   (a) Compare and contrast, using suitable diagrams, the outside vapor phase oxidation (OVPO) process and the modified chemical vapor deposition (MCVD) technique for the preparation of low loss optical fibers.
     (b) Briefly describe the salient features of vapor axial deposition (VAD) and the plasma-activated chemical vapor deposition (PCVD) when applied to the preparation of optical fibers.

4.4   Discuss the drawing of optical fibers from prepared glasses with regard to:
     (a) multicomponent glass fibers;
     (b) silica-rich fibers.

4.5   List the various optical fiber types currently on the market indicating their important features. Hence briefly describe the general areas of application for each type.

4.6   Briefly describe the major reasons for the cabling of optical fibers which are to be placed in a field environment. Thus state the functions of the optical fiber cable.

4.7   Explain how the Griffith theory is developed in order to predict the fracture stress of an optical fiber with an elliptical crack.
     Silica has a Young's modulus of 9×10^{10} N m^{-2} and a surface energy of 2.29 J. Estimate the fracture stress in psi for a silica optical fiber with a dominant elliptical crack of depth 0.5 μm. Also determine the strain at the break for the fiber (1 psi $\equiv$ 6894.76 N m^{-2}).

4.8   Another length of the optical fiber described in problem 4.7 is found to break at 1% strain. The failure is due to a single dominant elliptical crack. Estimate the depth of this crack.

4.9   Describe the effects of stress corrosion on optical fiber strength and durability.
     It is found that a 20 m length of fused silica optical fiber may be extended to 24 m at liquid nitrogen temperatures (i.e. little stress corrosion) before failure occurs. Estimate the fracture stress in psi for the fiber under these conditions. Young's modulus for silica is 9×10^{10} N m^{-2} and 1 psi $\equiv$ 6894.76 N m^{-2}.

4.10 Discuss optical fiber cable design with regard to:
     (a) fiber buffering;
     (b) cable strength and structural members;
     (c) cable sheath and water barrier.
     Further, compare and contrast possible cable designs for multifiber cables.

4.11 State the two major categories of fiber–fiber joint, indicating the differences between them. Briefly discuss the problem of Fresnel reflection at all types of optical fiber joint, and indicate how it may be avoided.
     A silica multimode step index fiber has a core refractive index of 1.46. Determine the optical loss in decibels due to Fresnel reflection at a fiber joint with:
     (a) a small air gap;
     (b) an index matching epoxy which has a refractive index of 1.40.

It may be assumed that the fiber axes and end faces are perfectly aligned at the joint.

4.12 The Fresnel reflection at a butt joint with an air gap in a multimode step index fiber is 0.46 dB. Determine the refractive index of the fiber core.

4.13 Describe the three types of fiber misalignment which may contribute to insertion loss at an optical fiber joint.

A step index fiber with a 200 μm core diameter is butt jointed. The joint which is index matched has a lateral offset of 10 μm but no longitudinal or angular misalignment. Using two methods, estimate the insertion loss at the joint assuming the uniform illumination of all guided modes.

4.14 A graded index fiber has a characteristic refractive index profile (α) of 1.85, and a core diameter of 60 μm. Estimate the insertion loss due to a 5 μm lateral offset at an index matched fiber joint assuming the uniform illumination of all guided modes.

4.15 A graded index fiber with a parabolic refractive index profile (α = 2) has a core diameter of 40 μm. Determine the difference in the estimated insertion losses at an index matched fiber joint with a lateral offset of 1 μm (no longitudinal or angular misalignment). When performing the calculation assume (a) the uniform illumination of only the guided modes and (b) the uniform illumination of both guided and leaky modes.

4.16 A graded index fiber with a 50 μm core diameter has a characteristic refractive index profile (α) of 2.25. The fiber is jointed with index matching and the connection exhibits an optical loss of 0.62 dB. This is found to be solely due to a lateral offset of the fiber ends. Estimate the magnitude of the lateral offset assuming the uniform illumination of all guided modes in the fiber core.

4.17 A step index fiber has a core refractive index of 1.47, a relative refractive index difference of 2% and a core diameter of 80 μm. The fiber is jointed with a lateral offset of 2 μm, an angular misalignment of the core axes of 3° and a small air gap (no longitudinal misalignment). Estimate the total insertion loss at the joint which may be assumed to comprise the sum of the misalignment losses.

4.18 Describe what is meant by the fusion splicing of optical fibers. Discuss the advantages and drawbacks of this jointing technique.

A multimode step index fiber with a core refractive index of 1.52 is fusion spliced. The splice exhibits an insertion loss of 0.8 dB. This insertion loss is found to be entirely due to the angular misalignment of the fiber core axes which is 7°. Determine the numerical aperture of the fiber.

4.19 Describe, with the aid of suitable diagrams, three common techniques used for the mechanical splicing of optical fibers.

A mechanical splice in a multimode step index fiber has a lateral offset of 16% of the fiber core radius. The fiber core has a refractive index of 1.49, and an index matching fluid with a refractive index of 1.45 is inserted in the splice between the butt jointed fiber ends. Assuming no longitudinal or angular misalignment, estimate the insertion loss of the splice.

4.20 Discuss the principles of operation of the two major categories of demountable optical fiber connector. Describe in detail a common technique for achieving a butt jointed fiber connector.

A butt jointed fiber connector used on a multimode step index fiber with a core refractive index of 1.42 and a relative refractive index difference of 1% has an angular misalignment of 9°. There is no longitudinal or lateral misalignment but there is a small air gap between the fibers in the connector. Estimate the insertion loss of the connector.

4.21 Briefly describe the types of demountable connector that may be used with single mode fibers. Further, indicate the problems involved with the connection of single mode fibers.

A single mode fiber connector is used with a 6 μm core diameter silica (refractive index 1.46) step index fiber which has a normalized frequency of 2.2 and a numerical aperture of 0.09. The connector has a lateral offset of 0.7 μm and an angular misalignment of 0.8°. Estimate the total insertion loss of the connector assuming that the joint is index matched and that there is no longitudinal misalignment.

4.22 A 10 μm core diameter single mode fiber has a normalized frequency of 2.0. A fusion splice at a point along its length exhibits an insertion loss of 0.15 dB. Assuming only lateral misalignment contributes to the splice insertion loss, estimate the magnitude of the lateral misalignment.

4.23 A 5 μm core diameter single mode step index fiber has a normalized frequency of 1.7, a core refractive index of 1.48 and a numerical aperture of 0.14. The loss in decibels due to angular misalignment at a fusion splice with a lateral offset of 0.4 μm is twice that due to the lateral offset. Estimate the magnitude in degrees of the angular misalignment.

4.24 Given the following parameters for a single mode step index fiber with a fusion splice estimate (a) the fiber core diameter; and (b) the numerical aperture for the fiber.

> Fiber normalized frequency = 1.9
> Fiber core refractive index = 1.46
> Splice lateral offset = 0.5 μm
> Splice lateral offset loss = 0.05 dB
> Splice angular misalignment = 0.3°
> Splice angular misalignment loss = 0.04 dB

Answers to Numerical Problems

4.7	7.43×10^4 psi, 0.6%	**4.16**	4.0 μm
4.8	0.2 μm	**4.17**	0.71 dB
4.9	2.61×10^6 psi	**4.18**	0.35
4.11	(a) 0.31 dB;	**4.19**	0.47 dB
	(b) 3.8×10^{-4} dB	**4.20**	1.51 dB
4.12	1.59	**4.21**	0.54 dB
4.13	0.29 dB	**4.22**	1.2 μm
4.14	0.67 dB	**4.23**	0.65°
4.15	(a) 0.19 dB;	**4.24**	(a) 7.0 μm;
	(b) 0.17 dB; difference 0.02 dB		(b) 0.10

REFERENCES

1 S. Tanaka, K. Inada, T. Akimoko and M. Kozima, 'Silicone-clad fused-silica-core fiber', *Electron. Lett.*, **11**(7), pp. 153–154, 1975.

2 K. J. Beales and C. R. Day, 'A review of glass fibers for optical communications', *Phys. and Chem. of Glass*, **21**(1), pp. 5–21, 1980.

3 T. Yamazuki and M. Yoshiyagawa, 'Fabrication of low-loss, multicomponent glass fibers with graded index and pseudo-step-index Borosilicate compound glass fibers', *Digest of International Conference on Integrated Optics and Optical Fiber Communication*, Osaka (Tokyo, IEEE, Japan), pp. 617–620, 1977.

4 K. J. Beales, C. R. Day, W. J. Duncan, J. E. Midwinter and G. R. Newns, 'Preparation of sodium borosilicate glass fibers for optical communication', *Proc. IEE (London)*, **123**, pp. 591–595, 1976.

5 G. R. Newns, P. Pantelis, J. L. Wilson, R. W. J. Uffen and R. Worthington, 'Absorption losses in glasses and glass fiber waveguides', *Opto-Electron*, **5**, pp. 289–296, 1973.

6 B. Scott and H. Rawson, 'Techniques for producing low loss glasses for optical fibre communications system', *Glass Technology*, **14**(5), pp. 115–124, 1973.

7 C. E. E. Stewart, D. Tyldesley, B. Scott, H. Rawson and G. R. Newns, 'High-purity glasses for optical-fibre communication', *Electron. Lett.*, **9**(21), pp. 482–483, 1973.

8 B. Scott and H. Rawson, 'Preparation of low loss glasses for optical fiber communication', *Opto-Electronics*, **5**(4), pp. 285–288, 1973.

9 N. S. Kapany, *Fiber Optics*, Academic Press, 1967.

10 A. M. Reid, W. W. Harper and A. Forbes, British Patent 50543, 1967.

11 B. P. Pal, 'Optical communication, fiber waveguide fabrication: a review', *Fiber Int. Opt.*, **2**(2), pp. 195–252, 1979.

12 G. R. Newns, 'Compound glass optical fibres', *2nd European Conference on Optical Fiber Communication* (Paris), pp. 21–26, 1976.

13 K. J. Beales, C. R. Day, W. J. Duncan, A. G. Dunn, P. L. Dunn, G. R. Newns and J. V. Wright, 'Low loss graded index fiber by the double crucible technique', *5th European Conference on Optical Fiber Communication* (Amsterdam), paper 3.2, 1979.

14 K. J. Beales, C. R. Day, W. J. Duncan and G. R. Newns, 'Low-loss compound-glass optical fibre', *Electron. Lett.*, **13**(24), pp. 755–756, 1977.

15 G. A. C. M. Spierings, T. P. M. Meeuwsen, F. Meyer, P. J. W. Severin and C. M. G. Jochem, 'Some aspects of the preparation of alkali lime germosilicate optical fibres', *Phys. Chem. Glasses* (GB), **21**(1), pp. 30–31, 1980.

16 H. Lydtin and F. Meyer, 'Review of techniques applied in optical fibre preparation', *Acta Electron.*, **22**(3), pp. 225–235, 1979.

17 J. F. Hyde, US Patent 2 272 342, 1942.

18 F. P. Kapron, D. B. Keck and R. D. Maurer, 'Radiation losses in optical waveguides', *Appl. Phys. Lett.*, **10**, pp. 423–425, 1970.

19 B. Bendow and S. S. Mitra, *Fiber Optics*, Plenum Press, 1979.

20 D. B. Keck and R. Bouillie, 'Measurements on high-bandwidth optical waveguides', *Optics Commun.*, **25**, pp. 43–48, 1978.

21 B. S. Aronson, D. R. Powers and R. Sommer, 'Chlorine drying of doped deposited silica preform simultaneous to consolidation', *Technical Digest of Topical Meeting on Optical Fiber Communication*, Washington, DC, p. 42, 1979.

22 T. Izawa, T. Miyashita and F. Hanawa, US Patent 4 062 665, 1977.

23 S. Sudo, M. Kawachi, M. Edahiro, T. Izawa, T. Shoida and H. Gotoh, 'Low-OH-content optical fiber fabricated by vapor-phase axial-deposition method', *Electron. Lett.*, **14**(17), pp. 534–535, 1978.

24 T. Izawa, S. Sudo and F. Hanawa, 'Continuous fabrication process for high-silica fiber preforms (vapor phase axial deposition)', *Trans. Inst. Electron. Commun. Eng. Jpn. Section E* (Japan), **E62**(11), pp. 779–785, 1979.

25 D. B. Keck and P. C. Schultz, US Patent 3 711 262, 1973.

26 W. G. French, J. B. MacChesney, P. B. O'Conner and G. W. Tasker, 'Optical waveguides with very low losses', *Bell Syst. Tech. J.*, **53**, pp. 951–954, 1974.

27 D. N. Payne and W. A. Gambling, 'New silica-based low-loss optical fibres', *Electron. Lett.*, **10**(15), pp. 289–290, 1974.

28 T. Miya, Y. Terunuma, T. Mosaka and T. Miyashita, 'Ultimate low-loss single-mode fibre at 1.55 µm', *Electron. Lett.*, **15**(4), pp. 106–108, 1979.

29 D. Gloge, 'The optical fibre as a transmission medium', *Rep. Prog. Phys.*, **42**, pp. 1778–1824, 1979.

30 S. R. Nagel, J. B. MacChesney and K. L. Walker, 'An overview of the modified chemical vapour deposition (MCVD) process and performance', *IEEE J. Quantum Electron.*, **QE-18**(4), pp. 459–477, 1982.

31 C. Lin, P. L. Lin, T. P. Lee, C. A. Burrus, F. T. Stone and A. J. Ritger, 'Measuring high bandwidth fibres in the 1.3 µm region with picosecond InGa injection lasers and ultrafast InGaAs detectors', *Electron. Lett.*, **17**(13), pp. 438–440, 1981.

32 D. Kuppers and J. Koenings, 'Preform fabrication by deposition of thousands of layers with the aid of plasma activated CVD', *2nd European Conference on Optical Fiber Communication* (Paris), p. 49, 1976.

33 R. E. Jaeger, J. B. MacChesney and T. J. Miller, 'The preparation of optical waveguide preforms by plasma deposition', *Bell Syst. Tech. J.*, **57**, pp. 205–210, 1978.

34 J. Irven and A. Robinson, 'Optical fibres prepared by plasma augmented vapour deposition', *Electron. Lett.*, **15**(9), pp. 252–254, 1979.

35 N. Nobukazu, 'Recent progress in glass fibers for optical communication', *Jap. J. Appl. Phys.*, **20**(8), pp. 1347–1360, 1981.

36 W. G. French, R. E. Jaeger, J. B. MacChesney, S. R. Nagel, K. Nassau and A. D. Pearson, 'Fiber preform preparation', in S. E. Miller and A. G. Chynoweth (Eds.), *Optical Fiber Telecommunications*, pp. 233–261, Academic Press, 1979.

37 W. A. Gambling, A. H. Hartog and C. M. Ragdale, 'Optical fibre transmission lines', *Radio Electron. Eng. (IERE J.)*, **51**(7/8), pp. 313–325, 1981.

38 P. W. Black, J. Irven and J. Titchmarsh, 'Fabrication of optical fibre waveguides', in C. P. Sandbank (Ed.), *Optical Fibre Communication Systems*, pp. 42–69, John Wiley, 1980.

39 J. B. MacChesney, 'Materials and processes for preform fabrication—modified chemical vapour deposition and plasma chemical vapour deposition', *Proc. IEEE*, **68**(10), pp. 1181–1184, 1980.

40 J. McDermott, 'Fiber-optic-cable choices expand to fill design needs', *EDN*, pp. 95–99, May 1981.

41 S. Ohr and S. Adlerstein, 'Fiber optics is growing strong—better connectors, cable will speed things up', *Electronic Design*, **27**(23), pp. 42–50, 52, 1979.

42 *Fibre-optics Summary Catalogue*, STC Electro-optic Components, UK.

43 *Fibre Optic Components and Systems*, Belling Lee Limited, Electronic Components Group, UK.

44 C. K. Koa, 'Optical fibres and cables', in M. J. Howes and D. V. Morgan (Eds.), *Optical Fibre Communications, Devices, Circuits and Systems*, pp. 189–249, John Wiley, 1980.

45 K. J. Beales, C. R. Day, A. G. Dunn and S. Partington, 'Multicomponent glass fibers for optical communications', *Proc. IEEE*, **68**(10), pp. 1191–1194, 1980.
46 G. De Loane, 'Optical fibre cables', *Telecomm. J. (Eng. Ed.) Switzerland*, **48**(11), pp. 649–656, 1981.
47 G. Galliano and F. Tosco, 'Optical fibre cables', *Optical Fibre Communications*, by Technical Staff of CSELT, pp. 501–540, McGraw-Hill, 1981.
48 T. Nakahara and N. Uchida, 'Optical cable design and characterization in Japan', *Proc. IEEE*, **68**(10), pp. 1220–1226, 1980.
49 A. A. Griffith, 'Phenomena of rupture and flow in solids', *Phil. Trans. R. Soc. Lond. Ser. A*, **221**, pp. 163–168, 1920.
50 W. Weibull, 'A statistical theory of the strength of materials', *Proc. R. Swedish Inst. Res.* No. 151, publication no. 4, 1939.
51 M. H. Reeve, 'Optical fibre cables', *Radio Electron. Eng. (IERE J.)*, **51**(7/8), pp. 327–332, 1981.
52 P. R. Bank and D. O. Lawrence, 'Emerging standards in fiber optic telecommunications cable', *Proc. Soc. Photo-optical Instrum. Eng.*, **224**, pp. 149–158, 1980.
53 J. C. Harrison, 'The metal foil/polyethylene cable sheath and its use in the Post Office', Institution of Post Office Engineers Paper No. 229, 1968.
54 J. A. Olszewski, G. H. Foot and Y. Y. Huang, 'Development and Installation of an optical-fiber cable for communications', *IEEE Trans. Commun.*, **COM-26**(7), pp. 991–998, 1978.
55 M. I. Schwartz, P. F. Gagen and M. R. Santana, 'Fiber cable design and characterization', *Proc. IEEE*, **68**(10), pp. 1214–1219, 1980.
56 J. E. Midwinter, 'Studies of monomode long wavelength fiber systems at the British Telecom Research Laboratories', *IEEE J. Quantum Electron.*, **QE-17**(6), pp. 911–918, 1981.
57 M. Born and W. Wolf, *Principles of Optics*, (6th Edn.), Pergamon Press, 1980.
58 P. Mossman, 'Connectors for optical fibre systems', *Radio Electron. Eng. (J. IERE)*, **51**(7/8), pp. 333/340, 1981.
59 J. S. Leach, M. A. Matthews and E. Dalgoutte, 'Optical fibre cable connections', in C. P. Sandbank (Ed.), *Optical Fibre Communication Systems*, pp. 86–105, John Wiley, 1980.
60 F. L. Thiel and R. M. Hawk, 'Optical waveguide cable connection', *Appl. Opt.* **15**(11), pp. 2785–2791, 1976.
61 K. Miyazaki *et al.*, 'Theoretical and experimental considerations of optical fiber connector', OSA Topical Meeting on Opt. Fiber Trans. Williamsburg, Va, paper WA 4-1, 1975.
62 H. Tsuchiya, H. Nakagome, N. Shimizu and S. Ohara, 'Double eccentric connectors for optical fibers', *Appl. Opt.*, **16**(5), pp. 1323–1331, 1977.
63 K. J. Fenton and R. L. McCartney, 'Connecting the thread of light', *Electronic Connector Study Group Symposium, 9th Annual Symposium Proc.*, p. 63, Cherry Hill, NJ, 1976.
64 C. M. Miller, Transmission vs transverse offset for parabolic-profile fiber splices with unequal core diameters', *Bell Syst. Tech. J.*, **55**(7), pp. 917–927, 1976.
65 D. Gloge, 'Offset and tilt loss in optical fiber splices', *Bell Syst. Tech. J.*, **55**(7), pp. 905–916, 1976.
66 C. M. Miller and S. C. Mettler, 'A loss model for parabolic-profile fiber splices', *Bell Syst. Tech. J.*, **57**(9), pp. 3167–3180, 1978.
67 J. S. Cook, W. L. Mammal and R. J. Crow, 'Effect of misalignment on coupling efficiency of single-mode optical fibre butt joints', *Bell Syst. Tech. J.*, **52**(8), pp. 1439–1448, 1973.
68 D. Marcuse, 'Loss analysis of single-mode fiber splices', *Bell Syst. Tech. J.*, **56**(5), pp. 703–718, 1977.

69 W. A. Gambling, H. Matsumura and A. G. Cowley, 'Jointing loss in single-mode fibres', *Electron. Lett.*, **14**(3), pp. 54–55, 1978.

70 W. A. Gambling, H. Matsumura and C. M. Ragdale, 'Joint loss in single-mode fibres', *Electron. Lett.*, **14**(15), pp. 491–493, 1978.

71 J. J. Esposito, 'Optical connectors, couplers and switches', in H. F. Wolf (Ed.), *Handbook of Fiber Optics, Theory and Applications*, pp. 241–303, Granada, 1979.

72 J. F. Dalgleish, 'Connections', *Electronics*, pp. 96–98, 5 Aug. 1976.

73 D. Botez and G. J. Herskowitz, 'Components for optical communications systems: a review', *Proc. IEEE*, **68**(6), pp. 689–731, 1980.

74 Y. Ushui, T. Ohshima, Y. Toda, Y. Kato and M. Tateda, 'Exact splice loss prediction for single-mode fiber', *IEEE J. Quantum Electron.*, **QE-18**(4), pp. 755–757, 1982.

75 G. Coppa and P. Di Vita, 'Length dependence of joint losses in multimode optical fibres', *Electron. Lett.*, **18**(2), pp. 84–85, 1982.

76 K. Petermann, 'Nonlinear distortions due to fibre connectors', *Proceedings of 6th European Conference on Optical Communication* (UK), pp. 80–83, 1980.

77 K. Petermann, 'Wavelength-dependent transmission at fibre connectors', *Electron. Lett.*, **15**(22), pp. 706–708, 1979.

78 M. Ikeda, Y. Murakami and K. Kitayama, 'Mode scrambler for optical fibers', *Appl. Opt.*, **16**(4), pp. 1045–1049, 1977.

79 N. Kashima and N. Uchida, 'Relation between splice loss and mode conversion in a graded-index optical fibre', *Electron. Lett.*, **15**(12), pp. 336–338, 1979.

80 R. B. Kummer, 'Precise characterization of long nonidentical-fiber splice loss effects', *Proceedings of 6th European Conference on Optical Communication* (UK), pp. 302–304, 1980.

81 A. H. Cherin and J. F. Dalgleish, 'Splices and connectors for optical fibre communications', *Telecommun. J. (Eng. Ed.) Switzerland*, **48**(11), pp. 657–665, 1981.

82 J. E. Midwinter, *Optical Fibers for Transmission*, John Wiley, 1979.

83 E. A. Lacy, *Fiber Optics*, Prentice-Hall, 1982.

84 R. Jocteur and A. Tardy, 'Optical fiber splicing with plasma torch and oxyhydric microburner', *2nd European Conference on Optical Fibre Communication* (Paris), 1976.

85 I. Hatakeyama and H. Tsuchiya, 'Fusion splices for single-mode optical fibers', *IEEE J. Quantum Electron.*, **QE-14**(8), pp. 614–619, 1978.

86 M. Hirai and N. Uchida, 'Melt splice of multimode optical fibre with an electric arc', *Electron. Lett.*, **13**(5), pp. 123–125, 1977.

87 M. Tsuchiya and I. Hatakeyama, 'Fusion splices for single-mode optical fibres', *Optical Fiber Transmission II*, Williamsburg, pp. PD1, 1–4, Feb. 1977.

88 F. Esposto and E. Vezzoni, 'Connecting and splicing techniques', *Optical Fibre Communication*, by Technical Staff of CSELT, pp. 541–643, McGraw-Hill, 1981.

89 D. B. Payne, D. J. McCartney and P. Healey, 'Fusion splicing of a 31.6 km monomode optical fibre system', *Electron. Lett.*, **18**(2), pp. 82–84, 1982.

90 D. R. Briggs and L. M. Jayne, 'Splice losses in fusion-spliced optical waveguide fibers with different core diameters and numerical apertures', *Proceedings of 27th International Wire and Cable Symposium*, pp. 356–361, 1978.

91 I. Hatakeyama, M. Tachikura and H. Tsuchiya, 'Mechanical strength of fusion-spliced optical fibres', *Electron. Lett.*, **14**(19), pp. 613–614, 1978.

92 C. K. Pacey and J. F. Dalgleish, 'Fusion splicing of optical fibres', *Electron. Lett.*, **15**(1), pp. 32–34, 1978.

93 J. Cook and P. K. Runge, 'Optical fiber connectors', in S. E. Miller (Ed.), *Optical Fiber Telecommunications*, pp. 483–497, Academic Press, 1979.

94 T. G. Giallorenzi, 'Optical communications research and technology', *Proc. IEEE*, **66**(7), pp. 744–780, 1978.

95 K. Nawata, Y. Iwahara and N. Suzuki, 'Ceramic capillary splices for optical fibres', *Electron. Lett.*, **15**(15), pp. 470–472, 1979.

96 C. M. Miller, 'Loose tube splice for optical fibres', *Bell Syst. Tech. J.*, **54**(7), pp. 1215–1225, 1975.

97 D. Gloge, A. H. Cherin, C. M. Miller and P. W. Smith, 'Fiber splicing', in S. E. Miller (Ed.), *Optical Fiber Telecommunications*, pp. 455–482, Academic Press, 1979.

98 D. G. Dalgoutte, 'Collapsed sleeve splices for field jointing of optical fibre cable', *Proceedings of 3rd European Conference on Optical Communication* (Munich), pp. 106–108, 1977.

99 P. Hensel, J. C. North and J. H. Stewart, 'Connecting optical fibers', *Electron. Power*, **23**(2), pp. 133–135, 1977.

100 A. R. Tynes and R. M. Derosier, 'Low-loss splices for single-mode fibres,' *Electron. Lett.*, **13**(22), pp. 673–674, 1977.

101 Y. Toda, O. Watanabe, M. Ogai and S. Seika, 'Low-loss fusion splice of single mode fibre', *Internat. Commun. Conf. (IEEE)*, paper 27.7, 1981.

102 E. L. Chinnock, D. Gloge, D. L. Bisbee and P. W. Smith, 'Preparation of optical fiber ends for low-loss tape splices', *Bell Syst. Tech. J.*, **54**, pp. 471–477, 1975.

103 C. M. Miller, 'Fiber optic array splicing with etched silicon chips', *Bell Syst. Tech. J.*, **57**(1), pp. 75–90, 1978.

104 G. Le Noane, 'Optical fibre cable and splicing technique', *2nd European Conference on Optical Fiber Communication* (Paris), pp. 247–252, 1976.

105 W. C. Young, P. Kaiser, N. K. Cheung, L. Curtis, R. E. Wagner and D. M. Folkes, 'A transfer molded biconic connector with insertion losses below 0.3 dB without index match', *Proceedings of 6th European Conference on Optical Communication*, pp. 310–313, 1980.

106 G. Le Noane, 'Low-loss optical-fibre connection systems', *Electron. Lett.*, **15**(1), pp. 12–13, 1979.

107 N. Suzuki and K. Nawata, 'Demountable connectors for optical fibre transmission equipment', *Rev. Elect. Commun. Labs (NTT)*, **27**(11–12), pp. 999–1009, 1979.

108 N. Suzuki, Y. Iwahara, M. Saruwatari and K. Nawata, 'Ceramic capillary connector for 1.3 μm single-mode fibres', *Electron. Lett.*, **15**(25), pp. 809–810, 1979.

109 P. Hensel, 'Triple-ball connector for optical fibres', *Electron. Lett.*, **13**(24), pp. 734–735, 1977.

110 N. Shimuzu and H. Tsuchiya, 'Single-mode fibre connection', *Electron. Lett.*, **14**(19), pp. 611–613, 1978.

111 N. Shimizu, H. Tsuchiya and T. Izawa, 'Low-loss single-mode fibre connectors', *Electron. Lett.*, **15**(1), pp. 28–29, 1978.

112 P. W. Smith, D. L. Bisbee, D. Gloge and E. L. Chinnock, 'A moulded-plastic technique for connecting and splicing optical fibre tapes and cables', *Bell Syst. Tech. J.*, **54**(6) pp. 971–984, 1975.

113 Y. Fujii, J. Minowa and N. Suzuki, 'Demountable multiple connector with precise V-grooved silicon', *Electron. Lett.*, **15**(14), pp. 424–425, 1979.

114 A. Nicia, 'Practical low-loss lens connector for optical fibres', *Electron. Lett.*, **14**(16), pp. 511–512, 1978.

115   D. B. Payne and C. A. Miller, 'Triple-ball connector using fibre-bead location', *Electron. Lett.*, **16**(1), pp. 11–12, 1980.

116   C. A. Miller and D. B. Payne, 'Monomode fibre connector using fibre bead location', *Proceedings of 6th European Conference on Optical Communication* (UK), pp. 306–309, 1980.

117   D. B. Payne and D. J. McCartney, 'Splicing and connectors for single-mode fibres', *Internat. Commun. Conf. (IEEE)*, paper 27.6, 1981.

118   M. A. Bedgood, J. Leach and M. Matthews, 'Demountable connectors for optical fiber systems', *Elect. Commun.*, **51**(2), pp. 85–91, 1976.

119   S. Nagasawa and H. Murata, 'Optical fibre connectors using a fused and drawn multi-glass-rod arrangement', *Electron. Lett.*, **17**(7), pp. 268–270, 1981.

120   G. Khoe, H. G. Kock, D. Kuppers, J. H. F. M. Poulissen and H. M. De Vrieze, 'Progress in monomode optical-fiber interconnection devices', *J. Lightwave Technol.*, **LT-2** (3), pp. 217–227, 1984.

5

Optical Fiber Measurements

5.1 INTRODUCTION

In this chapter we are primarily concerned with measurements on optical fibers which characterize the fiber. These may be split into three main areas:

(a) transmission characteristics;
(b) geometrical and optical characteristics;
(c) mechanical characteristics.

Data in these three areas are usually provided by the optical fiber manufacturer with regard to specific fibers. Hence fiber measurements are generally performed in the laboratory and techniques have been developed accordingly. This information is essential for the optical communication system designer in order that suitable choices of fibers, materials and devices may be made with regard to the system application. However, although the system designer and system user do not usually need to take fundamental measurements of the fiber characteristics there is a requirement for field measurements in order to evaluate overall system performance, and for functions such as fault location. Therefore we also include some discussion of field measurements which take into account the effects of cabled fiber, splice and connector losses, etc.

There are a number of major techniques used for the laboratory measurement of the various fiber characteristics. The transmission characteristics of greatest interest are those of optical attenuation and dispersion (bandwidth), whereas the important geometrical and optical characteristics include size (core and cladding diameters), numerical aperture and refractive index profile. Measurements of the mechanical characteristics such as tensile strength and durability were outlined in Section 4.6.1 and are therefore pursued no further in this chapter.

When attention is focused on the measurement of the transmission properties of multimode fibers, problems emerge regarding the large number of modes propagating in the fiber. The various modes show individual differences with regard to attenuation and dispersion within the fiber. Moreover, mode coupling occurs giving transfer of energy from one mode to another (see Section 2.3.7). The mode coupling which is associated with perturbations in the fiber composition or geometry, and external factors such as microbends or splices, is for instance responsible for the increased attenuation (due to radiation) of the higher order modes. These multimode propagation effects mean that both the fiber loss and bandwidth are not uniquely defined parameters but depend upon the fiber excitation conditions and environmental factors such as cabling, bending, etc. Also these transmission parameters may vary along the fiber length (i.e. they are not necessarily linear functions) due to the multimode propagation effects, making extrapolation of measured data to different fiber lengths less than meaningful.

It is therefore important that transmission measurements on multimode fibers are performed in order to minimize these uncertainties. In the laboratory, measurements are usually taken on continuous lengths of uncabled fiber in order to reduce the influence of external factors on the readings (this applies to both multimode and single mode fibers). However, this does mean that the system designer must be aware of the possible deterioration in the fiber transmission characteristics within the installed system. The multimode propagation effects associated with fiber perturbations may be accounted for by allowing or encouraging the mode distribution to reach a steady-state (equilibrium) distribution. This distribution occurs automatically after propagation has taken place over a certain fiber length (coupling length) depending upon the strength of the mode coupling within the particular fiber. At equilibrium the mode distribution propagates unchanged and hence the fiber attenuation and dispersion assume well-defined values. These values of the transmission characteristics are considered especially appropriate for the interpretation of measurements to long-haul links and do not depend on particular launch conditions.

The equilibrium mode distribution may be achieved by launching the optical signal through a long (dummy) fiber to the fiber under test. This technique has been used to good effect [Ref. 1] but may require several kilometers of dummy fiber and is therefore not suitable for dispersion measurements. Alternatively there are a number of methods of simulating the equilibrium mode distribution with a much shorter length of fiber. Mode equilibrium may be achieved using an optical source with a mode output which corresponds to the steady-state mode distribution of the fiber under test. This technique may be realized experimentally using an optical arrangement which allows the numerical aperture of the launched beam to be varied (using diaphragms) as well as the spot size of the source (using pinholes). In this case the input light beam is given an angular width which is equal to the equilibrium distribution numerical aperture of the fiber and the source spot size on the fiber input face is matched to the optical power distribution in a cross section of the fiber at equilibrium.

Other techniques involve the application of strong mechanical perturbations on a short section of the fiber in order to quickly induce mode coupling and hence equilibrium mode distribution within 1 m. These devices which simulate mode equilibrium over a short length of fiber are known as mode scramblers or mode mixers. A simple method [Ref. 2] is to sandwich the fiber between two sheets of abrasive paper (i.e. sandpaper) placed on wooden blocks in order to provide a suitable pressure. Two slightly more sophisticated techniques are illustrated in Fig. 5.1 [Refs. 3 and 4].

Figure 5.1(a) shows mechanical perturbations induced by enclosing the fiber with metal wires and applying pressure by use of a surrounding heat shrinkable tube. A method which allows adjustment and therefore an improved probability of repeatable results is shown in Fig. 5.1(b). This technique involves inserting the fiber between a row of equally spaced pins, subjecting it to sinusoidal bends. Hence the variables are the number of pins giving the number of periods, the pin diameter d and the pin spacing s.

In order to test that a particular mode scrambler gives an equilibrium mode distribution within the test fiber, it is necessary to check the insensitivity of the far field radiation pattern (this is related to the mode distribution, see Section 2.3.6) from the fiber with regard to changes in the launch conditions. It is also useful to compare the far field patterns from the mode scrambler and a separate long length at the test fiber for coincidence [Ref. 1]. However, it must be noted that at present mode scramblers tend to give only an approximate equilibrium mode distribution and their effects vary with different fiber types. Hence measurements involving the use of different mode scrambling methods can be subject to discrepancies. Nevertheless, the majority of laboratory measurement techniques to ascertain the transmission characteristics of multi-mode optical fibers use some form of equilibrium mode simulation in order to give values representative of long transmission lines.

We commence the discussion of optical fiber measurements in Section 5.2 by dealing with the major techniques employed in the measurement of fiber attenuation. These techniques include measurement of both total fiber attenuation and the attenuation resulting from individual mechanisms within the fiber (e.g. material absorption, scattering). In Section 5.3 fiber dispersion measurements in both the time and frequency domains are discussed. Various techniques for the measurement of the fiber refractive index profile are then con-

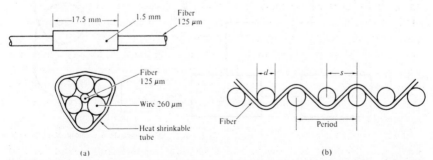

Fig. 5.1   Mode scramblers; (a) heat shrinking technique [Ref. 3]; (b) bending technique [Ref. 4].

sidered in Section 5.4. In Section 5.5 we discuss two simple methods for measuring the fiber numerical aperture. Measurement of the fiber outer diameter is then dealt with in Section 5.6. Finally, field measurements which may be performed on optical fiber links, together with examples of measurement instruments, are discussed in Section 5.7. Particular attention is paid in this concluding section to optical time domain reflectometry (OTDR).

5.2 FIBER ATTENUATION MEASUREMENTS

Fiber attenuation measurements techniques have been developed in order to determine the total fiber attenuation of the relative contributions to this total from both absorption losses and scattering losses. The overall fiber attenuation is of greatest interest to the system designer, but the relative magnitude of the different loss mechanisms is important in the development and fabrication of low loss fibers. Measurement techniques to obtain the total fiber attenuation give either the spectral loss characteristic (see Fig. 3.3) or the loss at a single wavelength (spot measurement).

5.2.1 Total Fiber Attenuation

A commonly used technique for determining the total fiber attenuation per unit length is the cut back or differential method. Figure 5.2 shows a schematic

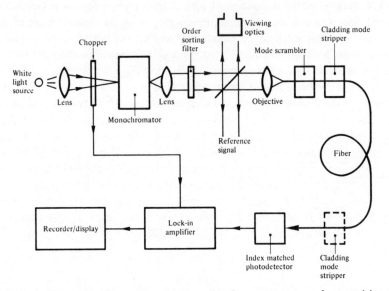

Fig. 5.2   A typical experimental arrangement for the measurement of spectral loss in optical fibers using the cut back technique.

diagram of the typical experimental set-up for measurement of the spectral loss to obtain the overall attenuation spectrum for the fiber. It consists of a 'white' light source, usually a tungsten halogen or xenon arc lamp. The focused light is mechanically chopped at a low frequency of a few hundred hertz. This enables the lock-in amplifier at the receiver to perform phase-sensitive detection. The chopped light is then fed through a monochromator which utilizes a prism or diffraction grating arrangement to select the required wavelength at which the attenuation is to be measured. Hence the light is filtered before being focused onto the fiber by means of a microscope objective lens. A beam splitter may be incorporated before the fiber to provide light for viewing optics and a reference signal used to compensate for output power fluctuations. As indicated in Section 5.1, when the measurement is performed on multimode fibers it is very dependent on the optical launch conditions. Therefore unless the launch optics are arranged to give the steady-state mode distribution at the fiber input, or a dummy fiber is used, then a mode scrambling device is attached to the fiber within the first meter. The fiber is also usually put through a cladding mode stripper, which may consist of an S-shaped groove cut in the Teflon and filled with glycerin. This device removes light launched into the fiber cladding through radiation into the index matched (or slightly higher refractive index) glycerin. A mode stripper can also be included at the fiber output end to remove any optical power which is scattered from the core into the cladding down the fiber length. This tends to be pronounced when the fiber cladding consists of a low refractive index silicone resin.

The optical power at the receiving end of the fiber is detected using a p–i–n or avalanche photodiode. In order to obtain reproducible results the photodetector surface is usually index matched to the fiber output end face using epoxy resin or an index matching cell [Ref. 5]. Finally the electrical output from the photodetector is fed to a lock-in amplifier; the output of which is recorded.

The cut back method involves taking a set of optical output power measurements over the required spectrum using a long length of fiber (usually at least a kilometer). This fiber is generally uncabled having only a primary protective coating. Increased losses due to cabling (see Section 4.6.2) do not tend to change the shape of the attenuation spectrum as they are entirely radiative, which for multimode fibers are almost wavelength independent. The fiber is then cut back to a point a few meters (e.g. 3 m) from the input end and, maintaining the same launch conditions, another set of power output measurements are taken. The following relationship for the optical attenuation per unit length α_{dB} for the fiber may be obtained from Eq. (3.3):

$$\alpha_{dB} = \frac{10}{L_1 - L_2} \log_{10} \frac{P_{02}}{P_{01}} \tag{5.1}$$

L_1 and L_2 are the original and cut back fiber lengths respectively, and P_{01} and

P_{02} are the corresponding output optical powers at a specific wavelength from the original and cut back fiber lengths. Hence when L_1 and L_2 are measured in kilometers, α_{dB} has units of dB km^{-1}.

Furthermore Eq. (5.1) may be written in the form:

$$\alpha_{dB} = \frac{10}{L_1 - L_2} \log_{10} \frac{V_2}{V_1} \tag{5.2}$$

where V_1 and V_2 correspond to output voltage readings from the original fiber length and the cut back fiber length respectively. The electrical voltages V_1 and V_2 may be directly substituted for the optical powers P_{01} and P_{02} of Eq. (5.1) as they are directly proportional to these optical powers (see Section 7.4.3). The accuracy of the results obtained for α_{dB} using this method is largely dependent on constant optical launch conditions and the achievement of the equilibrium mode distribution within the fiber. It is indicated [Refs. 6 and 7] that for constant launch conditions α_{dB} may be determined with a precision of around ±0.1 dB over 1 km lengths of fiber. Alternatively the total uncertainty in the measured attenuation is quoted [Ref. 8] as $\pm0.2/(L_1 - L_2)$ dB km^{-1} where L_1 and L_2 are in kilometers. Hence the approximate uncertainty for a 1 km fiber length is ±0.2 dB km^{-1}.

Example 5.1

A 2 km length of multimode fiber is attached to apparatus for spectral loss measurement. The measured output voltage from the photoreceiver using the full 2 km fiber length is 2.1 V at a wavelength of 0.85 μm. When the fiber is then cut back to leave a 3 m length the output voltage increases to 10.5 V. Determine the attenuation per kilometer for the fiber at a wavelength of 0.85 μm and estimate the accuracy of the result.

Solution: The attenuation per kilometer may be obtained from Eq. (5.2) where:

$$\alpha_{dB} = \frac{10}{L_1 - L_2} \log_{10} \frac{V_2}{V_1} = \frac{10}{1.997} \log_{10} \frac{10.5}{2.1}$$

$$= 3.5 \text{ dB km}^{-1}$$

The uncertainty in the measured attenuation may be estimated using:

$$\text{Uncertainty} = \frac{\pm0.2}{L_1 - L_2} = \frac{\pm0.2}{1.997} \approx \pm0.1 \text{ dB}$$

The dynamic range of the measurements that may be taken depends upon the exact configuration of the apparatus utilized, the optical wavelength and the fiber core diameter. However, a typical dynamic range is in the region 30–40 dB when using a white light source at a wavelength of 0.85 μm and multimode fiber with a core diameter around 50 μm. This may be increased to around 60 dB by use of a laser source operating at the same wavelength. It must be noted that a laser source is only suitable for making a single wavelength (spot) measurement as it does not emit across a broad band of spectral wavelengths.

Spot measurements may be performed on an experimental set-up similar to that shown in Fig. 5.2. However interference filters are frequently used instead of the monochromator in order to obtain a measurement at a particular optical wavelength. These provide greater dynamic range (10–15 dB improvement) than the monochromator but are of limited use for spectral measurements due to the reduced number of wavelengths that are generally available for measurement. A typical optical configuration for spot attenuation measurements is shown in Fig. 5.3. The interference filters are located on a wheel to allow measurement at a selection of different wavelengths. In the experimental arrangement shown in Fig. 5.3 the source spot size is defined by a pinhole and the beam angular width is varied by using different diaphragms. However, the electronic equipment utilized with this set-up is similar to that used for the spectral loss measurements illustrated in Fig. 5.2. Therefore determination of the optical loss per unit length for the fiber at a particular wavelength is performed in exactly the same manner, using the cut back method. Spot attenuation measurements are sometimes utilized after fiber cabling in order to obtain information on any degradation in the fiber attenuation resulting from the cabling process.

Although widely used, the cut back measurement method has the major drawback of being a destructive technique. Therefore, although suitable for

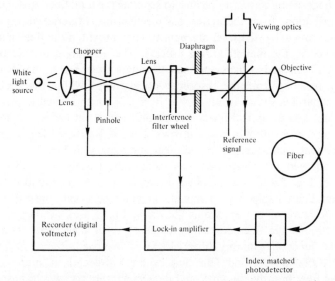

Fig. 5.3 An experimental arrangement for making spot (single wavelength) attenuation measurements using interference filters and employing the cut back technique.

laboratory measurement it is far from ideal for attenuation measurements in the field. Several nondestructive techniques exist which allow the fiber losses to be calculated through a single reading of the optical output power at the far end of the fiber after determination of the near end power level. The simplest is the insertion loss technique which utilizes the same experimental configuration as the cut back method. However, the fiber to be tested is spliced, or connected by means of a demountable connector, to a fiber with a known optical output at the wavelength of interest. When all the optical power is completely coupled between the two fibers, or when the insertion loss of the splice or connector are known, then the measurement of the optical output power from the second fiber gives the loss resulting from the insertion of this second fiber into the system. Hence the insertion loss due to the second fiber provides measurement of its attenuation per unit length. Unfortunately the accuracy of this measurement method is dependent on the coupling between the two fibers and is therefore somewhat uncertain.

The most popular nondestructive attenuation measurement technique for both laboratory and field use only requires access to one end of the fiber. It is the backscatter measurement method which uses optical time domain reflectometry and also provides measurement of splice and connector losses as well as fault location. Optical time domain reflectometry finds major use in field measurements and therefore is discussed in detail in Section 5.7.1.

5.2.2 Fiber Absorption Loss Measurement

It was indicated in the previous section that there is a requirement for the optical fiber manufacturer to be able to separate the total fiber attenuation into the contributions from the major loss mechanisms. Material absorption loss measurements allow the level of impurity content within the fiber material to be checked in the manufacturing process. The measurements are based on calorimetric methods which determine the temperature rise in the fiber or bulk material resulting from the absorbed optical energy within the structure.

The apparatus shown in Fig. 5.4(a) [Ref. 12] which is used to measure the absorption loss in optical fibers was modified from an earlier version which measured the absorption losses in bulk glasses [Ref. 13]. This temperature measurement technique, illustrated diagrammatically in Fig. 5.4(b), has been widely adopted for absorption loss measurements. The two fiber samples shown in Fig. 5.4(b) are mounted in capillary tubes surrounded by a low refractive index liquid (e.g. methanol) for good electrical contact, within the same enclosure of the apparatus shown in Fig. 5.4(a). A thermocouple is wound around the fiber containing capillary tubes using one of them as a reference junction (dummy fiber). Light is launched from a laser source (Nd:YAG or krypton ion depending on the wavelength of interest) through the main fiber (not the dummy), and the temperature rise due to absorption is measured by the thermocouple and indicated on a nanovoltmeter. Electrical

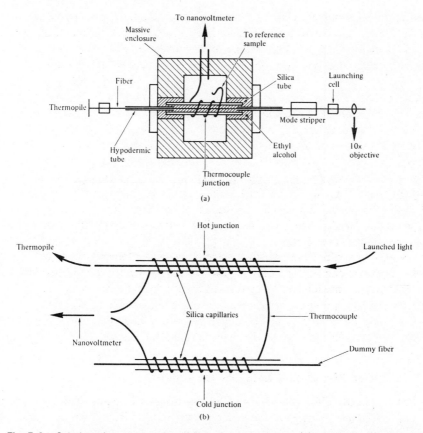

Fig. 5.4 Calorimetric measurement of fiber absorption losses: (a) schematic diagram of a version of the apparatus [Ref. 12]; (b) the temperature measurement technique using a thermocouple.

calibration may be achieved by replacing the optical fibers with thin resistance wires and by passing known electrical power through one. Independent measurements can then be made using the calorimetric technique and with electrical measurement instruments.

The calorimetric measurements provide the heating and cooling curve for the fiber sample used. A typical example of this curve is illustrated in Fig. 5.5(a). The attenuation of the fiber due to absorption α_{abs} may be determined from this heating and cooling characteristic. A time constant t_c can be obtained from a plot of $(T_\infty - T_t)$ on a logarithmic scale against the time t, an example of which shown in Fig. 5.5(c) was obtained from the heating characteristic displayed in Fig. 5.5(b) [Ref. 13]. T_∞ corresponds to the maximum temperature rise of the fiber under test and T_t is the temperature rise at a time t. It may be observed from Fig. 5.5(a) that T_∞ corresponds to a steady state temperature for the fiber when the heat loss to the surrounding

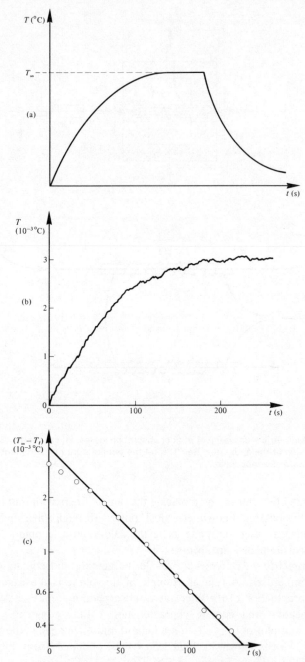

Fig. 5.5   (a) A typical heating and cooling curve for a glass fiber sample. (b) A heating curve and (c) the corresponding plot of $(T_\infty - T_t)$ against time for a sample glass rod (bulk material measurement). Reproduced with permission from K. I. White and. J. E. Midwinter, *Opto-electronics*, **5**, p. 323, 1973.

balances the heat generated in the fiber resulting from absorption at a particular optical power level. The time constant t_c may be obtained from the slope of the straight line plotted in Fig. 5.5(c) as:

$$t_c = \frac{t_2 - t_1}{\ln(T_\infty - T_{t_1}) - \ln(T_\infty - T_{t_2})} \tag{5.3}$$

where t_1 and t_2 indicate two points in time and t_c is a constant for the calorimeter which is inversely proportional to the rate of heat loss from the device.

From detailed theory it may be shown |Ref. 13| that the fiber attenuation due to absorption is given by:

$$\alpha_{abs} = \frac{CT_\infty}{P_{opt} t_c} \text{ dB km}^{-1} \tag{5.4}$$

where C is proportional to the thermal capacity per unit length of the silica capillary and the low refractive index liquid surrounding the fiber, and P_{opt} is the optical power propagating in the fiber under test. The thermal capacity per unit length may be calculated, or determined by the electrical calibration utilizing the thin resistance wire. Usually the time constant for the calorimeter t_c is obtained using a high absorption fiber which gives large temperature differences and greater accuracy. Once t_c is determined, the absorption losses of low loss test fibers may be calculated from their maximum temperature rise T_∞, using Eq. (5.4). The temperatures are measured directly in terms of the thermocouple output (microvolts), and the optical input to the test fiber is obtained by use of thermocouple or optical power meter.

Example 5.2

Measurements are made using a calorimeter and thermocouple experimental arrangement as shown in Fig. 5.4 in order to determine the absorption loss of an optical fiber sample. Initially a high absorption fiber is utilized to obtain a plot of $(T_\infty - T_t)$ on a logarithmic scale against t. It is found from the plot that the readings of $(T_\infty - T_t)$ after 10 and 100 seconds are 0.525 and 0.021 µV respectively.

The test fiber is then inserted in the calorimeter and gives a maximum temperature rise of 4.3×10^{-4} °C with a constant measured optical power of 98 mW at a wavelength of 0.75 µm. The thermal capacity per kilometer of the silica capillary and fluid is calculated to be 1.64×10^4 J °C^{-1}.

Determine the absorption loss in dB km^{-1}, at a wavelength of 0.75 µm, for the fiber under test.

Solution: Initially, the time constant for the calorimeter is determined from the measurements taken on the high absorption fiber using Eq. (5.3) where:

$$t_c = \frac{t_2 - t_1}{\ln(T_\infty - T_{t_1}) - \ln(T_\infty - T_{t_2})}$$

$$= \frac{100 - 10}{\ln (T_\infty - T_{10}) - \ln (T_\infty - T_{100})}$$

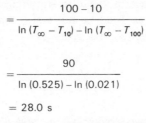

$$= \frac{90}{\ln (0.525) - \ln (0.021)}$$

$$= 28.0 \text{ s}$$

Then the absorption loss of the test fiber may be obtained using Eq. (5.4) where:

$$\alpha_{abs} = \frac{CT_\infty}{P_{opt} t_c} = \frac{1.64 \times 10^4 \times 4.3 \times 10^{-4}}{98 \times 10^{-3} \times 28.0}$$

$$= 2.6 \text{ dB km}^{-1}$$

Hence direct measurement of the contribution of absorption losses to the total fiber attenuation may be achieved. However, fiber absorption losses are often obtained indirectly from measurement of the fiber scattering losses (see the next section) by subtraction from the total fiber attenuation, measured by one of the techniques discussed in Section 5.2.1.

5.2.3 Fiber Scattering Loss Measurement

The usual method of measuring the contribution of the losses due to scattering within the total fiber attenuation is to collect the light scattered from a short length of fiber and compare it with the total optical power propagating within the fiber. Light scattered from the fiber may be detected in a scattering cell as illustrated in the experimental arrangement shown in Fig. 5.6. This may consist of a cube of six square solar cells (Tynes cell [Ref. 14]) or an integrating sphere

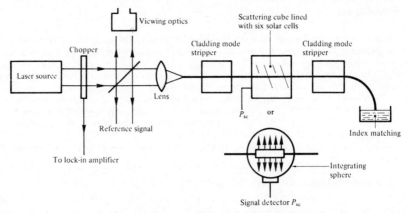

Fig. 5.6 An experimental set-up for measurement of fiber scattering loss illustrating both the solar cell cube and integrating sphere scattering cells.

and detector [Ref. 15]. The solar cell cube which contains index matching fluid surrounding the fiber gives measurement of the scattered light, but careful balancing of the detectors is required in order to achieve a uniform response. This problem is overcome in the integrating sphere which again usually contains index matching fluid but responds uniformly to different distributions of scattered light. However, the integrating sphere does exhibit high losses from internal reflections. Other variations of the scattering cell include the internally reflecting cell [Ref. 16] and the sandwiching of the fiber between two solar cells [Ref. 17].

A laser source (i.e. He–Ne, Nd:YAG, krypton ion) is utilized to provide sufficient optical power at a single wavelength together with a suitable instrument to measure the response from the detector. In order to avoid inaccuracies in the measurement resulting from scattered light which may be trapped in the fiber, cladding mode strippers (see Section 5.2.1) are placed before and after the scattering cell. These devices remove the light propagating in the cladding so that the measurements are taken only using the light guided by the fiber core. Also to avoid reflections contributing to the optical signal within the cell, the output fiber end is index matched using either a fluid or suitable surface.

The loss due to scattering α_{sc} following Eq. (3.3) is given by:

$$\alpha_{sc} = \frac{10}{l\,(km)} \log_{10} \left(\frac{P_{opt}}{P_{opt} - P_{sc}} \right) \text{ dB km}^{-1} \qquad (5.5)$$

where l (km) is the length of the fiber in km contained within the scattering cell, P_{opt} is the optical power propagating within the fiber at the cell and P_{sc} is the optical power scattered from the short length of fiber l within the cell. As $P_{opt} \gg P_{sc}$, then the logarithm in Eq. (5.5) may be expanded to give:

$$\alpha_{sc} = \frac{4.343}{l\,(km)} \left(\frac{P_{sc}}{P_{opt}} \right) \text{ dB km}^{-1} \qquad (5.6)$$

Since the measurements of length are generally in centimeters and the optical power is normally registered in volts, Eq. (5.6) can be written as:

$$\alpha_{sc} = \frac{4.343 \times 10^5}{l\,(cm)} \left(\frac{V_{sc}}{V_{opt}} \right) \text{ dB km}^{-1} \qquad (5.7)$$

where V_{sc} and V_{opt} are the voltage readings corresponding to the scattered optical power and the total optical power within the fiber at the cell. The relative experimental accuracy (i.e. repeatability) for scatter loss measurements are quoted as ± 0.2 dB [Ref. 6] using the solar cell cube and around 5% [Ref. 8] with the integrating sphere. However, it must be noted that the absolute accuracy of the measurements is somewhat poorer, being dependent on the calibration of the scattering cell and the mode distribution within the fiber.

Example 5.3

A He–Ne laser operating at a wavelength of 0.63 μm was used with a solar cell cube to measure the scattering loss in an optical fiber sample. With a constant optical output power the reading from the solar cell cube was 6.14 nV. The optical power measurement at the cube without scattering was 153.38 μV. The length of the fiber in the cube was 2.92 cm. Determine the loss due to scattering in dB km^{-1} for the fiber at a wavelength of 0.63 μm.

Solution: The scattering loss in the fiber at a wavelength of 0.63 μm may be obtained directly using Eq. (5.7) where:

$$\alpha_{sc} = \frac{4.343 \times 10^5}{l\,(cm)} \left(\frac{V_{sc}}{V_{opt}} \right)$$

$$= \frac{4.343 \times 10^5}{2.92} \left(\frac{6.14 \times 10^{-9}}{153.38 \times 10^{-6}} \right)$$

$$= 6.0 \text{ dB km}^{-1}$$

5.3 FIBER DISPERSION MEASUREMENTS

Dispersion measurements give an indication of the distortion to optical signals as they propagate down optical fibers. The delay distortion which, for example, leads to the broadening of transmitted light pulses, limits the information-carrying capacity of the fiber. Hence as shown in Section 3.7 the measurement of dispersion allows the bandwidth of the fiber to be determined. Therefore besides attenuation, dispersion is the most important transmission characteristic of an optical fiber. As discussed in Section 3.7 there are three major mechanisms which produce dispersion in optical fibers (material dispersion, waveguide dispersion and intermodal dispersion). The importance of these different mechanisms to the total fiber dispersion is dictated by the fiber type.

For instance, in multimode fibers (especially step index), intermodal dispersion tends to be the dominant mechanism, whereas in single mode fibers intermodal dispersion is nonexistent as only a single mode is allowed to propagate. In the single mode case the dominant dispersion mechanism is intramodal (i.e. material dispersion). The dominance of intermodal dispersion in multimode fibers makes it essential that dispersion measurements on these fibers are only performed when the equilibrium mode distribution has been established within the fiber, otherwise inconsistent results will be obtained. Therefore devices such as mode scramblers must be utilized in order to simulate the steady-state mode distribution.

Dispersion effects may be characterized by taking measurements of the impulse response of the fiber in the time domain, or by measuring the baseband frequency response in the frequency domain. If it is assumed that the fiber response is linear with regard to power [Ref. 19], a mathematical description in the time domain for the optical output power $P_o(t)$ from the fiber may be obtained by convoluting the power impulse response $h(t)$ with the optical input power $P_i(t)$ as:

$$P_o(t) = h(t) * P_i(t) \qquad (5.8)$$

where the asterisk * denotes convolution. The convolution of $h(t)$ with $P_i(t)$ shown in Eq. (5.8) may be evaluated using the convolution integral [Ref. 20] where:

$$P_o(t) = \int_{-\infty}^{\infty} P_i(t-x)h(x)\,\mathrm{d}x \qquad (5.9)$$

In the frequency domain the power transfer function $H(\omega)$ is the Fourier transform of $h(t)$ and therefore by taking the Fourier transforms of all the functions in Eq. (5.8) we obtain,

$$\mathcal{P}_o(\omega) = H(\omega)\mathcal{P}_i(\omega) \qquad (5.10)$$

where ω is the baseband angular frequency. The frequency domain representation given in Eq. (5.10) is the least mathematically complex, and by performing the Fourier transformation (or the inverse Fourier transformation) it is possible to switch between the time and frequency domains (or vice versa) by mathematical means. Hence, independent measurement of either $h(t)$ or $H(\omega)$ allows determination of the overall dispersive properties of the optical fiber. Thus fiber dispersion measurements can be made in either the time or frequency domains.

5.3.1   Time Domain Measurement

The most common method for time domain measurement of pulse dispersion in optical fibers is illustrated in Fig. 5.7 [Ref. 21]. Short optical pulses

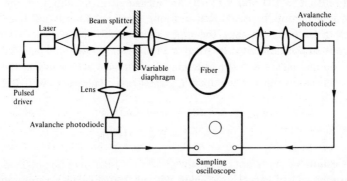

Fig. 5.7 Experimental arrangement for making fiber dispersion measurements in the time domain [Ref. 21].

(100–400 ps) are launched into the fiber from a suitable source (e.g. AlGaAs injection laser) using fast driving electronics. The pulses travel down the length of fiber under test (around 1 km) and are broadened due to the various dispersion mechanisms. However, it is possible to take measurements of an isolated dispersion mechanism by, for example, using a laser with a narrow spectral width when testing a multimode fiber. In this case the intramodal dispersion is negligible and the measurement thus reflects only intermodal dispersion. The pulses are received by a high speed photodetector (i.e. avalanche photodiode) and are displayed on a fast sampling oscilloscope. A beam splitter is utilized for triggering the oscilloscope and for input pulse measurement.

Alternatively, after the initial measurement of output pulse width, the long fiber length may be cut back to a short length (1–2 m) and the measurement repeated in order to obtain the effective input pulse width. The fiber dispersion is obtained from the two pulse width measurements which are taken at any convenient fraction of their amplitude. However, unlike the considerations of dispersion in Sections 3.7–3.10 where rms pulse widths are used, dispersion measurements are normally made on pulses using the half maximum amplitude or 3 dB points. If $P_i(t)$ and $P_o(t)$ of Eq. (5.8) are assumed to have a Gaussian shape then Eq. (5.8) may be written in the form [Ref. 20]:

$$\tau_o^2(3 \text{ dB}) = \tau^2(3 \text{ dB}) + \tau_i^2(3 \text{ dB}) \tag{5.11}$$

where $\tau_i(3 \text{ dB})$ and $\tau_o(3 \text{ dB})$ are the 3 dB pulse widths at the fiber input and output respectively and $\tau(3 \text{ dB})$ is the width of the fiber impulse response again measured at half the maximum amplitude. Hence the pulse dispersion in the fiber (commonly referred to as the pulse broadening when considering the 3 dB pulse width) in ns km^{-1} is given by:

$$\tau(3 \text{ dB}) = \frac{(\tau_o^2(3 \text{ dB}) - \tau_i^2(3 \text{ dB}))^{\frac{1}{2}}}{L} \text{ ns km}^{-1} \tag{5.12}$$

where $\tau(3 \text{ dB})$, $\tau_i(3 \text{ dB})$ and $\tau_o(3 \text{ dB})$ are measured in ns and L is the fiber length in km. It must be noted that if a long length of fiber is cut back to a short length in order to take the input pulse width measurement then L corresponds to the difference between the two fiber lengths in km. When the launched optical pulses and the fiber impulse response are Gaussian then the 3 dB optical bandwidth for the fiber B_{opt} may be calculated using [Ref. 22]:

$$\begin{aligned} B_{opt} \times \tau(3 \text{ dB}) &= 0.44 \text{ GHz ns} \\ &= 0.44 \text{ MHz ps} \end{aligned} \tag{5.13}$$

Hence estimates of the optical bandwidth for the fiber may be obtained from the measurements of pulse broadening without resorting to rigorous mathematical analysis.

Example 5.4

Pulse dispersion measurements are taken over a 1.2 km length of partially graded fiber. The 3 dB widths of the optical input pulses are 300 ps, and the corresponding 3 dB widths for the output pulses are found to be 12.6 ns. Assuming the pulse shapes and fiber impulse response are Gaussian calculate:

(a) the 3 dB pulse broadening for the fiber in ns km^{-1};
(b) the fiber bandwidth–length product.

Solution: (a) The 3 dB pulse broadening may be obtained using Eq. (5.12) where:

$$\tau(3 \text{ dB}) = \frac{(12.6^2 - 0.3^2)^{\frac{1}{2}}}{1.2} = \frac{(158.76 - 0.09)^{\frac{1}{2}}}{1.2}$$

$$= 10.5 \text{ ns km}^{-1}$$

(b) The optical bandwidth for the fiber is given by Eq. (5.13) as:

$$B_{\text{opt}} = \frac{0.44}{\tau(3 \text{ dB})} = \frac{0.44}{10.5} \text{ GHz km}$$

$$= 41.9 \text{ MHz km}$$

The value obtained for B_{opt} corresponds to the bandwidth–length product for the fiber because the pulse broadening in part (a) was calculated over a 1 km fiber length. Also it may be noted that in this case the narrow input pulse width makes little difference to the calculation of the pulse broadening. The input pulse width becomes significant when measurements are taken on low dispersion fibers (e.g. single mode).

The dispersion measurement technique discussed above is reasonably simple and gives a direct measurement of the pulse broadening. However, it has some drawbacks when performing measurements on low dispersion fibers. At present the time resolution of commercially available detectors is limited to a few tenths of a nanosecond. Hence inaccuracies occur in the measurement when the pulse broadening along the fiber is less than 0.5 ns. For low dispersion fibers this may dictate the testing of a considerable length of fiber.

A more convenient method of measuring the temporal dispersion of an optical pulse within a fiber which does not require a long fiber length is the shuttle pulse technique. The experimental set-up reported by Cohen [Ref. 23] is shown in Fig. 5.8. Both ends of a short fiber length are terminated with partially transparent mirrors and a pulse launched from a GaAs injection laser travels through one mirror into the fiber and then shuttles back and forth between the fiber ends. This technique has an added advantage in that it allows the length dependence of the impulse response to be studied by sampling the pulse after each $2N - 1$ (where $N = 1, 2, 3$, etc.) transits. The pulse at the output end is displayed on a sampling oscilloscope through the partially transparent mirror. Hence the pulse broadening may be measured by comparing the

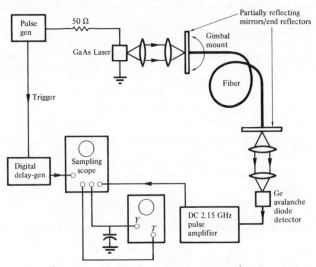

Fig. 5.8   Schematic diagram showing the apparatus used in the shuttle pulse technique for time domain dispersion measurements in optical fibers [Ref. 23].

widths of the observed output pulses. To ensure symmetrical reflection at the mirrors it is important that they are perpendicular to the fiber axis. The mirrors are therefore mounted in cylindrical holders which have grooves precisely milled to accurately machined faces. An index matching fluid is also utilized between the fiber end faces and the mirrors in order to achieve optimum optical transmission. Nevertheless the technique does have drawbacks including optical loss at each end reflection. Therefore accurate measurements over multiple reflections cannot always be ensured.

5.3.2   Frequency Domain Measurement

Frequency domain measurement is the preferred method for acquiring the bandwidth of optical fibers. This is because the baseband frequency response $H(\omega)$ of the fiber may be obtained directly from these measurements using Eq. (5.10) without the need for any assumptions of Gaussian shape, or alternatively, the mathematically complex deconvolution of Eq. (5.8) which is necessary with measurements in the time domain. Thus the optical bandwidth of a fiber is best obtained from frequency domain measurements.

One of two frequency domain measurement techniques is generally used. The first utilizes a similar pulsed source to that employed for the time domain measurements shown in Fig. (5.7). However, the sampling oscilloscope is replaced by a spectrum analyzer which takes the Fourier transform of the pulse in the time domain and hence displays its constituent frequency components. The experimental arrangement is illustrated in Fig. 5.9.

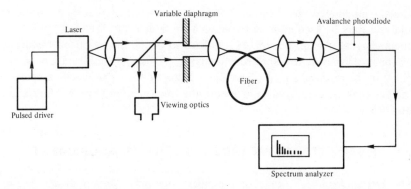

Fig. 5.9 Experimental set-up for making fiber dispersion measurements in the frequency domain using a pulsed laser source.

Comparison of the spectrum at the fiber output $\mathcal{P}_o(\omega)$ with the spectrum at the fiber input $\mathcal{P}_i(\omega)$ provides the baseband frequency response for the fiber under test following Eq. (5.10) where:

$$H(\omega) = \frac{\mathcal{P}_o(\omega)}{\mathcal{P}_i(\omega)} \qquad (5.14)$$

The second technique involves launching a sinusoidally modulated optical signal at different selected frequencies using a sweep oscillator. Therefore the signal energy is concentrated in a very narrow frequency band in the baseband region, unlike the pulse measurement method where the signal energy is spread over the entire baseband region. A possible experimental arrangement for this swept frequency measurement method is shown in Fig. 5.10 [Ref. 24]. The optical source can be an LED or an injection laser, both of which may be

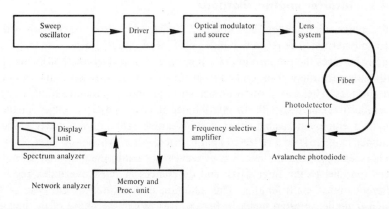

Fig. 5.10 Block schematic showing an experimental arrangement for the swept frequency measurement method to provide fiber dispersion measurements in the frequency domain [Ref. 24].

directly modulated (see Section 7.5) from the sweep oscillator. A spectrum analyzer may be used in order to obtain a continuous display of the swept frequency signal. Again, Eq. (5.14) is utilized to obtain the baseband frequency response. However, the spectrum analyzer provides no information on the phase of the received signal. Therefore a vector voltmeter or ideally a network analyzer can be employed to give both the frequency and phase information [Ref. 7].

5.4   FIBER REFRACTIVE INDEX PROFILE MEASUREMENTS

The refractive index profile of the fiber core plays an important role in characterizing the properties of optical fibers. It allows determination of the fiber's numerical aperture and the number of modes propagating within the fiber core, whilst largely defining any intermodal and/or profile dispersion caused by the fiber. Hence a detailed knowledge of the refractive index profile enables the impulse response of the fiber to be predicted. Also as the impulse response and consequently the information-carrying capacity of the fiber is strongly dependent on the refractive index profile, it is essential that the fiber manufacturer is able to produce particular profiles with great accuracy, especially in the case of graded index fibers (i.e. optimum profile). There is therefore a requirement for accurate measurement of the refractive index profile. These measurements may be performed using a number of different techniques each of which exhibit certain advantages and drawbacks. In this section we will discuss some of the more popular methods which may be relatively easily interpreted theoretically, without attempting to review all the possible techniques which have been developed.

5.4.1   Interferometric Methods

Interference microscopes (e.g. Mach–Zehnder, Michelson) have been widely used to determine the refractive index profiles of optical fibers. The technique usually involves the preparation of a thin slice of fiber (slab method) which has both ends accurately polished to obtain square (to the fiber axes) and optically flat surfaces. The slab is often immersed in an index matching fluid, and the assembly is examined with an interference microscope. Two major methods are then employed; using either a transmitted light interferometer (Mach–Zehnder [Ref. 25]) or a reflected light interferometer (Michelson [Ref. 26]). In both cases light from the microscope travels normal to the prepared fiber slice faces (parallel to the fiber axis), and differences in refractive index result in different optical path lengths. This situation is illustrated in the case of the Mach–Zehnder interferometer in Fig. 5.11(a). When the phase of the incident light is compared with the phase of the emerging light, a field of parallel

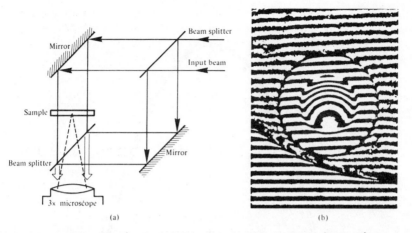

Fig. 5.11   (a) The principle of the Mach–Zehnder interferometer [Ref. 25]. (b) The interference fringe pattern obtained with an interference microscope from a graded index fiber. Reproduced with permission from L. G. Cohen, P. Kaiser, J. B. MacChesney, P. B. O'Conner and H. M. Presby, *Appl. Phys. Lett.*, **26**, p. 472, 1975.

interference fringes are observed. A photograph of the fringe pattern may then be taken, an example of which is shown in Fig. 5.11(b) [Ref. 28].

The fringe displacements for the points within the fiber core are then measured using as reference the parallel fringes outside the fiber core (in the fiber cladding). The refractive index difference between a point in the fiber core (e.g. the core axis) and the cladding can be obtained from the fringe shift q, which corresponds to a number of fringe displacements. This difference in refractive index δn is given by [Ref. 6]:

$$\delta n = \frac{q\lambda}{x} \qquad (5.15)$$

where x is the thickness of the fiber slab and λ is the incident optical wavelength. The slab method gives an accurate measurement of the refractive index profile, although computation of the individual points is somewhat tedious unless an automated technique is used [Ref. 28]. Figure 5.12 [Ref. 28] shows the refractive index profile obtained from the fringe pattern indicated in Fig. 5.11(b).

A limitation of this method is the time required to prepare the fiber slab. However, another interferometric technique has been developed [Ref. 30] which requires no sample preparation. In this method the light beam is incident to the fiber perpendicular to its axis; this is known as transverse shearing interferometry. Again fringes are observed from which the fiber refractive index profile may be obtained.

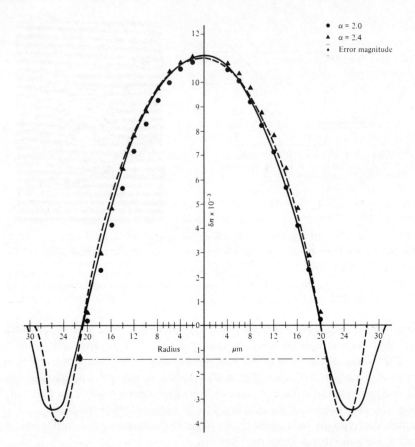

Fig. 5.12   The fiber refractive index profile computed from the interference pattern shown in Fig. 5.11(b). Reproduced with permission from L. G. Cohen, P. Kaiser, J. B. MacChesney, P. B. O'Conner and H. M. Presby, *Appl. Phys. Lett.*, **26**, p. 472, 1975.

5.4.2   Near Field Scanning Method

The near field scanning method utilizes the close resemblance that exists between the near field intensity distribution and the refractive index profile, for a fiber with all the guided modes equally illuminated. It provides a reasonably straightforward and rapid method for acquiring the refractive index profile. When a diffuse Lambertian source (e.g. tungsten filament lamp or LED) is used to excite all the guided modes then the near field optical power density at a radius r from the core axis $P_D(r)$ may be expressed as a fraction of the core axis near field optical power density $P_D(0)$ following [Ref. 31]:

$$\frac{P_D(r)}{P_D(0)} = C(r, z)\left[\frac{n_1^2(r) - n_2^2}{n_1^2(0) - n_2^2}\right] \tag{5.16}$$

where $n_1(0)$ and $n_1(r)$ are the refractive indices at the core axis and at a distance r from the core axis respectively, n_2 is cladding refractive index and $C(r, z)$ is a correction factor. The correction factor which is incorporated to compensate for any leaky modes present in the short test fiber may be determined analytically. A set of normalized correction curves is, for example, given in Ref. 33.

An experimental configuration is shown in Fig. 5.13. The output from a Lambertian source is focused onto the end of the fiber using a microscope objective lens. A magnified image of the fiber output end is displayed in the plane of a small active area photodetector (e.g. silicon $p–i–n$ photodiode). The photodetector which scans the field transversely receives amplification from the phase sensitive combination of the optical chopper and lock-in amplifier. Hence the profile may be plotted directly on an $X–Y$ recorder. However, the profile must be corrected with regard to $C(r, z)$ as illustrated in Fig. 5.14(a) which is very time consuming. Both the scanning and data acquisition can be automated with the inclusion of a minicomputer [Ref. 31].

The test fiber is generally less than 1 m in length to eliminate any differential mode attenuation and mode coupling. A typical refractive index profile for a practical step index fiber measured by the near field scanning method is shown in Fig. 5.14(b). It may be observed that the profile dips in the center at the fiber core axis. This results from the collapse of the fiber preform before the fiber is drawn during the manufacturing process (see Sections 4.3 and 4.4).

Measurements of the refractive index profile may also be obtained from the far field pattern produced by laser light scattered by the fiber under test. This technique, generally known as the scattered pattern method, requires complex analysis of the forward or backward patterns in order to determine the refractive index profile [Ref. 31]. Therefore, it is pursued no further in this text.

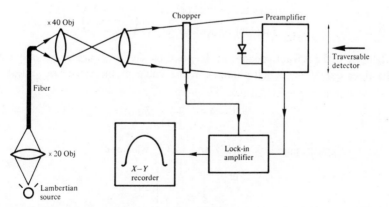

Fig. 5.13 Experimental set-up for the near field scanning measurement of the refractive index profile [Ref. 32].

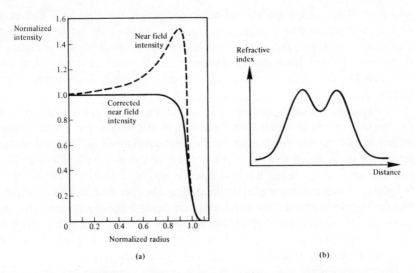

Fig. 5.14   (a) The refractive index profile of a step index fiber measured using the near field scanning method, showing the near field intensity and the corrected near field intensity. Reproduced with permission from F. E. M. Sladen, D. N. Payne and M. J. Adams, *Appl. Phys. Lett.*, **28**, p. 225, 1976. (b) The refractive index profile of a practical step index fiber measured by the near field scanning method [Ref. 31].

5.4.3 End Reflection Method

The refractive index at any point in the cross section of an optical fiber is directly related to the reflected power from the fiber surface in air at that point following the Fresnel reflection formula of Eq. (4.12). Hence the fraction of light reflected at the air–fiber interface is given by:

$$r = \frac{\text{reflected optical power}}{\text{incident optical power}} = \left(\frac{n_1 - 1}{n_1 + 1} \right)^2 \tag{5.17}$$

where n_1 is the refractive index at the point on the fiber surface (usually within the fiber core). For small changes in the value of the refractive index:

$$\delta r = 4 \frac{(n_1 - 1)}{(n_1 + 1)^3} \delta n_1 \tag{5.18}$$

Therefore combining Eqs. (5.18) and (5.17) we have:

$$\frac{\delta r}{r} = \left[\frac{4}{n_1^2 - 1} \right] \delta n_1 \tag{5.19}$$

Equation (5.19) gives the relative change in the Fresnel reflection coefficient

r which corresponds to the change of refractive index at the point of measurement. However, when the measurement is performed in air the small changes in refractive index δn_1 that must be measured give only very small changes in r, as demonstrated in example 5.5.

Example 5.5

Refractive index measurements are to be made using the end reflection method on a step index fiber in air. The fiber core refractive index is nominally 1.5. Estimate the percentage change in the Fresnel reflection coefficient that must be measured in order to allow a change in refractive index δn_1 of 0.001 to be resolved.

Solution: The relative change in the Fresnel reflection coefficient is given by Eq. (5.19), where:

$$\frac{\delta r}{r} = \left[\frac{4}{n_1^2 - 1} \right] \delta n_1$$

Therefore when n_1 is 1.5 and the requirement for δn_1 is 0.001:

$$\frac{\delta r}{r} = \left[\frac{4}{1.25} \right] 0.001 = 0.0032$$

Hence the change in the Fresnel reflection coefficient which must be measured is 0.32%.

It is clear from example 5.5 that for a fiber in air very accurate measurement of r is needed to facilitate a moderate resolution of δn_1. However, this problem may be overcome by immersing the fiber in a suitable index matching oil, as illustrated in example 5.6.

Example 5.6

The step index fiber of example 5.5 is immersed in oil with a refractive index of 1.45. Estimate the percentage change in the Fresnel reflection coefficient which must be measured in order to obtain the resolution in δn_1 required in example 5.5 (0.001).

Solution: In this case we must refer back to the Fresnel reflection formula of Eq. (4.12) where:

$$r = \left(\frac{n_1 - n}{n_1 + n} \right)^2 = \left(\frac{n_1 - 1.45}{n_1 + 1.45} \right)^2$$

Hence

$$\delta r = 4 \frac{(n_1 - 1.45)}{(n_1 + 1.45)^3} \delta n_1$$

and

$$\frac{\delta r}{r} = \left[\frac{4}{(n_1 + 1.45)(n_1 - 1.45)} \right] \delta n_1 = \left[\frac{4}{n_1^2 - 1.45^2} \right] \delta n_1$$

so

$$\frac{\delta r}{r} = \left[\frac{4}{2.25 - 2.103} \right] 0.001 = 0.0272$$

Therefore, the change in the Fresnel reflection coefficient which must be measured for the fiber in the oil is 2.72%.

The result from example 5.6 illustrates the increased sensitivity in the measurement with index matching over that calculated in example 5.5 (0.32%) when no index matching was used. Also the sensitivity may be increased further with improved index matching giving very accurate profile measurements.

Two experimental arrangements for performing end reflection measurements are illustrated in Fig. 5.15 [Refs. 34 and 35]. Both techniques utilize a focused laser beam incident on the fiber end face in order to provide the necessary spatial resolution. Figure 5.15(a) shows end reflection measurements without index matching of the fiber input end face. The laser beam is initially directed through a polarizer and a $\lambda/4$ plate in order to prevent feedback of the reflected optical power from both the fiber end face and the intermediate optics, causing modulation of the laser output through interference. The circularly polarized light beam from the $\lambda/4$ plate is then spatially filtered and expanded to provide a suitable spot size. A beam splitter is used to provide both a reference from the input light beam which is monitored with a solar cell, and two beams from the fiber end face reflection. The reflected beams are used for measurement via a p–i–n photodiode, lock-in amplifier combination, and for visual check of the alignment on the fiber end face using a screen. Focusing on the fiber end face is achieved with a microscope objective lens, and the fiber end is scanned slowly across the focal spot using precision translation stages. The reflected optical power is monitored as a function of the fiber linear position on an X–Y recorder and the refractive index profile may be obtained directly using Eq. (5.19). Possible reflections from the other fiber end face are avoided by immersing it in an index matching liquid.

The experimental arrangement shown in Fig. 5.15(b) provides increased sensitivity by immersing the fiber in an index matching oil as demonstrated in example 5.6. In this case the laser beam, which is again incident on a polarizer, and $\lambda/4$ plate is deflected vertically using a mirror. An oil immersion objective is utilized to focus the beam onto the immersed fiber end. This apparatus has shown sensitivity comparable with the near field method. However, there is a need for careful alignment of the apparatus in order to avoid stray reflections.

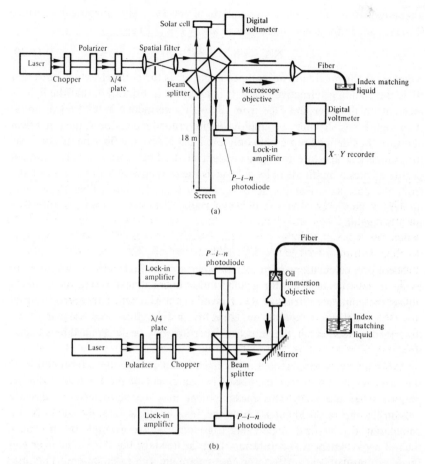

Fig. 5.15 Experimental arrangements for end reflection measurement of fiber refractive index profile: (a) without index matching of fiber input end face [Ref. 34]; (b) with index matching of fiber input end face [Ref. 35].

Also in both techniques it is essential that the fiber end face should be perfectly flat (cleaved but not polished), because the reflected power is severely affected by surface irregularities.

5.5 FIBER NUMERICAL APERTURE MEASUREMENTS

The numerical aperture is an important optical fiber parameter as it affects characteristics such as the light-gathering efficiency and the normalized frequency of the fiber (V). This in turn dictates the number of modes propagating within the fiber (also defining the single mode region) which has consequent effects on both the fiber dispersion (i.e. intermodal) and, possibly, the fiber attenuation (i.e. differential attenuation of modes). The numerical aperture (NA) is defined for a step index fiber in air by Eq. (2.8) as:

$$NA = \sin \theta_a = (n_1^2 - n_2^2)^{\frac{1}{2}} \tag{5.20}$$

where θ_a is the maximum acceptance angle, n_1 is the core refractive index and n_2 is the cladding refractive index. It is assumed in Eq. (5.20) that the light is incident on the fiber end face from air with a refractive index (n_0) of unity. Although Eq. (5.20) may be employed with graded index fibers, the numerical aperture thus defined represents only the local NA of the fiber on its core axis (the numerical aperture for light incident at the fiber core axis). The graded profile creates a multitude of local NAs as the refractive index changes radially from the core axis. For the general case of a graded index fiber these local numerical apertures NA(r) at different radial distances r from the core axis may be defined by:

$$NA(r) = \sin \theta_a(r) = (n_1^2(r) - n_2^2)^{\frac{1}{2}} \tag{5.21}$$

Therefore, calculations of numerical aperture from refractive index data are likely to be less accurate for graded index fibers than for step index fibers unless the complete refractive index profile is considered. However, if refractive index data is available on either fiber type from the measurements described in Section 5.4, the numerical aperture may be determined by calculation.

Alternatively, a simple, commonly used technique for the determination of the fiber numerical aperture involves measurement of the far field radiation pattern from the fiber. This measurement may be performed by directly measuring the far field angle from the fiber using a rotating stage, or by calculating the far field angle using trigonometry. An example of an experimental arrangement with a rotating stage is shown in Fig. 5.16. The fiber end faces are prepared in order to ensure square smooth terminations. The fiber output end is then positioned on the rotating stage with its end face parallel to the plane of the photodetector input, and so that its output is perpendicular to

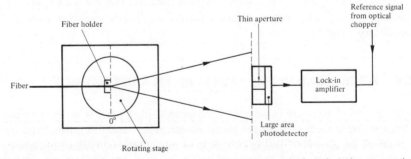

Fig. 5.16   Fiber numerical aperture measurement using a scanning photodetector and a rotating stage.

the axis of rotation. Light is launched into the fiber at all possible angles (overfilling the fiber) using an optical system similar to that used in the spot attenuation measurements (Fig. 5.3).

The photodetector, which may be either a small area device or an apertured large area device, is placed 10–20 cm from the fiber and positioned in order to obtain a maximum signal with no rotation (0°). Hence when the rotating stage is turned the limits of the far field pattern may be recorded. The output power is monitored and plotted as a function of angle; the maximum acceptance angle being obtained when the power drops a predetermined amount (e.g. 10%). Thus the numerical aperture of the fiber can be obtained from Eq. (5.20). This far field scanning measurement may also be performed with the photodetector located on a rotational stage and the fiber positioned at the center of rotation [Ref. 7]. A complementary technique utilizes a plane wave input to the fiber, which is then rotated around the input beam axis whilst its output is directly monitored.

A less precise measurement of the numerical aperture can be obtained from the far field pattern by trigonometric means. The experimental apparatus is shown in Fig. 5.17, where the end prepared fiber is located on an optical base plate or slab. Again light is launched into the fiber under test over the full range of its numerical aperture, and the far field pattern from the fiber is displayed on a screen which is positioned a known distance D from the fiber output end face. The test fiber is then aligned so that the optical intensity on the screen is maximized. Finally, the pattern size on the screen A is measured using a calibrated vernier caliper. The numerical aperture can be obtained from simple trigonometrical relationships where:

$$\text{NA} = \sin \theta_a = \frac{A/2}{[(A/2)^2 + D^2]^{\frac{1}{2}}} = \frac{A}{(A^2 + 4D^2)^{\frac{1}{2}}} \qquad (5.22)$$

Example 5.7

A trigonometrical measurement is performed in order to determine the numerical aperture of a step index fiber. The screen is positioned 10.0 cm from the fiber end face. When illuminated from a wide angled visible source the measured output pattern size is 6.2 cm. Calculate the approximate numerical aperture of the fiber.

Solution: The numerical aperture may be determined directly, using Eq. (5.22) where:

$$\text{NA} = \frac{A}{(A^2 + 4D^2)^{\frac{1}{2}}} = \frac{6.2}{(38.44 + 400)^{\frac{1}{2}}} = 0.30$$

It must be noted that the accuracy of this measurement technique is dependent upon the visual assessment of the far field pattern from the fiber.

The above measurement techniques are generally employed with multimode

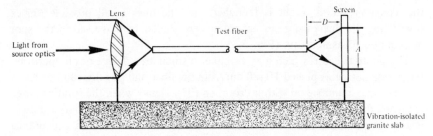

Fig. 5.17   Apparatus for trigonometric fiber numerical aperture measurement.

fibers only, as the far field patterns from single mode fibers are affected by diffraction phenomena. These are caused by the small core diameters of single mode fibers which tend to invalidate simple geometric optics measurements. However, more detailed analysis of the far field pattern allows determination of the normalized frequency and core radius for single mode fibers, from which the numerical aperture may be calculated using Eq. (2.69) |Ref. 36|.

Far field pattern measurements with regard to multimode fibers are dependent on the length of the fiber tested. When the measurements are performed on short fiber lengths (around 1 m) the numerical aperture thus obtained corresponds to that defined by Eqs. (5.20) or (5.21). However, when a long fiber length is utilized which gives mode coupling and the selective attenuation of the higher order modes, the measurement yields a lower value for the numerical aperture. It must also be noted that the far field measurement techniques give an average (over the local NAs) value for the numerical aperture of graded index fibers. Hence, alternative methods must be employed if accurate determination of the fiber's NA is required |Ref. 37|.

5.6   FIBER DIAMETER MEASUREMENTS

5.6.1   Outer Diameter

It is essential during the fiber manufacturing process (at the fiber drawing stage) that the fiber outer diameter (cladding diameter) is maintained constant to within 1%. Any diameter variations may cause excessive radiation losses and make accurate fiber–fiber connection difficult. Hence on-line diameter measurement systems are required which provide accuracy better than 0.3% at a measurement rate greater than 100 Hz (i.e. a typical fiber drawing velocity is 1 m s^{-1}). Use is therefore made of noncontacting optical methods such as fiber image projection and scattering pattern analysis.

The most common on-line measurement technique uses fiber image projection (shadow method) and is illustrated in Fig. 5.18 |Ref. 38|. In this method a laser beam is swept at a constant velocity transversely across the fiber and a

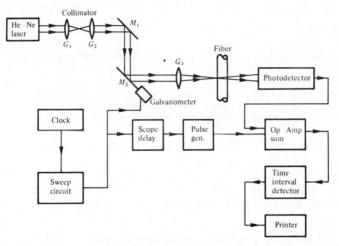

Fig. 5.18   The shadow method for the on-line measurement of the fiber outer diameter [Ref. 38].

measurement is made of the time interval during which the fiber intercepts the beam and casts a shadow on a photodetector. In the apparatus shown in Fig. 5.18 the beam from a laser operating at a wavelength of 0.6328 μm is collimated using two lenses (G_1 and G_2). It is then reflected off two mirrors (M_1 and M_2), the second of which (M_2) is driven by a galvanometer which makes it rotate through a small angle at a constant angular velocity before returning to its original starting position. Therefore, the laser beam which is focused in the plane of the fiber by a lens (G_3) is swept across the fiber by the oscillating mirror, and is incident on the photodetector unless it is blocked by the fiber. The velocity ds/dt of the fiber shadow thus created at the photodetector is directly proportional to the mirror velocity $d\phi/dt$ following:

$$\frac{ds}{dt} = l\frac{d\phi}{dt} \tag{5.23}$$

where l is the distance between the mirror and the photodetector.

Furthermore the shadow is registered by the photodetector as an electrical pulse of width W_e which is related to the fiber outer diameter d_o as:

$$d_o = W_e\frac{ds}{dt} \tag{5.24}$$

Thus the fiber outer diameter may be quickly determined and recorded on the printer. The measurement speed is largely dictated by the inertia of the mirror rotation and its accuracy by the risetime of the shadow pulse.

Example 5.8

The shadow method is used for the on-line measurement of the outer diameter of an optical fiber. The apparatus employs a rotating mirror with an angular velocity of 4 rad s^{-1} which is located 10 cm from the photodetector. At a particular instant in time a shadow pulse of width 300 μs is registered by the photodetector. Determine the outer diameter of the optical fiber in μm at this instant in time.

Solution: The shadow velocity may be obtained from Eq. (5.23) where:

$$\frac{ds}{dt} = l\frac{d\phi}{dt} = 0.1 \times 4 = 0.4 \text{ m s}^{-1}$$

$$= 0.4 \text{ μm μs}^{-1}$$

Hence the fiber outer diameter d_0 in μm is given by Eq. (5.24):

$$d_0 = W_e \frac{ds}{dt} = 300 \text{ μs} \times 0.4 \text{ μm μs}^{-1}$$

$$= 120 \text{ μm}$$

Other on-line measurement methods, enabling faster diameter measurements, involve the analysis of forward or backward far field patterns which are produced when a plane wave is incident transversely on the fiber. These techniques generally require measurement of the maxima in the center portion of the scattered pattern from which the diameter can be calculated after detailed mathematical analysis [Refs. 39–42]. They tend to give good accuracy (e.g. ±0.25 μm [Ref. 42]) even though the theory assumes a perfectly circular fiber cross section. Also for step index fibers the analysis allows determination of the core diameter, and core and cladding refractive indices.

Measurements of the fiber outer diameter after manufacture (off-line) may be performed using a micrometer or dial gage. These devices can give accuracies in the order of ±0.5 μm. Alternatively, off-line diameter measurements can be made with a microscope incorporating a suitable calibrated micrometer eyepiece.

5.6.2 Core Diameter

The core diameter for step index fibers is defined by the step change in the refractive index profile at the core–cladding interface. Therefore the techniques employed for determining the refractive index profile (interferometric, near field, end reflection, etc.) may be utilized to measure the core diameter. Graded index fibers present a more difficult problem as, in general, there is a continuous transition between the core and the cladding. In this case it is necessary to define the core as an area with a refractive index above a certain predetermined value if refractive index profile measurements are used to obtain the core diameter.

Core diameter measurement is also possible from the near field pattern of a suitably illuminated (all guided modes excited) fiber. The measurements may be taken using a microscope equipped with a micrometer eyepiece similar to that employed for off-line outer diameter measurements. However, the core–cladding interface for graded index fibers is again difficult to identify due to fading of the light distribution towards the cladding, rather than the sharp boundary which is exhibited in the step index case.

A second possible definition of the fiber core diameter is that it may be considered as the dimension of an area in which a certain fraction (e.g. 90 or 95%) of the total transmitted optical power is propagated. Hence measurements may be performed to establish the optical power distribution within the fiber either using near field scanning techniques [Ref. 7], or by measuring the power transmitted through a calibrated variable diameter iris positioned upon an enlarged image of the fiber. Unfortunately these transmission measurements are dependent on the fiber length and the optical launch conditions which tend to make the measurement an effective rather than true diameter reading. Other possible techniques include selectively etching the core material [Ref. 7], and the analysis of the far field scattering patterns (for step index fibers) as indicated when the measurement of fiber outer diameter was discussed.

5.7 FIELD MEASUREMENTS

The measurements discussed in the previous sections are primarily suited to the laboratory environment where quite sophisticated instrumentation may be used. However, there is a requirement for the measurement of the transmission characteristics of optical fibers when they are located in the field within an optical communication system. It is essential that optical fiber attenuation and dispersion measurements, connector and splice loss measurements and fault location be performed on optical fiber links in the field. Although information on fiber attenuation and dispersion is generally provided by the manufacturer, this is not directly applicable to cabled, installed fibers which are connected in series within an optical fiber system. Effects such as microbending (see Section 4.6.2) with the resultant mode coupling (see Section 2.3.7) affect both the fiber attenuation and dispersion. It is also found that the simple summation of the transmission parameters with regard to individual connected lengths of fiber cable does not accurately predict the overall characteristics of the link [Ref. 43]. Hence test equipment has been developed which allows these transmission measurements to be performed in the field.

In general, field test equipment differs from laboratory instrumentation in a number of aspects as it is required to meet the exacting demands of field measurement. Therefore the design criteria for field measurement equipment include:

(a) Sturdy and compact encasement which must be portable.
(b) The ready availability of electrical power must be ensured by the incorporation of batteries or by connection to a generator. Hence the equipment should maintain accuracy under conditions of varying supply voltage and/or frequency.
(c) In the event of battery operation, the equipment must have a low power consumption.
(d) The equipment must give reliable and accurate measurements under extreme environmental conditions of temperature, humidity and mechanical load.
(e) Complicated and involved fiber connection arrangements should be avoided. The equipment must be connected to the fiber in a simple manner without the need for fine or critical adjustment.
(f) The equipment cannot usually make use of external triggering or regulating circuits between the transmitter and receiver due to their wide spacing on the majority of optical links.

Even if the above design criteria are met, it is likely that a certain amount of inaccuracy will have to be accepted with field test equipment. For example, it may not be possible to include adjustable launching conditions (i.e. variation in spot size and numerical aperture) in order to create the optimum. Also, because of the large dynamic range required to provide measurements over long fiber lengths, lossy devices such as mode scramblers may be omitted. Therefore measurement accuracy may be impaired through inadequate simulation of the equilibrium mode distribution.

A number of portable, battery-operated, optical power meters are commercially available. These devices often measure absolute optical power in dBm and dBµ (i.e. 0 dBm is equivalent to 1 mW and 0 dBµ is equivalent to 1 µW; see example 5.9) over a specified spectral range (e.g. 0.4–1.15 µm). In a number of cases the spectral range may be altered by the incorporation of different demountable sensor heads (photodetectors). However, it must be noted that although these devices often take measurements over a certain spectral range this simply implies that they may be adjusted to be compatible with the center emission frequency of particular optical sources so as to obtain the most accurate reading of optical power. Therefore these devices do not generally give spectral attenuation measurements unless the source optical output frequency is controlled or filtered to achieve single wavelength operation. A typical example is the United Detector Technology S 550 fiber optics power meter shown in Fig. 5.19. This device may be used for measurement of the absolute optical attenuation on a fiber link by employing the cut back technique. Other optical system parameters which may also be obtained using this type of power meter are the measurement of individual splice and con-

Fig. 5.19 The United Detector Technology S 550 fiber optics power meter. (Courtesy of United Detector Technology.)

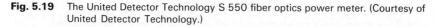

nector losses, the determination of the absolute optical output power emitted from the source (see Sections 6.5.3 and 7.4.1) and the measurement of the responsivity or the absolute photocurrent of the photodetector in response to particular levels of input optical power (see Section 8.6).

Example 5.9

An optical power meter records optical signal power in either dBm or dBμ.

(a) Convert the optical signal powers of 5 mW and 20 μW to dBm.
(b) Convert optical signal powers of 0.3 mW and 80 nW to dBμ.

Solution: The optical signal power can be expressed in decibels using:

$$dB = 10 \log_{10} \left(\frac{P_o}{P_r} \right)$$

where P_o is the received optical signal power and P_r is a reference power level.

(a) For a 1 mW reference power level:

$$dBm = 10 \log_{10} \left(\frac{P_o}{1 \text{ mW}} \right)$$

Hence an optical signal power of 5 mW is equivalent to

Optical signal power $= 10 \log_{10} 5 = 6.99$ dBm

and an optical signal power of 20 μW is equivalent to:

Optical signal power $= 10 \log_{10} 0.02$

$= -16.99$ dBm

(b) For a 1 µW reference power level:

$$dB\mu = 10 \log_{10} \left(\frac{P_o}{1 \ \mu W} \right)$$

Therefore an optical signal power of 0.3 mW is equivalent to:

$$\text{Optical signal power} = 10 \log_{10} \left(\frac{P_o}{1 \ \mu W} \right) = 10 \log_{10} 30$$

$$= 14.77 \ dB\mu$$

and an optical signal power of 800 nW is equivalent to:

$$\text{Optical signal power} = 10 \log_{10} 0.8$$

$$= -0.97 \ dB\mu$$

There are a number of portable measurement test sets specifically designed for fiber attenuation measurements which require access to both ends of the optical link. These devices tend to use the cut back measurement technique unless correction is made for any difference in connector losses between the link and a short length of similar reference cable. A block schematic of an optical attenuation meter consisting of a transmitter and receiver unit is shown in Fig. 5.20 [Ref. 43]. Reproducible readings may be obtained by keeping the launched optical power from the light source absolutely constant. A constant optical output power is achieved with the equipment illustrated in Fig. 5.20 using an injection laser and a regulating circuit which is driven from a reference output of the source derived from a photodiode. Hence any variations in the laser output power are rectified by automatic adjustment of the modulating voltage, and therefore current, from the pulse generator. A large area photodiode is utilized in the receiver to eliminate any effects from differing fiber end faces. It is generally found that when a measurement is made on multimode fiber a short cut back reference length of a few meters is insufficient to obtain an equilibrium mode distribution. Hence unless a mode scrambling device together with a mode stripper are used, it is likely that a reference length of around 500 m or more will be required if reasonably accurate measurements are to be made. When measurements are made without a steady-state mode distribution in the reference fiber a significantly higher loss value is obtained which may be as much as 1 dB km^{-1} above the steady-state attenuation [Refs. 22 and 44].

Several field test sets are available for making dispersion measurements on optical fiber links. These devices generally consist of transmitter and receiver units which take measurements in the time domain. Short light pulses ($\simeq$200 ns) are generated from an injection laser and are broadened by

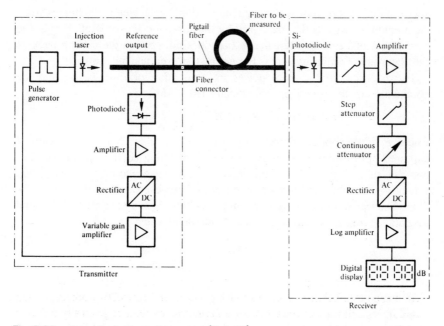

Fig. 5.20   An optical attenuation meter [Ref. 43].

transmission down the optical link before being received by a fast response photodetector (i.e. avalanche photodiode) and displayed on a sampling oscilloscope. This is similar to the dispersion measurements in the time domain discussed in Section 5.3. If it is assumed that the pulses have a near Gaussian shape, Eq. (5.12) may be utilized to determine the pulse broadening on the link, and hence the 3 dB optical bandwidth may be obtained.

5.7.1   Optical Time Domain Reflectometry (OTDR)

A measurement technique which is far more sophisticated and which finds wide application in both the laboratory and the field is the use of optical time domain reflectometry (OTDR). This technique is often called the backscatter measurement method. It provides measurement of the attenuation on an optical link down its entire length giving information on the length dependence of the link loss. In this sense it is superior to the optical attenuation measurement methods discussed previously (Section 5.2) which only tend to provide an averaged loss over the whole length measured in dB km^{-1}. When the attenuation on the link varies with length, the averaged loss information is inadequate. OTDR also allows splice and connector losses to be evaluated as well as the location of any faults on the link. It relies upon the measurement and analysis of the fraction of light which is reflected back within the fiber's numerical aperture due to Rayleigh scattering (see Section 3.4.1). Hence the backscattering method which was first described by Barnoski and Jensen |Ref. 45| has the advantages of being nondestructive (i.e. does not require the cutting back of the fiber) and of requiring access to one end of the optical link only.

The backscattered optical power as a function of time $P_{Ra}(t)$ may be obtained from the following relationship |Ref. 46|:

$$P_{Ra}(t) = \tfrac{1}{2}P_i S\gamma_R W_o v_g \exp(-\gamma v_g t) \qquad (5.25)$$

where P_i is the optical power launched into the fiber, S is the fraction of captured optical power, γ_R is the Rayleigh scattering coefficient (backscatter loss per unit length), W_o is the input optical pulse width, v_g is the group velocity in the fiber and γ is the attenuation coefficient per unit length for the fiber. The fraction of captured optical power S is given by the ratio of the solid acceptance angle for the fiber to the total solid angle as:

$$S \simeq \frac{\pi NA^2}{4\pi n_i^2} = \frac{NA^2}{4n_i^2} \qquad (5.26)$$

It must be noted that the relationship given in Eq. (5.26) applies to step index fibers and the parameter S for a graded index fiber is generally a factor of 2/3 lower than for a step index fiber with the same numerical aperture |Ref. 47|. Hence using Eqs. (5.25) and (5.26) it is possible to determine the backscattered optical power from a point along the link length in relation to the forward optical power at that point.

Example 5.10

An optical fiber link consists of step index fiber which has a numerical aperture of 0.2 and a core refractive index of 1.5. The Rayleigh scattering coefficient for the fiber is 0.7 km^{-1}. When light pulses of 50 ns duration are launched into the fiber, calculate the ratio in decibels of the backscattered optical power to the forward optical power at the fiber input. The velocity of light in a vacuum is 2.998 $\times$ 10^8 m s^{-1}.

Solution: The backscattered optical power $P_{Ra}(t)$ is given by Eq. (5.25) where:

$$P_{Ra}(t) = \tfrac{1}{2}P_o S\gamma_R W_o v_g \exp(-\gamma v_g t)$$

At the fiber input $t = 0$; hence the power ratio is:

$$\frac{P_{Ra}(0)}{P_i} = \tfrac{1}{2}S\gamma_R W_o v_g$$

Substituting for S from Eq. (5.26) gives:

$$\frac{P_{Ra}(0)}{P_i} = \frac{1}{2}\left[\frac{NA^2 \gamma_R W_o v_g}{4n_i^2}\right]$$

The group velocity in the fiber v_g is defined by Eq. (2.32) as:

$$v_g = \frac{c}{N_1} \simeq \frac{c}{n_1}$$

Therefore

$$\frac{P_{Ra}(0)}{P_i} = \frac{1}{2}\left[\frac{NA^2 \gamma_R W_0 c}{4n_1^3}\right]$$

$$= \frac{1}{2}\left[\frac{(0.02)^2 0.7 \times 10^{-3} \times 50 \times 10^{-9} \times 2.998 \times 10^8}{4(1.5)^3}\right]$$

$$= 1.555 \times 10^{-5}$$

In decibels

$$\frac{P_{Ra}(0)}{P_i} = 10 \log_{10} 1.555 \times 10^{-5}$$

$$= -48.1 \text{ dB}$$

Hence in example 5.10 the backscattered optical power at the fiber input is 48.1 dB down on the forward optical power. The backscattered optical power should not be confused with any Fresnel reflection at the fiber input end face resulting from a refractive index mismatch. This could be considerably greater than the backscattered light from the fiber, presenting measurement problems with OTDR if it is allowed to fall onto the receiving photodetector of the equipment described below.

A block schematic of the backscatter measurement method is shown in Fig. 5.21 [Ref. 49]. A light pulse is launched into the fiber in the forward direction

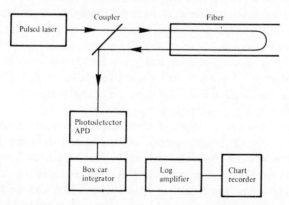

Fig. 5.21   Optical time domain reflectometry or the backscatter measurement method.

from an injection laser using either a directional coupler or a system of external lenses with a beam splitter (usually only in the laboratory). The back-scattered light is detected using an avalanche photodiode receiver which drives an integrator in order to improve the received signal to noise ratio by giving an arithmetic average over a number of measurements taken at one point within the fiber. This is necessary as the received optical signal power from a particular point along the fiber length is at a very low level compared with the forward power at that point by some 45–60 dB (see example 5.10), and is also swamped with noise. The signal from the integrator is fed through a logarithmic amplifier and averaged measurements for successive points within the fiber are plotted on a chart recorder. This provides location-dependent attenuation values which give an overall picture of the optical loss down the link. A possible backscatter plot is illustrated in Fig. 5.22 [Ref. 50] which shows the initial pulse caused by reflection and backscatter from the input coupler followed by a long tail caused by the distributed Rayleigh scattering from the input pulse as it travels down the link. Also shown in the plot is a pulse corresponding to the discrete reflection from a fiber joint, as well as a discontinuity due to excessive loss at a fiber imperfection or fault. The end of the fiber link is indicated by a pulse corresponding to the Fresnel reflection incurred at the output end face of the fiber. Such a plot yields the attenuation per unit length for the fiber by simply computing the slope of the curve over the length required. Also the location and insertion losses of joints and/or faults can be obtained from the power drop at their respective positions on the link. Finally the overall link length can be determined from the time difference between reflections from the fiber input and output end faces. Hence optical time domain reflectometry is a very powerful technique for field measurement on optical fiber links.

A number of optical time domain reflectometers are commercially available for operation in the shorter wavelength region below 1.0 μm. These devices emit a series of short (10–100 ns), intense optical pulses (100–500 mW) from which the backscattered light is received, analyzed and displayed on an oscilloscope, or plotted on a chart recorder. A typical example which will operate over a dynamic range of 40 dB two-way optical loss (often quoted as 2×20 dB since the single way optical loss is 20 dB) with location and attenuation accuracies of ± 4 m and $\pm 10\%$ respectively is shown in Fig. 5.23. In addition this device is capable of detecting reflecting breaks (i.e. from the 4% Fresnel reflection) over a single way dynamic range of up to 38 dB.

A major drawback of this technique, especially when using commercial optical time domain reflectometers, is the limited dynamic range of the measurement system. As indicated above this is currently around 40 dB (2×20 dB) for high performance devices. Hence, depending upon the fiber and coupling losses, the length of optical link which can be fully tested is restricted to at very best around 15 km. However, a method of optical time domain reflectometry by photon counting [Ref. 51] has shown some promise

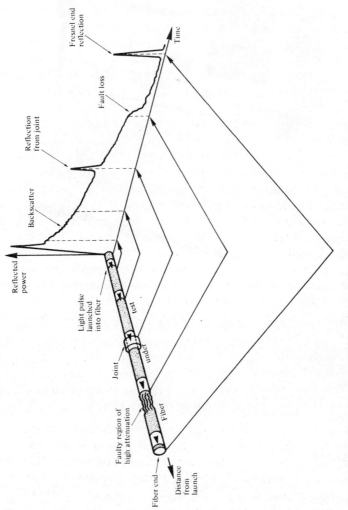

Fig. 5.22 An illustration of a possible backscatter plot from a fiber under test [Ref. 50].

Fig. 5.23   The STC OFR3 optical time domain reflectometer. (Courtesy of STC Components.)

especially when used as a diagnostic tool for fault location. In this method the avalanche photodiode is operated in a Geiger tube breakdown mode [Ref. 52] by biassing the device above its normal operating voltage where it can detect a single photon. Experimental [Ref. 51] measurements using this technique have demonstrated its ability to cope with single way losses of up to 40 dB (i.e. a two-way dynamic range of 80 dB) when detecting reflecting breaks in multimode fibers.

PROBLEMS

5.1   Describe what is meant by 'equilibrium mode distribution' and 'cladding mode stripping' with regard to transmission measurements in optical fibers. Briefly outline methods by which these conditions may be achieved when optical fiber measurements are performed.

5.2   Discuss with the aid of a suitable diagram the cut back technique used for the measurement of the total attenuation in an optical fiber. Indicate the differences in the apparatus utilized for spectral loss and spot attenuation measurement.

A spot measurement of fiber attenuation is performed on a 1.5 km length of optical fiber at a wavelength of 1.1 μm. The measured optical output power

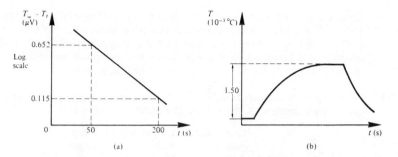

Fig. 5.24   Fiber absorption for measurements for problem 5.3: (a) plot of $(T_\infty - T_t)$ against time for a high absorption fiber; (b) the heating and cooling curve for the test fiber.

from the 1.5 km length of fiber is 50.1 μW. When the fiber is cut back to a 2 m length, the measured optical output power is 385.4 μW. Determine the attenuation per kilometer for the fiber at a wavelength of 1.1 μm, and estimate the accuracy of the result.

5.3   Briefly outline the principle behind the calorimetric methods used for the measurement of absorption loss in optical fibers.

A high absorption optical fiber was used to obtain the plot of $(T_\infty - T_t)$ (on a logarithmic scale) against time shown in Fig. 5.24(a). The measurements were achieved using a calorimeter and thermocouple experimental arrangement. Subsequently a different test fiber was passed three times through the same calorimeter before further measurements were taken. Measurements on the test fiber produced the heating and cooling curve shown in Fig. 5.24(b) when a constant 76 mW of optical power, at a wavelength of 1.06 μm, was passed through it. The constant C for the experimental arrangement was calculated to be 2.32×10^4 J °C^{-1}. Calculate the absorption loss in decibels per kilometer, at a wavelength of 1.06 μm, for the fiber under test.

5.4   Discuss the measurement of fiber scattering loss by describing the use of two common scattering cells.

A Nd:YAG laser operating at a wavelength of 1.064 μm is used with an integrating sphere to measure the scattering loss in an optical fiber sample. The optical power propagating within the fiber at the sphere is 98.45 μW and 5.31 nW of optical power is scattered within the sphere. The length of fiber in the sphere is 5.99 cm. Determine the optical loss due to scattering for the fiber at a wavelength of 1.064 μm in decibels per kilometer.

5.5   Fiber scattering loss measurements are taken at a wavelength of 0.75 μm using a solar cell cube. The reading of the input optical power to the cube is 7.78 V with a gain setting of 10^5. The corresponding reading from the scattering cell which incorporates a 4.12 cm length of fiber is 1.56 V with a gain setting of 10^9. Previous measurements of the total fiber attenuation at a wavelength of 0.75 μm gave a value of 3.21 dB km^{-1}. Calculate the absorption loss for the fiber at a wavelength of 0.75 μm in decibels per kilometer.

5.6   Discuss with the aid of suitable diagrams the measurement of dispersion in optical fibers. Consider both time and frequency domain measurement techniques.

Pulse dispersion measurements are taken on a graded index fiber in the time domain. The 3 dB width of the optical output pulses from a 950 m fiber length is 827 ps. When the fiber is cut back to a 2 m length the 3 dB width of the optical output pulses becomes 234 ps. Determine the optical bandwidth for a kilometer length of the fiber assuming Gaussian pulse shapes.

5.7 Pulse dispersion measurements in the time domain are taken on a multimode and a single mode step index fiber. The results recorded are:

	Input pulse width (3 dB)	Output pulse width (3 dB)	Fiber length (km)
(a) Multimode fiber	400 ps	31.20 ns	1.13
(b) Single mode fiber	200 ps	425 ps	2.35

Calculate the optical bandwidth over one kilometer for each fiber assuming Gaussian pulse shapes.

5.8 Describe the end reflection method for the detailed measurement of the refractive index profile of an optical fiber. Indicate how the resolution of this technique can be improved.

The end reflection technique is to be used in order to measure the refractive index profile of a graded index fiber in air. The fiber has a core axis refractive index of 1.46 and a relative index difference of 1%. It is envisaged that 20 point measurements will be made between the core axis and the cladding (inclusive). Estimate the percentage change in the Fresnel reflection coefficient which must be measured in order to facilitate these readings, assuming no index matching at the fiber input end face.

5.9 The fraction of light reflected at an air–fiber interface r is given by:

$$r = \left(\frac{n_1 - 1}{n_1 + 1} \right)^2$$

where n_1 is the fiber core refractive index at the point of reflection. Show that the fractional change in the core refractive index $\delta n_1/n_1$ may be expressed in terms of the fractional change in the reflection coefficient $\delta r/r$ following:

$$\frac{\delta n_1}{n_1} = \left(\frac{r^{\frac{1}{2}}}{1-r} \right) \frac{\delta r}{r}$$

Hence, show that for a step index fiber with n_1 of 1.5, a 5% change in r corresponds to only a 1% change in n_1.

5.10 A step index fiber has a nominal core refractive index of 1.48. The fiber input end face is immersed in oil with a refractive index of 1.51 prior to taking refractive index measurements using the end reflection method. Determine the anticipated resolution in the core refractive index measurement for a 2% change in the Fresnel reflection coefficient.

5.11 Compare and contrast two simple techniques used for the measurement of the numerical aperture of optical fibers.

Numerical aperture measurements are performed on an optical fiber. The angular limit of the far field pattern is found to be 26.1° when the fiber is rotated from a center zero point. The far field pattern is then displayed on a screen where its size is measured as 16.7 cm. Determine the numerical aperture for the fiber and the distance of the fiber output end face from the screen.

5.12 Describe, with the aid of a suitable diagram, the shadow method used for the on-line measurement of the outer diameter of an optical fiber.

The shadow method is used for the measurement of the outer diameter of an optical fiber. A fiber outer diameter of 347 μm generates a shadow pulse of 550 μs when the rotating mirror has an angular velocity of 3 rad s^{-1}. Calculate the distance between the rotating mirror and the optical fiber.

5.13 Outline the major design criteria of an optical fiber power meter for use in the field. Suggest any problems associated with field measurements using such a device.

Convert the following optical power meter readings to numerical values of power: 25 dBm, −5.2 dBm, 3.8 dBμ.

5.14 Describe what is meant by optical time domain reflectometry. Discuss how the technique may be used to take field measurements on optical fibers. Indicate the advantages of this technique over other measurement methods to determine attenuation in optical fibers.

A backscatter plot for an optical fiber link provided by OTDR is shown in Fig. 5.25. Determine:

(a) the attenuation of the optical link for the regions indicated *A*, *B* and *C* in decibels per kilometer.
(b) the insertion loss of the joint at the point *X*.

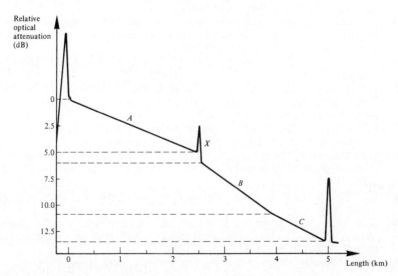

Fig. 5.25 The backscatter plot for the optical link of problem 5.14.

5.15   Discuss the sensitivity of OTDR in relation to commercial reflectometers. Comment on an approach which may lead to an improvement in the sensitivity of this measurement technique.

The Rayleigh scattering coefficient for a silica single mode step index fiber at a wavelength of 0.80 μm is 0.46 km^{-1}. The fiber has a refractive index of 1.46 and a numerical aperture of 0.14. When a light pulse of 60 ns duration at a wavelength of 0.80 μm is launched into the fiber, calculate the level in decibels of the backscattered light compared with the Fresnel reflection from a clean break in the fiber. It may be assumed that the fiber is surrounded by air.

Answers to Numerical Problems

5.2	5.92 dB km^{-1}, $\pm$0.13 dB	**5.10**	4.5 × 10^{-4}
5.3	1.77 dB km^{-1}	**5.11**	0.44, 17.0 cm
5.4	3.91 dB km^{-1}	**5.12**	21.0 cm
5.5	1.10 dB km^{-1}	**5.13**	316.2 mW, 302 μW, 2.40 μW
5.6	525.9 MHz km	**5.14**	(a) 2.0 dB km^{-1}, 3.0 dB km^{-1},
5.7	(a) 15.9 MHz km;		2.5 dB km^{-1};
	(b) 7.3 GHz km		(b) 1.0 dB
5.8	0.26%	**5.15**	−37.3 dB
5.9	1.0%		

REFERENCES

1   M. Tateda, T. Horiguchi, M. Tokuda and N. Uchida, 'Optical loss measurement in graded index fiber using a dummy fiber', *Appl. Opt.*, **18**(19), pp. 3272–3275, 1979.

2   M. Eve, A. M. Hill, D. J. Malyon, J. E. Midwinter, B. P. Nelson, J. R. Stern and J. V. Wright, 'Launching independent measurements of multimode fibres', *2nd European Conference on Optical Fiber Communication* (Paris), pp. 143–146, 1976.

3   M. Ikeda, Y. Murakami and K. Kitayama, 'Mode scrambler for optical fibres', *Appl. Opt.*, **16**(4), pp. 1045–1049, 1977.

4   S. Seikai, M. Tokuda, K. Yoshida and N. Uchida, 'Measurement of baseband frequency response of multimode fibre by using a new type of mode scrambler', *Electron. Lett.*, **13**(5), pp. 146–147, 1977.

5   J. P. Dakin, W. A. Gambling and D. N. Payne, 'Launching into glass–fibre waveguide', *Opt. Commun.*, **4**(5), pp. 354–357, 1972.

6   L. C. Cohen, P. Kaiser, P. D. Lazay and H. M. Presby, 'Fiber characterization', in S. E. Miller and A. G. Chynoweth (Eds.), *Optical Fiber Telecommunications*, pp. 343–400, Academic Press, 1979.

7   B. Costa and B. Sordo, 'Fibre characterization', *Optical Fibre Communication*, Technical Staff of CSELT, pp. 145–308, McGraw-Hill, 1981.

8   G. J. Cannell, R. Worthington and K. C. Byron, 'Measurement techniques', in C. P. Sandbank (Ed.), *Optical Fibre Communications Systems*, pp. 106–55, John Wiley, 1980.

9   D. Charlton and P. R. Reitz, 'Making fiber measurements', *Laser Focus*, pp. 52–64, Sept. 1979.

10   R. Bouillie and L. Jeunhomme, 'Measurement techniques for physical characteristics of optical fibers', *ICC International Conference on Communication* (Boston, USA), IEEE Pt 3, pp. 1–4, 1979.

11   J. E. Midwinter, *Optical Fibers for Transmission*, John Wiley, 1979.

12 K. I. White, 'A calorimetric method for the measurement of low optical absorption losses in optical communication fibres', *Opt. Quantum Electron.*, **8**, pp. 73–75, 1976.

13 K. I. White and J. E. Midwinter, 'An improved technique for the measurement of low optical absorption losses in bulk glass', *Opto-electronics*, **5**, pp. 323–334, 1973.

14 A. R. Tynes, 'Integrating cube scattering detector', *Appl. Opt.*, **9**(12), pp. 2706–2710, 1970.

15 F. W. Ostermayer and W. A. Benson, 'Integrating sphere for measuring scattering loss in optical fiber waveguides', *Appl. Opt.*, **13**(8), pp. 1900–1905, 1974.

16 S. de Vito and B. Sordo, 'Misure di attenuazione e diffusione in fibre ottiche multi-modo', LXXV Riuniuone AEI, Rome, 15–21 Sept. 1974.

17 J. P. Dakin, 'A simplified photometer for rapid measurement of total scattering attenuation of fibre optical waveguides', *Opt. Commun.*, **12**(1), pp. 83–88, 1974.

18 L. G. Cohen, P. Kaiser and C. Lin, 'Experimental techniques for evaluation of fiber transmission loss and dispersion', *Proc. IEEE*, **68**(10), pp. 1203–1208, 1980.

19 S. D. Personick, 'Baseband linearity and equalization in fiber optic digital communication systems', *Bell Syst. Tech. J.*, **52**(7), pp. 1175–1194, 1973.

20 B. P. Lathi, *Random Signals and Communication Theory*, International Textbook Company, 1968.

21 D. Gloge, E. L. Chinnock and T. P. Lee, 'Self pulsing GaAs laser for fiber dispersion measurement', *IEEE J. Quantum Electron.*, **QE-8**, pp. 844–846, 1972.

22 F. Krahn, W. Meininghaus and D. Rittich, 'Measuring and test equipment for optical cable', *Phillips Telecommun. Rev.*, **37**(4), pp. 241–249, 1979.

23 L. G. Cohen, 'Shuttle pulse measurements of pulse spreading in an optical fibre', *Appl. Opt.*, **14**(6), pp. 1351–1356, 1975.

24 I. Kokayashi, M. Koyama and K. Aoyama, 'Measurement of optical fibre transfer functions by swept frequency technique and discussion of fibre characteristics', *Electron. Commun. Jpn*, **60-C**(4), pp. 126–133, 1977.

25 W. E. Martin, 'Refractive index profile measurements of diffused optical waveguides', *Appl. Opt.*, **13**(9), pp. 2112–2116, 1974.

26 H. M. Presby, W. Mammel and R. M. Derosier, 'Refractive index profiling of graded index optical fibers', *Rev. Sci. Instr.*, **47**(3), pp. 348–352, 1976.

27 B. Costa and G. De Marchis, 'Test methods (optical fibres)', *Telecomm. J. (Engl. Ed.) Switzerland*, **48**(11), pp. 666–673, 1981.

28 L. G. Cohen, P. Kaiser, J. B. MacChesney, P. B. O'Conner and H. M. Presby, 'Transmission properties of a low-loss near-parabolic-index fiber', *Appl. Phys. Lett.*, **26**(8), pp. 472–474, 1975.

29 C. Lin, 'Measurement techniques in fiber optics', *IFOC Int. Fiber Opt. Commun. (USA)*, **2**(3), pp. 18–20, 52–53, 1981.

30 M. E. Marhic, P. S. Ho and M. Epstein, 'Nondestructive refractive index profile measurement of clad optical fibers', *Appl. Phys. Lett.*, **26**(10), pp. 574–575, 1975.

31 P. L. Chu, 'Measurements in optical fibres', *Proc. IEEE Australia*, **40**(4), pp. 102–114, 1979.

32 F. E. M. Sladen, D. N. Payne and M. J. Adams, 'Determination of optical fibre refractive index profile by near field scanning technique', *Appl. Phys. Lett.*, **28**(5), pp. 255–258, 1976.

33 M. J. Adams, D. N. Payne and F. M. E. Sladen, 'Correction factors for determination of optical fibre refractive-index profiles by near-field scanning techniques', *Electron. Lett.*, **12**(11), pp. 281–283, 1976.

34 W. Eickhoff and E. Weidel, 'Measuring method for the refractive index profile of optical glass fibres', *Opt. Quantum Electron.*, **7**, pp. 109–113, 1975.

35 B. Costa and B. Sordo, 'Measurements of the refractive index profile in optical fibres, comparison between different techniques', *2nd European Conference on Optical Fiber Communication* (Paris), pp. 81–86, 1976.

36 W. A. Gambling, D. N. Payne and H. Matsumura, 'Propagation studies on single mode phosphosilicate fibres', *2nd European Conference on Optical Fiber Communication* (Paris), pp. 95–100, 1976.

37 F. T. Stone, 'Rapid optical fibre delta measurement by refractive index tuning', *Appl. Opt.*, **16**(10), pp. 2738–2742, 1977.

38 L. G. Cohen and P. Glynn, 'Dynamic measurement of optical fibre diameter', *Rev. Sci. Instrum.*, **44**(12), pp. 1745–1752, 1973.

39 H. M. Presby, 'Refractive index and diameter measurements of unclad optical fibres', *J. Opt. Soc. Am.*, **64**(3), pp. 280–284, 1974.

40 P. L. Chu, 'Determination of diameters and refractive indices of step-index optical fibres', *Electron. Lett.*, **12**(7), pp. 150–157, 1976.

41 H. M. Presby and D. Marcuse, 'Refractive index and diameter determinations of step index optical fibers and preforms', *Appl. Opt.*, **13**(12), pp. 2882–2885, 1974.

42 D. Smithgall, L. S. Wakins and R. E. Frazee, 'High-speed noncontact fibre-diameter measurement using forward light scattering', *Appl. Opt.*, **16**(9), pp. 2395–2402, 1977.

43 F. Krahn, W. Meininghaus and D. Rittich, 'Field and test measurement equipment for optical cables', *Acta Electronica*, **23**(3), pp. 269–275, 1979.

44 R. Olshansky, M. G. Blankenship and D. B. Keck, 'Length-dependent attenuation measurements in graded-index fibres', *Proceedings of 2nd European Conference on Optical Communication* (Paris), pp. 111–113, 1976.

45 M. K. Barnoski and S. M. Jensen, 'Fiber waveguides: a novel technique for investigating attenuation characteristics', *Appl. Opt.*, **15**(9), pp. 2112–2115, 1976.

46 S. D. Pensonick, 'Photon probe, an optical fibre time-domain reflectometer', *Bell Syst. Tech. J.*, **56**(3), pp. 355–366, 1977.

47 E. G. Newman, 'Optical time domain reflectometer: comment', *Appl. Opt.*, **17**(11), p. 1675, 1978.

48 E. A. Lacy, *Fiber Optics*, Prentice-Hall, 1982.

49 M. K. Barnoski and S. D. Personick, 'Measurements in fiber optics', *Proc. IEEE*, **66**(4), pp. 429–440, 1978.

50 J. D. Archer, *Manual of Fibre Optics Communication*, STC Components Group, UK, 1981.

51 P. Healey, 'Optical time domain reflectometry by photon counting', *6th European Conference on Optical Communication* (UK), pp. 156–159, 1980.

52 P. P. Webb *et al.*, 'Single photon detection with avalanche photodiodes', *Bull. Am. Phys. Soc. II*, **15**, p. 813, 1970.

53 P. Healey, 'OTDR by photon counting', *IEE Colloquium on Test Equipment for Optical Fibre Commun. Syst.* (London), paper 4/1, May 1981.

6

Optical Sources 1: The Laser

6.1 INTRODUCTION

The optical source is often considered to be the active component in an optical fiber communication system. Its fundamental function is to convert electrical energy in the form of a current into optical energy (light) in an efficient manner which allows the light output to be effectively launched or coupled into the optical fiber. Three main types of optical light source are available. These are:

(a) wideband 'continuous spectra' sources (incandescent lamps);
(b) monochromatic incoherent sources (light emitting diodes' LEDs);
(c) monochromatic coherent sources (lasers).

To aid consideration of the sources currently in major use the historical aspect must be mentioned. In the early stages of optical fiber communication the most powerful narrowband coherent light sources were necessary due to severe attenuation and dispersion in the fibers. Therefore initially gas lasers (helium–neon) were utilized. However, the development of the semiconductor injection laser and the LED, together with the substantial improvement in the properties of optical fibers, has given prominence to these two specific sources.

To a large extent these two sources fulfil the major requirements for an optical fiber emitter which are outlined below.

(a) A size and configuration compatible with launching light into an optical fiber. Ideally the light output should be highly directional.
(b) Must accurately track the electrical input signal to minimize distortion and noise. Ideally the source should be linear.
(c) Should emit light at wavelengths where the fiber has low losses and low dispersion and where the detectors are efficient.
(d) Preferably capable of simple signal modulation (i.e. direct—see Section 7.5) over a wide bandwidth extending from audio frequencies to beyond the gigahertz range.
(e) Must couple sufficient optical power to overcome attenuation in the fiber plus additional connector losses and leave adequate power to drive the detector.

(f) Should have a very narrow spectral bandwidth (linewidth) in order to minimize dispersion in the fiber.

(g) Must be capable of maintaining a stable optical output which is largely unaffected by changes in ambient conditions (e.g. temperature).

(h) It is essential that the source is comparatively cheap and highly reliable in order to compete with conventional transmission techniques.

In order to form some comparison between these two types of light source the historical aspect must be enlarged upon. The first generation optical communication sources were designed to operate between 0.8 and 0.9 μm (ideally around 0.85 μm) because initially the properties of the semiconductor materials used lent themselves to emission at this wavelength. Also as suggested in (c) this wavelength avoided the loss incurred in many fibers near 0.9 μm due to the OH ion (see Section 3.3.2). These early systems utilized multimode step index fibers which required the superior performance of semiconductor lasers for links of reasonable bandwidth (tens of megahertz) and distances (several kilometers). The LED (being a lower power source generally exhibiting little spatial or temporal coherence) was not suitable for long distance wideband transmission, although it found use in more moderate applications.

However, the role of the LED as a source for optical fiber communications was enhanced following the development of multimode graded index fiber. The substantial reduction in intermodal dispersion provided by this fiber type over multimode step index fiber allowed incoherent LEDs emitting in the 0.8–0.9 μm wavelength band to be utilized for applications requiring wider bandwidths. This position was further consolidated with the development of second generation optical fiber sources operating at wavelengths between 1.1 and 1.6 μm where both material losses and dispersion are greatly reduced. In this wavelength region wideband graded index fiber systems utilizing LED sources may be operated over long distances without the need for intermediate repeaters. Furthermore, LEDs offer the advantages of relatively simple construction and operation with the inherent effects of these factors on cost and extended, trouble-free life.

In parallel with these later developments in multimode optical propagation came advances in single mode fiber construction. This has stimulated the development of single mode laser sources to take advantage of the extremely low dispersion offered by single mode fibers. These systems are ideally suited to extra wideband, very long-haul applications and are currently under intensive investigation for long-distance telecommunications. On the other hand, light is usually emitted from the LED in many spatial modes which cannot be readily focused and coupled into single mode fiber. Hence to date the LED has been utilized almost exclusively as a multimode source which will only give adequate coupling efficiency into multimode fiber. However, in this role the LED has become a primary multimode source which is extensively used for increasingly wider bandwidth, longer-haul applications. Therefore at present the LED is chosen for many applications using multimode fibers and the injection laser diode (ILD) tends to find more use as a single mode device in single mode fiber systems. Although other laser types (e.g. Nd:YAG laser, see Section 6.11) as well as the injection laser may eventually find limited use in optical fiber communications, this chapter and the following one will deal primarily with major structures and configurations of semiconductor sources (ILD and LED) taking into account recent developments and possible future advances.

We begin by describing in Section 6.2 the basic principles of laser operation which may be applied to all laser types. Immediately following in Section 6.3 is a discussion of optical emission from semiconductors in which we concentrate on the fundamental operating principles, the structure and the materials for the semiconductor laser. Aspects of practical semiconductor injection lasers are then considered in Section 6.4 prior to a more specific discussion of the structure and operation of multimode devices in Section 6.5. Following in Section 6.6 is a brief discussion of the single mode injection laser which provides a basis for the description of the major single mode structures presented in Section 6.7. As the preceding sections have primarily dealt with injection lasers operating in the shorter wavelength region (0.8–0.9 µm), a brief account of longer wavelength (1.1–1.6 µm) devices is given in Section 6.8. In Section 6.9 we describe the operating characteristics which are common to all injection laser types before a short discussion of injection laser to optical fiber coupling together with device packaging is presented in Section 6.10. Finally, in Section 6.11 nonsemiconductor lasers are briefly considered, the discussion concentrating on the neodymium-doped yttrium–aluminum–garnet (Nd:YAG) device.

6.2 BASIC CONCEPTS

To gain an understanding of the light-generating mechanisms within the major optical sources used in optical fiber communications it is necessary to consider both the fundamental atomic concepts and the device structure. In this context the requirements for the laser source are far more stringent than those for the LED. Unlike the LED, strictly speaking, the laser is a device which amplifies light. Hence the derivation of the term LASER as an acronym for Light Amplification by Stimulated Emission of Radiation. Lasers, however, are seldom used as amplifiers since there are practical difficulties in relation to the achievement of high gain whilst avoiding oscillation from the required energy feedback. Thus the practical realization of the laser is as an optical oscillator. The operation of the device may be described by the formation of an electromagnetic standing wave within a cavity (or optical resonator) which provides an output of monochromatic highly coherent radiation. By contrast the LED provides optical emission without an inherent gain mechanism. This results in incoherent light output.

In this section we elaborate on the basic principles which govern the operation of both these optical sources. It is clear, however, that the operation of the laser must be discussed in some detail in order to provide an appreciation of the way it functions as an optical source. Hence we concentrate first on the general principles of laser action.

6.2.1 Absorption and Emission of Radiation

The interaction of light with matter takes place in discrete packets of energy or quanta, called photons. Furthermore the quantum theory suggests that atoms exist only in certain discrete energy states such that absorption and emission of light causes them to make a transition from one discrete energy state to another. The frequency of the absorbed or emitted radiation f is related to the difference in energy E between the higher energy state E_2 and the lower energy state E_1 by the expression:

$$E = E_2 - E_1 = hf \tag{6.1}$$

where $h = 6.626 \times 10^{-34}$ J s is Planck's constant. These discrete energy states for the atom may be considered to correspond to electrons occurring in particular energy levels relative to the nucleus. Hence different energy states for the atom correspond to different electron configurations, and a single electron transition between two energy levels within the atom will provide a change in energy suitable for the absorption or emission of a photon. It must be noted, however, that modern quantum theory [Ref. 1] gives a probabilistic description which specifies the energy levels in which electrons are most likely to be found. Nevertheless, the concept of stable atomic energy states and electron transitions between energy levels is still valid.

Figure 6.1(a) illustrates a two energy state or level atomic system where an atom is initially in the lower energy state E_1. When a photon with energy $(E_2 - E_1)$ is incident on the atom it may be excited into the higher energy state E_2 through absorption of the photon. This process is sometimes referred to as stimulated absorption. Alternatively when the atom is initially in the higher energy state E_2 it can make a transition to the lower energy state E_1 providing the emission of a photon at a frequency corresponding to Eq. (6.1). This emission process can occur in two ways:

(a) by spontaneous emission in which the atom returns to the lower energy state in an entirely random manner;

(b) by stimulated emission when a photon having an energy equal to the energy difference between the two states $(E_2 - E_1)$ interacts with the atom in the upper energy state causing it to return to the lower state with the creation of a second photon.

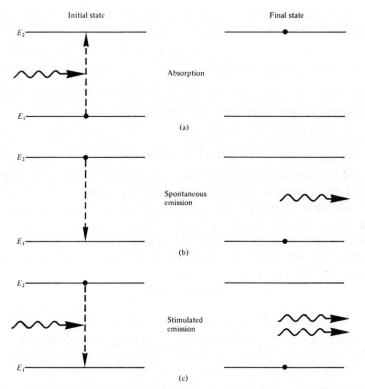

Fig. 6.1   Energy state diagram showing: (a) absorption; (b) spontaneous emission; (c) stimulated emission. The black dot indicates the state of the atom before and after a transition takes place.

These two emission processes are illustrated in Figs. 6.1(b) and (c) respectively. The random nature of the spontaneous emission process where light is emitted by electronic transitions from a large number of atoms gives incoherent radiation. A similar emission process in semiconductors provides the basic mechanism for light generation within the LED (see Section 6.3.2).

It is the stimulated emission process, however, which gives the laser its special properties as an optical source. Firstly the photon produced by stimulated emission is generally* of an identical energy to the one which caused it and hence the light associated with them is of the same frequency. Secondly the light associated with the stimulating and stimulated photon is in phase and has the same polarization. Therefore, in contrast to spontaneous emission, coherent radiation is obtained. Furthermore this means that when an

---

* A photon with energy hf will not necessarily always stimulate another photon with energy hf. Photons may be stimulated over a small range of energies around hf providing an emission which has a finite frequency or wavelength spread (linewidth).

atom is stimulated to emit light energy by an incident wave, the liberated energy can add to the wave in a constructive manner, providing amplification.

6.2.2   The Einstein Relations

Prior to discussion of laser action in semiconductors it is useful to consider optical amplification in the two level atomic system shown in Fig. 6.1. In 1917 Einstein [Ref. 2] demonstrated that the rates of the three transition processes of absorption, spontaneous emission and stimulated emission were related mathematically. He achieved this by considering the atomic system to be in thermal equilibrium such that the rate of the upward transitions must equal the rate of the downward transitions. The population of the two energy levels of such a system are described by Boltzmann statistics which give:

$$\frac{N_1}{N_2} = \frac{g_1 \exp(-E_1/KT)}{g_2 \exp(-E_2/KT)} = \frac{g_1}{g_2} \exp(E_2 - E_1/KT)$$

$$= \frac{g_1}{g_2} \exp(hf/KT) \tag{6.2}$$

where N_1 and N_2 represent the density of atoms in energy levels E_1 and E_2 respectively with g_1 and g_2 being the corresponding degeneracies* of the levels, K is Boltzmann's constant and T is the absolute temperature.

As the density of atoms in the lower or ground energy state E_1 is N_1, the rate of upward transition or absorption is proportional to both N_1 and the spectral density ρ_f of the radiation energy at the transition frequency f. Hence the upward transition rate R_{12} (indicating an electron transition from level 1 to level 2) may be written as:

$$R_{12} = N_1 \rho_f B_{12} \tag{6.3}$$

where the constant of proportionality B_{12} is known as the Einstein coefficient of absorption.

By contrast atoms in the higher or excited energy state can undergo electron transitions from level 2 to level 1 either spontaneously or through stimulation by the radiation field. For spontaneous emission the average time an electron exists in the excited state before a transition occurs is known as the spontaneous lifetime τ_{21}. If the density of atoms within the system with energy E_2 is N_2, then the spontaneous emission rate is given by the product of N_2 and

---

* In many cases the atom has several sublevels of equal energy within an energy level which is then said to be degenerate. The degeneracy parameters g_1 and g_2 indicate the number of sublevels within the energy levels E_1 and E_2 respectively. If the system is not degenerate then g_1 and g_2 may be set to unity [Ref. 1].

$1/\tau_{21}$. This may be written as $N_2 A_{21}$ where A_{21}, the Einstein coefficient of spontaneous emission, is equal to the reciprocal of the spontaneous lifetime.

The rate of stimulated downward transition of an electron from level 2 to level 1 may be obtained in a similar manner to the rate of stimulated upward transition. Hence the rate of stimulated emission is given by $N_2 \rho_f B_{21}$, where B_{21} is the Einstein coefficient of stimulated emission. The total transition rate from level 2 to level 1, R_{21}, is the sum of the spontaneous and stimulated contributions. Hence:

$$R_{21} = N_2 A_{21} + N_2 \rho_f B_{21} \tag{6.4}$$

For a system in thermal equilibrium, the upward and downward transition rates must be equal and therefore $R_{12} = R_{21}$, or

$$N_1 \rho_f B_{12} = N_2 A_{21} + N_2 \rho_f B_{21} \tag{6.5}$$

It follows that:

$$\rho_f = \frac{N_2 A_{21}}{N_1 B_{12} - N_2 B_{21}}$$

and

$$\rho_f = \frac{A_{21}/B_{21}}{(B_{12} N_1 / B_{21} N_2) - 1} \tag{6.6}$$

Substituting Eq. (6.2) into Eq. (6.6) gives

$$\rho_f = \frac{A_{21}/B_{21}}{[(g_1 B_{12}/g_2 B_{21}) \exp (hf/KT)] - 1} \tag{6.7}$$

However, since the atomic system under consideration is in thermal equilibrium it produces a radiation density which is identical to black body radiation. Planck showed that the radiation spectral density for a black body radiating within a frequency range f to $f + df$ is given by [Ref. 3]:

$$\rho_f = \frac{8\pi h f^3}{c^3} \left(\frac{1}{\exp (hf/KT) - 1} \right) \tag{6.8}$$

Comparing Eq. (6.8) with Eq. (6.7) we obtain the Einstein relations:

$$B_{12} = \left(\frac{g_2}{g_1} \right) B_{21} \tag{6.9}$$

and

$$\frac{A_{21}}{B_{21}} = \frac{8\pi h f^3}{c^3} \tag{6.10}$$

It may be observed from Eq. (6.9) that when the degeneracies of the two levels are equal ($g_1 = g_2$) then the probabilities of absorption and stimulated emission are equal. Furthermore, the ratio of the stimulated emission rate to the spontaneous emission rate is given by:

$$\frac{\text{stimulated emission rate}}{\text{spontaneous emission rate}} = \frac{B_{21} \, \rho_f}{A_{21}} = \frac{1}{\exp{(hf/KT)} - 1} \qquad (6.11)$$

Example 6.1

Calculate the ratio of the stimulated emission rate to the spontaneous emission rate for an incandescent lamp operating at a temperature of 1000 K. It may be assumed that the average operating wavelength is 0.5 μm.

Solution: The average operating frequency is given by:

$$f = \frac{c}{\lambda} = \frac{2.998 \times 10^8}{0.5 \times 10^{-6}} \simeq 6.0 \times 10^{14} \text{ Hz}$$

Using Eq. (6.11) the ratio is:

$$\frac{\text{stimulated emission rate}}{\text{spontaneous emission rate}} = \frac{1}{\exp\left(\dfrac{6.626 \times 10^{-34} \times 6 \times 10^{14}}{1.381 \times 10^{-23} \times 1000}\right)}$$

$$= \exp{(-28.8)}$$

$$= 3.1 \times 10^{-13}$$

The result obtained in example 6.1 indicates that for systems in thermal equilibrium spontaneous emission is by far the dominant mechanism. Furthermore it illustrates that the radiation emitted from ordinary optical sources in the visible spectrum occurs in a random manner, proving these sources are incoherent.

It is apparent that in order to produce a coherent optical source and amplification of a light beam the rate of stimulated emission must be increased far above the level indicated by example 6.1. From consideration of Eq. (6.5) it may be noted that for stimulated emission to dominate over absorption and spontaneous emission in a two level system both the radiation density and the population density of the upper energy level N_2 must be increased in relation to the population density of the lower energy level N_1.

6.2.3 Population Inversion

Under the conditions of thermal equilibrium given by the Boltzmann distribution (Eq. (6.2)) the lower energy level E_1 of the two level atomic system contains more atoms than the upper energy level E_2. This situation which is normal for structures at room temperature is illustrated in Fig. 6.2(a). However, to achieve optical amplification it is necessary to create a nonequilibrium distribution of atoms such that the population of the upper energy level is greater than that of the lower energy level (i.e. $N_2 > N_1$). This condition which is known as population inversion is illustrated in Fig. 6.2(b).

In order to achieve population inversion it is necessary to excite atoms into the upper energy level E_2 and hence obtain a nonequilibrium distribution. This process is achieved using an external energy source and is referred to as 'pumping'. A common method used for pumping involves the application of intense radiation (e.g. from an optical flash tube or high frequency radio field). In the former case atoms are excited into the higher energy state through stimulated absorption. However, the two level system discussed above does not lend itself to suitable population inversion. Referring to Eq. (6.9), when the two levels are equally degenerate (or not degenerate) then $B_{12} = B_{21}$. Thus the probabilities of absorption and stimulated emission are equal, providing at best equal populations in the two levels.

Population inversion, however, may be obtained in systems with three or four energy levels. The energy level diagrams for two such systems which correspond to two nonsemiconductor lasers are illustrated in Fig. 6.3. To aid attainment of population inversion both systems display a central metastable state in which the atoms spend an unusually long time. It is from this metastable level that the stimulated emission or lasing takes place. The three level system (Fig. 6.3(a)) consists of a ground level E_0, a metastable level E_1 and a third level above the metastable level E_2. Initially the atomic distribution will follow the Boltzmann law. However, with suitable pumping the electrons in

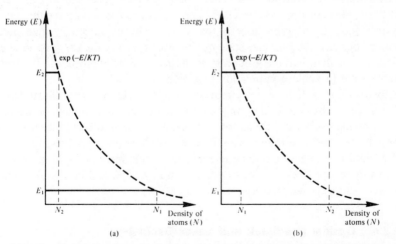

Fig. 6.2   Populations in a two energy level system: (a) Boltzmann distribution for a system in thermal equilibrium; (b) a nonequilibrium distribution showing population inversion.

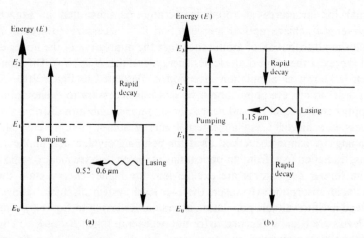

Fig. 6.3 Energy level diagrams showing population inversion and lasing for two non-semiconductor lasers: (a) three level system—ruby (crystal) laser; (b) four level system—He–Ne (gas) laser.

some of the atoms may be excited from the ground state into the higher level E_2. Since E_2 is a normal level the electrons will rapidly decay by nonradiative processes to either E_1 or directly to E_0. Hence empty states will always be provided in E_2. The metastable level E_1 exhibits a much longer lifetime than E_2 which allows a large number of atoms to accumulate at E_1. Over a period the density of atoms in the metastable state N_1 increases above those in the ground state N_0 and a population inversion is obtained between these two levels. Stimulated emission and hence lasing can then occur creating radiative electron transitions between levels E_1 and E_0. A drawback with the three level system such as the ruby laser is that it generally requires very high pump powers because the terminal state of the laser transition is the ground state. Hence more than half the ground state atoms must be pumped into the metastable state to achieve population inversion.

By contrast a four level system such as the He–Ne laser illustrated in Fig. 6.3(b) is characterized by much lower pumping requirements. In this case the pumping excites the atoms from the ground state into energy level E_3 and they decay rapidly to the metastable level E_2. However, since the populations of E_3 and E_1 remain essentially unchanged a small increase in the number of atoms in energy level E_2 creates population inversion, and lasing takes place between this level and level E_1.

6.2.4 Optical Feedback and Laser Oscillation

Light amplification in the laser occurs when a photon colliding with an atom in the excited energy state causes the stimulated emission of a second photon and then both these photons release two more. Continuation of this process effectively creates avalanche multiplication, and when the electromagnetic waves associated with these photons are in phase, amplified coherent emission is obtained. To achieve this laser action it is necessary to contain photons within the laser medium and maintain the conditions for coherence. This is accomplished by placing or forming mirrors (plane or curved) at either end of the amplifying medium as illustrated in Fig. 6.4. The optical cavity formed is more analogous to an oscillator than an amplifier as it provides positive feedback of the photons by reflection at the mirrors at either end of the cavity. Hence the optical signal is fed back many times whilst receiving amplification as it passes through the medium. The structure therefore acts as a Fabry–Perot resonator. Although the amplification of the signal from a single pass through the medium is quite small, after multiple passes the net gain can be large. Furthermore, if one mirror is made partially transmitting, useful radiation may escape from the cavity.

A stable output is obtained at saturation when the optical gain is exactly matched by the losses incurred in the amplifying medium. The major losses result from factors such as absorption and scattering in the amplifying medium, absorption, scattering and diffraction at the mirrors and non-useful transmission through the mirrors.

Oscillations occur in the laser cavity over a small range of frequencies where the cavity gain is sufficient to overcome the above losses. Hence the device is not a perfectly monochromatic source but emits over a narrow spectral band. The central frequency of this spectral band is determined by the mean energy level difference of the stimulated emission transition. Other oscillation frequencies within the spectral band result from frequency variations due to the thermal motion of atoms within the amplifying medium (known as Doppler broadening*) and by atomic collisions†. Hence the amplification within the

Fig. 6.4 The basic laser structure incorporating plane mirrors.

---

* Doppler broadening is referred to as an inhomogeneous broadening mechanism since individual groups of atoms in the collection have different apparent resonance frequencies.
† Atomic collisions provide homogeneous broadening as every atom in the collection has the same resonant frequency and spectral spread.

laser medium results in a broadened laser transition or gain curve over a finite spectral width as illustrated in Fig. 6.5. The spectral emission from the device therefore lies within the frequency range dictated by this gain curve.

Since the structure forms a resonant cavity, when sufficient population inversion exists in the amplifying medium the radiation builds up and becomes established as standing waves between the mirrors. These standing waves exist only at frequencies for which the distance between the mirrors is an integral number of half wavelengths. Thus when the optical spacing between the mirrors is L the resonance condition along the axis of the cavity is given by [Ref. 4]:

$$L = \frac{\lambda q}{2n} \qquad (6.12)$$

where λ is the emission wavelength, n is the refractive index of the amplifying medium and q is an integer. Alternatively discrete emission frequencies f are defined by

$$f = \frac{qc}{2nL} \qquad (6.13)$$

where c is the velocity of light. The different frequencies of oscillation within the laser cavity are determined by the various integer values of q and each constitutes a resonance or mode. Since Eqs. (6.12) and (6.13) apply for the case when L is along the longitudinal axis of the structure (Fig. 6.4) the frequencies given by Eq. (6.13) are known as the longitudinal or axial modes. Furthermore from Eq. (6.13) it may be observed that these modes are separated by a frequency interval δf where:

$$\delta f = \frac{c}{2nL} \qquad (6.14)$$

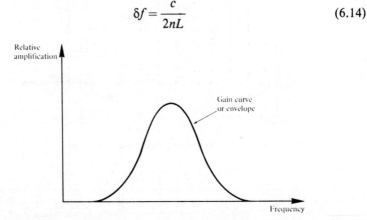

Fig. 6.5   The relative amplification in the laser amplifying medium showing the broadened laser transition line or gain curve.

Example 6.2

A ruby laser contains a crystal length 4 cm with a refractive index of 1.78. The peak emission wavelength from the device is 0.55 μm. Determine the number of longitudinal modes and their frequency separation.

Solution: The number of longitudinal modes supported within the structure may be obtained from Eq. (6.12) where:

$$q = \frac{2nL}{\lambda} = \frac{2 \times 1.78 \times 0.04}{0.55 \times 10^{-6}} = 2.6 \times 10^5$$

Using Eq. (6.14) the frequency separation of the modes is:

$$\delta f = \frac{2.998 \times 10^8}{2 \times 1.78 \times 0.04} = 2.1 \text{ GHz}$$

Although the result of example 6.2 indicates that a large number of modes may be generated within the laser cavity, the spectral output from the device is defined by the gain curve. Hence the laser emission will only include the longitudinal modes contained within the spectral width of the gain curve. This situation is illustrated in Fig. 6.6 where several modes are shown to be present in the laser output. Such a device is said to be multimode.

Laser oscillation may also occur in a direction which is transverse to the axis of the cavity. This gives rise to resonant modes which are transverse to the

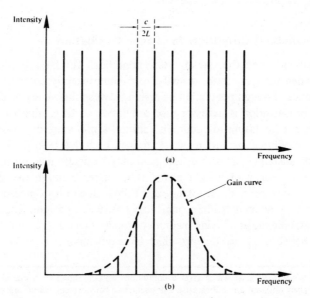

Fig. 6.6 (a) The modes in the laser cavity. (b) The longitudinal modes in the laser output.

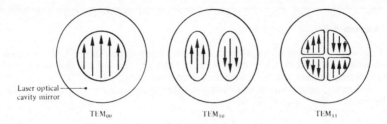

Fig. 6.7   The lower order transverse modes of a laser.

direction of propagation. These transverse electromagnetic modes are designated in a similar manner to transverse modes in waveguides (Section 2.3.2) by TEM_{lm} where the integers l and m indicate the number of transverse modes (see Fig. 6.7). Unlike the longitudinal modes which contribute only a single spot of light to the laser output, transverse modes may give rise to a pattern of spots at the output. This may be observed from the low order transverse mode patterns shown in Fig. 6.7 on which the direction of the electric field is also indicated. In the case of the TEM_{00} mode all parts of the propagating wavefront are in phase. This is not so, however, with higher order modes (TEM_{10}, TEM_{11}, etc.) where phase reversals produce the various mode patterns. Thus the greatest degree of coherence together with the highest level of spectral purity may be obtained from a laser which operates in only the TEM_{00} mode. Higher order transverse modes only occur when the width of the cavity is sufficient for them to oscillate. Consequently they may be eliminated by suitable narrowing of the laser cavity.

6.2.5   Threshold Condition for Laser Oscillation

It has been indicated that steady state conditions for laser oscillation are achieved when the gain in the amplifying medium exactly balances the total losses.* Hence although population inversion between the energy levels providing the laser transition is necessary for oscillation to be established, it is not alone sufficient for lasing to occur. In addition a minimum or threshold gain within the ampiifying medium must be attained such that laser oscillations are initiated and sustained. This threshold gain may be determined by considering the change in energy of a light beam as it passes through the amplifying medium. For simplicity all the losses except those due to transmission through the mirrors may be included in a single loss coefficient per unit length $\bar{a}$ cm^{-1}. Again we assume the amplifying medium occupies a length L completely filling the region between the two mirrors which have reflectivities r_1 and r_2. On each

---

*This applies to CW laser which gives a continuous output, rather than pulsed devices for which slightly different conditions exist. For oscillation to commence the fractional gain and loss must be matched.

round trip the beam passes through the medium twice. Hence the fractional loss incurred by the light beam is:

$$\text{Fractional loss} = r_1 r_2 \exp(-2\bar{\alpha}L) \qquad (6.15)$$

Furthermore it is found that the increase in beam intensity resulting from stimulated emission is exponential [Ref. 4]. Therefore if the gain coefficient per unit length produced by stimulated emission is $\bar{g}$ cm^{-1}, the fractional round trip gain is given by

$$\text{Fractional gain} = \exp(2\bar{g}l) \qquad (6.16)$$

Hence

$$\exp(2\bar{g}l) \times r_1 r_2 \exp(-2\bar{\alpha}L) = 1$$

and

$$r_1 r_2 \exp 2(\bar{g}-\bar{\alpha})L = 1 \qquad (6.17)$$

The threshold gain per unit length may be obtained by rearranging the above expression to give:

$$\bar{g}_{th} = \bar{\alpha} + \frac{1}{2L} \ln \frac{1}{r_1 r_2} \qquad (6.18)$$

The second term on the right hand side of Eq. (6.18) represents the transmission loss through the mirrors.*

For laser action to be easily achieved it is clear that a high threshold gain per unit length is required in order to balance the losses from the cavity. However it must be noted that the parameters displayed in Eq. (6.18) are totally dependent on the laser type.

6.3 OPTICAL EMISSION FROM SEMICONDUCTORS

6.3.1 The p–n Junction

To allow consideration of semiconductor optical sources it is necessary to review some of the properties of semiconductor materials, especially with regard to the p–n junction. A perfect semiconductor crystal containing no impurities or lattice defects is said to be intrinsic. The energy band structure [Ref. 1] of an intrinsic semiconductor is illustrated in Fig. 6.8(a) which shows the valence and conduction bands separated by a forbidden energy gap or bandgap E_g, the width of which varies for different semiconductor materials.

---

*This term is sometimes expressed in the form $1/L \ln 1/r$, where r, the reflectivity of the mirrored ends, is equal to $\sqrt{(r_1 r_2)}$.

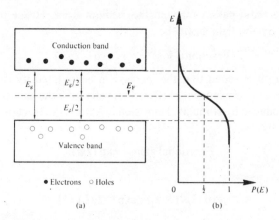

Fig. 6.8   (a) The energy band structure of an intrinsic semiconductor at a temperature above absolute zero showing an equal number of electrons and holes in the conduction band and the valence band respectively. (b) The Fermi–Dirac probability distribution corresponding to (a).

Figure 6.8(a) shows the situation in the semiconductor at a temperature above absolute zero where thermal excitation raises some electrons from the valence band into the conduction band leaving empty hole states in the valence band. These thermally excited electrons in the conduction band and the holes left in the valence band allow conduction through the material, and are called carriers.

For a semiconductor in thermal equilibrium the energy level occupation is described by the Fermi–Dirac distribution function (rather than the Boltzmann). Consequently the probability $P(E)$ that an electron gains sufficient thermal energy at an absolute temperature T that it will be found occupying a particular energy level E, is given by the Fermi–Dirac distribution [Ref. 1]:

$$P(E) = \frac{1}{1 + \exp{(E - E_F)/KT}} \qquad (6.19)$$

where K is Boltzmann's constant and E_F is known as the Fermi energy or Fermi level. The Fermi level is only a mathematical parameter but it gives an indication of the distribution of carriers within the material. This is shown in Fig. 6.8(b) for the intrinsic semiconductor where the Fermi level is at the center of the bandgap, indicating that there is a small probability of electrons occupying energy levels at the bottom of the conduction band and a corresponding number of holes occupying energy levels at the top of the valence band.

To create an extrinsic semiconductor the material is doped with impurity atoms which either create more free electrons (donor impurity) or holes (acceptor impurity). These two situations are shown in Fig. 6.9 where the donor impurities form energy levels just below the conduction band whilst acceptor impurities form energy levels just above the valence band.

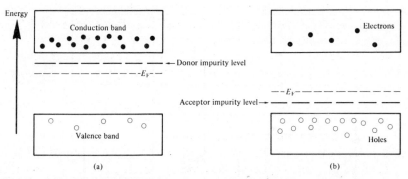

Fig. 6.9   Energy band diagrams: (a) *n* type semiconductor; (b) *p* type semiconductor.

When donor impurities are added, thermally excited electrons from the donor levels are raised into the conduction band to create an excess of negative charge carriers and the semiconductor is said to be *n* type, with the majority carriers being electrons. The Fermi level corresponding to this carrier distribution is raised to a position above the center of the bandgap as illustrated in Fig. 6.9(a). When acceptor impurities are added as shown in Fig. 6.9(b) thermally excited electrons are raised from the valence band to the acceptor impurity levels leaving an excess of positive charge carriers in the valence band and creating a *p* type semiconductor where the majority carriers are holes. In this case the Fermi level is lowered below the centre of the bandgap.

The *p–n* junction diode is formed by creating adjoining *p* and *n* type semiconductor layers in a single crystal as shown in Fig. 6.10(a). A thin depletion region or layer is formed at the junction through carrier recombination which effectively leaves it free of mobile charge carriers (both electrons and holes). This establishes a potential barrier between the *p* and *n* type regions which restricts the interdiffusion of majority carriers from their respective regions as illustrated in Fig. 6.10(b). In the absence of an externally applied voltage no current flows as the potential barrier prevents the net flow of carriers from one region to another. When the junction is in this equilibrium state the Fermi level for the *p* and *n* type semiconductor is the same as shown in Fig. 6.10(b).

The width of the depletion region and thus the magnitude of the potential barrier is dependent upon the carrier concentrations (doping) in the *p* and *n* type regions, and any external applied voltage. When an external positive voltage is applied to the *p* type region with respect to the *n* type, both the depletion region width and the resulting potential barrier are reduced and the diode is said to be forward biassed. Electrons from the *n* type region and holes from the *p* type region can flow more readily across the junction into the opposite type region. These minority carriers are effectively injected across the junction by the application of the external voltage and form a current flow through the

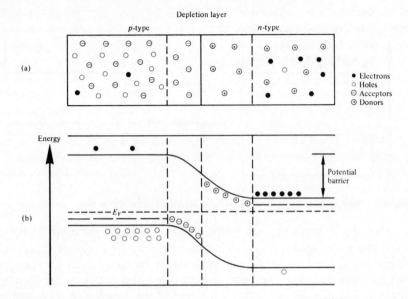

Fig. 6.10 (a) The impurities and charge carriers at a $p–n$ junction. (b) The energy band diagram corresponding to (a).

device as they continuously diffuse away from the interface. However, this situation in suitable semiconductor materials allows carrier recombination with the emission of light.

6.3.2 Spontaneous Emission

The increased concentration of minority carriers in the opposite type region in the forward biassed $p–n$ diode leads to the recombination of carriers across the bandgap. This process is shown in Fig. 6.11 for a direct bandgap (see Section

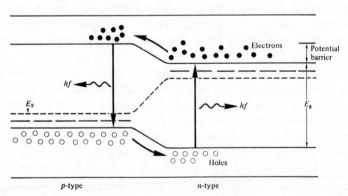

Fig. 6.11 The $p–n$ junction with forward bias giving spontaneous emission of photons.

6.3.3) semiconductor material where the normally empty electron states in the conduction band of the *p* type material and the normally empty hole states in the valence band of the *n* type material are populated by injected carriers which recombine across the bandgap. The energy released by this electron–hole recombination is approximately equal to the bandgap energy E_g. Excess carrier population is therefore decreased by recombination which may be radiative or nonradiative.

In nonradiative recombination the energy released is dissipated in the form of lattice vibrations and thus heat. However, in band to band radiative recombination the energy is released with the creation of a photon (see Fig. 6.11) with a frequency following Eq. (6.20) where the energy is approximately equal to the bandgap energy E_g and therefore:

$$E_g \simeq hf \tag{6.20}$$

This spontaneous emission of light from within the diode structure is known as electroluminescence.* The light is emitted at the site of carrier recombination which is primarily close to the junction, although recombination may take place throughout the diode structure as carriers diffuse away from the junction region (see Fig. 6.12). However, the amount of radiative recombination and the emission area within the structure is dependent upon the semiconductor materials used and the fabrication of the device.

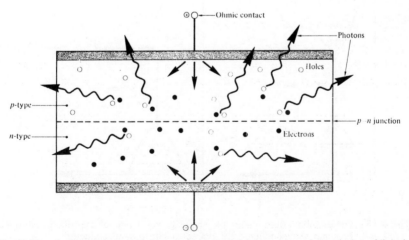

Fig. 6.12   An illustration of carrier recombination giving spontaneous emission of light in a *p–n* junction diode.

---

* The term electroluminescence is used when the optical emission results from the application of an electric field.

6.3.3   Carrier Recombination

6.3.3.1   *Direct and Indirect Bandgap Semiconductors*

In order to encourage electroluminescence it is necessary to select an appropriate semiconductor material. The most useful materials for this purpose are direct bandgap semiconductors in which electrons and holes on either side of the forbidden energy gap have the same value of crystal momentum and thus direct recombination is possible. This process is illustrated in Fig. 6.13(a) with an energy-momentum diagram for a direct bandgap semiconductor. It may be observed that the energy maximum of the valence band occurs at the same (or very nearly the same) value of electron crystal momentum* as the energy minimum of the conduction band. Hence when electron–hole recombination occurs the momentum of the electron remains virtually constant and the energy released, which corresponds to the bandgap energy E_g, may be emitted as light. This direct transition of an electron across the energy gap provides an efficient mechanism for photon emission and the average time the minority carrier remains in a free state before recombination (the minority carrier lifetime) is short (10^{-8}–10^{-10} s). Some commonly used direct bandgap semiconductor materials are shown in Table 6.1 [Ref. 3].

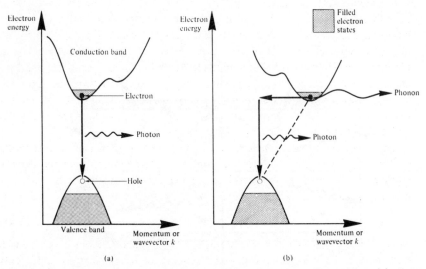

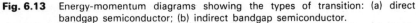

Fig. 6.13   Energy-momentum diagrams showing the types of transition: (a) direct bandgap semiconductor; (b) indirect bandgap semiconductor.

---

* The crystal momentum p is related to the wavevector k for an electron in a crystal by $p = 2\pi hk$, where h is Planck's constant [Ref. 1]. Hence the abscissa of Fig. 6.13 is often shown as the electron wavevector rather than momentum.

Table 6.1   Some direct and indirect bandgap semiconductors with calculated recombination coefficients

Semiconductor material	Energy bandgap (eV)	Recombination coefficient B_r (cm^3 s^{-1})
GaAs	Direct: 1.43	7.21×10^{-10}
GaSb	Direct: 0.73	2.39×10^{-10}
InAs	Direct: 0.35	$8.5 \ \times 10^{-11}$
InSb	Direct: 0.18	4.58×10^{-11}
Si	Indirect: 1.12	1.79×10^{-15}
Ge	Indirect: 0.67	5.25×10^{-14}
GaP	Indirect: 2.26	5.37×10^{-14}

In indirect bandgap semiconductors, however, the maximum and minimum energies occur at different values of crystal momentum (Fig. 6.13(b)). For electron–hole recombination to take place it is essential that the electron loses momentum such that it has a value of momentum corresponding to the maximum energy of the valence band. The conservation of momentum requires the emission or absorption of a third particle, a phonon. This three particle recombination process is far less probable than the two particle process exhibited by direct bandgap semiconductors. Hence, the recombination in indirect bandgap semiconductors is relatively slow (10^{-2}–10^{-4} s). This is reflected by a much longer minority carrier lifetime together with a greater probability of nonradiative transitions. The competing nonradiative recombination processes which involve lattice defects and impurities (e.g. precipitates of commonly used dopants) become more likely as they allow carrier recombination in a relatively short time in most materials. Thus the indirect bandgap emitters such as silicon and germanium shown in Table 6.1 give insignificant levels of electroluminescence. This disparity is further illustrated in Table 6.1 by the values of the recombination coefficient B_r given for both the direct and indirect bandgap recombination semiconductors shown.

The recombination coefficient is obtained from the measured absorption coefficient of the semiconductor, and for low injected minority carrier density relative to the majority carriers it is related approximately to the radiative minority carrier lifetime* τ_r by [Ref. 4]:

$$\tau_r = [B_r(N + P)]^{-1} \qquad (6.21)$$

where N and P are the respective majority carrier concentrations in the n and p type regions. The significant difference between the recombination coefficients for the direct and indirect bandgap semiconductors shown, underlines the importance of the use of direct bandgap materials for electroluminescent

---

* The radiative minority carrier lifetime is defined as the average time a minority carrier can exist in a free state before radiative recombination takes place.

sources. Direct bandgap semiconductor devices in general have a much higher internal quantum efficiency. This is the ratio of the number of radiative recombinations (photons produced within the structure) to the number of injected carriers which is often expressed as a percentage.

Example 6.3

Compare the approximate radiative minority carrier lifetimes in gallium arsenide and silicon when the minority carriers are electrons injected into the p type region which has a hole concentration of 10^{18} cm^{-3}. The injected electron density is small compared with the majority carrier density.

Solution: Equation (6.21) gives the radiative minority carrier lifetime τ_r as

$$\tau_r \simeq [B_r (N + P)]^{-1}$$

In the p type region the hole concentration determines the radiative carrier lifetime as $P \gg N$. Hence,

$$\tau_r \simeq [B_r N]^{-1}$$

Thus for gallium arsenide:

$$\tau_r \simeq [7.21 \times 10^{-10} \times 10^{18}]^{-1}$$
$$= 1.39 \times 10^{-9}$$
$$= 1.39 \text{ ns}$$

For silicon:

$$\tau_r \simeq [1.79 \times 10^{-15} \times 10^{18}]^{-1}$$
$$= 5.58 \times 10^{-4}$$
$$= 0.56 \text{ ms}$$

Thus the direct bandgap gallium arsenide has a radiative carrier lifetime factor of around 2.5×10^{-6} less than the indirect bandgap silicon.

6.3.3.2 *Other Radiative Recombination Processes*

In the previous sections only full bandgap transitions have been considered to give radiative recombination. However energy levels may be introduced into the bandgap by impurities or lattice defects within the material structure which may greatly increase the electron–hole recombination (effectively reduce the carrier lifetime). The recombination process through such impurity or defect centers may be either radiative or nonradiative. Major radiative recombination processes at 300 K other than band to band transitions are shown in Fig. 6.14. These are band to impurity center or impurity center to band, donor level to acceptor level and recombination involving isoelectronic impurities.

Hence an indirect bandgap semiconductor may be made into a more useful electroluminescent material by the addition of impurity centers which will effectively convert it into a direct bandgap material. An example of this is the introduction of nitrogen as an impurity into gallium phosphide. In this case the

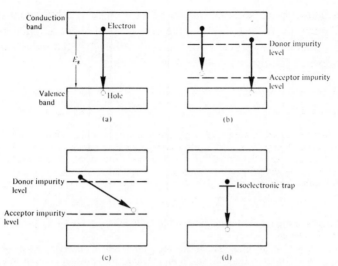

Fig. 6.14   Major radiative recombination processes at 300 K: (a) conduction to valence band (band to band) transition; (b) conduction band to acceptor impurity, and donor impurity to valence band transition; (c) donor impurity to acceptor impurity transition; (d) recombination from an isoelectronic impurity to the valence band.

nitrogen forms an isoelectronic impurity as it has the same number of valence (outer shell) electrons as phosphorus but with a different covalent radius and higher electronegativity [Ref. 1]. The nitrogen impurity center thus captures an electron and acts as an isoelectronic trap which has a large spread of momentum. This trap then attracts the oppositely charged carrier (a hole) and a direct transition takes place between the impurity center and the valence band. Hence gallium phosphide may become an efficient light emitter when nitrogen is incorporated. However, such conversion of indirect to direct bandgap transitions is only readily achieved in materials where the direct and indirect bandgaps have a small energy difference. This is the case with gallium phosphide but not with silicon or germanium.

6.3.4   Stimulated Emission and Lasing

The general concept of stimulated emission via population inversion was indicated in Section 6.2.3. Carrier population inversion is achieved in an intrinsic (undoped) semiconductor by the injection of electrons into the conduction band of the material. This is illustrated in Fig. 6.15 where the electron energy and the corresponding filled states are shown. Figure 6.15(a) shows the situation at absolute zero when the conduction band contains no electrons. Electrons injected into the material fill the lower energy states in the conduction band up to the injection energy or the quasi Fermi level for electrons. Since charge neutrality is conserved within the material an equal density of

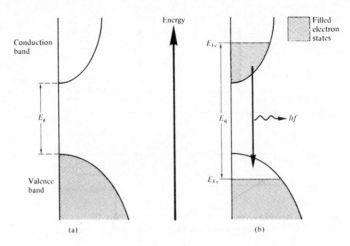

Fig. 6.15   The filled electron states for an intrinsic direct bandgap semiconductor at absolute zero [Ref. 5]: (a) in equilibrium; (b) with high carrier injection.

holes is created in the top of the valence band by the absence of electrons as shown in Fig. 6.15(b) [Ref. 5].

Incident photons with energy E_g but less than the separation energy of the quasi Fermi levels $E_q = E_{Fc} - E_{Fv}$ cannot be absorbed because the necessary conduction band states are occupied. However, these photons can induce a downward transition of an electron from the filled conduction band states into the empty valence band states thus stimulating the emission of another photon. The basic condition for stimulated emission is therefore dependent on the quasi Fermi level separation energy as well as the bandgap energy and may be defined as:

$$E_{Fc} - E_{Fv} > hf > E_g \qquad (6.22)$$

However, it must be noted that we have described an ideal situation whereas at normal operating temperatures the distribution of electrons and holes is less well defined but the condition for stimulated emission is largely maintained.

Population inversion may be obtained at a $p–n$ junction by heavy doping (degenerative doping) of both the p and n type material. Heavy p type doping with acceptor impurities causes a lowering of the Fermi level or boundary between the filled and empty states into the valence band. Similarly degenerative n type doping causes the Fermi level to enter the conduction band of the material. Energy band diagrams of a degenerate $p–n$ junction are shown in Fig. 6.16. The position of the Fermi level and the electron occupation (shading) with no applied bias are shown in Fig. 6.16(a). Since in this case the junction is in thermal equilibrium, the Fermi energy has the same value throughout the material. Figure 6.16(b) shows the $p–n$ junction when a forward bias nearly equal to the bandgap voltage is applied and hence there is

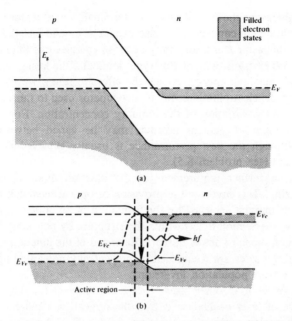

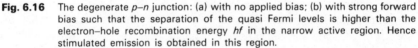

Fig. 6.16   The degenerate p–n junction: (a) with no applied bias; (b) with strong forward bias such that the separation of the quasi Fermi levels is higher than the electron–hole recombination energy hf in the narrow active region. Hence stimulated emission is obtained in this region.

direct conduction. At high injection carrier density* in such a junction there exists an active region near the depletion layer that contains simultaneously degenerate populations of electrons and holes (sometimes termed doubly degenerate). For this region the condition for stimulated emission of Eq. (6.22) is satisfied for electromagnetic radiation of frequency $E_g/h < f < (E_{Fc} - E_{Fv})/h$. Therefore any radiation of this frequency which is confined to the active region will be amplified. In general the degenerative doping distinguishes a p–n junction which provides stimulated emission from one which gives only spontaneous emission as in the case of the LED.

Finally it must be noted that high impurity concentration within a semiconductor causes differences in the energy bands in comparison with an intrinsic semiconductor. These differences are particularly apparent in the degeneratively doped p–n junctions used for semiconductor lasers. For instance at high donor level concentrations in gallium arsenide, the donor impurity levels form a band that merges with the conduction band. These energy states, sometimes referred to as 'bandtail' states [Ref. 7] extend into the forbidden energy gap. The laser transition may take place from one of these

---

* This may be largely considered to be electrons injected into the p–n region because of their greater mobility.

states. Furthermore the transitions may terminate on acceptor states which because of their high concentration also extend as a band into the energy gap. In this way the lasing transitions may occur at energies less than the bandgap energy E_g. When transitions of this type dominate, the lasing peak energy is less than the bandgap energy. Hence the effective lasing wavelength can be varied within the electroluminescent semiconductor used to fabricate the junction laser through variation of the impurity concentration. For example, the lasing wavelength of gallium arsenide may be varied between 0.85 and 0.95 μm although the best performance is usually achieved in the 0.88 to 0.91 μm band (see problem 6.5).

However, a further requirement of the junction diode is necessary to establish lasing. This involves the provision of optical feedback to give laser oscillation. It may be achieved by the formation of an optical cavity (Fabry–Perot cavity, see Section 6.2.4) within the structure by polishing the end faces of the junction diode to act as mirrors. Each end of the junction is polished or cleaved and the sides are roughened to prevent any unwanted light emission and hence wasted population inversion.

In common with all other laser types a requirement for the initiation and maintenance of laser oscillation is that the optical gain matches the optical losses within the cavity (see Section 6.2.5). For the p–n junction or semiconductor laser this occurs at a particular photon energy within the spectrum of spontaneous emission (usually near the peak wavelength of spontaneous emission). Thus when extremely high currents are passed through the device (i.e. injection levels of around 10^{18} carriers cm^{-3}), spontaneous emission with a wide spectrum (linewidth) becomes lasing (when a current threshold is passed) and the linewidth subsequently narrows.

An idealized optical output power against current characteristic (also called the light output against current characteristic) for a semiconductor laser is shown in Fig. 6.17. The current threshold is indicated and it may be observed that the device gives little light output in the region below the threshold current

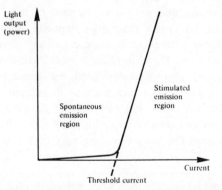

Fig. 6.17   The ideal light output against current caracteristic for an injection laser.

which corresponds to spontaneous emission only within the structure. However, after the threshold current is reached, the light output increases substantially for small increases in current through the device. This corresponds to the region of stimulated emission when the device is acting as an amplifier of light.

For strongly confined structures the threshold current density for stimulated emission J_{th} is to a fair approximation [Ref. 4] related to the threshold gain coefficient $\bar{g}_{th}$ for the laser cavity through:

$$\bar{g}_{th} = \bar{\beta} J_{th} \tag{6.23}$$

where the gain factor $\bar{\beta}$ is a constant appropriate to specific devices. Detailed discussion of the more exact relationship is given in Ref. 4.

Substituting for $\bar{g}_{th}$ from Eq. (6.18) and rearranging we obtain:

$$J_{th} = \frac{1}{\bar{\beta}} \left[\bar{\alpha} + \frac{1}{2L} \ln \frac{1}{r_1 r_2} \right] \tag{6.24}$$

Since for the semiconductor laser the mirrors are formed by a dielectric plane and often uncoated, the mirror reflectivities r_1 and r_2 may be calculated using the Fresnel reflection relationship of Eq. (4.12).

Example 6.4

A GaAs injection laser has an optical cavity of length 250 μm and width 100 μm. At normal operating temperature the gain factor $\bar{\beta}$ is 21×10^{-3} A cm^{-3} and the loss coefficient $\bar{\alpha}$ per cm is 10. Determine the threshold current density and hence the threshold current for the device. It may be assumed that the cleaved mirrors are uncoated and that the current is restricted to the optical cavity. The refractive index of GaAs may be taken as 3.6.

Solution: The reflectivity for normal incidence of a plane wave on GaAs–air interface may be obtained from Eq. (4.12) where:

$$r_1 = r_2 = r = \left(\frac{n-1}{n+1} \right)^2$$

$$= \left(\frac{3.6-1}{3.6+1} \right)^2 \simeq 0.32$$

The threshold current density may be obtained from Eq. (6.24) where:

$$J_{th} = \frac{1}{\bar{\beta}} \left[\bar{\alpha} + \frac{1}{L} \ln \frac{1}{r} \right]$$

$$= \frac{1}{21 \times 10^{-3}} \left[10 + \frac{1}{250 \times 10^{-4}} \ln \frac{1}{0.32} \right]$$

$$= 2.65 \times 10^3 \text{ A cm}^{-2}$$

The threshold current I_{th} is given by:

$$I_{th} = J_{th} \times \text{area of the optical cavity}$$

$$= 2.65 \times 10^3 \times 250 \times 100 \times 10^{-8}$$

$$\simeq 663 \text{ mA}$$

Therefore the threshold current for this device is 663 mA if the current flow is restricted to the optical cavity.

As the stimulated emission minority carrier lifetime is much shorter (typically 10^{-11} s) than that due to spontaneous emission, further increases in input current above the threshold will result almost entirely in stimulated emission, giving a high internal quantum efficiency (50–100%). Also, whereas incoherent spontaneous emission has a linewidth of tens of nanometers, stimulated coherent emission has a linewidth of a nanometer or less.

6.3.5 Heterojunctions

The previous sections have considered the photoemissive properties of a single *p–n* junction fabricated from a single crystal semiconductor material. This is known as a homojunction. However the radiative properties of a junction diode may be improved by the use of heterojunctions. A heterojunction is an interface between two adjoining single crystal semiconductors with different bandgap energies. Devices which are fabricated with heterojunctions are said to have heterostructure.

Heterojunctions are classified into either an isotype (*n–n* or *p–p*) or an anisotype (*p–n*). The isotype heterojunction provides a potential barrier within the structure which is useful for the confinement of minority carriers to a small active region (carrier confinement). It effectively reduces the carrier diffusion length and thus the volume within the structure where radiative recombination may take place. This technique is widely used for the fabrication of injection lasers and high radiance LEDs. Isotype heterojunctions are also extensively used in LEDs to provide a transparent layer close to the active region which substantially reduces the absorption of light emitted from the structure.

Alternatively anisotype heterojunctions with sufficiently large bandgap differences improve the injection efficiency of either electrons or holes. Both types of heterojunction provide a dielectric step due to the different refractive indices at either side of the junction. This may be used to provide radiation confinement to the active region (i.e. the walls of an optical waveguide). The efficiency of the containment depends upon the magnitude of the step which

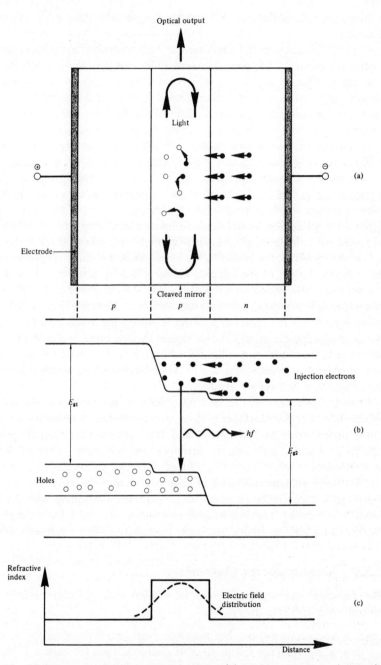

Fig. 6.18   The double heterojunction injection laser: (a) the layer structure, shown with an applied forward bias; (b) energy band diagram indicating a *p–p* heterojunction on the left and a *p–n* heterojunction on the right; (c) the corresponding refractive index diagram and electric field distribution.

is dictated by the difference in bandgap energies and the wavelength of the radiation.

It is useful to consider the application of heterojunctions in the fabrication of a particular device. They were first used to provide potential barriers in injection lasers. When a double heterojunction (DH) structure was implemented, the resulting carrier and optical confinement reduced the threshold currents necessary for lasing by a factor of around 100. Thus stimulated emission was obtained with relatively small threshold currents (50–200 mA). The layer structure and an energy band diagram for a DH injection laser are illustrated in Fig. 6.18. A heterojunction is shown either side of the active layer for laser oscillation. The forward bias is supplied by connecting a positive electrode of a supply to the p side of the structure and a negative electrode to the n side. When a voltage which corresponds to the bandgap energy of the active layer is applied, a large number of electrons (or holes) are injected into the active layer and laser oscillation commences. These carriers are confined to the active layer by the energy barriers provided by the heterojunctions which are placed within the diffusion length of the injected carriers. It may also be observed from Fig. 6.18(c) that a refractive index step (usually a difference of 5–10%) at the heterojunctions provides radiation containment to the active layer. In effect the active layer forms the center of a dielectric waveguide which strongly confines the electroluminescence within this region as illustrated in Fig. 6.18(c). The refractive index step shown is the same for each heterojunction which is desirable in order to prevent losses due to lack of waveguiding which can occur if the structure is not symmetrical.

Careful fabrication of the heterojunctions is also important in order to reduce defects at the interfaces such as misfit dislocations or inclusions which cause nonradiative recombination and thus reduce the internal quantum efficiency. Lattice matching is therefore an important criterion for the materials used to form the interface. Ideally heterojunctions should have a very small lattice parameter mismatch of no greater than 0.1%. However, it is often not possible to obtain such good lattice parameter matching with the semiconductor materials required to give emission at the desired wavelength and therefore much higher lattice parameter mismatch is often tolerated ($\simeq 0.6\%$).

6.3.6 Semiconductor Materials

The semiconductor materials used for optical sources must broadly fulfill several criteria. These are:

(a) p–n junction formation. The materials must lend themselves to the formation of p–n junctions with suitable characteristics for carrier injection.

(b) Efficient electroluminescence. The devices fabricated must have a high probability of radiative transitions and therefore a high internal quantum efficiency. Hence the materials utilized must be either direct bandgap semiconductors or indirect bandgap semiconductors with appropriate impurity centers.

(c) Useful emission wavelength. The materials must emit light at a suitable wavelength to be utilized with current optical fibers and detectors (0.8–1.7 μm). Ideally they should allow bandgap variation with appropriate doping and fabrication in order that emission at a desired specific wavelength may be achieved.

Initial investigation of electroluminescent materials for LEDs in the early 1960s centered around the direct bandgap III–V alloy semiconductors including the binary compounds gallium arsenide (GaAs) and gallium phosphide (GaP) and the ternary gallium arsenide phosphide ($GaAs_x P_{1-x}$). Gallium arsenide gives efficient electroluminescence over an appropriate wavelength band (0.88–0.91 μm) and for the first generation optical fiber communication systems was the first material to be fabricated into homojunction semiconductor lasers operating at low temperature [Ref. 8]. It was quickly realized that improved devices could be fabricated with heterojunction structures which through carrier and radiation confinement would give enhanced light output for drastically reduced device currents. These heterostructure devices were first fabricated using liquid phase epitaxy (LPE) to produce $GaAs/Al_x Ga_{1-x} As$ single heterojunction lasers. This process involves the precipitation of material from a cooling solution onto an underlying substrate. When the substrate consists of a single crystal and the lattice constant or parameter of the precipitating material is the same or very similar to that of the substrate (i.e. the unit cells within the two crystalline structures are of a similar dimension), the precipitating material forms an epitaxial layer on the substrate surface. Subsequently the same technique was used to produce double heterojunctions consisting of $Al_x Ga_{1-x} As/GaAs/Al_x Ga_{1-x} As$ epitaxial layers, which gave continuous (CW) operation at room temperature [Refs. 9 and 10]. Some of the common material systems now utilized for double heterojunction device fabrication together with their useful wavelength ranges are shown in Table 6.2.

Table 6.2 Some common material systems used in the fabrication of electroluminescent sources for optical fiber communications

Material systems active layer/confining layers	Useful wavelength range (μm)	Substrate
$GaAs/Al_x Ga_{1-x} As$	0.8–0.9	GaAs
$GaAs/In_x Ga_{1-x} P$	0.9	GaAs
$Al_y Ga_{1-y} As/Al_x Ga_{1-x} As$	0.65–0.9	GaAs
$In_y Ga_{1-y} As/In_x Ga_{1-x} P$	0.85–1.1	GaAs
$GaAs_{1-x} Sb_x/Ga_{1-y} Al_y As_{1-x} Sb_x$	0.9–1.1	GaAs
$Ga_{1-y} Al_y As_{1-x} Sb_x/GaSb$	1.0–1.7	GaSb
$In_{1-x} Ga_x As_y P_{1-y}/InP$	0.92–1.7	InP

The GaAs/AlGaAs DH system is currently by far the best developed and is used for fabricating both lasers and LEDs for the shorter wavelength region. The bandgap in this material may be 'tailored' to span the entire 0.8–0.9 μm wavelength band by changing the AlGa composition. Also there is very little lattice mismatch (0.017%) between the AlGaAs epitaxial layer and the GaAs substrate which gives good internal quantum efficiency. In the longer wavelength region (1.1–1.6 μm) a number of III–V alloys have been investigated which are compatible with GaAs, InP and GaSb substrates. These include ternary alloys such as $GaAs_{1-x}Sb_x$ and $In_xGa_{1-x}As$ grown on GaAs.

However, although the ternary alloys allow bandgap tailoring they have a fixed lattice parameter. Therefore, quaternary alloys which allow both bandgap tailoring and control of the lattice parameter (i.e. a range of lattice parameters is available for each bandgap) appear to be of more use for the longer wavelength region. The most advanced are $In_{1-x}Ga_xAs_yP_{1-y}$ lattice matched to InP and $Ga_{1-y}Al_yAs_{1-x}Sb_x$ lattice matched to GaSb. Both these material systems allow emission over the entire 1.0–1.7 μm wavelength band. At present the InGaAsP/InP material system is the most favorable for both long wavelength light sources and detectors. This is due to the ease of fabrication with lattice matching on InP which is also a suitable material for the active region with a bandgap energy of 1.35 eV at 300 K. Hence InP/InGaAsP (active/confining) devices may be fabricated. Conversely GaSb is a low bandgap material (0.78 eV at 300 K) and the quaternary alloy must be used for the active region in the GaAlAsSb/GaSb system. Thus compositional control must be maintained for three layers in this system in order to minimize lattice mismatch in the active region whereas it is only necessary for one layer in the InP/InGaAsP system.

6.4 THE SEMICONDUCTOR INJECTION LASER

The electroluminescent properties of the forward biassed *p–n* junction diode have been considered in the previous sections. Stimulated emission by the recombination of the injected carriers is encouraged in the semiconductor injection laser (often called the injection laser diode (ILD) or simply the injection laser) by the provision of an optical cavity in the crystal structure in order to provide the feedback of photons. This gives the injection laser several major advantages over other semiconductor sources (e.g. LEDs) that may be used for optical communications. These are:

(a) High radiance due to the amplifying effect of stimulated emission. Injection lasers will generally supply milliwatts of optical output power.
(b) Narrow linewidth of the order of 1 nm (10 Å) or less which is useful in minimizing the effects of material dispersion.

(c) Modulation capabilities which at present extend up into the gigahertz range and will undoubtedly be improved upon.

(d) Relative temporal coherence which is considered essential to allow heterodyne (coherent) detection in high capacity systems, but at present is primarily of use in single mode systems.

(e) Good spatial coherence which allows the output to be focused by a lens into a spot which has a greater intensity than the dispersed unfocused emission. This permits efficient coupling of the optical output power into the fiber even for fibers with low numerical aperture. The spatial field matching to the optical fiber which may be obtained with the laser source is not possible with an incoherent emitter and consequently coupling efficiencies are much reduced.

These advantages, together with the compatibility of the injection laser with optical fibers (e.g. size) led to the early developments of the device in the 1960s. Early injection lasers had the form of a Fabry–Perot cavity often fabricated in gallium arsenide which was the major III–V compound semiconductor with electroluminescent properties at the appropriate wavelength for first generation systems. The basic structure of this homojunction device is shown in Fig. 6.19, where the cleaved ends of the crystal act as partial mirrors in order to encourage stimulated emission in the cavity when electrons are injected into the p type region. However, as mentioned previously these devices had a high threshold current density (greater than 10^4 A cm^{-2}) due to their lack of carrier containment and proved inefficient light sources. The high current densities required dictated that these devices when operated at 300 K were largely utilized in a pulsed mode in order to minimize the junction temperature and thus avert damage.

Improved carrier containment and thus lower threshold current densities (around 10^3 A cm^{-2}) were achieved using heterojunction structures (see Section 6.3.5). The double heterojunction injection laser fabricated from lattice matched III–V alloys provided both carrier and optical confinement on both

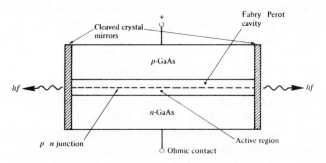

Fig. 6.19 Schematic diagram of a GaAs homojunction injection laser with a Fabry–Perot cavity.

sides of the p–n junction giving the injection laser a greatly enhanced performance. This enabled these devices with the appropriate heat sinking to be operated in a continuous wave (CW) mode at 300 K with obvious advantages for optical communications (e.g. analog transmission). However, in order to provide reliable CW operation of the DH injection laser it was necessary to provide further carrier and optical confinement which led to the introduction of stripe geometry DH laser configurations. Prior to discussion of this structure, however, it is useful to consider the efficiency of the semiconductor injection laser as an optical source.

6.4.1 Efficiency

There are a number of ways in which the operational efficiency of the semiconductor laser may be defined. A useful definition is that of the differential external quantum efficiency η_D which is the ratio of the increase in photon output rate for a given increase in the number of injected electrons. If P_e is the optical power emitted from the device, I is the current, e is the charge on an electron, and hf is the photon energy, then:

$$\eta_D = \frac{dP_e/hf}{dI/e} \simeq \frac{dP_e}{dI(E_g)} \qquad (6.25)$$

where E_g is the bandgap energy expressed in electronvolts. It may be noted that η_D gives a measure of the rate of change of the optical output power with current and hence defines the slope of the output characteristic (Fig. 6.17) in the lasing region, for a particular device. Hence η_D is sometimes referred to as the slope quantum efficiency. For a CW semiconductor laser it usually has values in the range 40–60%. Alternatively the internal quantum efficiency of the semiconductor laser η_i, which was defined in Section 6.3.3.1 as:

$$\eta_i = \frac{\text{number of photons produced in the laser cavity}}{\text{number of injected electrons}} \qquad (6.26)$$

may be quite high with values usually in the range 50–100%. It is related to the differential external quantum efficiency by the expression [Ref. 4]:

$$\eta_D = \eta_i \left[\frac{1}{1 + (2\bar{a}L/\ln(1/r_1 r_2))} \right] \qquad (6.27)$$

where $\bar{a}$ is the loss coefficient of the laser cavity, L is the length of the laser cavity and r_1, r_2 are the cleaved mirror reflectivities.

Another parameter is the total efficiency (external quantum efficiency) η_T which is efficiency defined as:

$$\eta_T = \frac{\text{total number of output photons}}{\text{total number of injected electrons}} \qquad (6.28)$$

$$= \frac{P_e/hf}{I/e} \simeq \frac{P_e}{IE_g} \tag{6.29}$$

As the power emitted P_e changes linearly when the injection current I is greater than the threshold current I_{th}, then:

$$\eta_T \simeq \eta_D \left(1 - \frac{I_{th}}{I} \right) \tag{6.30}$$

For high injection current (e.g. $I = 5I_{th}$) then $\eta_T \simeq \eta_D$, whereas for lower currents ($I \simeq 2I_{th}$) the total efficiency is lower and around 15–25%.

The external power efficiency of the device (or device efficiency) η_{ep} in converting electrical input to optical output is given by:

$$\eta_{ep} = \frac{P_e}{P} \times 100 = \frac{P_e}{IV} \times 100\% \tag{6.31}$$

where $P = IV$ is the d.c. electrical input power.

Using Eq. (6.29) for the total efficiency we find:

$$\eta_{ep} = \eta_T \left(\frac{E_g}{V} \right) \times 100\% \tag{6.32}$$

Example 6.5

The total efficiency of an injection laser with a GaAs active region is 18%. The voltage applied to the device is 2.5 V and the bandgap energy for GaAs is 1.43 eV. Calculate the external power efficiency of the device.

Solution: Using Eq. (6.32), the external power efficiency is given by:

$$\eta_{ep} = 0.18 \left(\frac{1.43}{2.5} \right) \times 100 \simeq 10\%$$

This result indicates the possibility of achieving high overall power efficiencies from semiconductor lasers which are much larger than for other laser types.

6.4.2 Stripe Geometry

The DH laser structure provides optical confinement in the vertical direction through the refractive index step at the heterojunction interfaces, but lasing takes place across the whole width of the device. This situation is illustrated in Fig. 6.20 which shows the broad area DH laser where the sides of the cavity are simply formed by roughening the edges of the device in order to reduce unwanted emission in these directions and limit the number of horizontal transverse modes. However, the broad emission area creates several problems including difficult heat sinking, lasing from multiple filaments in the relatively

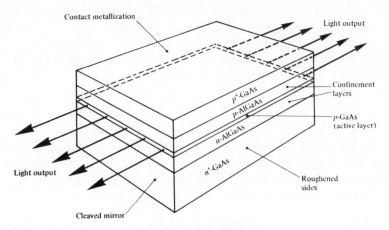

Fig. 6.20   A broad area GaAs/AlGaAs DH injection laser.

wide active area and unsuitable light output geometry for efficient coupling to the cylindrical fibers.

To overcome these problems whilst also reducing the required threshold current, laser structures in which the active region does not extend to the edges of the device were developed. A common technique involved the introduction of stripe geometry to the structure to provide optical containment in the horizontal plane. The structure of a DH stripe contact laser is shown in Fig. 6.21 where the major current flow through the device and hence the active region is within the stripe. Generally the stripe is formed by the creation of high resistance areas on either side by techniques such as proton bombardment [Ref. 9] or oxide isolation [Ref. 10]. The stripe therefore acts as a guiding mechanism which overcomes the major problems of the broad area device. However, although the active area width is reduced the light output is still not

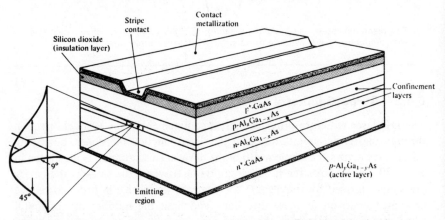

Fig. 6.21   Schematic representation of an oxide stripe AlGaAs DH injection laser.

particularly well collimated due to isotropic emission from a small active region and diffraction within the structure. The optical output and far field emission pattern are also illustrated in Fig. 6.21. The output beam divergence is typically 45° perpendicular to the plane of the junction and 9° parallel to it. Nevertheless this is a substantial improvement on the broad area laser.

The stripe contact device also gives, with the correct balance of guiding, single transverse (in a direction parallel to the junction plane) mode operation whereas the broad area device tends to allow multimode operation in this horizontal plane. Numerous stripe geometry laser structures have been investigated with stripe widths ranging from 2 to 65 μm, and the DH stripe geometry structure is universally utilized for optical fiber communications.

6.5 MULTIMODE INJECTION LASERS

6.5.1 Laser Modes

The typical output spectrum for a broad area injection laser is shown in Fig. 6.22(a). It does not consist of a single wavelength output but a series of wavelength peaks corresponding to different longitudinal (in the plane of the junction, along the optical cavity) modes within the structure. As indicated in Section 6.2.4 the spacing of these modes is dependent on the optical cavity length as each one corresponds to an integral number of lengths. They are generally separated by a few tenths of a nanometer, and the laser is said to be a multimode device. However, Fig. 6.22(a) also indicates some broadening of the longitudinal mode peaks due to subpeaks caused by higher order horizontal

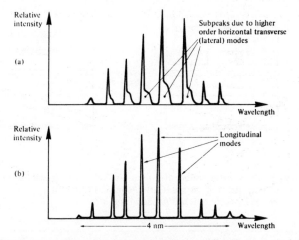

Fig. 6.22 Output spectra for multimode injection lasers: (a) broad area device with multi-transverse modes; (b) stripe geometry device with single transverse mode.

transverse modes.* These higher order lateral modes may exist in the broad area device due to the unrestricted width of the active region. The correct stripe geometry inhibits the occurrence of the higher order lateral modes by limiting the width of the optical cavity leaving only a single lateral mode which gives the output spectrum shown in Fig. 6.22(b) where only the longitudinal modes may be observed. This represents the typical output spectrum for a good multimode injection laser.

6.5.2 Structures

Fabrication of multimode injection lasers with a single lateral mode is achieved by the use of stripe geometry. The constriction of the current flow to the stripe is realized in the structure either by implanting the regions outside the stripe with protons (proton isolated stripe) to make them highly resistive, or by oxide or p–n junction isolation. The structure for an aluminum gallium arsenide oxide isolated stripe DH laser is shown in Fig. 6.21. It has an active region of gallium arsenide bounded on both sides by aluminum gallium arsenide regions. This technique has been widely applied especially for laser structures used in the shorter wavelength region. The current is confined by etching a narrow stripe in a silicon dioxide film.

The other two basic techniques are illustrated in Figs. 6.23(a) and (b) which show the proton isolated stripe and the p–n junction isolated stripe structures respectively. In Fig. 6.24(a) the resistive region formed by the proton bombardment gives better current confinement than the simple oxide stripe and has

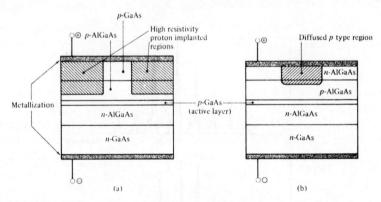

Fig. 6.23 Schematic representation of structures for stripe geometry injection lasers: (a) proton isolated stripe GaAs/AlGaAs laser; (b) p–n junction isolated diffused planar stripe) GaAs/AlGaAs laser.

---

* Transverse modes in the plane of the junction are often called lateral modes, transverse mode being reserved for modes perpendicular to the junction plane.

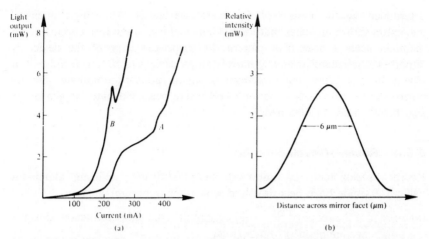

Fig. 6.24 (a) The light output against current characteristic for an injection laser with nonlinearities or a kink in the stimulated emission region. (b) A typical near field intensity distribution (pattern) in the plane of the junction for an injection laser.

superior thermal properties due to the absence of the silicon dioxide layer. *p–n* junction isolation involves a selective diffusion through the *n* type surface region in order to reach the *p* type layers as illustrated in Fig. 6.24(b). None of these structures confines all the radiation and current to the stripe region and spreading occurs on both sides of the stripe.

6.5.3 Optical Output Power

The optical output power against current characteristic for the ideal semiconductor laser was illustrated in Fig. 6.17. However, with many practical laser diodes this characteristic is not linear in the stimulated emission region but exhibits kinks. These kinks may be classified into two broad categories.

The first type of kink results from changes in the dominant lateral mode of the laser as the current is changed. The output characteristic for laser *A* in Fig. 6.24(a) illustrates this type of kink where lasing from the device changes from the fundamental lateral mode to a higher order lateral mode (second order) in a current region corresponding to a change in slope. The second type of kink involves a 'spike' as observed for laser *B* of Fig. 6.24(a). These spikes have been shown to be associated with filamentary behavior within the active region of the device [Ref. 4]. The filaments result from defects within the crystal structure.

Both these mechanisms affect the near and far field intensity distributions (patterns) obtained from the laser. A typical near field intensity distribution corresponding to a single optical output power level in the plane of the junction is shown in Fig. 6.24(b). As this distribution is in the lateral direction, it is determined by the nature of the lateral waveguide. The single intensity maximum shown indicates that the fundamental lateral mode is dominant. To maintain such a near field pattern the stripe geometry of the device is important. In general relatively narrow stripe devices ($\leqslant 10 \ \mu$m) formed by a planar process allow the fundamental lateral mode to dominate. This is especially the case at low power levels where near-field patterns similar to Fig. 6.24(b) may be obtained.

6.5.4 Recent Developments

Recent developments in multimode laser fabrication utilizing aluminum gallium arsenide have been involved with three important areas:

(a) to reduce the required threshold current at room temperature and thus the power consumption of the devices;
(b) to obtain a stable, narrow, near field intensity distribution for efficient coupling of the emitted light to the optical fibers and to remove or reduce any kinks in the light output against current characteristic;
(c) to increase the reliability of the devices.

Various stripe widths have been utilized in an attempt to optimize these factors. The 20 μm wide oxide stripe laser may be fabricated so that its light output against current characteristic is relatively free of kinks [Ref. 11] (see Fig. 6.25), but it requires a fairly high threshold current of the order of 180 mA. If attempts are made to lower the threshold current by shortening the cavity then the first order longitudinal mode has a lower output power. Also the light output against current characteristic, although free from major kinks, tends to be nonlinear in the lasing region and is therefore of little use for analog modulation.

The narrow stripe laser which utilizes an oxide-defined stripe of less than 5 μm appears largely to overcome these difficulties. The light output against

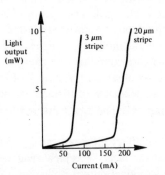

Fig. 6.25 The light output against current characteristics for a 20 μm wide and a 3 μm wide oxide stripe injection laser [Refs. 11 and 12].

current characteristic for a 3 μm stripe laser [Ref. 12] is also shown in Fig.
6.25 and it may be observed that the threshold current is of the order of
70 mA. This device also has a very linear light output against current
characteristic. In addition it provides an emission of many longitudinal modes
which tends to reduce modal noise (see Section 3.11) in the fiber.

Further reduction in threshold current to 27 mA has been achieved with a
12 μm wide oxide-defined stripe $Al_x Ga_{1-x} As$ laser [Ref. 13] by reducing the
cavity length to 100 μm and limiting the current spread by a thin p type
gallium arsenide cap. Also a dielectric stack was added to the back mirror
facet. This provided substantially improved reflectivity from this facet ensuring
that virtually all the optical power was emitted from the other facet thus
maintaining a good external efficiency with a reduced current density through
the device. These devices have nearly doubled the useful output efficiencies to
values as high as 14%, and similar laser structures have been operated in CW
mode to light output levels above 30 mW. Optical output power levels as high
as 130 mW have been achieved with wide (65 μm) oxide-defined stripe lasers
[Ref. 14]. However, at these light levels reliability becomes a problem as the
degradation mechanisms are accelerated (see Section 6.9.6).

6.6 SINGLE MODE INJECTION LASERS

These devices are becoming increasingly important in optical fiber communications. The development of high radiance LEDs for use with low loss,
low dispersion graded index fibers has diminished the requirement for multimode laser sources. Early optical fiber system design utilized LED sources
only for short-haul, low-bandwidth applications with step index fibers. More
recently advances made in both LED fabrication (see Section 7.3) and graded
index fibers have led to the use of LED sources in medium-haul, medium-
bandwidth systems. However, for long-haul, high-bandwidth applications it is
necessary to use single mode fiber where the only current suitable source is the
single mode laser. Hence there has been much activity in the area of single
mode laser fabrication in recent years along with the developments in low
attenuation single mode fiber.

This has proved to be of special interest in the longer wavelength region,
around 1.3 and 1.55 μm, where low fiber attenuation may be achieved. However,
this does not mean that the shorter wavelength region has been neglected with
regard to single mode lasers. The expertise and knowledge of fabricating with
the GaAs/AlGaAs system has led to the development of a number of
structures which allow single mode CW operation. The well-proven and
successful AlGaAs single mode laser structures will therefore be discussed
before consideration of the application of similar structures to the longer
wavelength region.

6.6.1 Single Mode Operation

For single mode operation, the optical output from a laser must contain only a single longitudinal and single transverse mode. Hence the spectral width of the emission from the single mode device is far smaller than the broadened transition linewidth discussed in Section 6.2.4. It was indicated that an inhomogeneously broadened laser can support a number of longitudinal and transverse modes simultaneously giving a multimode output. Single transverse mode operation, however, may be obtained by reducing the aperture of the resonant cavity such that only the TEM_{00} mode is supported. To obtain single mode operation it is then necessary to eliminate all but one of the longitudinal modes.

One method of achieving single longitudinal mode operation is to reduce the length L of the cavity until the frequency separation of the adjacent modes given by Eq. (6.14) as $\delta f = c/2nL$ is larger than the laser transition line width or gain curve. Then only the single mode which falls within the transition linewidth can oscillate within the laser cavity. However, it is clear that rigid control of the cavity parameters is essential to provide the mode stabilization necessary to achieve and maintain this single mode operation.

6.6.2 Mode Stabilization

The structures required to give mode stability have been discussed with regard to the multimode injection laser (see Section 6.5.2) and similar techniques are required to produce a laser emitting a single longitudinal and transverse mode. The correct DH structure restricts the vertical width of the waveguiding region to less than 0.4 μm allowing only the fundamental transverse mode to be supported and removing any interference of the higher order transverse modes on the emitted longitudinal modes.

The lateral modes (in the plane of the junction) are confined by the restrictions on the current flow provided by the stripe geometry. In general only the lower order modes are excited which appear as satellites to each of the longitudinal modes. However, as mentioned previously (Section 6.5.3), stripe contact devices often have instabilities and strong nonlinearities (e.g. kinks) in their light output against current characteristics. Tight current confinement as well as good waveguiding are therefore essential in order to achieve only the required longitudinal modes which form between the mirror facets in the plane of the junction. Finally, as indicated in the previous section, single mode operation may be obtained through control of the optical cavity length such that only a single longitudinal mode falls within the gain bandwidth of the device. Figure 6.26 shows a typical output spectrum for a single mode device.

However, injection lasers with short cavity lengths (around 50 μm) are difficult to handle and have not been particularly successful. A number of

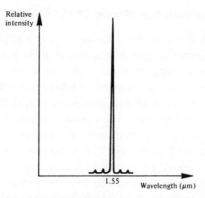

Fig. 6.26 Typical single longitudinal mode output spectrum from a single mode injection laser.

other structures are available which give electrical and optical containment and allow single mode operation.

6.7 SINGLE MODE STRUCTURES

6.7.1 Buried Heterostructure (BH) Laser

The BH laser is obtained by etching a narrow mesa stripe (as small as 1 μm in width) in DH semiconductor material and effectively burying it in high resistivity, lattice matched n type material with an appropriate bandgap and low refractive index. This process involves a rather complicated double liquid phase epitaxial (LPE) growth to give the structure illustrated in Fig. 6.27. These devices may have very small active regions to allow single mode operation. The small active region also gives low threshold currents of 10 mA or less [Refs. 15 and 16] and good linearity without kinks. Wide modulation bandwidths of 2 GHz have also been obtained [Ref. 16] but the maximum reliable optical output power is restricted. In practice the CW optical power must be kept below 1 mW/facet where lifetimes in excess of 2000 hours at 70 °C have been reported [Ref. 16]. The other major drawback is the beam divergence in the junction plane (40–50°) from the small active region.

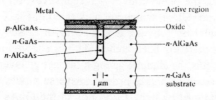

Fig. 6.27 Schematic representation of the structure of a GaAs/AlGaAs buried heterostructure laser.

6.7.2   Transverse Junction Stripe (TJS) Laser

This structure is one of the most promising for single mode operation. It has had substantial development since its conception in the mid-1970s and is now fabricated on a semi-insulating substrate which has largely removed the earlier problem of excessive temperature-dependent leakage current. A recent form of the device is shown in Fig. 6.28 and consists of a DH multilayer in n type semiconductor material. The lateral waveguide is achieved by two consecutive zinc diffusions in order that the structure is confined by p^+–n and p–n gallium arsenide homojunctions. Carrier injection is obtained laterally across these homojunctions in the central layer and the DH structure confines the carriers to the central active region. The device has good characteristics for single longitudinal and transverse mode operation with typical threshold currents of the order of 20 mA giving CW optical output power of around 3 mW [Ref. 18].

Extrapolated device lifetimes of 10^6 hours at room temperature have been reported [Ref. 19] for the AlGaAs structure which appear to be aided by the moderate junction temperature rise (approximately 10 °C) due to the low values of drive current. However, the structure has a strong threshold current dependence on temperature (see Section 6.9.1) which makes control of the optical emission more difficult at elevated temperatures.

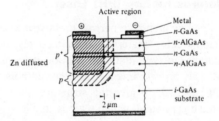

Fig. 6.28   The structure of a GaAs/AlGaAs transverse junction stripe laser.

6.7.3   Channelled Substrate Lasers

There are several single mode structures which rely on the growth by liquid phase epitaxy over channelled substrates. These include the channelled substrate planar (CSP), the plano-convex waveguide (PCW) and the constricted double heterojunction (CDH) which are illustrated in Fig. 6.29. The CSP laser structure is fabricated by growing a DH layer on a substrate into which a shallow channel has already been etched. This is shown in Fig. 6.29(a) where the n type AlGaAs fills the channel giving a flat active layer. Mode selection is thought to be the result of higher order transverse modes undergoing a large propagation loss induced by light absorption in the GaAs substrate on either side of the channel. Single mode operation is dependent on the achievement of

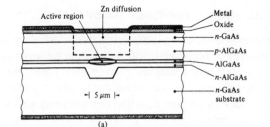

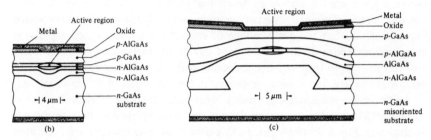

Fig. 6.29 Schematic representations of single mode AlGaAs laser structures: (a) channeled substrate planar; (b) plano-convex waveguide; (c) constricted double-heterojunction.

a thin flat layer over the channelled substrate and on deep zinc diffusions (wider than the channel) in order to create a uniform current across the channel. Typically these devices have a threshold current around 70 mA and a median lifetime of 780 hours at 70 °C [Refs. 18 and 22].

The PCW laser structure utilizes a lens-shaped lateral waveguide grown by LPE over the channel as shown in Fig. 6.29(b). Lateral confinement of both current and radiation is provided by the variation in thickness over the lens shape which also tends to focus the light giving a narrow active region of 2–3 μm. A stripe contact is used to restrict the current flow to the active region. Single mode operation with current thresholds around 40 mA giving CW optical output to 10 mW and linear light output against current characteristics have been obtained [Ref. 23].

The CDH laser structure is grown with a 'double-dovetail' channel configuration as illustrated in Fig. 6.29(c). The resulting constricted active region is defined by a stripe contact approximately 10 μm wide and provides the single mode operation of the device. Tight current confinement is not required with this structure as the optical cavity is on the least resistive path. This offers a distinct advantage over the CSP and PCW laser structures. The devices operate with threshold currents in the range 40–70 mA giving CW single longitudinal mode output to twice current threshold (10–15 mW optical output power) [Refs. 18 and 24].

6.7.4 Distributed Feedback (DFB) Lasers

These devices consist of a complex structure which determines the wavelength of the longitudinal mode emission rather than the material composition as in the more conventional cleaved mirror structures. An optical grating is incorporated into the heterostructure waveguide to provide periodic variations in refractive index along the direction of wave propagation so that feedback of optical energy is obtained through Bragg scattering (see Section 11.8.3). The corrugated grating may be applied over the whole active length of the device where it gives what is known as distributed feedback and eliminates the need for end mirrors. Originally the corrugations were applied directly to the active layer [Ref. 25] but the performance of such devices deteriorated rapidly at temperatures above 80 K. It is believed this resulted from excessive nonradiative recombination from defects in contact with the active layer introduced by the processes (e.g. ion milling) employed in the formation of the corrugations.

An improved mesa stripe geometry DFB laser structure fabricated with the GaAs/AlGaAs system which employs a separate confinement heterostructure is shown in Fig. 6.30 [Ref. 26]. In this device the corrugations are separated from the active layer and formed in an $Al_{0.07}Ga_{0.93}As$ layer on the p side of the junction. The n type $Al_{0.3}Ga_{0.7}As$ and p type $Al_{0.17}Ga_{0.83}As$ layers confine the optical field to the p type GaAs active layer. However, sufficient optical power leaks across the thin p type ($\simeq 0.1$ μm) $Al_{0.17}Ga_{0.83}As$ buffered layer into the corrugated region such that distributed feedback is obtained. Furthermore the

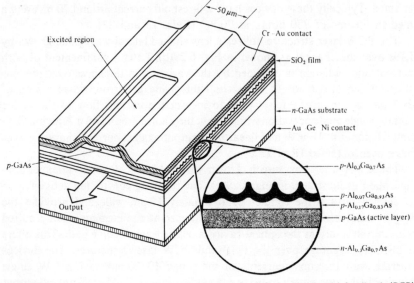

Fig. 6 30   The structure of a mesa stripe geometry AlGaAs distributed feedback (DFB) laser [Ref. 26].

p type $Al_{0.17}Ga_{0.83}As$ layer also acts as a barrier for injected electrons confining them to the active layer, thus avoiding excessive nonradiative recombination in the corrugation region.

The emission frequency from the structure is determined by the corrugation period (see Section 11.8.3). Hence the DFB structure can provide superior longitudinal mode discrimination over conventional Fabry–Perot structures where single longitudinal mode operation is dependent on the gain spectrum of the optical cavity. When suitably fabricated and operating with a single longitudinal and transverse mode such DFB lasers can give narrow emission linewidths of less than 1 nm in comparison with 1–2 nm for conventional DH injection lasers. Fabrication, however, on the channelled substrate requires great accuracy to ensure longitudinal mode selection and transverse mode control because the cleaved edges of the device are not part of the optical cavity.

A further advantage of the DFB structure over the Fabry–Perot cavity design is that it exhibits a reduced wavelength sensitivity to changes in temperature and injection current. The emission from a Fabry–Perot injection laser follows the temperature dependence of the energy gap whereas the lasing from the DFB structure follows the smaller temperature dependence of the refractive index. The typical wavelength shift with temperature for a DFB laser is 0–0.05 nm K^{-1} while the ordinary DH structure gives a typical shift of 0.2–0.5 nm K^{-1} [Ref. 27]. Therefore, despite the constructional complexity the DFB laser offers interesting possibilities for application in optical fiber communications. This is especially the case in relation to integrated optical techniques (Section 11.7) which are starting to be used in the fabrication of components and circuits for optical fiber systems.

6.7.5   Large Optical Cavity (LOC) Lasers

Ideally, long distance wideband applications require light sources with high power single mode CW optical outputs in the range 15–30 mW. To obtain such high power operation with current technology it is necessary to increase the lasing spot size (active region) both transversely and laterally whilst maintaining the single mode selectivity. Two methods of achieving this increase in the transverse direction are by use of very thin active layers of 0.05–0.06 μm instead of the more typical value around 0.2 μm [Refs. 18 and 31], or by use of the large optical cavity (LOC) technique. The LOC laser utilizes an additional guide layer with a refractive index intermediate between that of the active layer and of the n type AlGaAs layer as illustrated in Fig. 6.31 [Ref. 18]. It is within this guide layer that the optical mode is mainly propagated whilst obtaining optical gain from the active layer.

The LOC technique has been applied to BH, PCW and CDH devices, with the BH and CDH structures giving increased spot size in both the transverse

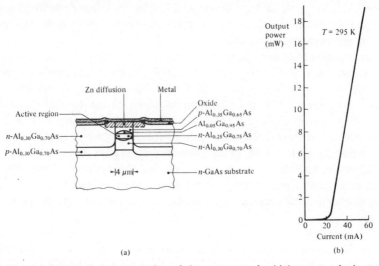

(a)   (b)

Fig. 6.31   (a) Schematic representation of the structure of a high power, single mode buried heterostructure large optical cavity laser. (b) The light output against current characteristic for the BH LOC laser. Reproduced with permission from D. Botez, *Proc. SPIE (USA)*, **224**, p. 102, 1980.

and lateral planes [Ref. 32]. The structure shown in Fig. 6.31(a) is for a BH LOC laser. However, a problem with the LOC technique is that the guide layer allows carrier leakage as well as the spread of the optical mode. This leakage is reduced in the structure of Fig. 6.31(a) by the reverse biassed *p–n* junction which is formed during the second LPE growth. The resulting current confinement gives low threshold currents around 20 mA as may be seen from a typical characteristic shown in Fig. 6.31(b). It may also be noted that the device gives CW optical output power approaching 20 mW. Bandwidths in the range of 2 GHz have also been reported [Ref. 16] for these devices which exhibit a virtually flat frequency response.

6.8   LONGER WAVELENGTH INJECTION LASERS

Semiconductor materials currently under investigation for the longer wavelength (1.1–1.6 μm) region were outlined in Section 6.3.6. The work has centered on the InGaAsP/InP and AlGaAsSb/GaSb systems, the former prepared by both liquid phase and vapor phase epitaxial techniques; the latter generally prepared by LPE. Especially promising characteristics have been obtained from devices utilizing InGaAsP prepared by both techniques [Ref. 14].

Figure 6.32(a) shows the structure of a 12 μm stripe contact InGaAsP/InP CW injection laser grown by VPE and emitting at 1.25 μm [Ref. 36]. The

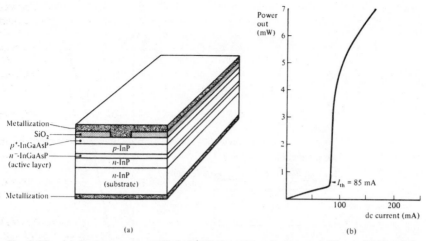

Fig. 6.32 (a) Structure of an InGaAsP/InP laser for emission at 1.25 μm. (b) The light output against current characteristic for the device in (a). Reproduced with permission from G. H. Olsen, C. J. Neuse and M. Ettenburg, *Appl. Phys. Lett.*, **34**, p. 262, 1979.

device has a current threshold of 85 mA which may be seen from Fig. 6.32(b) and the light output against current characteristic is linear up to 4 mW where a kink occurs. It exhibits single longitudinal mode optical output up to this power level which indicates that the radiation is essentially confined to the contact stripe.

However, improved single mode operation may be obtained using the mode-confining techniques discussed in the previous section. BH and TJS laser structures have been found to give consistent single mode operation with the InGaAsP system. Threshold currents as low as 22 mA have been reported [Ref. 37] for BH devices with a 1 μm wide active region and single mode operation to 8 mW has been observed in wider structures of this type [Ref. 38]. The structure of a BH InGaAsP/InP laser which operates at 1.5 μm is illustrated in Fig. 6.33(a) [Ref. 39]. It comprises a semi-insulating Fe-doped InP substrate onto which are grown successively an n type InP confining layer, an InGaAsP active layer, a p type InP confining layer and a p type InGaAsP cap layer. The buried heterostructure so formed provides an optical cavity of length 250 μm and width 2.5 μm. In addition, the epitaxial layers are mesa etched down to the n type InP layer (right hand side of Fig. 6.33a), leaving a 5 μm length for the n type electrode. The p type electrode is formed on the p type InGaAsP cap layer. Hence, the two electrodes are provided on the same side of the laser chip. Although the previously reported devices operating in the 1.5–1.7 μm wavelength band have exhibited fairly high threshold currents (200– 300 mA) [Ref. 40], this device was reported to have a threshold current as low as 38 mA for CW operation at 25 °C. This is demonstrated by the light output against current characteristic for the device shown in Fig. 6.33(b). It may also

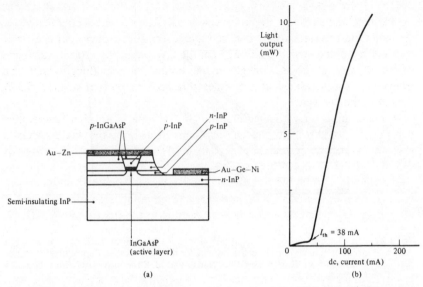

Fig. 6.33 (a) Schematic representation of the structure of a BH InGaAsP/InP laser on a semi-insulating substrate. (b) The light output against current characteristic for the BH laser under CW operation at 25°C. Reproduced with permission from T. Matsuoka, K. Takahei, Y. Noguchi and H. Nagai, *Electron. Lett.*, **17**, p. 12, 1981.

be noted that this characteristic is kink free to an optical output power of 10 mW. However, single mode operation was obtained only at 1.4 times threshold or about 2 mW optical output power.

Another InGaAsP/InP DH structure emitting near the 1.5 μm wavelength is the separated multiclad layer (SML) stripe laser [Ref. 41]. This device has a novel waveguide and internal current confinement structure which is illustrated in Fig. 6.34. The structure is four layered with the stripe region separating the multiclad layer. A coupled waveguide perpendicular to the junction plane is formed by the active layer and the *n* type InGaAsP layer. There is a difference in the propagation constant between the mode within the stripe and the modes

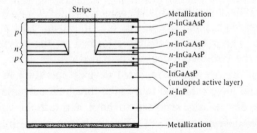

Fig. 6.34 Structure of an InGaAsP/InP separated multiclad layer stripe laser [Ref. 41].

of the coupled waveguide outside the confines of the stripe which contains the optical field and stabilizes the transverse mode. The current is confined to the stripe by the reverse biassed *p–n* heterojunction. For a device with a stripe width of 6 μm and cavity length 250 μm the CW threshold current was found to be 90 mA at 25 °C. Single mode optical output (longitudinal and transverse) was obtained at 1.5 times threshold current or around 5 mW optical output power.

A major problem with the InGaAsP/InP system is its high temperature dependence on threshold current in comparison with the GaAs/AlGaAs shorter wavelength system (see Section 6.9.1). This temperature sensitivity dictates the use of proper heat sinking so that the device temperature does not rise much above room temperature. It is often necessary to use thermoelectric cooling in order to maintain a specified working point. However, as demand for longer wavelength injection lasers rises due to the tremendous interest in long-haul, high-bandwidth systems, it is likely that device performance will improve. A reduction in temperature dependence may be brought about through improved electrical contacts and device mounting techniques. Also lower threshold currents, enhanced modal stability and improved dynamic response are to be expected as the technology for the fabrication of these devices matures. Hence the requirements of the external techniques for temperature stabilization may be reduced in the future.

6.9 INJECTION LASER CHARACTERISTICS

When considering the use of the injection laser for optical fiber communications it is necessary to be aware of certain of its characteristics which may affect its efficient operation. The following sections outline the major operating characteristics of the device (the ones which have not been dealt with in detail previously) which apply to all the various materials and structures previously discussed although there is substantial variation in behavior between them.

6.9.1 Threshold Current Temperature Dependence

Figure 6.35 shows the typical variation in threshold current with temperature for an injection laser. The threshold current tends to increase with temperature, the temperature dependence of the threshold current density J_{th} being approximately exponential [Ref. 4] for most common structures. It is given by:

$$J_{th} \propto \exp \frac{T}{T_0} \tag{6.33}$$

where T is the device absolute temperature and T_0 is the threshold temperature coefficient which is a characteristic temperature describing the quality of the

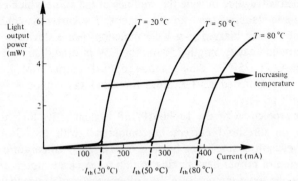

Fig. 6.35   A typical light output against current characteristic for an injection laser showing the variation in threshold current with temperature.

material, but which is also affected by the structure of the device. For AlGaAs devices, T_0 is usually in the range 120–190 K, whereas for InGaAsP devices it is between 40 and 75 K [Ref. 42]. This emphasizes the stronger temperature dependence of InGaAsP structures which is illustrated in example 6.6. The increase in threshold current with temperature for AlGaAs devices can be accounted for with reasonable accuracy by consideration of the increasing energy spread of electrons and holes injected into the conduction and valence bands. For InGaAsP lasers it appears that the high temperature sensitivity results from an additional large and rapidly rising nonradiative component of the recombination current in the active layer [Ref. 4].

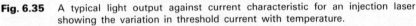

Example 6.6

Compare the ratio of the threshold current densities at 20 °C and 80 °C for a AlGaAs injection laser with $T_0 = 160$ K and the similar ratio for an InGaAsP device with $T_0 = 55$ K.

   Solution: From Eq. (6.33) the threshold current density:

$$J_{th} \propto \exp \frac{T}{T_0}$$

For the AlGaAs device:

$$J_{th} \ (20\ °C) \propto \exp \frac{293}{160} = 6.24$$

$$J_{th} \ (80\ °C) \propto \exp \frac{353}{160} = 9.08$$

Hence the ratio of the current densities:

$$\frac{J_{th} \ (80\ °C)}{J_{th} \ (20\ °C)} = \frac{9.08}{6.24} = 1.46$$

For the InGaAsP device:

$$J_{th} \ (20 \ °C) \propto \exp \frac{293}{55} = 205.88$$

$$J_{th} \ (80 \ °C) \propto \exp \frac{353}{55} = 612.89$$

Hence the ratio of the current densities:

$$\frac{J_{th} \ (80 \ °C)}{J_{th} \ (20 \ °C)} = \frac{612.89}{205.88} = 2.98$$

Thus in example 6.6 the threshold current density for the AlGaAs device increases by a factor of 1.5 over the temperature range, whereas the threshold current density for the InGaAsP device increases by a factor of 3. Hence the stronger dependence of threshold current on temperature for InGaAsP structures is shown in this comparison of two average devices. It may also be noted that it is important to obtain high values of T_0 for the devices in order to minimize temperature dependence.

However, carrier leakage due to low potential barriers within the laser structure causes T_0 to be reduced. Hence T_0 for the best AlGaAs TJS devices is only 95 K, and for LOC structures it is around 100 K rather than the higher values quoted for the standard DH structures. Conversely the threshold current variation for AlGaAs CDH structures appears much less than the standard DH devices with T_0 of around 250 K reported [Ref. 18]. In all cases adequate heat sinking along with consideration of the working environment are essential in order that the devices operate reliably over the anticipated current range.

6.9.2 Dynamic Response

The dynamic behavior of the injection laser is critical especially when it is used in high bit rate (wideband) optical fiber communication systems. The application of a current step to the device results in a switch on delay often followed by high frequency (of the order of 10 GHz) damped oscillations known as relaxation oscillations (RO). These transient phenomena occur whilst the electron and photon populations within the structure come into equilibrium and are illustrated in Fig. 6.36. The switch-on delay t_d may last for 0.5 ns and the RO for perhaps twice that period. At data rates above 100 Mbits s^{-1} this behavior can produce a serious deterioration in the pulse shape. Thus reducing t_d and damping the relaxation oscillations is highly desirable.

The switch-on delay is caused by the initial build-up of photon density resulting from stimulated emission. It is related to the minority carrier lifetime

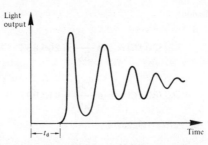

Fig. 6.36 The possible dynamic behavior of an injection laser showing the switch-on delay and relaxation oscillations.

and the current through the device [Ref. 7]. The current term, and hence the switch-on delay, may be reduced by biassing the laser near threshold (prebiassing). However, damping of the ROs is less straightforward. They are a basic laser phenomenon which vary with device structure and operating conditions; however, RO damping has been observed, and is believed to be due to several mechanisms including lateral carrier diffusion [Refs. 43 and 44], feeding of the spontaneous emission into the lasing mode [Ref. 45] and gain nonlinearities [Ref. 46]. Narrow stripe geometry DH lasers and all the mode stabilized devices (i.e. BH, TJS, CSP, CDH and LOC lasers) give RO damping, but it tends to coincide with a relatively slow increase in output power. This is thought to be the result of lateral carrier diffusion due to lack of lateral carrier confinement. However, it appears that RO damping and fast response may be obtained in BH structures with stripe widths less than the carrier diffusion length (i.e. less than 3 μm) [Ref. 47].

6.9.3 Self Pulsations

Injection lasers also exhibit another type of output fluctuation (apart from ROs) when operating in the CW mode. These self pulsations which are observed in aged and poor-quality devices are a major problem as they can occur after several hundred hours operation. Both transient as well as continuous self-sustained pulsations occur in the frequency range 0.2–4 GHz. The pulsation frequency is related to both the spontaneous recombination lifetime of the carriers (approximately 2 ns) and the photon lifetime in the cavity (approximately 10 ps).

Mechanisms within the device structure believed to cause these self pulsations include quantum noise effects (see Section 9.2.3), defects and filament formation and temperature-induced changes. More recent work [Refs. 48 and 49] traces the pulsations to the absorbing dark line defects (see Section 6.9.6), and to regions of carrier depletion. Also there appears to be correlation between the occurrence of self pulsations and deep level traps [Ref. 51] as well as the presence of visible defects [Ref. 52]. It is likely that the presence of absorbing regions within the active layer of the device will be affected by any slight increase in photon density which will tend to reduce their absorption capability (i.e. excess electron–hole pairs created in the absorption region bringing it near to transparency). The subsequent decrease in loss in the optical cavity will enhance the net gain of the device and increase the photon density giving a peak in the output. However, stimulated emission from the device will reduce the amount of population inversion in the amplifying material and allow the laser to fall below threshold, thus switching the device off. The absorption of the absorbing regions will therefore again be enhanced, and the amount of population inversion will be increased by the continued injection current. Hence the threshold will be reached again and the whole cycle repeated.

External means may be adopted to suppress these self-sustained pulsations in the laser response. These include optical injection [Ref. 53], feedback by an external mirror [Ref. 54] and changes in drive circuit reactance [Ref. 55]. However, it is likely that these pulsations will cause difficulties until a better understanding of the nonlinear absorption and nonlinear gain mechanisms within injection lasers is obtained.

6.9.4 Noise

Another important characteristic of injection laser operation involves the noise behavior of the device. This is especially the case when considering analog transmission.

The sources of noise are:

(a) quantum noise (see Section 9.2.3);
(b) instabilities in operation such as kinks in the light output against current characteristic and self pulsation;
(c) reflection of light back into the device;
(d) partition noise in multimode devices.

It is possible to reduce, if not remove, (b), (c) and (d) by using mode stabilized devices and optical isolators. Quantum noise, however, is an intrinsic property of all laser types. It results from the discrete and random spontaneous or stimulated transitions which cause intensity fluctuations in the optical emission. For injection lasers operating at frequencies less than 100 MHz quantum noise levels are usually low (signal to noise ratios less than −80 dB) unless the device is biassed within 10% of threshold. Over this region the noise spectrum is flat. However, for wideband systems when the laser is operating above threshold quantum noise becomes more pronounced. This is especially the case with multimode devices (signal to noise ratios of around −60 dB). The higher noise level results from a peak in the noise spectrum due to a relaxation resonance which typically occurs between 200 MHz and 1 GHz [Ref. 7]. Single mode lasers have shown [Ref. 55] greater noise immunity by as much as 30 dB when the current is raised above threshold. This is a further

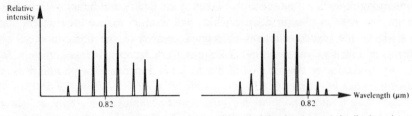

Fig. 6.37   The effect of partition noise in a multimode injection laser. It is displayed as a variation in the distribution of the various longitudinal modes emitted from the device.

advantage of the use of single mode lasers as sources for high data rate systems.

Partition noise is a phenomenon which occurs in multimode lasers when the modes are not well stabilized. Temperature changes can cause a variation in the distribution of the various longitudinal modes as illustrated in Fig. 6.37. This leads to increased dispersion on the link and hence, if not allowed for, may cause errors on a digital channel.

6.9.5   Mode Hopping

The single longitudinal mode output spectrum of a single mode laser is illustrated in Fig. 6.38(a). Mode hopping to a longer wavelength as the current is increased above threshold is demonstrated by comparison with the output spectrum shown in Fig. 6.38(b). This behavior occurs in all single mode injection lasers and is a consequence of increases in temperature of the device junction. The transition (hopping) from one mode to another is not a continuous function of the drive current but occurs suddenly over only 1–2 mA. Mode hopping alters the light output against current characteristics of the laser, and is responsible for the kinks observed in the characteristics of many single mode devices.

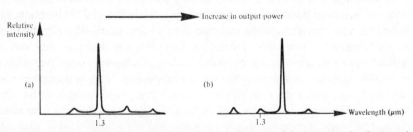

Fig. 6.38   Mode hopping in a single mode injection laser: (a) single longitudinal mode optical output; (b) mode hop to a longer peak emission wavelength at an increased optical output power.

Between hops the mode tends to shift slightly with temperature in the range 0.05–0.08 nm K^{-1}. Stabilization against mode hopping and mode shift may be obtained with adequate heat sinking or thermoelectric cooling. However, at constant heat sink temperature, shifts due to thermal increases can only be fully controlled by the use of feedback from external or internal grating structures (see Section 10.2.3).

6.9.6 Reliability

Device reliability has been a major problem with injection lasers and although it has been extensively studied, not all aspects of the failure mechanisms are fully understood. Nevertheless, much progress has been made since the early days when device lifetimes were very short (a few hours).

The degradation behavior may be separated into two major processes known as 'catastrophic' and 'gradual' degradation. Catastrophic degradation is the result of mechanical damage of the mirror facets and leads to the partial or complete laser failure. It is caused by the average optical flux density within the structure at the facet and therefore may be limited by using the device in a pulsed mode. However, its occurrence may severely restrict the operation (to low optical power levels) and lifetime of CW devices.

Gradual degradation results from internal damage caused by the energy released by nonradiative carrier recombination. It is generally accepted [Refs. 4 and 57] that this energy enhances point defect (e.g. vacancies and interstitials) displacement leading to the accumulation of defects in the active region. Hence if nonradiative electron–hole recombination occurs, for instance at the damaged surface of a laser where it has been roughened, this accelerates the diffusion of the point defect into the active region of the device. The emission characteristics of the active region therefore gradually deteriorate through the accumulation of point defects until the device is no longer useful. Mobile impurities formed by the precipitation process such as oxygen, copper or interstitial beryllium or zinc atoms may also be displaced into the active region. These atoms tend to cluster around existing dislocations encouraging high local absorption of photons. This causes 'dark lines' in the output spectrum which are a major problem with gradual degradation.

Over recent years techniques have evolved to reduce, if not eliminate, the introduction of defects into the injection laser active region. These include the use of substrates with low dislocation densities (i.e. less than 10^3 cm^{-2}), passivating the mirror facets to avoid surface-related effects and mounting with soft solders to avoid external strain. Together with improvements in crystal growth, device fabrication and material selection this has led to CW injection lasers with reported mean lifetimes in excess of 10^6 hours, or more than 100 years. These projections have been reported [Ref. 58] for a variety of GaAs/AlGaAs laser structures. In the longer wavelength region where techniques are not as well advanced reported [Refs. 41 and 59] extrapolated lifetimes for CW InGaAsP/InP structures are around 10^5 hours. These predictions bode well for the future where it is clear that injection lasers will no longer be restricted in their application on the grounds of reliability.

6.10 INJECTION LASER COUPLING AND PACKAGING

Although injection lasers are relatively directional the divergence of the beam must be considered when coupling the device to an optical fiber. A lens assembly is therefore required to direct the beam within the numerical aperture of the fiber if reasonable coupling efficiencies are to be obtained. A method of achieving this is by use of a fiber lens as illustrated in Fig. 6.39 [Refs. 57 and 59]. The fiber lens is mounted in a V-groove directly in front of the laser chip and at right angles to the fiber pigtail and gives a coupling efficiency of about 30%.

The whole assembly is incorporated in a hermetically sealed composite package (for increased reliability) with a monitor detector for feedback control (see Section 10.2.3) along with possibly some of the control and drive circuits. Optical power is taken out of the package by means of the optical fiber pigtail which itself passes through the hermetic seal.

The monitor detector is mounted behind the laser chip and monitors the optical power emerging from the rear face of the device. Alternatively a specially fabricated beam splitter may be used in order to obtain optical power from the mean output to the monitor detector. The whole assembly has the dimensions of a large scale integrated circuit and can therefore be placed on a

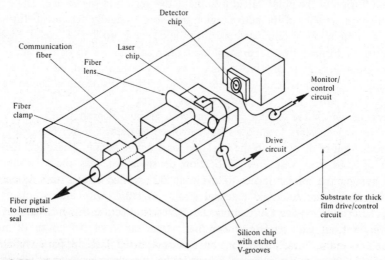

Fig. 6.39 Schematic diagram of a composite package for an injection laser with a fiber lens giving increased coupling efficiency [Ref. 57].

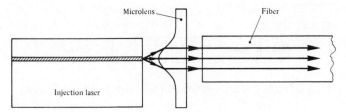

Fig. 6.40 Coupling the light output from an injection laser to the optical fiber using a cylindrical microlens [Ref. 61].

printed circuit board along with other components. This type of packaging also uses low inductance drive leads and the compact nature of the drive circuit reduces stray capacitance which eases the problems of driving the laser with large current pulses at high bit rates.

More recently similar composite packages have been utilized with cylindrical microlenses for coupling the laser emission to the optical fiber [Ref. 61]. This coupling is illustrated in Fig. 6.40, and experimental versions of this configuration have already succeeded in launching more than 1 mW of optical power at 1.3 μm wavelength into single mode fiber using an 8 μm diameter microlens. Also hemispherical lenses grown on the fiber end may be utilized to direct the laser emission into the fiber [Ref. 42], and the resulting action gives a coupling efficiency of around 20%. These various techniques are still under investigation and as yet no standard method has emerged.

6.11 NONSEMICONDUCTOR LASERS

Although at present injection lasers are the major lasing source for optical fiber communications it is possible that in the future certain nonsemiconductor sources will find application in high-bandwidth, long-haul systems. Solid state lasers doped with neodymium which emit in the 1.05–1.3 μm range appear to be attractive sources even though their application to optical fiber communications is at an early stage. The most advanced of these promising solid state sources is the neodymium-doped yttrium–aluminum–garnet (Nd : YAG) laser. This structure has several important properties which may enhance its use as an optical fiber communication source:

(a) Single mode operation near 1.064 and 1.32 μm, making it a suitable source for single mode systems.
(b) A narrow linewidth (⩽0.01 nm) which is useful for reducing dispersion on optical links.
(c) A potentially long lifetime, although comparatively little data are available.
(d) The possibility that the dimensions of the laser may be reduced to match those of the single mode step index fiber.

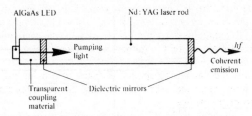

Fig. 6.41   Schematic diagram of an end pumped Nd:YAG laser.

However, the Nd:YAG laser also has several drawbacks which are common to all neodymium doped solid state devices:

(a) The device must be optically pumped. However, long lifetime AlGaAs LEDs may be utilized which improve the overall lifetime of the laser.

(b) A long fluorescence lifetime of the order of 10^{-4} seconds which only allows direct modulation (see Section 7.5) of the device at very low bandwidths. Thus an external optical modulator is necessary if the laser is to be usefully utilized in optical fiber communications.

(c) The device cannot take advantage of the well-developed technology associated with semiconductors and integrated circuits.

(d) The above requirements (i.e. pumping and modulation) tend to give a cost disadvantage in comparison with semiconductor lasers.

An illustration of a typical end pumped Nd:YAG laser is shown in Fig. 6.41. The Nd:YAG laser is a four level system (see Section 6.2.3) with a number of pumping bands and fluorescent transitions. The strongest pumping bands are at wavelengths of 0.75 and 0.81 μm giving major useful lasing transitions at 1.064 and 1.32 μm.

More complex neodymium-based compounds (rather than doped) are also under investigation for solid state optical sources. These include neodymium pentaphosphate (NdP_4O_{14}) and lithium neodymium tetraphosphate ($LiNdP_4O_{12}$) which may give high power single mode emission with less optical pumping. Other types of laser system (e.g. gas) do not appear useful for optical fiber communication systems owing to problems of size, fragility and high operating voltages. However, these devices are often used for laboratory evaluations of optical fibers and components where their high optical output power is an asset.

PROBLEMS

6.1   Briefly outline the general requirements for a source in optical fiber communications.

Discuss the areas in which the injection laser fulfils these requirements, and comment on any drawbacks of using this device as an optical fiber communication source.

6.2   Briefly describe the two processes by which light can be emitted from an atom. Discuss the requirement for population inversion in order that stimulated emission may dominate over spontaneous emission. Illustrate your answer with an energy level diagram of a common nonsemiconductor laser.

6.3   Discuss the mechanism of optical feedback to provide oscillation and hence amplification within the laser. Indicate how this provides a distinctive spectral output from the device.

  The longitudinal modes of a gallium arsenide injection laser emitting at a wavelength of 0.87 μm are separated in frequency by 278 GHz. Determine the length of the optical cavity and the number of longitudinal modes emitted. The refractive index of gallium arsenide is 3.6.

6.4   When GaSb is used in the fabrication of an electroluminescent source, estimate the necessary hole concentration in the p type region in order that the radiative minority carrier lifetime is 1 ns.

6.5   The energy bandgap for lightly doped gallium arsenide at room temperature is 1.43 eV. When the material is heavily doped (degenerative) it is found that the lasing transitions involve 'bandtail' states which effectively reduce the bandgap transition by 8%. Determine the difference in the emission wavelength of the light between the lightly doped and this heavily doped case.

6.6   With the aid of suitable diagrams, discuss the principles of operation of the injection laser.

  Outline the semiconductor materials used for emission over the wavelength range 0.8–1.7 μm and give reasons for their choice.

6.7   Describe the techniques used to give both electrical and optical confinement in multimode injection lasers.

6.8   A DH injection laser has an optical cavity of length 50 μm and width 15 μm. At normal operating temperature the loss coefficient is 10 cm^{-1} and the current threshold is 50 mA. When the mirror reflectivity at each end of the optical cavity is 0.3, estimate the gain factor $\bar{\beta}$ for the device. It may be assumed that the current is confined to the optical cavity.

6.9   The coated mirror reflectivity at either end of the 350 μm long optical cavity of an injection laser is 0.5 and 0.65. At normal operating temperature the threshold current density for the device is 2×10^3 A cm^{-2} and the gain factor β is 22×10^{-3} cm A^{-1}. Estimate the loss coefficient in the optical cavity.

6.10 A gallium arsenide injection laser with a cavity of length 500 μm has a loss coefficient of 20 cm^{-1}. The measured differential external quantum efficiency of the device is 45%. Calculate the internal quantum efficiency of the laser. The refractive index of gallium arsenide is 3.6.

6.11 Describe, with the aid of suitable diagrams, the major structures utilized in the fabrication of single mode injection lasers. Give reasons for the current interest in these devices.

6.12 Compare the ideal light output against current characteristics for the injection laser with one from a more typical device. Discuss the major points on the characteristics and indicate why the two differ.

6.13 The threshold current density for a stripe geometry AlGaAs laser is $3000 \, \text{A cm}^{-1}$ at a temperature of 15 °C. Estimate the required threshold current at a temperature of 60 °C when the threshold temperature coefficient T_0 for the device is 180 K, and the contact stripe is $20 \times 100 \, \mu\text{m}$.

6.14 Briefly describe what is meant by the following terms when they are used in relation to injection lasers:
(a) relaxation oscillations;
(b) self pulsations;
(c) mode hopping;
(d) partition noise.

6.15 Discuss degradation mechanisms in injection lasers. Comment on these with regard to the CW lifetime of the devices.

Answers to Numerical Problems

6.3	150 μm, 1241	**6.9**	28 cm^{-1}
6.4	4.2×10^{18} cm^{-3}	**6.10**	84.5%
6.5	0.07 μm	**6.13**	77.0 mA
6.8	3.76×10^{-2} cm A^{-1}		

REFERENCES

1   C. Kittel, *Introduction to Solid State Physics* (5th Edn.), John Wiley, 1976.

2   E. S. Yang, *Fundamentals of Semiconductor Devices*, McGraw-Hill, 1978.

3   Y. P. Varshni, 'Band to band radiative recombination in groups IV, VI and III–V semiconductor I', *Phys. Stat. Solidi* (Germany), **19**(2), pp. 459–514, 1967.

4   H. Kressel and J. K. Butler, *Semiconductor Lasers and Heterojunction LEDs*, Academic Press, 1977.

5   H. Kressel, 'Electroluminescent sources for fiber systems', in M. K. Barnoski (Ed.), *Fundamentals of Optical Fiber Communications*, pp. 109–141, Academic Press, 1976.

6   S. M. Sze, *Physics of semiconductor devices* (2nd Edn.), John Wiley, 1981.

7   H. C. Casey and M. B. Parish, *Heterostructure Lasers: Part A and B*, Academic Press, 1978.

8   R. N. Hall, G. E. Fenner, J. D. Kingsley, T. J. Soltys and R. O. Carlson, 'Coherent light emission from GaAs junctions', *Phys. Rev. Lett.*, **19**, p. 366, 1962.

9   J. C. Dyment, L. A. D'Asaro, J. C. Norht, B. I. Miller and J. E. Ripper, 'Proton-bombardment formation of stripe-geometry heterostructure lasers for 300 K CW operation', *Proc. IEEE*, **60**, pp. 726–728, 1972.

10 H. Kressel and M. Ettenburg, 'Low-threshold double, heterojunction AlGaAs/ GaAs laser diodes: theory and experiment', *J. Appl. Phys.*, **47**(8), pp. 3533–3537, 1976.

11 P. R. Selway, A. R. Goodwin and P. A. Kirby, 'Semiconductor laser light sources for optical fiber communications', in C. P. Sandbank (Ed.), *Optical Fiber Communication Systems*, pp. 156–183, John Wiley, 1980.

12 A. R. Goodwin, A. W. Davis, P. A. Kirby, R. E. Epworth and R. G. Plumb, 'Narrow stripe semiconductor laser for improved performance of optical communication systems', *Proceedings of the 5th European Conference on Optical Communications* (Amsterdam), paper 4–3, 1979.

13 M. Ettenburg and H. F. Lockwood, 'Low-threshold-current CW injection lasers, *Fiber Integ. Opt.*, **2**(1), pp. 47–60, 1979.

14 C. J. Neuse, 'Advances in heterojunction lasers for fiber optics applications', *Opt. Eng.*, **15**(1), pp. 20–24, 1979.

15 K. Saito, N. Shige, T. Kajimura, T. Tsukada, M. Maeda and R. Ito, 'Buried-heterostructure lasers as light sources in fiber optic communications', *Technical Digest of 1977 International Conference on Integrated Optics and Optical Fiber Communication (Tokyo, Japan)*, pp. 65–68, 1977.

16 K. Saito and R. Ito, 'Buried-heterostructure AlGaAs lasers', *IEEE J. Quantum Electron.*, **1 QE-16**(2), pp. 205–215, 1980.

17 N. Bar-chaim, J. Katz, I. Uray and A. Yariv, 'Buried heterostructure AlGaAs lasers on semi-insulating substrates', *Electron. Lett.*, **17**(3), pp. 108–109, 1981.

18 D. Botez, 'Single mode AlGaAs laser diodes', *Proc. Soc. Photo-Opt. Instrum. Eng. (USA)*, **224**, pp. 102–112, 1980.

19 S. Nita, H. Namizaki, S. Takamiya and W. Susaki, 'Single-mode junction-up TJS lasers with estimated lifetime 10^6 hours', *IEEE J. Quantum Electron.*, **QE-15**(11), pp. 1208–1210, 1979.

20 H. Kumbe, T. Tanaka and H. Narisaki, 'High TEM single-mode CW operation with junction laser', *Appl. Phys. Lett.*, **33**(1), pp. 38–39, 1978.

21 D. C. O'Shea, W. Russell Callen and W. T. Rhodes, *Introduction to Lasers and Their Applications*, Addison-Wesley, 1978.

22 K. Aiki, M. Nakamura, T. Kuroda, J. Umeda, R. Ito, N. Chinone and M. Maeda, 'Transverse mode stabilized $Al_x Ga_{1-x} As$, injection lasers with channel-substrate-planar structures', *IEEE J. Quantum Electron.*, **QE-14**(2), pp. 89–97, 1978.

23 T. Furuse, I. Sakuma, Y. Ide, K. Nishida and F. Saito, 'Transverse mode stabilized AlGaAs DH laser having built in plano-convex waveguide', *Proceedings of the 5th European Conference on Optical Communications* (Amsterdam), paper 2.2, 1979.

24 D. Botez, 'Single-mode CW operation of "double-dovetail" constricted DH (AlGa)As diode lasers', *Appl. Phys. Lett.*, **33**(10), pp. 872–874, 1978.

25 H. C. Casey Jr, S. Somekh and M. Ilegems, 'Room-temperature operation of low-threshold separate confinement heterostructure injection laser with distributed feedback', *Appl. Phys. Lett.*, **27**(3), pp. 142–144, 1975.

26 M. Nakamura, K. Aidi, J. Umeda and A. Yariv, 'CW operation of distributed-feedback, GaAs–GaAlAs diode lasers at temperatures up to 300 K', *Appl. Phys. Lett.*, **27**(7), pp. 403–405, 1975.

27 A. Yariv and M. Nakamura, 'Periodic structures for integrated optics', *IEEE J. Quantum Electron.*, **QE-13**(4), pp. 233–253, 1977.

28 A. Yariv, *Introduction to Optical Electronics* (2nd Edn), Holt, Rinehart and Winston, 1976.

29 C. A. Burrus, H. Craig Casey and T. Li, 'Optical sources', in S. E. Miller (Ed.), *Optical Fiber Telecommunications*, pp. 499–556, Academic Press, 1979.

30 L. R. Tomasetta, H. D. Law, K. Nukano and J. S. Harris, 'GaAlAsSb/GaSb lattice matched laser operating at 1.25–1.40 μm', *IEEE Internat. Semiconductor Laser Conf. (San Francisco, CA)*, paper 11–13, 1978.

31 D. Botez, 'Near and far field analytical approximations for the fundamental mode in symmetric waveguide DH lasers', *RCA Review*, **39**(4), pp. 577–603, 1978.

32 N. Chinone, K. Saito, R. Ito, K. Aiki and N. Shige, 'Highly efficient (GaAl)As buried heterostructure lasers with buried optical guide', *Appl. Phys. Lett.*, **35**(7), pp. 513–516, 1979.

33 H. F. Wolf, 'Optical sources', in H. F. Wolf (Ed.), *Handbook of Fiber Optics*, pp. 153–201, Granada, 1979.

34 R. C. Goodfellow and R. Davies, 'Optical source devices', in M. J. Howes and D. V. Morgan (Eds.), *Optical Fibre Communications*, pp. 27–106, John Wiley, 1980.

35 S. Akiba, K. Sakai, Y. Matsushima and T. Yamamoto, 'Room temperature CW operation of InGaAsP/InP heterostructure lasers emitting at 1.56 μm', *Electron. Lett.*, **15**(19), pp. 606–607, 1979.

36 G. H. Olsen, C. J. Neuse and M. Ettenburg, 'Low-threshold 1.25 μm vapour-grown InGaAsP CW lasers', *Appl. Phys. Lett.*, **34**, pp. 262–264, 1979.

37 M. Nakamura, 'Semiconductor injection lasers for long wavelength optical communications', *Technical Digest—1980 Integ., and Guided-Wave Opt., Opt. Soc. of America*, Paper MD–1, 1980.

38 H. Kans and K. Sugiyama, 'Operation characteristics of buried stripe GaInAsP/InP DH lasers by melt back method', *J. Appl. Phys.*, **50**, pp. 7934–7938, 1979.

39 T. Matsuoka, K. Takahei, Y. Noguchi and H. Nagai, '1.5 μm region InP/GaInAsP buried heterostructure lasers on semi-insulating substrates', *Electron. Lett.*, **17**(1), pp. 12–14, 1981.

40 G. H. Olsen, 'InGaAsP laser diodes', *Proc. Soc. Photo-Opt. Instrum. Eng.*, **224**, pp. 113–121, 1980.

41 H. Imai, H. Ishikawa, T. Tanahashi and M. Takusagawa, 'InGaAsP/InP separated multiclad layer stripe geometry lasers emitting at 1.5 μm', *Electron. Lett.*, **17**(1), pp. 17–19, 1981.

42 D. Botez and G. J. Herkskowitz, 'Components for optical communications systems: a review', *Proc. IEEE*, **68**(6), pp. 689–730, 1980.

43 T. Ikegami, 'Spectrum broadening and tailing effect in direct-modulated injection lasers', *Proceedings of 1st European Conference on Optical Fiber Communication* (London, UK), p. 111, 1975.

44 K. Furuya, Y. Suematsu and T. Hong, 'Reduction of resonance like peak in direct modulation due to carrier diffusion in injection laser', *Appl. Opt.*, **17**(12), pp. 1949–1952, 1978.

45 P. M. Boers, M. T. Vlaardingerbroek and M. Danielson, 'Dynamic behaviour of semiconductor lasers', *Electron. Lett.*, **11**(10), pp. 206–208, 1975.

46 D. J. Channin, 'Effect of gain saturation on injection laser switching', *J. Appl. Phys.*, **50**(6), pp. 3858–3860, 1979.

47 N. Chinane, K. Aiki, M. Nakamura and R. Ito, 'Effects of lateral mode and carrier density profile on dynamic behaviour of semiconductor lasers', *IEEE J. Quantum Electron.*, QE-14, pp. 625–631, 1977.

48 G. Arnold and K. Petermann, 'Self pulsing phenomena in (GaAl)As double-heterostructure injection lasers', *Opt. Quantum Electron.*, **10**, pp. 311–322, 1978.

49 R. W. Dixon and W. B. Joyce, 'A possible model for sustained oscillations (pulsations) in (AlGa)As double-heterostructure lasers', *IEEE J. Quantum Electron.*, QE-15, pp. 470–474, 1979.

50 G. A. Acket and K. Koelman, 'Recent developments in semiconductor injection lasers', *Acta Electron. (France)*, **22**(4), pp. 295–300, 1979.

51   R. L. Hartman and R. W. Dixon, 'Reliability of DH GaAs lasers at elevated temperatures', *Appl. Phys. Lett.*, **26**, pp. 239–242, 1975.

52   J. A. Copeland, 'Semiconductor-laser self pulsing due to deep level traps', *Electron. Lett.*, **14**(25), pp. 809–810, 1978.

53   R. Lang and K. Koboyashi, 'Suppression of the relaxation oscillation in the modulated output of semiconductor lasers', *IEEE J. Quantum Electron.*, **QE-12**(3), pp. 194–199, 1976.

54   N. Chinone, K. Aiki and R. Ito, 'Stabilization of semiconductor laser outputs by a mirror close to the laser facet', *Appl. Phys. Lett.*, **33**, pp. 990–992, 1978.

55   Y. Suematsu and T. Hong, 'Suppression of relaxation oscillations in light output of injection lasers by electrical resonance circuit', *IEEE J. Quantum Electron.*, **QE-13**(9), pp. 756–762, 1977.

56   R. E. Nahory, M. A. Pollock and J. C. De Winter, 'Temperature dependence of InGaAsP double-heterostructure laser characteristics', *Electron. Lett.*, **15**(21), pp. 695–696, 1979.

57   I. Garrett and J. E. Midwinter, 'Optical communication systems', in M. J. Howes and D. V. Morgan (Eds.), *Optical Fibre Communications*, pp. 251–299, John Wiley, 1980.

58   H. Kogelnik, 'Devices for optical communications', *Solid State Devices Research Conf. (ESSDERC)* and *4th Symposium on Solid Device Technology* (Munich, W. Germany), **53**, pp. 1–19, 1980.

59   T. Yamamoto, K. Sakai and S. Akiba, '10000-h continuous CW operation of $In_{1-x}Ga_xAs_yP_{1-y}/InP$ DH lasers at room temperature', *IEEE J. Quantum Electron.*, **QE-15**(8), pp. 684–687, 1979.

60   N. Chinone and H. Makashima, 'Semiconductor lasers with thin active layer', *Appl. Opt.*, **17**(2), pp. 311–315, 1978.

61   P. A. Kirby, 'Semiconductor laser sources for optical communication', *Radio Electron. Eng., J. IERE*, **51**(7/8), pp. 363–376, 1981.

62   J. Stone and C. A. Burrus, 'Self contained LED pumped single crystal Nd:YAG fiber laser', *Fiber Integ. Opt.*, **2**, p. 19, 1979.

63   J. K. Butler (Ed.), *Semiconductor Injection Lasers*, IEEE Press, 1980.

64   K. Shirahata, W. Susaki and H. Namizaki, 'Recent developments in fiber optic devices', *IEEE Trans. Microwave Theory and Techniques*, **MTT-30**(2), pp. 121–130, 1982.

7

Optical Sources 2: The Light Emitting Diode

7.1 INTRODUCTION

Spontaneous emission of radiation in the visible and infrared regions of the spectrum from a forward biassed *p–n* junction was discussed in Section 6.3.2. The normally empty conduction band of the semiconductor is populated by electrons injected into it by the forward current through the junction, and light is generated when these electrons recombine with holes in the valence band to emit a photon. This is the mechanism by which light is emitted from an LED, but stimulated emission is not encouraged as it is in the injection laser by the addition of an optical cavity and mirror facets to provide feedback of photons.

The LED can therefore operate at lower current densities than the injection laser, but the emitted photons have random phases and the device is an incoherent optical source. Also the energy of the emitted photons is only roughly equal to the bandgap energy of the semiconductor material, which gives a much wider spectral linewidth (possibly by a factor of 100) than the injection laser. The linewidth for an LED is typically $1–2 KT$, where K is Boltzmann's constant and T is the absolute temperature. This gives linewidths of 30–40 nm at room temperature. Thus the LED supports many optical modes within its structure and is generally a multimode source which is primarily utilized with multimode step index or graded index fiber.

At present LEDs have several further drawbacks in comparison with injection lasers. These include:

(a) generally lower optical power coupled into a fiber (microwatts);
(b) relatively small modulation bandwidth (often less than 50 MHz);
(c) harmonic distortion.

However, although these problems may initially appear to make the LED a far less attractive optical source than the injection laser, the device has a number of distinct advantages which have given it a prominent place in optical fiber communications:

(a) Simpler fabrication. There are no mirror facets and in some structures no striped geometry.

(b) Cost. The simpler construction of the LED leads to much reduced cost which is always likely to be maintained.
(c) Reliability. The LED does not exhibit catastrophic degradation and has proved far less sensitive to gradual degradation than the injection laser. It is also immune to self pulsation and modal noise problems.
(d) Less temperature dependence. The light output against current characteristic is less affected by temperature than the corresponding characteristic for the injection laser. Furthermore the LED is not a threshold device and therefore raising the temperature does not increase the threshold current above the operating point and hence halt operation.
(e) Simpler drive circuitry. This is due to the generally lower drive currents and reduced temperature dependance which makes temperature compensation circuits unnecessary.
(f) Linearity. Ideally the LED has a linear light output against current characteristic (see Section 7.4.1) unlike the injection laser. This can prove advantageous where analog modulation is concerned.

These advantages coupled with the development of high radiance medium bandwidth devices have made the LED a widely used optical source for communications applications.

Structures fabricated using the GaAs/AlGaAs material system are well advanced for the shorter wavelength region. There is also much interest in LEDs for the longer wavelength region especially around 1.3 μm where material dispersion in silica-based fibers goes through zero and where the wide linewidth of the LED imposes far less limitation on link length than intermodal dispersion within the fiber. Furthermore the reduced attenuation allows longer-haul LED systems. As with injection lasers InGaAsP/InP is the material structure currently favored in this region for the high radiance devices. These longer wavelength systems utilizing graded index fibers are likely to lead to the development of wider bandwidth devices as data rates of hundreds of Mbit s^{-1} are already feasible. It is therefore likely that in the near future injection lasers will only find major use as single mode devices within single mode fiber systems for the very long-haul, ultra-wide band applications whilst LEDs will become the primary source for all other system applications.

Having dealt with the basic operating principles for the LED in Section 6.3.2, we continue in Section 7.2 with a discussion of LED efficiency in relation to the launching of light into optical fibers. Moreover, at the end of this section we include a brief account of the operation of an efficient LED which employs a double heterostructure. This leads into a discussion in Section 7.3 of the major practical LED structures where again we have regard of their light coupling efficiency. The various operating characteristics and limitations on LED performance are described in Section 7.4. Finally, in Section 7.5 we include a brief discussion on the possible modulation techniques for semiconductor optical sources.

7.2 LED EFFICIENCY

The absence of optical amplification through stimulated emission in the LED tends to limit the internal quantum efficiency (ratio of photons generated to injected electrons) of the device. Reliance on spontaneous emission allows non-radiative recombination to take place within the structure due to crystalline imperfections and impurities giving at best an internal quantum efficiency of 50% for simple homojunction devices. However, as with injection lasers double heterojunction (DH) structures have been implemented which recombination lifetime measurements suggest [Ref. 1] give internal quantum efficiencies of 60–80%.

Although the possible internal quantum efficiency can be relatively high the radiation geometry for an LED which emits through a planar surface is essentially Lambertian in that the surface radiance (the power radiated from a unit area into a unit solid angle; given in W sr^{-1} m^{-2}) is constant in all directions. The Lambertian intensity distribution is illustrated in Fig. 7.1 where the maximum intensity I_0 is perpendicular to the planar surface but is reduced on the sides in proportion to the cosine of the viewing angle θ as the apparent area varies with this angle. This reduces the external power efficiency to a few per cent as most of the light generated within the device is trapped by total internal reflection (see Section 2.2.1) when it is radiated at greater than the critical angle for the crystal–air interface. As with the injection laser (see Section 6.4.1) the external power efficiency η_{ep} is defined as the ratio of the optical power emitted externally P_e to the electrical power provided to the device P or:

$$\eta_{ep} \simeq \frac{P_e}{P} \times 100\% \qquad (7.1)$$

Also the optical power emitted P_e into a medium of low refractive index n from the face of a planar LED fabricated from a material of refractive index n_x is given approximately by [Ref. 2]:

$$P_e = \frac{P_{int} F n^2}{4 n_x^2} \qquad (7.2)$$

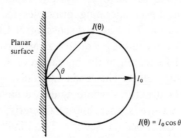

Planar
surface

$I(\theta)$

θ

I_0

$I(\theta) = I_0 \cos \theta$

Fig. 7.1   The Lambertian intensity distribution typical of a planar LED.

where P_{int} is the power generated internally and F is the transmission factor of the semiconductor–external interface. Hence it is possible to estimate the percentage of optical power emitted.

Example 7.1

A planar LED is fabricated from gallium arsenide which has a refractive index of 3.6.
(a) Calculate the optical power emitted into air as a percentage of the internal optical power for the device when the transmission factor at the crystal–air interface is 0.68.
(b) When the optical power generated internally is 50% of the electrical power supplied, determine the external power efficiency.
Solution: (a) The optical power emitted is given by Eq. (7.2), in which the refractive index n for air is 1.

$$P_e \simeq \frac{P_{int}Fn^2}{4n_x^2} = \frac{P_{int}\,0.68 \times 1}{4(3.6)^2} = 0.013\,P_{int}$$

Hence the power emitted is only 1.3% of the optical power generated internally.
(b) The external power efficiency is given by Eq. (7.1), where

$$\eta_{ep} = \frac{P_e}{P} \times 100 = 0.013\frac{P_{int}}{P} \times 100$$

Also the optical power generated internally $P_{int} = 0.5P$.
Hence

$$\eta_{ep} = \frac{0.013P_{int}}{2P_{int}} \times 100 = 0.65\%$$

A further loss is encountered when coupling the light output into a fiber. Considerations of this coupling efficiency are very complex; however, it is possible to use an approximate simplified approach [Ref. 3]. If it is assumed for step index fibers that all the light incident on the exposed end of the core within the acceptance angle θ_a is coupled, then for a fiber in air using Eq. (2.8),

$$\theta_a = \sin^{-1}(n_1^2 - n_2^2)^{\frac{1}{2}} = \sin^{-1}(NA) \tag{7.3}$$

Also incident light at angles greater than θ_a will not be coupled. For a Lambertian source, the radiant intensity at an angle θ, $I(\theta)$ is given by (see Fig. 7.1):

$$I(\theta) = I_0 \cos \theta \tag{7.4}$$

where I_0 is the radiant intensity along the line $\theta = 0$. Considering a source which is smaller than, and in close proximity to, the fiber core, and assuming cylindrical symmetry, the coupling efficiency η_c is given by:

$$\eta_c = \frac{\int_0^{\theta_a} I(\theta) \sin \theta \, d\theta}{\int_0^{\pi/2} I(\theta) \sin \theta \, d\theta} \tag{7.5}$$

Hence substituting from Eq. (7.4)

$$\eta_c = \frac{\int_0^{\theta_a} I_0 \cos \theta \sin \theta \, d\theta}{\int_0^{\pi/2} I_0 \cos \theta \sin \theta \, d\theta}$$

$$= \frac{\int_0^{\theta_a} I_0 \sin 2\theta \, d\theta}{\int_0^{\pi/2} I_0 \sin 2\theta \, d\theta}$$

$$\eta_c = \frac{[-I_0 \cos 2\theta/2]_0^{\theta_a}}{[-I_0 \cos 2\theta/2]_0^{\pi/2}}$$

$$= \sin^2 \theta_a \tag{7.6}$$

Furthermore from Eq. (7.3),

$$\eta_c = \sin^2 \theta_a = (NA)^2 \tag{7.7}$$

Equation (7.7) for the coupling efficiency allows estimates for the percentage of optical power coupled into the step index fiber relative to the amount of optical power emitted from the LED.

Example 7.2

The light output from the GaAs LED of example 7.1 is coupled into a step index fiber with a numerical aperture of 0.2, a core refractive index of 1.4 and a diameter larger than the diameter of the device. Estimate:

(a) The coupling efficiency into the fiber when the LED is in close proximity to the fiber core.
(b) The optical loss in decibels, relative to the power emitted from the LED, when coupling the light output into the fiber.
(c) The loss relative to the internally generated optical power in the device when coupling the light output into the fiber when there is a small air gap between the LED and the fiber core.

Solution: (a) From Eq. (7.7), the coupling efficiency is given by:

$$\eta_c = (NA)^2 = (0.2)^2 = 0.04$$

Thus about 4% of the externally emitted optical power is coupled into the fiber.

(b) Let the optical power coupled into the fiber be P_c. Then the optical loss in decibels relative to P_e when coupling the light output into the fiber is:

$$\text{Loss} = -10 \log_{10} \frac{P_c}{P_e}$$

$$= -10 \log_{10} \eta_c$$

Hence,

$$\text{Loss} = -10 \log_{10} 0.04$$
$$= 14.0 \text{ dB}$$

(c) When the LED is emitting into air, from example 7.1.

$$P_e = 0.013 P_{int}$$

Assuming a very small air gap (i.e. cylindrical symmetry unaffected); then from (a) the power coupled into the fiber is:

$$P_c = 0.04 P_e = 0.04 \times 0.013 P_{int}$$
$$= 5.2 \times 10^{-4} \, P_{int}$$

Hence in this case only about 0.05% of the internal optical power is coupled into the fiber.

The loss in decibels relative to P_{int} is:

$$\text{Loss} = -10 \log_{10} \frac{P_c}{P_{int}} = -10 \log_{10} 5.2 \times 10^{-4} = 32.8 \text{ dB}$$

If significant optical power is to be coupled from an incoherent LED into a low NA fiber the device must exhibit very high radiance. This is especially the case when considering graded index fibers where the Lambertian coupling efficiency with the same NA (same refractive index difference) and $\alpha \simeq 2$ (see Section 2.5) is about half that into step index fibers [Ref. 4]. To obtain the necessary high radiance, direct bandgap semiconductors (see Section 6.3.3.1) must be used fabricated with DH structures which may be driven at high current densities. The principle of operation of such a device will now be considered prior to discussion of various LED structures.

7.2.1 The Double Heterojunction LED

The principle of operation of the DH LED is illustrated in Fig. 7.2. The device shown consists of a p type GaAs layer sandwiched between a p type AlGaAs and an n type AlGaAs layer. When a forward bias is applied (as indicated in Fig. 7.2(a)) electrons from the n type layer are injected through the p–n junction into the p type GaAs layer where they become minority carriers. These minority carriers diffuse away from the junction [Ref. 5] recombining with majority carriers (holes) as they do so. Photons are therefore produced with energy corresponding to the bandgap energy of the p type GaAs layer. The injected electrons are inhibited from diffusing into the p type AlGaAs layer because of the potential barrier presented by the p–p heterojunction (see Fig. 7.2(b)). Hence electroluminescence only occurs in the GaAs junction layer, providing both good internal quantum efficiency and high radiance emission. Furthermore light is emitted from the device without reabsorption because the bandgap energy in the AlGaAs layer is large in comparison with that in GaAs. The DH structure is therefore used to provide the most efficient

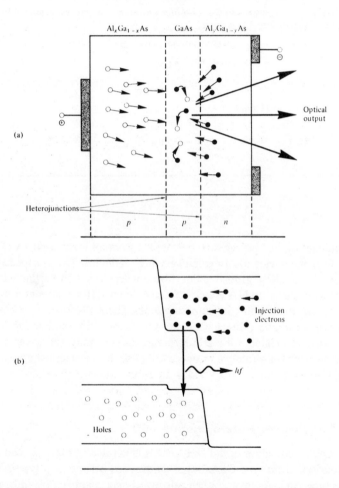

Fig. 7.2 The double heterojunction LED: (a) the layer structure, shown with an applied forward bias; (b) the corresponding energy band diagram.

incoherent sources for application within optical fiber communications. Nevertheless these devices generally exhibit the previously discussed constraints in relation to coupling efficiency to optical fibers. This and other LED structures are considered in greater detail in the following section.

7.3 LED STRUCTURES

There are four major types of LED structure although only two have found extensive use in optical fiber communications. These are the etched well surface emitter, often simply called the surface emitter, or Burrus (after the originator) LED, and the edge emitter. The other two structures, the planar and dome LEDs, find more application as cheap plastic encapsulated visible devices for use in such areas as intruder alarms, TV channel changes and industrial counting. However, infrared versions of these devices have been used in optical communications mainly with fiber bundles and it is therefore useful to consider them briefly before progressing on to the high radiance LED structures.

7.3.1 Planar LED

The planar LED is the simplest of the structures that are available and is fabricated by either liquid or vapor phase epitaxial processes over the whole surface of a GaAs substrate. This involves a *p* type diffusion into the *n* type substrate in order to create the junction illustrated in Fig. 7.3. Forward current flow through the junction gives Lambertian spontaneous emission and the device emits light from all surfaces. However, only a limited amount of light escapes the structure due to total internal reflection as discussed in Section 7.2, and therefore the radiance is low.

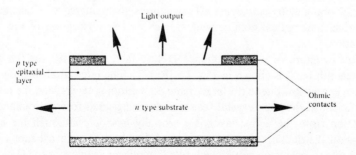

Fig. 7.3   The structure of a planar LED showing the emission of light from all surfaces.

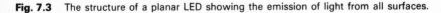

7.3.2   Dome LED

The structure of a typical dome LED is shown in Fig. 7.4. A hemisphere of n type GaAs is formed around a diffused p type region. The diameter of the dome is chosen to maximize the amount of internal emission reaching the surface within the critical angle of the GaAs–air interface. Hence this device has a higher external power efficiency than the planar LED. However, the geometry of the structure is such that the dome must be far larger than the active recombination area, which gives a greater effective emission area and thus reduces the radiance.

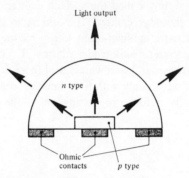

Fig. 7.4   The structure of a dome LED.

7.3.3   Surface Emitter (Burrus Type) LED

A method for obtaining high radiance is to restrict the emission to a small active region within the device. The technique pioneered by Burrus and Dawson [Ref. 6] with homostructure devices was to use an etched well in a GaAs substrate in order to prevent heavy absorption of the emitted radiation, and physically to accommodate the fiber. These structures have a low thermal impedance in the active region allowing high current densities and giving high radiance emission into the optical fiber. Furthermore considerable advantage may be obtained by employing DH structures giving increased efficiency from electrical and optical confinement as well as less absorption of the emitted radiation.

The structure of a high radiance DH surface emitter for the 0.8–0.9 μm wavelength band is shown in Fig. 7.5 [Ref. 7]. The internal absorption in this device is very low due to the larger bandgap confining layers, and the reflection coefficient at the back crystal face is high giving good forward radiance. The emission from the active layer is essentially isotropic although the external emission distribution may be considered Lambertian with a beam width of 120° due to refraction from a high to a low refractive index at the GaAs–fiber interface. The power coupled P_c into a step index fiber may be estimated from

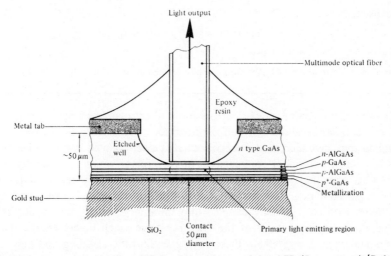

Fig. 7.5   The structure of an AlGaAs DH surface-emitting LED (Burrus type) [Ref. 7].

the relationship [Ref. 8]:

$$P_c = \pi(1 - r)AR_D(NA)^2 \qquad (7.8)$$

where r is the Fresnel reflection coefficient at the fiber surface, A is the smaller of the fiber core cross section or the emission area of the source and R_D is the radiance of the source. However, the power coupled into the fiber is also dependent on many other factors including the distance and alignment between the emission area and the fiber, the LED emission pattern and the medium between the emitting area and the fiber. For instance the addition of epoxy resin in the etched well tends to reduce the refractive index mismatch and increase the external power efficiency of the device. Hence, DH surface emitters often give more coupled optical power than predicted by Eq. (7.8). Nevertheless Eq. (7.8) may be used to gain an estimate of the power coupled although accurate results may only be obtained through measurement.

Example 7.3

A DH surface emitter which has an emission area diameter of 50 μm is butt jointed to an 80 μm core step index fiber with a numerical aperture of 0.15. The device has a radiance of 30 W sr^{-1} cm^{-2} at a constant operating drive current. Estimate the optical power coupled into the fiber if it is assumed that the Fresnel reflection coefficient at the index matched fiber surface is 0.01.

Solution: Using Eq. (7.8), the optical power coupled into the fiber P_c is given by:

$$P_c = \pi(1 - r)AR_D(NA)^2$$

In this case A represents the emission area of the source.
Hence,

$$A = \pi(25 \times 10^{-4})^2 = 1.96 \times 10^{-5} \text{ cm}^2$$

Thus,

$$P_c = \pi(1 - 0.01)1.96 \times 10^{-5} \times 30 \times (0.15)^2$$
$$= 41.1 \ \mu W$$

In this example around 41 μW of optical power is coupled into the step index fiber.

However, for graded index fiber optimum direct coupling requires that the source diameter be about one half the fiber core diameter. In both cases lens coupling may give increased levels of optical power coupled into the fiber but at the cost of additional complexity. Another factor which complicates the LED fiber coupling are the transmission characteristics of the leaky modes or large angle skew rays (see Section 2.3.6). Much of the optical power from an incoherent source is initially coupled into these large angle rays, which fall within the acceptance angle of the fiber but have much higher energy than meridional rays. Energy from these rays goes into the cladding and may be lost. Hence much of the light coupled into a multimode fiber from an LED is lost within a few hundred meters. It must therefore be noted that the effective optical power coupled into a short length of fiber significantly exceeds that coupled into a longer length.

7.3.4 Lens Coupling

It is apparent that much of the light emitted from LEDs is not coupled into the generally narrow acceptance angle of the fiber. Even with the etched well surface emitter where the low NA fiber is butted directly into the emitting aperture of the device, coupling efficiencies are poor (of the order of 1–2%). However, it has been found that greater coupling efficiency may be obtained if lenses are used to collimate the emission from the LED. There are several lens coupling configurations which include spherically polished structures not unlike the dome LED, spherical-ended fiber coupling, truncated spherical microlenses and integral lens structures.

A GaAs/AlGaAs spherical-ended fiber coupled LED is illustrated in Fig. 7.6 [Ref. 9]. It consists of a planar surface emitting structure with the spherical-ended fiber attached to the cap by epoxy resin. An emitting diameter of 35 μm was fabricated into the device and the light was coupled into fibers with core diameters of 75 and 110 μm. The geometry of the situation is such that it is essential that the active diameter of the device be substantially less (factor of 2) than the fiber core diameter if increased coupling efficiency is to be obtained. In this case good performance was obtained with coupling efficiencies around 6%. This is in agreement with theoretical [Ref. 10] and other experimental [Ref. 11] results which suggest an increased coupling efficiency of 2–5 times through the spherical fiber lens.

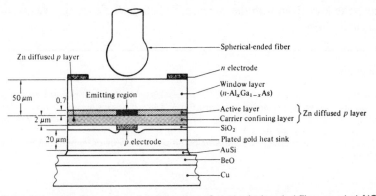

Fig. 7.6 Schematic illustration of the structure of a spherical-ended fiber coupled AlGaAs LED [Ref. 9].

Another common lens coupling technique employs a truncated spherical microlens. This configuration is shown in Fig. 7.7 for an etched well InGaAsP/InP DH surface emitter [Ref. 12] operating at a wavelength of 1.3 μm. Again a requirement for efficient coupling is that the emission region diameter is much smaller than the core diameter of the fiber. In this case the best results were obtained with a 14 μm active diameter and an 85 μm core diameter step index fiber with a numerical aperture of 0.16. The coupling efficiency was increased by a factor of 13, again supported by theory [Ref. 10] which suggests possible increases of up to 30 times.

However, the overall power conversion efficiency η_{pc} which is defined as the ratio of the optical power coupled into the fiber P_c to the electrical power applied at the terminals of the device P and is therefore given by:

$$\eta_{pc} = \frac{P_c}{P} \tag{7.9}$$

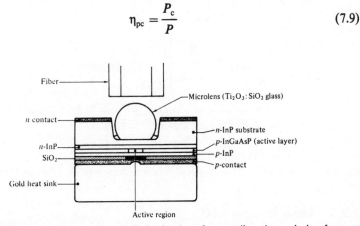

Fig. 7.7 The use of a truncated spherical microlens for coupling the emission from an InGaAsP surface-emitting LED to the fiber [Ref. 12].

is still quite low. Even with the increased coupling efficiency η_{pc} was found to be around 0.4%.

Example 7.4

A lens coupled surface-emitting LED launches 190 μW of optical power into a step index fiber when a forward current of 25 mA is flowing through the device. Determine the overall power conversion efficiency when the corresponding forward voltage across the diode is 1.5 V.
Solution: The overall power conversion efficiency may be obtained from Eq. (7.9) where,

$$\eta_{pc} = \frac{P_c}{P} = \frac{190 \times 10^{-6}}{25 \times 10^{-3} \times 1.5} = 5.1 \times 10^{-3}$$

Hence the overall power conversion efficiency is 0.5%.

The integral lens structure has perhaps the greatest potential for both a low current, high power source for small core fibers and an ultra-high power source for large core fibers. In this device a low absorption lens is formed directly in the semiconductor material as illustrated in Fig. 7.8 [Ref. 11], instead of being fabricated in glass and attached to the LED with epoxy. This technique eliminates the semiconductor, epoxy, lens interface thus increasing the theoretical coupling efficiency into the fiber. For optimized devices coupling efficiencies in excess of 15% are anticipated [Ref. 13].

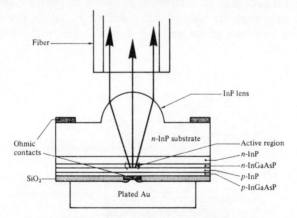

Fig. 7.8   An example of an integrated lens structure with an InGaAsP LED [Ref. 11].

7.3.5   Edge Emitter LED

The other basic high radiance LED structure currently used in optical communications is the stripe geometry DH edge emitter. This device has a similar geometry to a conventional injection laser as shown in Fig. 7.9. It takes advantage of transparent guiding layers with a very thin active layer (50–100 µm) in order that the light produced in the active layer spreads into the transparent guiding layers, reducing self-absorption in the active layer. The consequent waveguiding narrows the beam divergence to a half power width of around 30° in the plane perpendicular to the junction. However, the lack of waveguiding in the plane of the junction gives a Lambertian output with a half power width of around 120° as illustrated in Fig. 7.9.

Most of the propagating light is emitted at one end face only due to a reflector on the other end face and an antireflection coating on the emitting end face. The effective radiance at the emitting end face can be very high giving an increased coupling efficiency into small NA fiber compared with the surface emitter. However, surface emitters generally radiate more power into air (2.5–3 times) than edge emitters since the emitted light is less affected by reabsorption and interfacial recombination. Comparisons [Refs. 15–17] have shown that edge emitters couple more optical power into low NA (less than 0.3) than surface emitters, whereas the opposite is true for large NA (greater than 0.3).

The enhanced waveguiding of the edge emitter enables it in theory [Ref. 16] to couple 7.5 times more power into low NA fiber than a comparable surface emitter. However, in practice the increased coupling efficiency has been found to be slightly less than this (3.5–6 times) [Refs. 16 and 17]. Similar coupling efficiencies may be obtained into low NA fiber with surface emitters by the use of a lens. Furthermore it has been found that lens coupling with edge emitters may increase the coupling efficiencies by comparable factors (around 5 times).

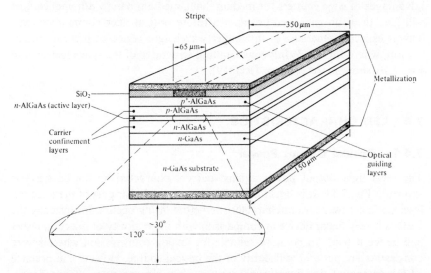

Fig. 7.9 Schematic illustration of the structure of a stripe geometry DH AlGaAs edge-emitting LED.

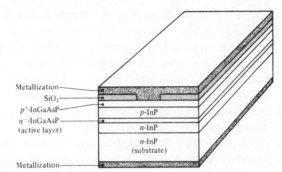

Fig. 7.10   The structure of an InGaAsP edge-emitting LED for operation at a wavelength of 1.3 μm [Ref. 19].

The stripe geometry of the edge emitter allows very high carrier injection densities for given drive currents. Thus it is possible to couple approaching a milliwatt of optical power into low NA (0.14) step index fiber with edge-emitting LEDs operating at high drive currents (500 mA) [Ref. 18].

Edge emitters have also been found to have a substantially better modulation bandwidth in the order of hundreds of megahertz than comparable surface-emitting structures with the same drive level [Ref. 17]. In general it is possible to construct edge-emitting LEDs with a narrower linewidth than surface emitters, but there are manufacturing problems with the more complicated structure (including difficult heat sinking geometry) which moderate the benefits of these devices. However, although surface emitters appear to be the favored incoherent sources at present it is likely that the significant advantages of edge emitters for medium-haul, medium-bandwidth applications will give them an enhanced position in future optical fiber communications. This is especially the case in the longer wavelength region at present around 1.3 μm, where InGaAsP/InP edge emitting structures of the type illustrated in Fig. 7.10 show more promise [Ref. 19].

7.4   LED CHARACTERISTICS

7.4.1   Optical Output Power

The ideal light output power against current characteristic for an LED is shown in Fig. 7.11. It is linear corresponding to the linear part of the injection laser optical power output characteristic before lasing occurs. Intrinsically the LED is a very linear device in comparison with the majority of injection lasers and hence it tends to be more suitable for analog transmission where severe constraints are put on the linearity of the optical source. However, in practice LEDs do exhibit significant nonlinearities which depend upon the configuration utilized. It is therefore often necessary to use some form of linearizing

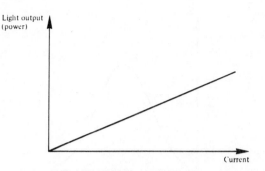

Fig. 7.11   An ideal light output against current characteristic for an LED.

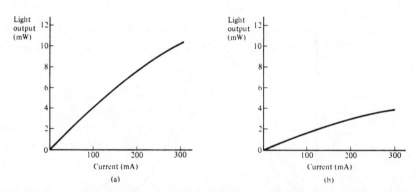

Fig. 7.12   Light output (power) into air against d.c. drive current for typically good LEDs [Ref. 17]: (a) an AlGaAs surface emitter with a 50 μm diameter dot contact; (b) an AlGaAs edge emitter with a 65 μm wide stripe and 100 μm length.

circuit technique (e.g. predistortion linearization or negative feedback) in order to ensure the linear performance of the device to allow its use in high quality analog transmission systems [Ref. 21]. Figures 7.12(a) and (b) show the light output against current characteristics for typically good surface and edge emitters respectively [Ref. 17]. It may be noted that the surface emitter radiates significantly more optical power into air than the edge emitter, and that both devices are reasonably linear at moderate drive currents.

7.4.2   Output Spectrum

The spectral linewidth of an LED operating at room temperature in the 0.8–0.9 μm wavelength band is usually between 25 and 40 nm at the half maximum intensity points (full width at half power (FWHP) points). For materials with smaller bandgap energies operating in the 1.1–1.7 μm wavelength region the linewidth tends to increase to around 50–100 nm. Examples of these two output spectra are shown in Fig. 7.13 [Refs. 3 and 22]. Also illustrated in Fig. 7.13(b) are the increases in linewidth due to increased doping

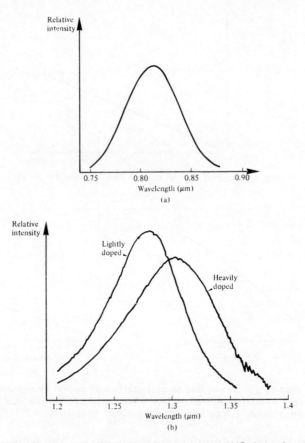

Fig. 7.13   LED output spectra: (a) output spectrum for an AlGaAs surface emitter with doped active region [Ref. 3]; (b) output spectra for an InGaAsP surface emitter showing both the lightly doped and heavily doped cases. Reproduced with permission from A. C. Carter, *The Radio and Electron. Eng.*, **51**, p. 341, 1981.

levels and the formation of bandtail states (see Section 6.3.4). This becomes apparent in the differences in the output spectra between surface- and edge-emitting LEDs where the devices have generally heavily doped and lightly doped (or undoped) active layers respectively. It may also be noted that there is a shift to lower peak emission wavelength (i.e. higher energy) through reduction in doping in Fig. 7.13(b), and hence the active layer composition must be adjusted if the same center wavelength is to be maintained.

The output spectra also tend to broaden with increases in temperature due to the greater energy spread in carrier distributions at higher temperatures. Increases in temperature of the junction affect the peak emission wavelength as well, and it is shifted by $+0.3$–0.4 nm $°C^{-1}$. It may therefore be necessary to utilize heat sinks with LEDs for certain optical fiber communication applications, although this is far less critical (normally insignificant compared with the linewidth) than the cooling requirement for injection lasers.

7.4.3   Modulation Bandwidth

The modulation bandwidth in optical communications may be defined in either electrical or optical terms. However, it is often more useful when considering the associated electrical circuitry in an optical fiber communication system to use the electrical definition where the electrical signal power has dropped to half its constant value due to the modulated portion of the optical signal. This corresponds to the electrical 3 dB point or the frequency at which the output electrical power is reduced by 3 dB with respect to the input electrical power. As optical sources operate down to d.c. we only consider the high frequency 3 dB point, the modulation bandwidth being the frequency range between zero and this high frequency 3 dB point.

Alternatively if the 3 dB bandwidth of the modulated optical carrier (optical bandwidth) is considered, we obtain an increased value for the modulation bandwidth. The reason for this inflated modulation bandwidth is illustrated in example 7.5 and Fig. 7.14. In considerations of bandwidth within the text the electrical modulation bandwidth will be assumed unless otherwise stated following current practice.

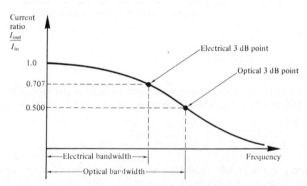

Fig. 7.14   The frequency response for an optical fiber system showing the electrical and optical bandwidths.

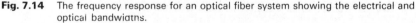

Example 7.5

Compare the electrical and optical bandwidths for an optical fiber communication system and develop a relationship between them.

Solution: In order to obtain a simple relationship between the two bandwidths it is necessary to compare the electrical current through the system. Current rather than voltage (which is generally used in electrical systems) is compared as both the optical source and optical detector (see Section 8.6) may be considered to have a linear relationship between light and current.

Electrical bandwidth: The ratio of the electrical output power to the electrical input power in decibels RE_{dB} is given by:

$$RE_{dB} = 10 \log_{10} \frac{\text{electrical power out (at the detector)}}{\text{electrical power in (at the source)}}$$

$$= 10 \log_{10} \frac{I_{out}^2/R_{out}}{I_{in}^2/R_{in}}$$

$$\propto 10 \log_{10} \left[\frac{I_{out}}{I_{in}}\right]^2$$

The electrical 3 dB points occur when the ratio of electrical powers shown above is $\frac{1}{2}$. Hence it follows that this must occur when:

$$\left[\frac{I_{out}}{I_{in}}\right]^2 = \frac{1}{2}, \quad \text{or} \quad \frac{I_{out}}{I_{in}} = \frac{1}{\sqrt{2}}$$

Thus in the electrical regime the bandwidth may be defined by the frequency when the output current has dropped to $1/\sqrt{2}$ or 0.707 of the input current to the system.

Optical bandwidth: The ratio of the optical output power to the optical input power in decibels RO_{dB} is given by,

$$RO_{dB} = 10 \log_{10} \frac{\text{optical power out (received at detector)}}{\text{optical power in (transmitted at source)}}$$

$$\propto 10 \log_{10} \frac{I_{out}}{I_{in}}$$

(due to the linear light/current relationships of the source and detector). Hence the optical 3 dB points occur when the ratio of the currents is equal to $\frac{1}{2}$, and:

$$\frac{I_{out}}{I_{in}} = \frac{1}{2}$$

Therefore in the optical regime the bandwidth is defined by the frequencies at which the output current has dropped to $\frac{1}{2}$ or 0.5 of the input current to the system. This corresponds to an electrical power attenuation of 6 dB.

The comparison between the two bandwidths is illustrated in Fig. 7.14 where it may be noted that the optical bandwidth is significantly greater than the electrical bandwidth. The difference between them (in frequency terms) depends on the shape of the frequency response for the system. However, if the system response is assumed to be Gaussian, then the optical bandwidth is a factor of $\sqrt{2}$ greater than the electrical bandwidth [Ref. 23].

The modulation bandwidth of LEDs is generally determined by three mechanisms. These are:

(a) the doping level in the active layer;
(b) the reduction in radiative lifetime due to the injected carriers;
(c) the parasitic capacitance of the device.

Assuming negligible parasitic capacitance, the speed at which an LED can be directly current modulated is fundamentally limited by the recombination lifetime of the carriers, where the optical output power $P_e(\omega)$ of the device (with constant peak current) and angular modulation frequency ω is given by [Ref. 24],

$$\frac{P_e(\omega)}{P_{dc}} = \frac{1}{[1 + (\omega\tau_i)^2]^{\frac{1}{2}}} \tag{7.10}$$

where τ_i is the injected (minority) carrier lifetime in the recombination region and P_{dc} is the d.c. optical output power for the same drive current.

Example 7.6

The minority carrier recombination lifetime for an LED is 5 ns. When a constant d.c. drive current is applied to the device the optical output power is 300 μW. Determine the optical output power when the device is modulated with an rms drive current corresponding to the d.c. drive current at frequencies of (a) 20 MHz; (b) 100 MHz.

It may be assumed that parasitic capacitance is negligible. Further determine the 3 dB optical bandwidth for the device and estimate the 3 dB electrical bandwidth assuming a Gaussian response.

Solution: (a) From Eq. (7.10), the optical output power at 20 MHz is:

$$P_e(20 \text{ MHz}) = \frac{P_{dc}}{[1 + (\omega\tau_i)^2]^{\frac{1}{2}}}$$

$$= \frac{300 \times 10^{-6}}{[1 + (2\pi \times 20 \times 10^6 \times 5 \times 10^{-9})^2]^{\frac{1}{2}}}$$

$$= \frac{300 \times 10^{-6}}{[1.39]^{\frac{1}{2}}}$$

$$= 254.2 \text{ μW}$$

(b) Again using Eq. (7.10):

$$P_e(100 \text{ MHz}) = \frac{300 \times 10^{-6}}{[1 + (2\pi \times 100 \times 10^6 \times 5 \times 10^{-9})^2]^{\frac{1}{2}}}$$

$$= \frac{300 \times 10^{-6}}{[10.87]^{\frac{1}{2}}}$$

$$= 90.9 \ \mu W$$

This example illustrates the reduction in the LED optical output power as the device is driven at higher modulating frequencies. It is therefore apparent that there is a somewhat limited bandwidth over which the device may be usefully utilized.

To determine the optical 3 dB bandwidth: the high frequency 3 dB point occurs when $P_e(\omega)/P_{dc} = 1/2$. Hence, using Eq. (7.10):

$$\frac{1}{[1 + (\omega\tau_i)^2]^{\frac{1}{2}}} = \frac{1}{2}$$

and $1 + (\omega\tau_i)^2 = 4$. Therefore $\omega\tau_i = \sqrt{3}$, and:

$$f = \frac{\sqrt{3}}{2\pi\tau} = \frac{\sqrt{3}}{\pi \times 10^{-8}} = 55.1 \ \text{MHz}$$

Thus the 3 dB optical bandwidth B_{opt} is 55.1 MHz as the device similar to all LEDs operates down to d.c.

Assuming a Gaussian frequency response, the 3 dB electrical bandwidth B will be:

$$B = \frac{55.1}{\sqrt{2}} = 39.0 \ \text{MHz}$$

Thus the corresponding electrical bandwidth is 39 MHz. However, it must be remembered that parasitic capacitance may reduce the modulation bandwidth below this value.

The carrier lifetime is dependent on the doping concentration, the number of injected carriers into the active region, the surface recombination velocity and the thickness of the active layer. All of these parameters tend to be interdependent and are adjustable within limits in present day technology. In general the carrier lifetime may be shortened by either increasing the active layer doping or by decreasing the thickness of the active layer. However, in surface emitters this can reduce the external power efficiency of the device due to the creation of an increased number of nonradiative recombination centers.

Edge-emitting LEDs have a very thin virtually undoped active layer and the carrier lifetime is controlled only by the injected carrier density. At high current densities the carrier lifetime decreases with injection level because of a bimolecular recombination process [Ref. 24]. This bimolecular recombination process allows edge-emitting LEDs with narrow recombination regions to have short recombination times, and therefore relatively high modulation capabilities at reasonable operating current densities. For instance, edge-emitting devices with electrical modulation bandwidths of 145 MHz have been achieved with moderate doping and extremely thin (approximately 50 nm) active layers [Ref. 25].

However, at present LEDs are fundamentally slower than injection lasers because of the longer lifetime of electrons in the donor region resulting from spontaneous recombination rather than stimulated emission. Thus at high modulation bandwidths the optical output power tends to decrease as demonstrated in example 7.6, and shown in Fig. 7.15 [Ref. 26]. The figure illustrates the decrease in optical output power with electrical modulation bandwidth for surface emitters (solid lines) whilst also indicating an edge-emitting device. It may be noted that at a modulation bandwidth of 120 MHz the edge-emitting LED provides more optical power into air than the surface-emitters.

Furthermore, homojunction LEDs fabricated using vapor phase epitaxial techniques which give modulation bandwidths to 1 GHz have been reported [Refs. 27 and 28] and it is likely that the modulation bandwidths for all device types will improve as the technology advances. At present commercially available LEDs are generally restricted to bandwidths of below 100 MHz.

Longer wavelength LEDs, especially those fabricated from the InGaAsP/InP system are also becoming commercially available. These devices which at present tend to operate in the 1.1–1.3 μm wavelength band take advantage of the reduced dispersion and attenuation at these wavelengths. They also exhibit

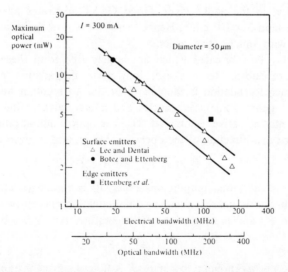

Fig. 7.15 Optical output power with a 300 mA drive current against bandwidth for AlGaAs surface emitters reported by Lee and Dentai [Ref. 1] and for AlGaAs surface and edge emitters reported by Botez and Ettenburg [Refs. 17 and 26]. Reproduced with permission from D. Botez and G. J. Herskowitz, 'Components for Optical Communications Systems: A Review', *Proc. IEEE*, **68**, p. 689, 1980. Copyright © 1980 IEEE.

larger modulation bandwidths (2–3 times) than the GaAs/AlGaAs system with the same doping density [Ref. 28], probably due to shorter carrier lifetimes as a consequence of a larger value of recombination coefficient and a larger number of nonradiating centers [Ref. 24]. Modulation bandwidths up to 300 MHz have been reported [Ref. 12] for these devices, and it is clear that there will be continued activity both in the 1.1–1.3 μm wavelength band as well as in the 1.3–1.6 μm band in order to achieve increased modulation bandwidths [Ref. 30]. It is therefore certain that as the technology for detectors over these wavelength bands becomes more established, then large data rate, longer wavelength LED systems will be brought into operation.

7.4.4 Reliability

LEDs are not generally affected by catastrophic degradation mechanisms which severely affect injection lasers (see Section 6.9.6). However, they do exhibit gradual degradation which may take the form of a rapid degradation mode or a slow degradation mode.

Rapid degradation in LEDs appears to be similar to that in injection lasers, and is due to the growth of dislocations in the active region giving rise to both dark line defects and dark spot defects (absorbing regions) under device ageing [Ref. 30]. The growth of these defects does not depend upon substrate orientation but on the injection current density, the temperature, and the impurity concentration in the active layer. Good GaAs substrates have dislocation densities around 5×10^4 cm^{-2}. Hence there is less probability of dislocations in devices with small active regions.

LEDs may be fabricated which are largely free from these defects and are therefore subject to a slower long term degradation process. This homogeneous degradation is thought to be due to recombination enhanced point defect generation (i.e. vacancies and interstitials), or the migration of impurities into the active region [Ref. 31]. The optical output power $P_e(t)$ may be expressed as a function of the operating time t, and is given by [Ref. 31]:

$$P_e(t) = P_{out} \exp (-\beta_r t) \tag{7.11}$$

where P_{out} is the initial output power and β_r is the degradation rate. The degradation rate is characterized by the activation energy of homogeneous degradation E_a and is a function of temperature. It is given by:

$$\beta_r = \beta_0 \exp [-E_a/KT] \tag{7.12}$$

where β_0 is a proportionality constant, K is Boltzmann's constant and T is the absolute temperature of the emitting region. The activation energy E_a is a variable which is dependent on the material system and the structure of the device. The value of E_a is in the range 0.56–0.6 eV, and 0.9–1.0 eV for surface-emitting GaAs/AlGaAs and InGaAsP/InP LEDs respectively [Ref. 7]. These values suggest $10^6 - 10^7$ hours (100–1000 years) CW operation at room temperature for AlGaAs devices and in excess of 10^9 hours for surface-emitting InGaAsP LEDs.

Example 7.7

An InGaAsP surface emitter has an activation energy of 1 eV with a constant of proportionality (β_0) of 1.84×10^7 h^{-1}. Estimate the CW operating lifetime for the LED with a constant junction temperature of 17 °C, if it is assumed that the device is no longer useful when its optical output power has diminished to 0.67 of its original value.

Solution: Initially it is necessary to obtain the degradation rate β_r, thus from Eq. (7.12)

$$\beta_r = \beta_0 \exp \left[-E_a / KT \right]$$

$$= 1.84 \times 10^7 \exp \left[\frac{-1 \times 1.602 \times 10^{-19}}{1.38 \times 10^{-23} \times 290} \right]$$

$$= 1.84 \times 10^7 \exp [-40]$$

$$= 7.82 \times 10^{-11} \text{ h}^{-1}$$

Now using Eq. (7.11)

$$\frac{P_e(t)}{P_{out}} = \exp (-\beta_r t) = 0.67$$

Therefore

$$\beta_r t = -\ln 0.67$$

and

$$t = \frac{\ln 0.67}{7.82 \times 10^{-11}} = \frac{0.40}{7.82 \times 10^{-11}}$$

$$= 5.1 \times 10^9 \text{ h}$$

Hence the estimated lifetime of the device under the specified conditions in example 7.7 is 5.1×10^9 hours. It must be noted that the junction temperature even for a device operating at room temperature is likely to be well in excess of room temperature when substantial drive currents are passed. Also the diminished level of optical output in the example is purely arbitrary and for many applications this reduced level may be unacceptable. Nevertheless even with more rigorous conditions the anticipated lifetime of such devices is excellent and is unlikely to cause problems in any optical fiber communication system.

Extrapolated accelerated lifetime tests are also in broad agreement with the theoretical estimates [Refs. 31 and 32]. Therefore as the LED is a reasonably simple compact structure which operates without a current threshold, it is likely that its operational life will always exceed that of the injection laser.

7.5 MODULATION

In order to transmit information via an optical fiber communication system it is necessary to modulate a property of the light with the information signal. This property may be intensity, frequency, phase or polarization (direction) with either digital or analog signals. The choices are indicated by the characteristics of the optical fiber, the available optical sources and detectors, and considerations of the overall system.

However, at present in optical fiber communications considerations of the above generally limit the options to some form of intensity modulation of the source. For instance, with optical heterodyne (coherent) detection it is necessary to demodulate the frequency or phase of the light. Although not impossible this is very difficult and requires a stable narrow linewidth single mode source as well as extremely high quality single mode fiber. Therefore direct (envelope) detection of the intensity modulated source is currently the favored method.

Intensity modulation is easy to implement with the electroluminescent sources available at present (LEDs and injection lasers). These devices can be directly modulated simply by variation of their drive currents at rates up to gigahertz. Thus direct modulation of the optical source is satisfactory for the modulation bandwidths currently under investigation. However, considering the recent interest in integrated optical devices (see Section 11.8) it is likely that external optical modulators [Ref. 33] may be utilized in the future in order to achieve greater bandwidths and to allow the use of nonsemiconductor sources (e.g. Nd:YAG laser) which cannot be directly modulated at high frequency (see Section 6.11). External optical modulators are active devices which tend to be used primarily to modulate the frequency or phase of the light, but may also be used for time division multiplexing and switching of optical signals. However, modulation considerations within this text will be almost exclusively concerned with the direct modulation of the intensity of the optical source.

Intensity modulation may be utilized with both digital and analog signals. Analog intensity modulation is usually easier to apply but requires comparatively large signal to noise ratios (see Section 9.2.5) and therefore it tends to be limited to relatively narrow bandwidth, short distance applications. Alternatively digital intensity modulation gives improved noise immunity but requires wider bandwidths, although these may be small in comparison with the available bandwidth. It is therefore ideally suited to optical fiber transmission where the available bandwidth is large. Hence at present most fiber systems in the medium to long distance range use digital intensity modulation.

PROBLEMS

7.1 Describe with the aid of suitable diagrams the mechanism giving the emission of light from an LED. Discuss the effects of this mechanism on the properties of the LED in relation to its use as an optical source for communications.

7.2 Briefly outline the advantages and drawbacks of the LED in comparison with the injection laser for use as a source in optical fiber communications.

7.3 Estimate the external power efficiency of a GaAs planar LED when the transmission factor of the GaAs–air interface is 0.68 and the internally generated optical power is 30% of the electrical power supplied. The refractive index of GaAs may be taken as 3.6.

7.4 The external power efficiency of an InGaAsP/InP planar LED is 0.75% when the internally generated optical power is 30 mW. Determine the transmission factor for the InP–air interface if the drive current is 37 mA and the potential difference across the device is 1.6 V. The refractive index of InP may be taken as 3.46.

7.5 A GaAs planar LED emitting at a wavelength of 0.85 μm has an internal quantum efficiency of 60% when passing a forward current of 20 mA s^{-1}. Estimate the optical power emitted by the device into air, and hence determine the external power efficiency if the potential difference across the device is 1 V. It may be assumed that the transmission factor at the GaAs–air interface is 0.68 and that the refractive index of GaAs is 3.6. Comment on any approximations made.

7.6 The external power efficiency of a planar GaAs LED is 1.5% when the forward current is 50 mA and the potential difference across its terminals is 2 V. Estimate the optical power generated within the device if the transmission factor at the coated GaAs–air interface is 0.8.

7.7 Outline the common LED structures for optical fiber communications discussing their relative merits and drawbacks. In particular, compare surface- and edge-emitting devices.

7.8 Derive an expression for the coupling efficiency of a surface-emitting LED into a step index fiber, assuming the device to have a Lambertian output. Determine the optical loss in decibels when coupling the optical power emitted from the device into a step index fiber with an acceptance angle of 14°. It may be assumed that the LED is smaller than the fiber core and that the two are in close proximity.

7.9 Considering the LED of problem 7.3, calculate:
(a) the coupling efficiency and optical loss in decibels of coupling the emitted light into a step index fiber with an NA of 0.15, when the device is in close proximity to the fiber and is smaller than the fiber core;
(b) the optical loss relative to the optical power generated internally if the device emits into a thin air gap before light is coupled into the fiber.

7.10 Estimate the optical power coupled into a 50 μm diameter core step index fiber with an NA of 0.18 from a DH surface emitter with an emission area diameter of 75 μm and a radiance of 60 W sr^{-1} cm^{-2}. The Fresnel reflection at index matched semiconductor–fiber interface considered negligible.

Further, determine the optical loss when coupling light into the fiber relative to the power emitted by the device into air if the Fresnel reflection at the semiconductor–air interface is 30%.

7.11 The Fresnel reflection coefficient at a fiber core of refractive index n_1 is given approximately from the classical Fresnel formulae by

$$r = \left[\frac{n_1 - n}{n_1 + n} \right]^2$$

where n is the refractive index of the surrounding medium.
(a) Estimate the optical loss due to Fresnel reflection at a fiber core from GaAs which have refractive indices of 1.5 and 3.6 respectively.
(b) Calculate the optical power coupled into a 200 μm diameter core step index fiber with an NA of 0.3 from a GaAs surface-emitting LED with an emission diameter of 90 μm and a radiance of 40 W sr^{-1} cm^{-2}. Comment on the result.
(c) Estimate the optical power emitted into air for the device in (b).

7.12 Determine the overall power conversion efficiency for the LED in problem 7.11 if it is operating with a drive current of 100 mA and a forward voltage of 1.9 V.

7.13 Discuss lens coupling of LEDs to optical fibers and outline the various techniques employed.

7.14 Discuss the relationship between the electrical and optical modulation bandwidths for an optical fiber communication system. Estimate the 3 dB optical bandwidth corresponding to a 3 dB electrical bandwidth of 50 MHz. A Gaussian frequency response may be assumed.

7.15 Determine the optical modulation bandwidth for the LED of problem 7.11 if the device emits 840 μW of optical power into air when modulated at a frequency of 150 MHz.

7.16 Estimate the electrical modulation bandwidth for an LED with a carrier recombination lifetime of 8 ns. The frequency response of the device may be assumed to be Gaussian.

7.17 Discuss the reliability of LEDs in comparison with injection lasers.

Estimate the CW operating lifetime for an AlGaAs LED with an activation energy of 0.6 eV and a constant of proportionality (β_0) of 2.3×10^3 h^{-1} when the junction temperature of the device is a constant of 50 °C. It may be assumed that the LED is no longer useful when its optical output power is 0.8 of its original value.

7.18   What is meant by the direct modulation of an optical source? Give reasons for the current use of direct intensity modulation of semiconductor optical sources and comment on possible alternatives.

Answers to Numerical Problems

7.3	0.4%	**7.11**	(a) 0.81 dB
7.4	0.70		(b) 600 μW;
7.5	230 μW, 1.15%		(c) 5.44 mW
7.6	97.2 mW	**7.12**	0.32%
7.8	12.3 dB	**7.14**	70.7 MHz
7.9	(a) 16.7 dB	**7.15**	40.6 MHz
	(b) 35.2 dB	**7.16**	24.4 MHz
7.10	0.12 mW, 16.9 dB	**7.17**	2.21 × 10^5 hours

REFERENCES

1   T. P. Lee and A. G. Dentai, 'Power and modulation bandwidth of GaAs–AlGaAs high radiance LEDs for optical communication systems', *IEEE J. Quantum Electron.*, **QE-14**(3), pp. 150–156, 1978.

2   R. C. Goodfellow and R. Davis, 'Optical source devices', in M. J. Howes and D. V. Morgan (Eds.), *Optical Fibre Communications*, pp. 27–106, John Wiley, 1980.

3   J. P. Wittke, M. Ettenburg and H. Kressel, 'High radiance LED for single fiber optical links', *RCA Rev.*, **37**(2), pp. 160–183, 1976.

4   T. G. Giallorenzi, 'Optical communications research and technology: fiber optics', *Proc. IEEE*, **66**, pp. 744–780, 1978.

5   A. A. Bergh and P. J. Dean, *Light-Emitting Diodes*, Oxford University Press, 1976.

6   C. A. Burrus and R. W. Dawson, 'Small area high-current density GaAs electroluminescent diodes and a method of operation for improved degradation characteristics', *Appl. Phys. Lett.*, **17**(3), pp. 97–99, 1970.

7   C. A. Burrus and B. I. Miller, 'Small-area double-heterostructure aluminumgallium arsenide electroluminescent diode sources for optical fiber transmission lines', *Opt. Commun.*, **4**, pp. 307–369, 1971.

8   T. P. Lee, 'Recent developments in light emitting diodes for optical fiber communication systems', *Proc. Soc. Photo Opt. Instrum. Eng. (USA)*, **224**, pp. 92–101, 1980.

9   M. Abe, I. Umebu, O. Hasegawa, S. Yamakoshi, T. Yamaoka, T. Kotani, H. Okada and H. Takamashi, 'Highly efficient long lived GaAlAs LEDs for fiber-optical communications', *IEEE Trans. Electron. Devices*, **ED-24**(7), pp. 990– 994, 1977.

10   R. A. Abram, R. W. Allen and R. C. Goodfellow, 'The coupling of light emitting diodes to optical fibres using sphere lenses', *J. Appl. Phys.*, **46**(8), pp. 3468–3474, 1975.

11   O. Wada, S. Yamakoshi, A. Masayuki, Y. Nishitani and T. Sakurai, 'High radiance InGaAsP/InP lensed LEDs for optical communication systems at 1.2–1.3 μm, *IEEE J. Quantum Electron.*, **QE-17**(2), pp. 174–178, 1981.

12   R. C. Goodfellow, A. C. Carter, I. Griffith and R. R. Bradley, 'GaInAsP/InP fast, high radiance, 1.05–1.3 μm wavelength LEDs with efficient lens coupling to small numerical aperture silica optical fibers', *IEEE Trans. Electron. Devices*, ED-26(8), pp. 1215–1220, 1979.

13 R. A. Abram and R. C. Goodfellow, 'Coupling efficiency calculations on an integrated LED sphere lens source for optical fibres', *Electron. Lett.*, 16(1), pp. 14–16, 1980.

14 C. A. Burrus, H. Craig Casey Jr and T. Li, 'Optical sources', in S. E. Miller and A. G. Chynoweth (Eds.), *Optical Fiber Telecommunications*, pp. 499–556, Academic Press, 1979.

15 D. Gloge, 'LED design for fibre system', *Electron. Lett.*, 13(4), pp. 399–400, 1977.

16 D. Marcuse, 'LED fundamentals: Comparison of front and edge emitting diodes', *IEEE J. Quantum Electron.*, QE-13(10), pp. 819–827, 1977.

17 D. Botez and M. Ettenburg, 'Comparison of surface and edge emitting LEDs for use in fiber-optical communications', *IEEE Trans. Electron. Devices*, ED-26(3), pp. 1230–1238, 1979.

18 M. Ettenburg, H. Kressel and J. P. Wittke, 'Very high radiance edge-emitting LED', *IEEE J. Quantum Electron.*, QE-12(6), pp. 360–364, 1979.

19 G. H. Olsen, F. Z. Hawrylo, D. J. Channin, D. Botez and M. Ettenburg, 'High performance 1.3 µm InGaAsP edge emitting LEDs', *IEEE 1980 Internat. Electron Devices Meeting Tech. Dig.* (Washington, DC, USA), pp. 530–533, 1980.

20 J. Straus, 'The nonlinearity of high-radiance light-emitting diodes', *IEEE J. Quantum Electron.*, QE-14(11), pp. 813–819, 1979.

21 J. Straus, 'Linearized transmitters for analog fiber links', *Laser Focus (USA)*, 14(10), pp. 54–61, 1978.

22 A. C. Carter, 'Light-emitting diodes for optical fibre systems', *Radio Electron. Eng. J. IERE*, 51(7/8), pp. 341–348, 1981.

23 I. Garrett and J. E. Midwinter, 'Optical communication systems', in M. J. Howes and D. V. Morgan (Eds.), *Optical Fibre Communications*, pp. 251–300, John Wiley, 1980.

24 H. Kressel and J. K. Butler, *Semiconductor Lasers and Heterojunction LEDs*, Academic Press, 1977.

25 H. F. Lockwood, J. P. Wittke and M. Ettenburg, 'LED for high data rate, optical communications', *Opt. Commun.*, 16, p. 193, 1976.

26 D. Botez and G. J. Herkowitz, 'Components for optical communications systems: a review', *Proc. IEEE*, 68(6), pp. 689–731, 1980.

27 R. C. Goodfellow and A. Mabbit, 'Wide band high radiance gallium arsenide LEDs for fibre optic communication', *Electron. Lett.*, 12(2), pp. 50–51, 1976.

28 A. C. Carter, R. C. Goodfellow and R. Davis, 'High speed GaAs and GaInAs high radiance LEDs', *Internat. Electron. Devices Meeting*, Washington DC (USA), pp. 577–581, 1977.

29 I. Umebu, O. Hasegawa and K. Akita, 'InGaAsP/InP DH LEDs for fibre-optical communication', *Electron. Lett.*, 14(16), pp. 499–500, 1978.

30 T. P. Lee, 'Recent development in light emitting diodes (LEDs) for optical fiber communications systems', *Proc. Soc. Photo-opt. Instrum. Eng. (USA)*, 340, pp. 22–31, 1982.

31 S. Yamakoshi, A. Masayuki, O. Wada, S. Komiya and T. Sakurai, 'Reliability of high radiance InGaAsP/InP LEDs operating in the 1.2–1.3 µm wavelength', *IEEE J. Quantum Electron.*, QE-17(2), pp. 167–173, 1981.

32 S. Yamakoshi, T. Sugahara, O. Hasegawa, Y. Toyama and H. Takanashi, 'Growth mechanism of $\langle 100 \rangle$ dark-line defects in high radiance GaAlAs LEDs', *International Electronic Devices Meeting*, pp. 642–645, 1978.

33   I. P. Kaminow and T. Li, 'Modulation techniques', in S. E. Miller (Ed.), *Optical Fiber Telecommunications*, pp. 557–591, Academic Press, 1979.

34   A. G. Dentai, T. P. Lee and C. A. Burrus, 'Small-area high radiance LEDs emitting at 1.2 to 1.3 μm', *Electron. Lett.*, **13**(16), pp. 484–485, 1977.

35   H. F. Wolf, 'Optical sources', in H. F. Wolf (Ed.), *Handbook of Fiber Optics*, pp. 153–201, Granada, 1981.

36   K. Iga, T. Kambayashi, K. Wakao, C. Kitahara and K. Moriki, 'GaInAsP/InP double-heterostructure planar LED's', *IEEE Trans. Electron. Devices*, **ED-26**(8), pp. 1227–1230, 1979.

37   A. C. Carter, R. C. Goodfellow and R. Davis, '1.3–1.6 μm GaInAsP LEDs and their application in long haul, high data rate fibre optic systems', *Internat. Conf. on Communications Pt. II*, Seattle (USA), IEEE, Pt 28.1, 1980.

8

Optical Detectors

8.1 INTRODUCTION

We are concerned in this chapter with photodetectors currently in use and under investigation for optical fiber communications.

The detector is an essential component of an optical fiber communication system and is one of the crucial elements which dictate the overall system performance. Its function is to convert the received optical signal into an electrical signal, which is then amplified before further processing. Therefore when considering signal attenuation along the link, the system performance is determined at the detector. Improvement of detector characteristics and performance thus allows the installation of fewer repeater stations and lowers both the capital investment and maintenance costs.

The role the detector plays demands that it must satisfy very stringent requirements for performance and compatibility. The following criteria define the important performance and compatibility requirements for detectors which are generally similar to the requirements for sources.

(a) High sensitivity at the operating wavelengths. The first generation systems have wavelengths between 0.8 and 0.9 μm (compatible with AlGaAs laser and LED emission lines). However, considerable advantage may be gained at the detector from second generation sources with operating wavelengths above 1.1 μm as both fiber attenuation and dispersion are reduced. There is much research activity at present in this longer wavelength region, especially concerning wavelengths around 1.3 μm where attenuation and material dispersion can be minimized. In this case semiconductor materials are currently under investigation (see Section 8.4.3) in order to achieve good sensitivity at normal operating temperatures (i.e. 300 K).

(b) High fidelity. To reproduce the received signal waveform with fidelity, for analog transmission the response of the photodetector must be linear with regard to the optical signal over a wide range.

(c) Large electrical response to the received optical signal. The photodetector should produce a maximum electrical signal for a given amount of optical power, i.e. the quantum efficiency should be high.

(d) Short response time to obtain a suitable bandwidth. Present systems extend into the hundreds of megahertz. However, it is predicted that future systems (single mode fiber) will operate in the gigahertz range, and possibly above.

(e) A minimum noise introduced by the detector. Dark currents, leakage currents and shunt conductances must be low. Also the gain mechanism within either the detector or associated circuitry must be of low noise.

(f) Stability of performance characteristics. Ideally, the performance characteristics of the detector should be independent of changes in ambient conditions. However, the detectors currently favored (photodiodes) have characteristics (sensitivity, noise, internal gain) which vary with temperature, and therefore compensation for temperature effects is often necessary.

(g) Small size. The physical size of the detector must be small for efficient coupling to the fiber and to allow easy packaging with the following electronics.

(h) Low bias voltages. Ideally the detector should not require excessive bias voltages or currents.

(i) High reliability. The detector must be capable of continuous stable operation at room temperature for many years.

(j) Low cost. Economic considerations are often of prime importance in any large scale communication system application.

We continue the discussion in Section 8.2 by briefly indicating the various types of device which could be employed for optical detection. From this discussion it is clear that semiconductor photodiodes currently provide the best solution for detection in optical fiber communications. Therefore in Sections 8.3 and 8.4 we consider the principles of operation of these devices, together with the characteristics of the semiconductor materials employed in their construction. Sections 8.5–8.7 then briefly outline the major operating parameters (quantum efficiency, responsivity, long wavelength cutoff) of such photodiodes. Following in Sections 8.8 and 8.9 we discuss the structure and operation of the major device types ($p–n$, $p–i–n$ and avalanche photodiode). Finally, in Section 8.10 we consider recent developments in phototransistors which could mean they may eventually find wider use as detectors for optical fiber communications.

8.2 DEVICE TYPES

To detect optical radiation (photons) in the near infrared region of the spectrum, both external and internal photoemission of electrons may be utilized. External photoemission devices typified by photomultiplier tubes and vacuum photodiodes meet some of the performance criteria but are too bulky, and require high voltages for operation. However, internal photoemission devices especially semiconductor photodiodes with or without internal (avalanche) gain provide good performance and compatibility with relatively low cost. These photodiodes are made from semiconductors such as silicon, germanium and an increasing number of III–V alloys, all of which satisfy in various ways most of the detector requirements. They are therefore used in all major current optical fiber communication systems.

The internal photoemission process may take place in both intrinsic and extrinsic semiconductors. With intrinsic absorption, the received photons excite electrons from the valence to the conduction bands in the semiconductor, whereas extrinsic absorption involves impurity centers created within the material. However, for fast response coupled with efficient absorption of photons, the intrinsic absorption process is preferred and at present all detectors for optical fiber communications use intrinsic photodetection.

Silicon photodiodes [Ref. 1] have high sensitivity over the 0.8–0.9 μm wavelength band with adequate speed (hundreds of megahertz), negligible shunt conductance, low dark current and long term stability. They are therefore widely used in first generation systems and are currently commercially available. Their usefulness is limited to the first generation wavelength region as silicon has an indirect bandgap energy (see Section 8.4.1) of 1.14 eV giving a loss in response above 1.09 μm. Thus for second generation systems in the longer wavelength range 1.1–1.6 μm research is devoted to the investigation of semiconductor materials which have narrower bandgaps. Interest has focused on germanium and III–V alloys which give a good response at the longer wavelengths.

In both wavelength bands two main device types are currently the topic of major study. These are the *p–i–n* and avalanche photodiodes. We shall therefore consider these devices in greater detail.

8.3 OPTICAL DETECTION PRINCIPLES

The basic detection process in an intrinsic absorber is illustrated in Fig. 8.1 which shows a *p–n* photodiode. This device is reverse biassed and the electric field developed across the *p–n* junction sweeps mobile carriers (holes and electrons) to their respective majority sides (*p* and *n* type material). A depletion region or layer is therefore created on either side of the junction. This barrier has the effect of stopping the majority carriers crossing the junction in the opposite direction to the field. However, the field accelerates minority carriers from both sides to the opposite side of the junction, forming the reverse leakage current of the diode. Thus intrinsic conditions are created in the depletion region.

A photon incident in or near the depletion region of this device which has an

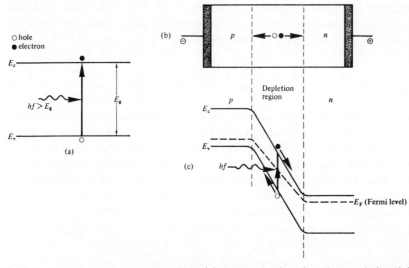

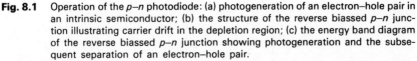

Fig. 8.1 Operation of the *p–n* photodiode: (a) photogeneration of an electron–hole pair in an intrinsic semiconductor; (b) the structure of the reverse biassed *p–n* junction illustrating carrier drift in the depletion region; (c) the energy band diagram of the reverse biassed *p–n* junction showing photogeneration and the subsequent separation of an electron–hole pair.

energy greater than or equal to the bandgap energy E_g of the fabricating material (i.e. $hf \geqslant E_g$) will excite an electron from the valence band into the conduction band. This process leaves an empty hole in the valence band and is known as the photogeneration of an electron–hole (carrier) pair as shown in Fig. 8.1(a). Carrier pairs so generated near the junction are separated and swept (drift) under the influence of the electric field to produce a displacement current in the external circuit in excess of any reverse leakage current (Fig. 8.1(b)). Photogeneration and the separation of a carrier pair in the depletion region of this reverse biassed *p–n* junction is illustrated in Fig. 8.1(c).

The depletion region must be sufficiently thick to allow a large fraction of the incident light to be absorbed in order to achieve maximum carrier pair generation. However, since long carrier drift times in the depletion region restrict the speed of operation of the photodiode it is necessary to limit its width. Thus there is a trade-off between the number of photons absorbed (sensitivity) and the speed of response.

8.4 ABSORPTION

8.4.1 Absorption Coefficient

The absorption of photons in a photodiode to produce carrier pairs and thus a photocurrent, is dependent on the absorption coefficient α_0 of the light in the semiconductor used to fabricate the device. At a specific wavelength and assuming only bandgap transitions (i.e. intrinsic absorber) the photocurrent I_p produced by incident light of optical power P_o is given by [Ref. 4]:

$$I_p = \frac{P_o e(1-r)}{hf} [1 - \exp(-\alpha_0 d)] \qquad (8.1)$$

where e is the charge on an electron, r is the Fresnel reflection coefficient at the semiconductor–air interface and d is the width of the absorption region.

The absorption coefficients of semiconductor materials are strongly dependent on wavelength. This is illustrated for some common semiconductors [Ref. 4] in Fig. 8.2. It may be observed that there is a variation between the absorption curves for the materials shown and that they are each suitable for

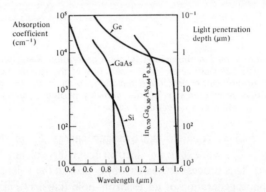

Fig. 8.2 Optical absorption curves for some common semiconductor photodiode materials (silicon, gemanium, gallium arsenide and indium gallium arsenide phosphide). Reproduced with permission from S. E. Miller and A. G. Chynoweth (Eds.), *Optical Fiber Telecommunications*, 1979, courtesy Academic Press Inc., Ltd.

Table 8.1 Bandgaps for some semiconductor photodiode materials at 300 K

	Bandgap (eV) at 300 K	
	Indirect	Direct
Si	1.14	4.10
Ge	0.67	0.81
GaAs	—	1.43
InAs	—	0.35
InP	—	1.35
GaSb	—	0.73
$In_{0.53}Ga_{0.47}As$	—	0.75
$In_{0.14}Ga_{0.86}As$	—	1.15
$GaAs_{0.88}Sb_{0.12}$	—	1.15

different wavelength applications. This results from their differing bandgaps energies as shown in Table 8.1. However, it must be noted that the curves depicted in Fig. 8.2 also vary with temperature.

8.4.2 Direct and Indirect Absorption: Silicon and Germanium

Table 8.1 indicates that silicon and germanium absorb light by both direct and indirect optical transitions. Indirect absorption requires the assistance of a phonon so that momentum as well as energy are conserved. This makes the transition probability less likely for indirect absorption than for direct absorption where no phonon is involved. In this context direct and indirect absorption may be contrasted with direct and indirect emission discussed in Section 6.3.3.1. Therefore as may be seen from Fig. 8.2 silicon is only weakly absorbing over the wavelength band of interest in optical fiber communications (i.e. first generation 0.8–0.9 μm). This is because transitions over this wavelength band in silicon are due only to the indirect absorption mechanism. As mentioned previously (Section 8.2) the threshold for indirect absorption occurs at 1.09 μm. The bandgap for direct absorption in silicon is 4.10 eV corresponding to a threshold of 0.30 μm in the ultraviolet and thus is well outside the wavelength range of interest.

Germanium is another semiconductor material for which the lowest energy absorption takes place by indirect optical transitions. However, the threshold for direct absorption occurs at 1.53 μm, below which germanium becomes strongly absorbing corresponding to the kink in the characteristic shown in Fig. 8.2. Thus germanium may be used in the fabrication of detectors over the whole of the wavelength range of interest (i.e. first and second generation 0.8–1.6 μm), especially considering that indirect absorption will occur up to a threshold of 1.85 μm.

Ideally a photodiode material should be chosen with a bandgap energy slightly less than the photon energy corresponding to the longest operating wavelength of the system. This gives a sufficiently high absorption coefficient to ensure a good response, and yet limits the number of thermally generated carriers in order to achieve a low dark current (i.e. displacement current generated with no incident light (see Fig. 3.5)). Germanium photodiodes have relatively large dark currents due to their narrow bandgaps in comparison to other semiconductor materials. This is a major disadvantage with the use of germanium photodiodes especially at shorter wavelengths (below 1.1 μm).

8.4.3 III–V Alloys

The drawback with germanium as a fabricating material for semiconductor photodiodes has led to increased investigation of direct bandgap III–V alloys for the longer wavelength region. These materials are potentially superior to germanium because their bandgaps can be tailored to the desired wavelength by changing the relative concentrations of their constituents, resulting in lower dark currents. They may also be fabricated in heterojunction structures (see Section 6.3.5) which enhances their high speed operations.

Ternary alloys such as InGaAs and GaAlSb deposited on GaSb substrates have been used to fabricate photodiodes for the 1.0–1.4 μm wavelength band. However, difficulties in growth of these alloys with lattice matching have led to defects which cause increased dark currents and microplasma sites (small areas with lower breakdown voltages than the rest of the junction).* These defects limit the performance of a device fabricated from ternary alloys. More encouraging results have been obtained with quaternary alloys such as InGaAsP grown on InP and GaAlAsSb grown on GaSb. These systems have the major advantage that the bandgap and lattice constant can be varied independently. This permits the bandgap tailoring whilst maintaining a lattice match to the substrate.

8.5 QUANTUM EFFICIENCY

The quantum efficiency η is defined as the fraction of incident photons which are absorbed by the photodetector and generate electrons which are collected at the detector terminals:

$$\eta = \frac{\text{number of electrons collected}}{\text{number of incident photons}} \qquad (8.2)$$

Hence,

$$\eta = \frac{r_e}{r_p} \qquad (8.3)$$

where r_p is the incident photon rate (photons per second) and r_e is the corresponding electron rate (electrons per second).

One of the major factors which determines the quantum efficiency is the absorption coefficient (see Section 8.4.1) of the semiconductor material used within the photodetector. The quantum efficiency is generally less than unity as not all of the incident photons are absorbed to create electron–hole pairs. Furthermore, it should be noted that it is often quoted as a percentage (e.g. a quantum efficiency of 75% is equivalent to 75 electrons collected per 100 incident photons). Finally, in common with the absorption coefficient, the quantum efficiency is a function of the photon wavelength and must therefore only be quoted for a specific wavelength.

---

* It should be noted that microplasmas are only of concern in avalanche photodiodes (see Section 8.9.1).

8.6 RESPONSIVITY

The expression for quantum efficiency does not involve photon energy and therefore the responsivity R is often of more use when characterizing the performance of a photodetector. It is defined as:

$$R = \frac{I_p}{P_o} \text{ A W}^{-1} \tag{8.4}$$

where I_p is the output photocurrent in amperes and P_o is the incident optical power in watts. The responsivity is a useful parameter as it gives the transfer characteristic of the detector (i.e. photocurrent per unit incident optical power).

The relationship for responsivity (Eq. (8.4)) may be developed to include quantum efficiency as follows. Considering Eq. (6.1) the energy of a photon $E = hf$. Thus the incident photon rate r_p may be written in terms of incident optical power and the photon energy as:

$$r_p = \frac{P_o}{hf} \tag{8.5}$$

In Eq. (8.3) the electron rate is given by:

$$r_e = \eta r_p \tag{8.6}$$

Substituting from Eq. (8.5) we obtain

$$r_e = \frac{\eta P_o}{hf} \tag{8.7}$$

Therefore, the output photocurrent is:

$$I_p = \frac{\eta P_o e}{hf} \tag{8.8}$$

where e is the charge on an electron. Thus from Eq. (8.4) the responsivity may be written as:

$$R = \frac{\eta e}{hf} \tag{8.9}$$

Equation (8.9) is a useful relationship for responsivity which may be developed a further stage to include the wavelength of the incident light.

The frequency f of the incident photons is related to their wavelength λ and the velocity of light in air c, by:

$$f = \frac{c}{\lambda} \tag{8.10}$$

Substituting into Eq. (8.9) a final expression for the responsivity is given by:

$$R = \frac{\eta e \lambda}{hc} \qquad (8.11)$$

It may be noted that the responsivity is directly proportional to the quantum efficiency at a particular wavelength.

The ideal responsivity against wavelength characteristic for a silicon photodiode with unit quantum efficiency is illustrated in Fig. 8.3. Also shown is the typical responsivity of a practical silicon device.

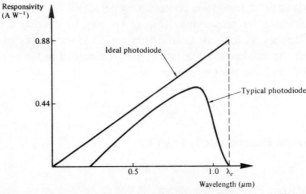

Fig. 8.3 Responsivity against wavelength characteristic for an ideal silicon photodiode. The responsivity of a typical device is also shown.

Example 8.1

When 3×10^{11} photons each with a wavelength of 0.85 µm are incident on a photodiode, on average 1.2×10^{11} electrons are collected at the terminals of the device. Determine the quantum efficiency and the responsivity of the photodiode at 0.85 µm.

Solution: From Eq. (8.2),

$$\text{quantum efficiency} = \frac{\text{number of electrons collected}}{\text{number of incident photons}}$$

$$= \frac{1.2 \times 10^{11}}{3 \times 10^{11}}$$

$$= 0.4$$

The quantum efficiency of the photodiode at 0.85 µm is 40%.
From Eq. (8.11),

$$\text{responsivity } R = \frac{\eta e \lambda}{hc}$$

$$= \frac{0.4 \times 1.602 \times 10^{-19} \times 0.85 \times 10^{-6}}{6.626 \times 10^{-34} \times 2.998 \times 10^{8}}$$

$$= 0.274 \ A \ W^{-1}$$

The responsivity of the photodiode at 0.85 μm is 0.27 A W^{-1}.

Example 8.2

A photodiode has a quantum efficiency of 65% when photons of energy 1.5 × 10^{-19} J are incident upon it.

(a) At what wavelength is the photodiode operating?
(b) Calculate the incident optical power required to obtain a photocurrent of 2.5 μA when the photodiode is operating as described above.

Solution: (a) From Eq. (6.1), the photon energy $E = hf = hc/\lambda$. Therefore

$$\lambda = \frac{hc}{E} = \frac{6.626 \times 10^{-34} \times 2.998 \times 10^{8}}{1.5 \times 10^{-19}}$$

$$= 1.32 \ \mu m$$

The photodiode is operating at a wavelength of 1.32 μm.
   (b) From Eq. (8.9),

$$\text{responsivity } R = \frac{\eta e}{hf} = \frac{0.65 \times 1.602 \times 10^{-19}}{1.5 \times 10^{-19}}$$

$$= 0.694 \ A \ W^{-1}$$

Also from Eq. (8.4),

$$R = \frac{I_p}{P_o}$$

Therefore

$$P_o = \frac{2.5 \times 10^{-6}}{0.694} = 3.60 \ \mu W$$

The incident optical power required is 3.60 μW.

8.7   LONG WAVELENGTH CUT OFF

It is essential when considering the intrinsic absorption process that the energy of incident photons be greater than or equal to the bandgap energy E_g of the material used to fabricate the photodetector. Therefore the photon energy

$$\frac{hc}{\lambda} \geqslant E_g \qquad\qquad (8.12)$$

giving

$$\lambda \leqslant \frac{hc}{E_g} \tag{8.13}$$

Thus the threshold for detection commonly known as the long wavelength cutoff point λ_c is:

$$\lambda_c = \frac{hc}{E_g} \tag{8.14}$$

The expression given in Eq. (8.14) allows the calculation of the longest wavelength of light to give photodetection for the various semiconductor materials used in the fabrication of detectors.

It is important to note that the above criterion is only applicable to intrinsic photodetectors. Extrinsic photodetectors violate the expression given in Eq. (8.12), but are not currently used in optical fiber communications.

Example 8.3

GaAs has a bandgap energy of 1.43 eV at 300 K. Determine the wavelength above which an intrinsic photodetector fabricated from this material will cease to operate.
Solution: From Eq. (8.14), the long wavelength cutoff:

$$\lambda_c = \frac{hc}{E_g} = \frac{6.626 \times 10^{-34} \times 2.998 \times 10^8}{1.43 \times 1.602 \times 10^{-19}}$$

$$= 0.867 \ \mu\text{m}$$

The GaAs photodetector will cease to operate above 0.87 μm.

8.8 SEMICONDUCTOR PHOTODIODES WITHOUT INTERNAL GAIN

Semiconductor photodiodes without internal gain generate a single electron hole pair per absorbed photon. This mechanism was outlined in Section 8.3, and in order to understand the development of this type of photodiode it is now necessary to elaborate upon it.

8.8.1 *p–n* Photodiode

Figure 8.4 shows a reverse biassed *p–n* photodiode with both the depletion and diffusion regions. The depletion region is formed by immobile positively charged donor atoms in the *n* type semiconductor material and immobile

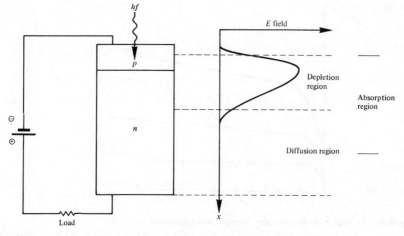

Fig. 8.4 *p–n* photodiode showing depletion and diffusion regions.

negatively charged acceptor atoms in the *p* type material, when the mobile carriers are swept to their majority sides under the influence of the electric field. The width of the depletion region is therefore dependent upon the doping concentrations for a given applied reverse bias (i.e. the lower the doping, the wider the depletion region). For the interested reader expressions for the depletion layer width are given Ref. 5.

Photons may be absorbed in both the depletion and diffusion regions as indicated by the absorption region in Fig. 8.4. The absorption region's position and width depends upon the energy of the incident photons and on the material from which the photodiode is fabricated. Thus in the case of the weak absorption of photons, the absorption region may extend completely throughout the device. Electron–hole pairs are therefore generated in both the depletion and diffusion regions. In the depletion region the carrier pairs separate and drift under the influence of the electric field, whereas outside this region the hole diffuses towards the depletion region in order to be collected. The diffusion process is very slow compared to drift and thus limits the response of the photodiode (see Appendix D).

It is therefore important that the photons are absorbed in the depletion region. Thus it is made as long as possible by decreasing the doping in the *n* type material. The depletion region width in a *p–n* photodiode is normally 1–3 μm and is optimized for the efficient detection of light at a given wavelength. For silicon devices this is in the visible spectrum (0.4–0.7 μm) and for germanium in the near infrared (0.7–0.9 μm).

Typical output characteristics for the reverse-biassed *p–n* photodiode are illustrated in Fig. 8.5. The different operating conditions may be noted moving from no light input to a high light level.

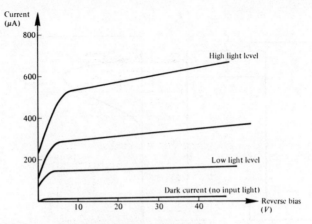

Fig. 8.5 Typical *p–n* photodiode output characteristics.

8.8.2 *p–i–n* Photodiode

In order to allow operation at longer wavelengths where the light penetrates more deeply into the semiconductor material a wider depletion region is necessary. To achieve this the *n* type material is doped so lightly that it can be considered intrinsic, and to make a low resistance contact a highly doped *n* type (n^+) layer is added. This creates a *p–i–n* (or PIN) structure as may be seen in Fig. 8.6 where all the absorption takes place in the depletion region.

Figure 8.7 shows the structures of two types of silicon *p–i–n* photodiode for operation in the shorter wavelength band below 1.09 μm. The front illuminated

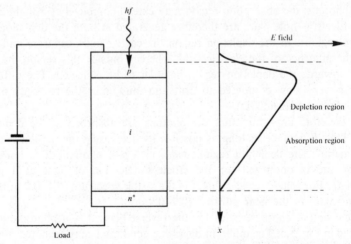

Fig. 8.6 *p–i–n* photodiode showing combined absorption and depletion region.

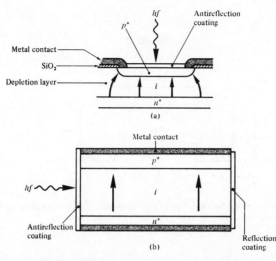

Fig. 8.7 (a) Structure of a front illuminated silicon p–i–n photodiode. (b) Structure of a side illuminated (parallel to junction) p–i–n photodiode.

photodiode when operating in the 0.8–0.9 μm band (Fig. 8.7(a)) requires a depletion region of between 20 and 50 μm in order to attain high quantum efficiency (typically 85%) together with fast response (less than 1 ns) and low dark current (1 nA). Dark current arises from surface leakage currents as well as generation–recombination currents in the depletion region in the absence of illumination. The side illuminated structure (Fig. 8.7(b)), where light is injected parallel to the junction plane, exhibits a large absorption width ($\simeq 500$ μm) and hence is particularly sensitive at wavelengths close to the bandgap limit (1.09 μm) where the absorption coefficient is relatively small.

Germanium p–i–n photodiodes which span the entire wavelength range of interest are also commercially available, but as mentioned previously (Section 8.4.2) the relatively high dark currents are a problem (typically 100 nA at 20 °C increasing to 1 μA at 40 °C). However, as outlined in Section 8.4.3 III–V alloys are under investigation for detection in the longer wavelength region. The two of particular interest in view of lattice matching are $In_{1-x}Ga_x As_y P_{1-y}$ grown on InP and $Ga_x Al_{1-x} As_y Sb_{1-y}$ grown on GaSb. The structure for a p–i–n photodiode [Ref. 7] of the former is shown in Fig. 8.8.

The quaternary wafer was grown by liquid phase epitaxy using a conventional sliding-boat technique. The photodiode formed a mesa structure in which the edge of the p–n junction was exposed to the environment or the material in the package. It operated at a wavelength of 1.26 μm with low dark current (less than 0.2 nA) and with a quantum efficiency around 60%. Also the response time was estimated at no greater than 100 ps. It is likely that photodiodes fabricated from these materials will find wide application within longer wavelength optical fiber systems.

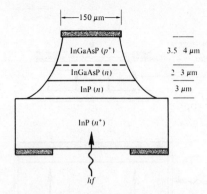

Fig. 8.8 Structure of an InGaAsP p–i–n photodiode. The InP base is transparent to the radiation absorbed in the quaternary layers [Ref. 7].

8.9 SEMICONDUCTOR PHOTODIODES WITH INTERNAL GAIN

8.9.1 Avalanche Photodiodes

The second major type of optical communications detector is the avalanche photodiode (APD). This has a more sophisticated structure than the p–i–n photodiode in order to create an extremely high electric field region (approximately 3×10^5 V cm^{-1}) as may be seen in Fig. 8.9(a). Therefore, as well as the depletion region where most of the photons are absorbed and the primary carrier pairs generated there is a high field region in which holes and electrons can acquire sufficient energy to excite new electron–hole pairs. This process is known as impact ionization and is the phenomenon that leads to avalanche breakdown in ordinary reverse biassed diodes. It requires very high

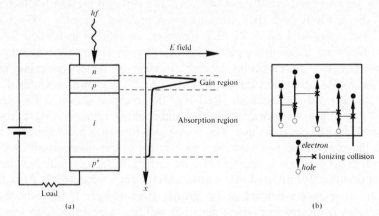

Fig. 8.9 (a) Avalanche photodiode showing high electric field (gain) region. (b) Carrier pair multiplication in the gain region.

reverse bias voltages (100–400 V) in order that the new carriers created by impact ionization can themselves produce additional carriers by the same mechanism as shown in Fig. 8.9(b).

Carrier multiplication factors as great as 10^4 may be obtained using defect-free materials to ensure uniformity of carrier multiplication over the entire photosensitive area. However, other factors affect the achievement of high gain within the device. Microplasmas, which are small areas with lower breakdown voltages than the remainder of the junction, must be reduced through the selection of defect-free materials together with careful device processing and fabrication [Ref. 14]. In addition, excessive leakage at the junction edges can be eliminated by the use of a guard ring structure as shown in Fig. 8.10. At present both silicon and germanium APDs are available.

Operation of these devices at high speed requires full depletion in the absorption region. As indicated in Section 8.8.1, when carriers are generated in undepleted material, they are collected somewhat slowly by the diffusion process. This has the effect of producing a long 'diffusion tail' on a short optical pulse. When the APD is fully depleted by employing electric fields in excess of 10^4 V m^{-1}, all the carriers drift at saturation-limited velocities. In this case the response time for the device is limited by three factors. These are:

(a) the transit time of the carriers across the absorption region (i.e. the depletion width); and

(b) the time taken by the carriers to perform the avalanche multiplication process.

(c) the RC time constant incurred by the junction capacitance of the diode and its load.

Often an asymmetric pulse shape is obtained from the APD which results from a relatively fast rise time as the electrons are collected and a fall time dictated by the transit time of the holes travelling at a slower speed. Hence, although the use of suitable materials and structures may give rise times between 150 and 200 ps, fall times of 1 ns or more are quite common which limit the overall response of the device.

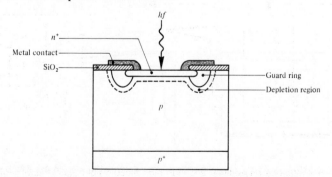

Fig. 8.10   Structure of a silicon avalanche photodiode with guard ring.

8.9.2 Silicon Reach Through Avalanche Photodiodes

To ensure carrier multiplication without excess noise within the APD it is necessary to reduce the ratio of the ionization coefficients for electrons and holes k (see Section 9.3.4). In silicon this ratio is a strong function of the electric field varying from around 0.1 at 3×10^5 V m^{-1} to 0.5 at 6×10^5 V m^{-1}. Hence for minimum noise, the electric field at avalanche breakdown must be as low as possible and the impact ionization should be initiated by electrons. To this end a 'reach through' structure has been implemented with the silicon avalanche photodiode. The silicon 'reach through' APD(RAPD) consists of $p^+ - \pi - p - n^+$ layers as shown in Fig. 8.11(a). As may be seen from the corresponding field plot in Fig. 8.11(b), the high field region where the avalanche multiplication takes place is relatively narrow and centered on the $p-n^+$ junction. Thus under low reverse bias most of the voltage is dropped across the $p-n^+$ junction.

When the reverse bias voltage is increased the depletion layer widens across the p region until it 'reaches through' to the nearly intrinsic (lightly doped) π region. Since the π region is much wider than the p region the field in the π region is much lower than that at the $p-n^+$ junction (see Fig. 8.11(b)). This has the effect of removing some of the excess applied voltage from the multiplication region to the π region giving a relatively slow increase in multiplication factor with applied voltage. Although the field in the π region is lower than in the multiplication region it is high enough (2×10^4 V cm^{-1}) when the photodiode is operating to sweep the carriers through to the multiplication region at their scattering limited velocity (10^7 cm s^{-1}). This limits the transit time and ensures a fast response (as short as 0.5 ns).

Measurements [Ref. 16] for a silicon RAPD for optical fiber communication applications at a wavelength of 0.825 µm have shown a quantum efficiency (without avalanche gain) of nearly 100% in the working region, as may be seen in Fig. 8.12. The dark currents for this photodiode are also low and depend only slightly on bias voltage.

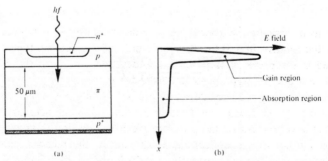

Fig. 8.11 (a) Structure of a silicon RAPD. (b) The field distribution in the RAPD showing the gain region across the $p-n^+$ junction.

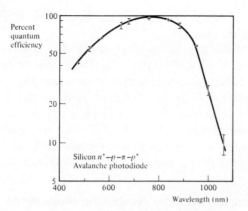

Fig. 8.12   Measurements of quantum efficiency against wavelength for a silicon RAPD. After Ref. 16. Reprinted with permission from *The Bell System Technical Journal.* © 1978, AT&T.

8.9.3   Germanium Avalanche Photodiodes

Work is also continuing to optimize germanium APDs in order to achieve low dark currents with reasonable multiplication factors over the whole of the wavelength range of interest. A low noise n^+-n-p germanium APD [Ref. 25] which will operate over the 0.8–1.5 μm wavelength band has gone some way to achieving this, although at a wavelength of 1.3 μm with a multiplication factor of 10 the device has a dark current of approximately 1 μA.

8.9.4   III–V Alloy Avalanche Photodiodes

Recently interest has focused on APDs fabricated from III–V alloys. Initially they were found to give excessively large dark currents at the biasses required

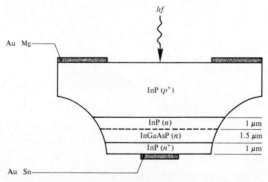

Fig. 8.13   Cross section of an InGaAsP/InP heterostructure APD with the high electric field (gain) region in the InP layer [Ref. 26].

to achieve gain. However, a structure has been reported which appears to reduce this problem [Ref. 26].

An example of this structure is given in Fig. 8.13 which shows an etched inverted mesa InGaAsP/InP APD. The interesting innovation with this structure which reduces the dark current to an acceptable level (200 pA for a multiplication factor of 10) is that the $p-n^+$ junction, and therefore the multiplication region is located in the InP substrate with the depletion region extending into the photosensitive InGaAsP. The structure therefore takes advantage of the low leakage, high gain properties of InP APDs and the longer wavelength response of the narrower bandgap InGaAsP (0.35–0.70 eV).

8.9.5 Drawbacks with the Avalanche Photodiode

APDs have a distinct advantage over photodiodes without internal gain for the detection of the very low light levels often encountered in optical fiber communications. However, they also have several drawbacks, which are:

(a) Fabrication difficulties due to their more complex structure and hence increased cost.
(b) The random nature of the gain mechanism which gives an additional noise contribution (see Section 9.3.3).
(c) The high bias voltages required (100–400 V) which are wavelength dependent.
(d) The variation of the gain (multiplication factor) with temperature as shown in Fig. 8.14 for a silicon RAPD [Ref. 16]. Thus temperature compensation is necessary to stabilize the operation of the device.

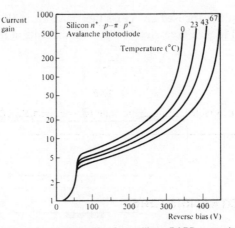

Fig. 8.14   Current gain against reverse bias for a silicon RAPD operating at a wavelength of 0.825 μm. After Ref. 16. Reprinted with permission from *The Bell System Technical Journal.* © 1978, AT&T.

8.9.6 Multiplication Factor

The multiplication factor M is a measure of the internal gain provided by the APD. It is defined as:

$$M = \frac{I}{I_p} \tag{8.15}$$

where I is the total output current at the operating voltage (i.e. where carrier multiplication occurs) and I_p is the initial or primary photocurrent (i.e. before carrier multiplication occurs).

Example 8.4

The quantum efficiency of a particular silicon RAPD is 80% for the detection of radiation at a wavelength of 0.9 μm. When the incident optical power is 0.5 μW, the output current from the device (after avalanche gain) is 11 μA. Determine the multiplication factor of the photodiode under these conditions.

Solution: From Eq. (8.11), the responsivity

$$R = \frac{\eta e \lambda}{hc} = \frac{0.8 \times 1.602 \times 10^{-19} \times 0.9 \times 10^{-6}}{6.626 \times 10^{-34} \times 2.998 \times 10^{8}}$$

$$= 0.581 \ A \ W^{-1}$$

Also from Eq. (8.4), the photocurrent

$$I_p = P_o R$$

$$= 0.5 \times 10^{-6} \times 0.581$$

$$= 0.291 \ \mu A$$

Finally using Eq. (8.15):

$$M = \frac{I}{I_p} = \frac{11 \times 10^{-6}}{0.291 \times 10^{-6}}$$

$$= 37.8$$

The multiplication factor of the photodiode is approximately 38.

8.10 PHOTOTRANSISTORS

The problems encountered with APDs for use in the longer wavelength region has stimulated a renewed interest in bipolar phototransistors. However, these devices have yet to find use in major optical fiber communication systems. In common with the APD the phototransistor provides internal gain of the photocurrent. This is achieved through transistor action rather than avalanche multiplication. A symbolic representation of the *n–p–n* bipolar phototransistor is shown in Fig. 8.15(a). It differs from the conventional bipolar transistor in that the base is unconnected, the base–collector junction being photosensitive to act as a light-gathering element. Thus absorbed light affects the base current giving multiplication of primary photocurrent through the device.

The structure of a recent *n–p–n* InGaAsP/InP heterojunction photo-transistor is shown in Fig. 8.15(b) [Ref. 30]. The three layer heterostructure (see Section 6.3.5) is grown on an InP substrate using liquid phase epitaxy (LPE). It consists of an *n* type InP collector layer followed by a thin (0.1 μm) *p* type InGaAsP base layer. The third layer is a wide bandgap *n* type InP emitter layer. Radiation incident on the device passes unattenuated through the wide bandgap emitter and is absorbed in the base, base–collector depletion region

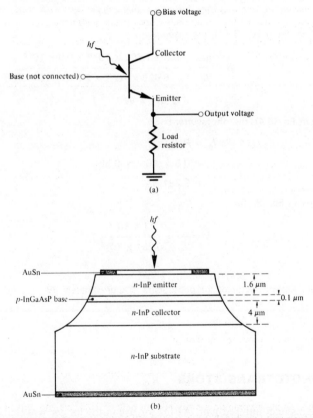

Fig. 8.15   (a) Symbolic representation of the *n–p–n* phototransistor showing the external connections. (b) Cross section of an *n–p–n* InGaAsP/InP heterojunction phototransistor [Ref. 30].

and the collector. A large secondary photocurrent between the emitter and collector is obtained as the photogenerated holes are swept into the base, increasing the forward bias on the device. The use of the heterostructure permits low emitter–base and collector–base junction capacitances together with low base resistance. This is achieved through low emitter and collector doping levels coupled with heavy doping of the base, and allows large current gain. In addition the potential barrier created by the heterojunction at the emitter–base junction effectively eliminates hole injection from the base when the junction is forward biassed. This gives good emitter base injection efficiency. The optical gain G_o of the device is given approximately by |Ref. 30|

$$G_o \simeq \eta h_{FE} = \frac{hf}{e} \frac{I_c}{P_o} \tag{8.16}$$

where η is the quantum efficiency of the base–collector photodiode, h_{FE} is the common emitter current gain, I_c is the collector current, P_o is the incident optical power, e is the electronic charge and hf is the photon energy.

The phototransistor shown in Fig. 8.15(b) is capable of operating over the 0.9–1.3 µm wavelength band giving optical gains in excess of 100 as demonstrated in example 8.5.

Example 8.5

The phototransistor of Fig. 8.15(b) has a collector current of 15 mA when the incident optical power at a wavelength of 1.26 µm is 125 µW. Estimate:

(a) the optical gain of the device under the above operating conditions;
(b) the common emitter current gain if the quantum efficiency of the base–collector photodiode at a wavelength of 1.26 µm is 40%.

Solution: (a) Using Eq. (8.16), the optical gain is given by:

$$G_o \simeq \frac{hf}{e} \frac{I_c}{P_o} = \frac{hc}{\lambda_e} \frac{I_c}{P_o}$$

$$= \frac{6.626 \times 10^{-34} \times 2.998 \times 10^8 \times 15 \times 10^{-3}}{1.26 \times 10^{-6} \times 1.602 \times 10^{-19} \times 125 \times 10^{-6}}$$

$$= 118.1$$

(b) The common emitter current gain is:

$$h_{FE} = \frac{G_o}{\eta} = \frac{118.1}{0.4} = 295.3$$

In this example a common emitter current gain of 295 gives an optical gain of 118. It is therefore possible that this type of device will become an alternative to the APD for optical detection at wavelengths above 1.1 µm [Refs. 33–35].

PROBLEMS

8.1 Outline the reasons for the adoption of the materials and devices used for photodetection in optical fiber communications. Discuss in detail the p–i–n photodiode with regard to performance and compatibility requirements in photodetectors.

8.2 A p–i–n photodiode on average generates one electron–hole pair per three incident photons at a wavelength of 0.8 μm. Assuming all the electrons are collected calculate:
  (a) the quantum efficiency of the device;
  (b) its maximum possible bandgap energy;
  (c) the mean output photocurrent when the received optical power is 10^{-7} W.

8.3 Explain the detection process in the p–n photodiode. Compare this device with the p–i–n photodiode.

8.4 Define the quantum efficiency and the responsivity of a photodetector.
  Derive an expression for the responsivity of an intrinsic photodetector in terms of the quantum efficiency of the device and the wavelength of the incident radiation.
  Determine the wavelength at which the quantum efficiency and the responsivity are equal.

8.5 A p–n photodiode has a quantum efficiency of 50% at a wavelength of 0.9 μm. Calculate:
  (a) its responsivity at 0.9 μm;
  (b) the received optical power if the mean photocurrent is 10^{-6} A;
  (c) the corresponding number of received photons at this wavelength.

8.6 When 800 photons per second are incident on a p–i–n photodiode operating at a wavelength of 1.3 μm they generate on average 550 electrons per second which are collected. Calculate the responsivity of the device.

8.7 Explain what is meant by the long wavelength cutoff point for an intrinsic photodetector, deriving any relevant expressions.
  Considering the bandgap energies given in Table 8.1, calculate the long wavelength cutoff points for both direct and indirect optical transitions in silicon and germanium.

8.8 A p–i–n photodiode ceases to operate when photons with energy greater than 0.886 eV are incident upon it; of which material is it fabricated?

8.9 Discuss the operation of the silicon RAPD, describing how it differs from the p–n photodiode.
  Outline the advantages and drawbacks with the use of the RAPD as a detector for optical fiber communications.

8.10 An APD with a multiplication factor of 20 operates at a wavelength of 1.5 μm. Calculate the quantum efficiency and the output photocurrent from the device if its responsivity at this wavelength is $0.6 \, A \, W^{-1}$ and 10^{10} photons of wavelength 1.5 μm are incident upon it per second.

8.11 Discuss the materials used in the fabrication of APDs and comment on their relative merits and drawbacks when employed in devices utilized for optical fiber communication.

8.12 Given that the following measurements were taken for an APD calculate the multiplication factor for the device.
Received optical power at 1.35 μm = 0.2 μW
Corresponding output photocurrent = 4.9 μA
(after avalanche gain)
Quantum efficiency at 1.35 μm   = 40%

8.13 An APD has a quantum efficiency of 45% at 0.85 μm. When illuminated with radiation of this wavelength it produces an output photocurrent of 10 μA after avalanche gain with a multiplication factor of 250. Calculate the received optical power to the device. How many photons per second does this correspond to?

8.14 When 10^{11} photons per second each with an energy of 1.28×10^{-19} J are incident on an ideal photodiode, calculate:
(a) the wavelength of the incident radiation;
(b) the output photocurrent;
(c) the output photocurrent if the device is an APD with a multiplication factor of 18.

8.15 A silicon RAPD has a multiplication factor of 10^3 when operating at a wavelength of 0.82 μm. At this operating point the quantum efficiency of the device is 90% and the dark current is 1 nA.
Determine the number of photons per second of wavelength 0.82 μm required in order to register a light input to the device corresponding to an output current (after avalanche gain) which is greater than the level of the dark current (i.e. $I > 1 \, nA$).

8.16 An InGaAsP heterojunction phototransistor has a common emitter current gain of 170 when operating at a wavelength of 1.3 μm with an incident optical power of 80 μW. The base collector quantum efficiency at this wavelength is 65%. Estimate the collector current in the device.

Answers to Numerical Problems

8.2	(a) 33%; (b) 24.8×10^{-20} J;	**8.10**	50%, 15.9 nA
	(c) 21.3 nW	**8.12**	24.1
8.4	1.24 μm	**8.13**	77.8 nW,
8.5	(a) 0.36 A W^{-1}; (b) 2.78 μW;		3.33×10^{11} photons s^{-1}
	(c) 1.26×10^{13} photons s^{-1}	**8.14**	(a) 1.55 μm; (b) 1.6 μA;
8.6	0.72 A W^{-1}		(c) 28.8 μA
8.7	0.3 μm, 1.09 μm, 1.53 μm,	**8.15**	6.94×10^6 photons s^{-1}
	1.85 μm	**8.16**	9.3 mA
8.8	$In_{0.7} Ga_{0.3} As_{0.64} P_{0.36}$		

REFERENCES

1 H. Melchior, M. B. Fisher and F. R. Arams, 'Photodetectors for optical communication systems', *Proc. IEEE*, **58**, pp. 1466–1486, 1970.

2 H. Melchior, 'Detectors for lightwave communications', *Phys. Today*, **30**, pp. 32–39, 1977.

3 S. D. Personick, 'Photodetectors for fiber systems', in M. K. Barnoski (Ed.), *Fundamentals of Optical Fiber Communications*, 2nd Ed., pp. 257–293, Academic Press, 1981.

4 T. P. Lee and T. Li, 'Photodetectors', in S. E. Miller and A. G. Chynoweth (Eds.), *Optical Fiber Telecommunications*, pp. 593–626, Academic Press, 1979.

5 S. M. Sze, *Physics of Semiconductor Devices* (2nd Ed.), John Wiley, 1981.

6 B. O. Seraphin and H. E. Bennett, 'Optical constants', in R. K. Willardson and A. C. Beer (Eds.), *Semiconductors and Semimetals*, Vol. 3, pp. 449–543, Academic Press, 1967.

7 C. A. Burrus, A. G. Dentai and T. P. Lee, 'InGaAsP $p-i-n$ photodiodes with low dark currents and small capacitance', *Electron. Lett.*, **15**(20), pp. 655–656, 1979.

8 T. Sukegawa, T. Hiraguchi, A. Tanaka and M. Haginer, 'Highly efficient $P-GaSb-N-Ga_{1-x}Al_xSb$ photodiodes', *Appl. Phys. Lett.*, **32**(6), pp. 376–378, 1978.

9 A. R. Clawson, W. Y. Lum, G. E. McWilliams and H. H. Wieder, 'Quaternary alloy InGaAsP/InP detectors', *Appl. Phys. Lett.*, **17**(11), pp. 2065–2066, 1978.

10 R. E. Leheny, R. E. Nahory and M. A. Pollack, '$In_{0.5}Ga_{0.47}As$, $p-i-n$ photodiodes for long wavelength fibre optic system', *Electron. Lett.*, **15**(22), pp. 713–715, 1979.

11 A. Kashinwazwa, A. Yamaguchu and M. Fuguwara, 'Silicon pin photodiodes', *Nat. Tech. Rep. (Jpn) EEE*, **25**(6), pp. 1180–1189, 1979.

12 T. Pearsall, 'Photodetectors for communication by optical fibres', in M. J. Howes and D. V. Morgan (Eds.), *Optical Fibre Communications*, pp. 107–164, John Wiley, 1980.

13 C. E. Hurwitz and J. J. Hsieh, 'GaInAsP/InP avalanche photodiodes', *Appl. Phys. Lett.*, **32**(8), pp. 487–489, 1978.

14 T. P. Lee, C. A. Burrus Jr and A. G. Dentai, 'InGaAsP/InP photodiodes microplasma-limited avalanche multiplication at $1-1.3$ μm wavelength', *IEEE J. Quantum Electron.*, **QE-15**, pp. 30–35, 1979.

15 P. P. Webb, R. J. McIntyre and J. Conradi, 'Properties of avalanche photodiodes', *RCA Rev.*, **35**, pp. 235–277, 1974.

16 A. R. Hartman, H. Melchior, D. P. Schinke and T. E. Seidel, 'Planar epitaxial silicon avalanche photodiode', *Bell Sys. Tech. J.*, **57**, pp. 1791–1807, 1978.

17 H. H. Weider, A. R. Clawson and G. E. McWilliams, '$In_xGa_{1-x}As_yP_{1-y}$/InP heterojunction photodiodes', *Appl. Phys. Lett.*, **31**, pp. 468–470, 1977.

18 G. E. Stillman and C. M. Wolfe, 'Avalanche photodiodes', in P. K. Willardson and A. C. Beers (Eds.), *Semiconductors and Semimetals*, Vol. 12, p. 291, Academic Press, 1977.

19 G. H. Oslen and H. Kressel, 'Vapour grown 1.3 μm InGaAsP/InP avalanche photodiodes', *Appl. Phys. Lett.*, **34**, pp. 581–583, 1979.

20 K. Nishida, K. Taguchi, Y. Matsumoto, 'InGaAsP heterostructure avalanche photodiodes with high avalanche gain', *Appl. Phys. Lett.*, **35**, pp. 251–253, 1979.

21 H. Melngailis, 'Photodiodes at 1.06–1.6 μm', IEEE Optical Fiber Communication Conf. Washington DC, USA Paper THA2/90, 1979.

22   R. G. Smith, 'Photodetectors for fiber transmission systems', *Proc. IEEE*, **68**(10), pp. 1247–1253, 1980.

23   D. Botez and C. J. Herskowitz, 'Components for optical communication systems: a review', *Proc. IEEE*, **68**(6), pp. 689–730, 1980.

24   H. Ando, H. Kanbe, T. Kimwa, T. Yamaska and T. Kaneda, 'Characteristics of germanium, avalanche photodiodes in the wavelength region 1–1.6 μm', *IEEE J. Quantum Electron.*, **QE-14**, pp. 804–809, 1978.

25   T. Mikawa, S. Kagawa, T. Kaneda, T. Sakwai, H. Ando and O. Mikami, 'A low-noise n^+np germanium avalanche photodiode', *IEEE J. Quantum Electron.* **QE-17**(2), pp. 210–216, 1981.

26   H. Kanbe, N. Susa, H. Nakagome and H. Ando, 'InGaAs avalanche photodiode with InP *p–n* junction, *Electron. Lett.*, **16**, pp. 163–165, 1980.

27   C. E. Hurwitz, 'Detectors for the 1.1–1.6 μm spectral region', *Proc. Soc. Photo-opt. Instrum. Eng. (USA)*, **224**, pp. 122–127, 1980.

28   H. Melchior and A. R. Hartman, 'Epitaxial silicon $n^+–p–\pi–p^+$ avalanche photodiode for optical fiber communications at 800 to 900 nanometers', *Tech. Dig. in Electronic Devices Meeting*, p. 412, 1976.

29   M. Tobe, Y. Amemiya, S. Sakai and M. Umeno, 'High-sensitivity InGaAsP/InP phototransistors', *Appl. Phys. Lett.*, **37**(1), pp. 73–75, 1980.

30   P. D. Wright, R. J. Nelson and T. Cella, 'High gain InGaAsP–InP heterojunction phototransistors', *Appl. Phys. Lett.*, **37**(2), pp. 192–194, 1980.

31   A. N. Saxena and H. F. Wolf, 'Optical detectors', in H. F. Wolf (Ed.), *Handbook of Fiber Optics, Theory and Applications*, pp. 203–240, Granada, 1981.

32   S. D. Personick, 'Fundamental limits in optical communication', *Proc. IEEE* **69**(2), pp. 262–266, 1981.

33   R. A. Milano, P. D. Dapkus and G. E. Stillman, 'Heterojunction phototransistors for fiber-optic communications', *Proc. Soc. Photo-opt. Instrum. Eng.*, **272**, pp. 43–50, 1981.

34   K. Tubatabaie-Alavi and C. G. Fonstad, 'Recent advances in InGaAs/InP phototransistors', *Proc. Soc. Photo-opt. Instrum. Eng.*, **272**, pp. 38–42, 1981.

35   G. E. Stillman, L. W. Cook, G. E. Bulman, N. Tabatabaie, R. Chin and P. D. Dapkus, 'Long-wavelength (1.3- to 1.6-μm) detectors for fiber-optical communications, *IEEE Trans. Electron. Dev.*, **ED-29**(9), pp. 1355–1371, 1982.

9

Receiver Noise Considerations

9.1 INTRODUCTION

The receiver in an optical fiber communication system essentially consists of the photodetector plus an amplifier with possibly additional signal processing circuits. Therefore the receiver initially converts the optical signal incident on the detector into an electrical signal, which is then amplified before further processing to extract the information originally carried by the optical signal.

The importance of the detector in the overall system performance was stressed in Chapter 8. However, it is necessary to consider the properties of this device in the context of the associated circuitry combined in the receiver. It is essential that the detector performs efficiently with the following amplifying and signal processing circuits. Inherent to this process is the separation of the information originally contained in the optical signal from the noise generated within the rest of the system and in the receiver itself, as well as any limitations on the detector response imposed by the associated circuits. These factors play a crucial role in determining the performance of the system.

In order to consider receiver design it is useful to regard the limit on the performance of the system set by the signal to noise ratio (SNR) at the receiver. It is therefore necessary to outline noise sources within optical fiber systems. The noise in these systems has different origins from that of copper-based systems. Both types of system have thermal noise generated in the receiver. However, although optical fiber systems exhibit little crosstalk the noise generated within the detector must be considered, as well as the noise properties associated with the electromagnetic carrier.

In Section 9.2 we therefore briefly review the major noise mechanisms which are present in optical fiber communication receivers prior to more detailed discussion of the limitations imposed by photon (or quantum) noise in both digital and analog transmission. This is followed in Section 9.3 with a more specific discussion of the noise associated with the two major receiver types (i.e. employing $p-i-n$ and avalanche photodiode detectors). Expressions for the SNRs of these two receiver types are also developed in this section. Section 9.4 considers the noise and bandwidth performance of common preamplifier structures utilized in the design of optical fiber receivers. Finally in

Section 9.5 we present a brief account of low noise field effect transistor (FET) preamplifiers which find wide use within optical fiber communication receivers. This discussion also includes consideration of p–i–n photodiode/FET (PIN–FET) hybrid receiver circuits which have been developed for optical fiber communications.

9.2 NOISE

Noise is a term generally used to refer to any spurious or undesired disturbances that mask the received signal in a communication system. In optical fiber communication systems we are generally concerned with noise due to spontaneous fluctuations rather than erratic disturbances which may be a feature of copper-based systems (due to electromagnetic interference, etc.).

There are three main types of noise due to spontaneous fluctuations in optical fiber communication systems: thermal noise, dark current noise and quantum noise.

9.2.1 Thermal Noise

This is the spontaneous fluctuation due to thermal interaction between, say, the free electrons and the vibrating ions in a conducting medium, and it is especially prevalent in resistors at room temperature.

The thermal noise current i_t in a resistor R may be expressed by its mean square value [Ref. 1] and is given by:

$$\overline{i_t^2} = \frac{4KTB}{R} \tag{9.1}$$

where K is Boltzmann's constant, T is the absolute temperature and B is the post-detection (electrical) bandwidth of the system (assuming the resistor is in the optical receiver).

9.2.2 Dark Current Noise

When there is no optical power incident on the photodetector a small reverse leakage current still flows from the device terminals. Thus dark current (see Section 8.4.2) contributes to the total system noise and gives random fluctuations about the average particle flow of the photocurrent. It therefore manifests itself as shot noise [Ref. 1] on the photocurrent. Thus the dark current noise $\overline{i_d^2}$ is given by:

$$\overline{i_d^2} = 2eBI_d \tag{9.2}$$

where e is the charge on an electron and I_d is the dark current. It may be reduced by careful design and fabrication of the detector.

9.2.3 Quantum Noise

The quantum nature of light was discussed in Section 6.2.1 and the equation for the energy of this quantum or photon stated as $E = hf$. The quantum behavior of electromagnetic radiation must be taken into account at optical frequencies since $hf > KT$ and quantum fluctuations dominate over thermal fluctuations.

The detection of light by a photodiode is a discrete process since the creation of an electron–hole pair results from the absorption of a photon, and the signal emerging from the detector is dictated by the statistics of photon arrivals. Hence the statistics for monochromatic coherent radiation arriving at a detector follows a discrete probability distribution which is independent of the number of photons previously detected.

It is found that the probability $P(z)$ of detecting z photons in time period τ when it is expected on average to detect z_m photons obeys the Poisson distribution [Ref. 2]:

$$P(z) = \frac{z_m^z \exp(-z_m)}{z!} \tag{9.3}$$

where z_m is equal to the variance of the probability distribution. This equality of the mean and the variance is typical of the Poisson distribution. From Eq. (8.7) the electron rate r_e generated by incident photons is $r_e = \eta P_0/hf$. The number of electrons generated in time τ is equal to the average number of photons detected over this time period z_m. Therefore:

$$z_m = \frac{\eta P_0 \tau}{hf} \tag{9.4}$$

The Poisson distributions for $z_m = 10$ and $z_m = 1000$ are illustrated in Fig. 9.1 and represent the detection process for monochromatic coherent light.

Incoherent light is emitted by independent atoms and therefore there is no phase relationship between the emitted photons. This property dictates an

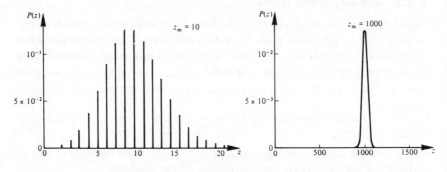

Fig. 9.1 Poisson distributions for $z_m = 10$ and $z_m = 1000$.

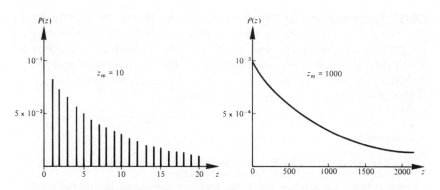

Fig. 9.2 Probability distributions indicating the statistical fluctuations of incoherent light for $z_m = 10$ and $z_m = 1000$.

exponential intensity distribution for incoherent light which if averaged over the Poisson distribution [Ref. 2] gives:

$$P(z) = \frac{z_m^z}{(1 + z_m)^{z+1}} \tag{9.5}$$

Equation (9.5) is identical to the Bose–Einstein distribution [Ref. 3] which is used to describe the random statistics of light emitted in black body radiation (thermal light). The statistical fluctuations for incoherent light are illustrated by the probability distributions shown in Fig. 9.2.

9.2.4 Digital Signalling Quantum Noise

For digital optical fiber systems it is possible to calculate a fundamental lower limit to the energy a pulse of light must contain in order to be detected with a given probability of error. The premise on which this analysis is based is that the ideal receiver has a sufficiently low amplifier noise to detect the displacement current of a single electron–hole pair generated within the detector (i.e. an individual photon may be detected). Thus in the absence of light, and neglecting dark current, no current will flow. Therefore the only way an error can occur is if a light pulse is present and no electron–hole pairs are generated. The probability of no pairs being generated when a light pulse is present may be obtained from Eq. (9.3) and is given by:

$$P(0/1) = \exp(-z_m) \tag{9.6}$$

Thus in the receiver described $P(0/1)$ represents the system error probability $P(e)$ and therefore:

$$P(e) = \exp(-z_m) \tag{9.7}$$

However, it must be noted that the above analysis assumes that the photodetector emits no electron–hole pairs in the absence of illumination. In this sense it is considered perfect. Equation (9.7) therefore represents an absolute receiver sensitivity and allows the determination of a fundamental limit in digital optical communications. This is the minimum pulse energy E_{min} required to maintain a given bit error rate (BER) which any practical receiver must satisfy and is known as the quantum limit.

Example 9.1

A digital optical fiber communication system operating at a wavelength of 1 µm requires a maximum bit error rate of 10^{-9}. Determine:
(a) the theoretical quantum limit at the receiver in terms of the quantum efficiency of the detector and the energy of an incident photon;
(b) the minimum incident optical power required at the detector in order to achieve the above bit error rate when the system is employing ideal binary signalling at 10 Mbits s^{-1}, and assuming the detector is ideal.

Solution: (a) From Eq. (9.7) the probability of error

$$P = \exp(-z_m) = 10^{-9}$$

and thus $z_m = 20.7$.

z_m corresponds to an average number of photons detected in a time period τ for a BER of 10^{-9}.
From Eq. (9.4):

$$z_m = \frac{\eta P_0 \tau}{hf} = 20.7$$

Hence the minimum pulse energy or quantum limit

$$E_{min} = P_0 \tau = \frac{20.7\, hf}{\eta}$$

Thus the quantum limit at the receiver to maintain a maximum BER of 10^{-9} is

$$\frac{20.7\, hf}{\eta}$$

(b) From part (a) the minimum pulse energy:

$$P_0 \tau = \frac{20.7\, hf}{\eta}$$

Therefore the average received optical power required to provide the minimum pulse energy is:

$$P_0 = \frac{20.7\, hf}{\tau \eta}$$

However for ideal binary signalling there are an equal number of ones and zeros (50% in the on state and 50% in the off state). Thus the average received optical power may be considered to arrive over two bit periods, and

$$P_0 \text{(binary)} = \frac{20.7\,hf}{2\tau\eta} = \frac{20.7\,hf\,B_T}{2\eta}$$

where B_T is the bit rate. At a wavelength of 1 μm, $f = 2.998 \times 10^{14}$ Hz, and assuming an ideal detector, $\eta = 1$.

Hence

$$P_0 \text{(binary)} = \frac{20.7 \times 6.626 \times 10^{-34} \times 2.998 \times 10^{14} \times 10^7}{2}$$

$$= 20.6 \text{ pW}$$

In decibels (dB)

$$P_0 \text{ in dB} = 10 \log_{10} \frac{P_0}{P_r}$$

where P_r is a reference power level.
When the reference power level is one watt:

$$P_0 = 10 \log_{10} P_0 \quad \text{where } P_0 \text{ is expressed in watts}$$
$$= 10 \log_{10} 2.06 \times 10^{-11}$$
$$= 3.14 - 110$$
$$= -106.9 \text{ dBW}$$

When the reference power level is one milliwatt

$$P_0 = 10 \log_{10} 2.06 \times 10^{-8}$$
$$= 3.14 - 80$$
$$= -76.9 \text{ dBm}$$

Therefore the minimum incident optical power required at the receiver to achieve an error rate of 10^{-9} with ideal binary signalling is 20.6 pW or −76.9 dBm.

The result of example 9.1 is a theoretical limit and in practice receivers are generally found to be at least 10 dB less sensitive.

9.2.5 Analog Transmission Quantum Noise

In analog optical fiber systems quantum noise manifests itself as shot noise which also has Poisson statistics [Ref. 1]. The shot noise current i_s on the photocurrent I_p is given by,

$$\overline{i_s^2} = 2eBI_p \tag{9.8}$$

Neglecting other sources of noise the SNR at the receiver may be written as:

$$\frac{S}{N} = \frac{I_p^2}{\overline{i_s^2}} \tag{9.9}$$

Substituting for $\overline{i_s^2}$ from Eq. (9.8) gives

$$\frac{S}{N} = \frac{I_p}{2eB} \qquad (9.10)$$

The expression for the photocurrent I_p given in Eq. (8.8) allows the SNR to be obtained in terms of the incident optical power P_o.

$$\frac{S}{N} = \frac{\eta P_o e}{hf\, 2eB} = \frac{\eta P_o}{2hfB} \qquad (9.11)$$

Equation (9.11) allows calculation of the incident optical power required at the receiver in order to obtain a specified SNR when considering quantum noise in analog optical fiber systems.

Example 9.2

An analog optical fiber system operating at a wavelength of 1 μm has a post detection bandwidth of 5 MHz. Assuming an ideal detector and considering only quantum noise on the signal, calculate the incident optical power necessary to achieve an SNR of 50 dB at the receiver.
Solution: From Eq. (9.11), the SNR is

$$\frac{S}{N} = \frac{\eta P_o}{2hfB}$$

Hence

$$P_o = \left(\frac{S}{N}\right) \frac{2hfB}{\eta}$$

For S/N = 50 dB, when considering signal and noise powers:

$$10 \log_{10} \frac{S}{N} = 50$$

and therefore S/N = 10^5

At 1 μm, $f = 2.998 \times 10^{14}$ Hz. For an ideal detector $\eta = 1$ and, thus the incident optical power:

$$P_o = \frac{10^5 \times 2 \times 6.626 \times 10^{-34} \times 2.998 \times 10^{14} \times 5 \times 10^6}{1}$$

$$= 198.6 \text{ nW}$$

In dBm

$$P_o = 10 \log_{10} 198.6 \times 10^{-6}$$
$$= -40 + 2.98$$
$$= -37.0 \text{ dBm}$$

Therefore the incident optical power required to achieve an SNR of 50 dB at the receiver is 198.6 nW which is equivalent to −37.0 dBm.

In practice receivers are less sensitive than example 9.2 suggests and thus in terms of the absolute optical power requirements analog transmission compares unfavorably with digital signalling.

However, it should be noted that there is a substantial difference in information transmission capacity between the digital and analog cases (over similar bandwidths) considered in examples 9.1 and 9.2. For example a 10 Mbit s^{-1} digital optical fiber communication system would provide only about 150 speech channels using standard baseband digital transmission techniques (see Section 10.5). In contrast a 5 MHz analog system, again operating in the baseband, could provide as many as 1250 similar bandwidth ($\simeq$3.4 kHz) speech channels. A comparison of signal to quantum noise ratios between the two transmission methods taking account of this information capacity aspect yields less disparity although digital signalling still proves far superior. For instance, applying the figures quoted above within examples 9.1 and 9.2, in order to compare two systems capable of transmitting the same number of speech channels (e.g. digital bandwidth of 10 Mbit s^{-1} and analog bandwidth of 600 kHz) gives a difference in absolute sensitivity in favor of digital transmission of approximately 31 dB. This indicates a reduction of around 9 dB on the 40 dB difference obtained by simply comparing the results over similar bandwidths. Nevertheless, it is clear that digital signalling techniques still provide a significant benefit in relation to quantum noise when employed within optical fiber communications.

9.3 RECEIVER NOISE

In order to investigate the optical receiver in greater detail it is necessary to consider the relative importance and interplay of the various types of noise mentioned in the previous section. This is dependent on both the method of demodulation and the type of device used for detection.

The conditions for coherent detection are not usually met in current optical fiber systems for the reasons outlined in Section 7.5. Thus heterodyne detection, which is very sensitive and provides excellent rejection of adjacent channels, is not used, as the optical signal arriving at the receiver tends to be incoherent. In practice all currently installed optical fiber communication

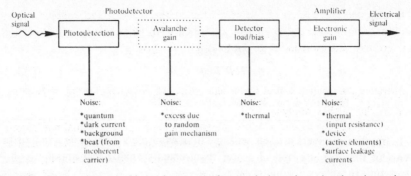

Fig. 9.3 Block schematic of the front end of an optical receiver showing the various sources of noise.

systems use incoherent or direct detection in which the variation of the optical power level is monitored and no information is carried in the phase or frequency content of the signal. Therefore the noise considerations in this section are based on a receiver employing direct detection of the modulated optical carrier which gives the same signal to noise ratio as an unmodulated optical carrier.

Figure 9.3 shows a block schematic of the front end of an optical receiver and the various noise sources associated with it. The majority of the noise sources shown apply to both main types of optical detector (p–i–n and avalanche photodiode). The noise generated from background radiation, which is important in atmospheric propagation and some copper-based systems, is negligible in both types of optical fiber receiver, and thus is often ignored. Also the beat noise generated from the various spectral components of the incoherent optical carrier can be shown to be insignificant [Ref. 4] with multimode propagation and hence will not be considered. It is necessary, however, to take into account the other sources of noise shown in Fig. 9.3.

The avalanche photodiode receiver is the most complex case as it includes noise resulting from the random nature of the internal gain mechanism (dotted in Fig. 9.3). It is therefore useful to consider noise in optical fiber receivers employing photodiodes without internal gain, before avalanche photodiode receivers are discussed.

9.3.1 p–n and p–i–n Photodiode Receiver

The two main sources of noise in photodiodes without internal gain are dark current noise and quantum noise, both of which may be regarded as shot noise on the photocurrent (i.e. effectively consider analog quantum noise). When the expressions for these noise sources given in Eqs. (9.2) and (9.4) are combined the total shot noise $\overline{i_{TS}^2}$ is given by:

$$\overline{i_{TS}^2} = 2eB(I_p + I_d) \qquad (9.12)$$

If it is necessary to take the noise due to the background radiation into account then the expression given in Eq. (9.12) may be expanded to include the background radiation induced photocurrent I_b, giving

$$\overline{i_{TS}^2} = 2eB(I_p + I_d + I_b) \tag{9.13}$$

However, as I_b is usually negligible the expression given in Eq. (9.12) will be used in the further analysis.

When the photodiode is without internal avalanche gain, thermal noise from the detector load resistor and from active elements in the amplifier tends to dominate. This is especially the case for wideband systems operating in the 0.8–0.9 μm wavelength band because the dark currents in well-designed silicon photodiodes can be made very small. The thermal noise $\overline{i_t^2}$ due to the load resistance R_L may be obtained from Eq. (9.1) and is given by:

$$\overline{i_t^2} = \frac{4KTB}{R_L} \tag{9.14}$$

The dominating effect of this thermal noise over the shot noise in photodiodes without internal gain may be observed in example 9.3.

Example 9.3

A silicon *p–i–n* photodiode incorporated into an optical receiver has a quantum efficiency of 60% when operating at a wavelength of 0.9 μm. The dark current in the device at this operating point is 3 nA and the load resistance is 4 kΩ.

The incident optical power at this wavelength is 200 nW and the post detection bandwidth of the receiver is 5 MHz. Compare the shot noise generated in the photodiode with the thermal noise in the load resistor at a temperature of 20 °C.

Solution: From Eq. (8.8) the photocurrent is given by:

$$I_p = \frac{\eta P_o e}{hf} = \frac{\eta P_o e\lambda}{hc}$$

Therefore

$$I_p = \frac{0.6 \times 200 \times 10^{-9} \times 1.602 \times 10^{-19} \times 0.9 \times 10^{-6}}{6.626 \times 10^{-34} \times 2.998 \times 10^8}$$

$$= 87.1 \text{ nA}$$

From Eq. (9.12) the total shot noise is:

$$\overline{i_{TS}^2} = 2eB(I_d + I_p)$$
$$= 2 \times 1.602 \times 10^{-19} \times 5 \times 10^6 \ [(3 + 87.1) \times 10^{-9}]$$
$$= 1.44 \times 10^{-19} \text{ A}^2$$

and the root mean square (rms) shot noise current is

$$(\overline{i_{TS}^2})^{\frac{1}{2}} = 3.79 \times 10^{-10} \text{ A}$$

The thermal noise in the load resistor is given by Eq. (9.14):

$$\overline{i_t^2} = \frac{4KTB}{R_L}$$

$$= \frac{4 \times 1.381 \times 10^{-23} \times 293 \times 5 \times 10^6}{4 \times 10^3}$$

$$= 2.02 \times 10^{-17} \ A^2$$

$(T = 20\ °C = 293\ K)$
Therefore the rms thermal noise current is

$$(\overline{i_t^2})^{\frac{1}{2}} = 4.49 \times 10^{-9}\ A$$

In this example the rms thermal noise current is a factor of 12 greater than the total rms shot noise current.

Example 9.3 does not include the noise sources within the amplifier, shown in Fig. 9.3. These noise sources, associated with both the active and passive elements of the amplifier, can be represented by a series voltage noise source $\overline{v_a^2}$ and a shunt current noise source $\overline{i_a^2}$.

Thus the total noise associated with the amplifier $\overline{i_{amp}^2}$ is given by:

$$i_{amp}^2 = \frac{1}{B}\int_0^B (\overline{i_a^2} + \overline{v_a^2}\,|Y|^2)df \qquad (9.15)$$

where Y is the shunt admittance (combines the shunt capacitances and resistances) and f is the frequency. An equivalent circuit for the front end of the receiver, including the effective input capacitance C_a and resistance R_a of the amplifier is shown in Fig. 9.4. The capacitance of the detector C_d is also shown and the noise resulting from C_d is usually included in the expression for i_{amp}^2 given in Eq. (9.15).

The SNR for the p–n or p–i–n photodiode receiver may be obtained by summing the noise contributions from Eqs. (9.12), (9.14) and (9.15). It is given by:

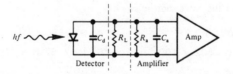

Fig. 9.4 The equivalent circuit for the front end of an optical fiber receiver.

$$\frac{S}{N} = \frac{I_p^2}{2eB(I_p + I_d) + \dfrac{4KTB}{R_L} + \overline{i_{amp}^2}} \tag{9.16}$$

The thermal noise contribution may be reduced by increasing the value of the load resistor R_L, although this reduction may be limited by bandwidth considerations which are discussed later. Also the noise associated with the amplifier $\overline{i_{amp}^2}$ may be reduced with low detector and amplifier capacitance.

However, when the noise associated with the amplifier $\overline{i_{amp}^2}$ is referred to the load resistor R_L, the noise figure F_n [Ref. 1] for the amplifier may be obtained. This allows $\overline{i_{amp}^2}$ to be combined with the thermal noise from the load resistor $\overline{i_t^2}$ to give:

$$\overline{i_t^2} + \overline{i_{amp}^2} = \frac{4KTBF_n}{R_L} \tag{9.17}$$

The expression for the SNR given in Eq. (9.16) can now be written in the form:

$$\frac{S}{N} = \frac{I_p^2}{2eB(I_p + I_d) + \dfrac{4KTBF_n}{R_L}} \tag{9.18}$$

Thus if the noise figure F_n for the amplifier is known, Eq. (9.18) allows the SNR to be determined.

Example 9.4

The receiver in example 9.3 has an amplifier with a noise figure of 3 dB. Determine the SNR at the output of the receiver under the same conditions as example 9.3.
  Solution: From example 9.3:

$$I_p = 87.1 \times 10^{-9} \text{ A}$$
$$\overline{i_{TS}^2} = 1.44 \times 10^{-19} \text{ A}^2$$
$$\overline{i_t^2} = 2.02 \times 10^{-17} \text{ A}^2$$

The amplifier noise figure

$$F_n = 3 \text{ dB}$$
$$= 10 \log_{10} 2$$

Thus F_n may be considered as $\times 2$.
  In Eq. (9.18) the SNR is given by:

$$\frac{S}{N} = \frac{I_p^2}{2eB(I_p + I_d) + \dfrac{4KTBF_n}{R_L}}$$

$$= \frac{i_p^2}{\overline{i_{TS}^2} + (\overline{i_t^2} \times F_n)}$$

$$= \frac{(87.1 \times 10^{-9})^2}{(1.44 \times 10^{-19}) + (2.02 \times 10^{-17} \times 2)}$$

$$= 1.87 \times 10^2$$

SNR in dB $= 10 \log_{10} 1.87 \times 10^2 = 22.72$ dB.

Alternatively it is possible to conduct the calculation in dB if we neglect the shot noise (say $\overline{i_{TS}^2} = 0$).

In dB:

$$I_p = 9.40 - 80 = -70.60$$

Hence

$$I_p^2 = -141.20 \text{ dB}$$

and

$$\overline{i_t^2} = 3.05 - 170 = -166.95 \text{ dB.}$$

The amplifier noise figure $F_n = 3$ dB.

Therefore the

$$SNR = -141.20 + 166.95 - 3$$

$$= 22.75 \text{ dB}$$

A slight difference in the final answer may be noted. This is due to the neglected shot noise term.

A quantity which is often used in the specification of optical detectors (or detector–amplifier combinations) is the noise equivalent power (NEP). It is defined as the amount of incident optical power P_o per unit bandwidth required to produce an output power equal to the detector (or detector–amplifier combination) output noise power. The NEP is therefore the value of P_o which gives an output SNR of unity. Thus the lower the NEP for a particular detector (or detector–amplifier combination), the less optical power is needed to obtain a particular SNR.

9.3.2 Receiver Capacitance

Considering the equivalent circuit shown in Fig. 9.4, the total capacitance for the front end of an optical receiver C_T is given by:

$$C_T = C_d + C_a \tag{9.19}$$

where C_d is the detector capacitance and C_a is the amplifier input capacitance. It is important that this total capacitance is minimized not only from the noise considerations discussed previously but also from the bandwidth penalty which is incurred due to the time constant of C_T and the load resistance R_L.

We assume here that R_L is the total loading on the detector and therefore have neglected the amplifier input resistance R_a. However, in practical receiver configurations R_a may have to be taken into account (see Section 9.4.1). The reciprocal of the time constant $2\pi R_L C_T$ must be greater than, or equal to, the post detection bandwidth B;

$$\frac{1}{2\pi R_L C_T} \geqslant B \qquad (9.20)$$

When the equality exists in Eq. (9.20) it defines the maximum possible value of B for the straightforward termination indicated in Fig. 9.4.

Assuming that the total capacitance may be minimized, then the other parameter which affects B is the load resistance R_L. To increase B it is necessary to reduce R_L. However, this introduces a thermal noise penalty as may be seen from Eq. (9.14) where both the increase in B and decrease in R_L contribute to an increase in the thermal noise. A trade-off therefore exists between the maximum bandwidth and the level of thermal noise which may be tolerated. This is especially important in receivers which are dominated by thermal noise.

Example 9.5

A photodiode has a capacitance of 6 pF. Calculate the maximum load resistance which allows an 8 MHz post detection bandwidth.

Determine the bandwidth penalty with the same load resistance when the following amplifier also has an input capacitance of 6 pF.

Solution: From Eq. (9.20) the maximum bandwidth is given by:

$$B = \frac{1}{2\pi R_L C_d}$$

Therefore the maximum load resistance

$$R_L(\text{max}) = \frac{1}{2\pi C_d B} = \frac{1}{2\pi \times 6 \times 10^{-12} \times 8 \times 10^6}$$

$$= 3.32 \text{ k}\Omega$$

Thus for an 8 MHz bandwidth the maximum load resistance is 3.32 kΩ.

Also, considering the amplifier capacitance, the maximum bandwidth

$$B = \frac{1}{2\pi R_L (C_d + C_a)} = \frac{1}{2\pi \times 3.32 \times 10^3 \times 12 \times 10^{-12}}$$

$$= 4 \text{ MHz}$$

As would be expected the maximum post detection bandwidth is halved.

9.3.3   Avalanche Photodiode (APD) Receiver

The internal gain mechanism in an APD increases the signal current into the amplifier and so improves the SNR because the load resistance and amplifier noise remain unaffected (i.e. the thermal noise and amplifier noise figure are unchanged). However, the dark current and quantum noise are increased by the multiplication process and may become a limiting factor. This is because the random gain mechanism introduces excess noise into the receiver in terms of increased shot noise above the level that would result from amplifying only the primary shot noise. Thus if the photocurrent is increased by a factor M (mean avalanche multiplication factor), then the shot noise is also increased by an excess noise factor M^x, such that the total shot noise $\overline{i_{SA}^2}$ is now given by:

$$\overline{i_{SA}^2} = 2eB(I_p + I_d)M^{2+x} \qquad (9.21)$$

where x is between 0.3 and 0.5 for silicon APDs and between 0.7 and 1.0 for germanium or III–V alloy APDs.

Equation (9.21) is often used as the total shot noise term in order to compute the SNR, although there is a small amount of shot noise current which is not multiplied through impact ionization. The shot noise current in the detector which is not multiplied is a device parameter and may be considered as an extra shot noise term. However it tends to be insignificant in comparison with the multiplied shot noise and is therefore neglected in the further analysis (i.e. all shot noise is assumed to be multiplied).

The SNR for the avalanche photodiode may be obtained by summing the combined noise contribution from the load resistor and the amplifier given in Eq. (9.17), which remains unchanged, with the modified noise term given in Eq. (9.21). Hence the SNR for the APD is:

$$\frac{S}{N} = \frac{M^2 I_p^2}{2eB(I_p + I_d)M^{2+x} + \dfrac{4KTBF_n}{R_L}} \qquad (9.22)$$

It is apparent from Eq. (9.22) that the relative significance of the combined thermal and amplifier noise term is reduced due to the avalanche multiplication of the shot noise term. When Eq. (9.22) is written in the form:

$$\frac{S}{N} = \frac{I_p^2}{2eB(I_p + I_d)M^x + \dfrac{4KTBF_n}{R_L}M^{-2}} \qquad (9.23)$$

it may be seen that the first term in the denominator increases with increasing M whereas the second term decreases. For low M the combined thermal and amplifier noise term dominates and the total noise power is virtually unaffected

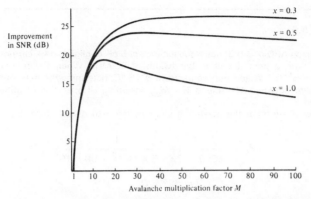

Fig. 9.5 The improvement in SNR as a function of avalanche multiplication factor M for different excess noise factors M^x. Reproduced with permission from I. Garrett, *The Radio and Electron. Eng.*, **51**, p. 349, 1981.

when the signal level is increased, giving an improved SNR. However, when M is large, the thermal and amplifier noise term becomes insignificant and the SNR decreases with increasing M at the rate of M^x. An optimum value of the multiplication factor M_{op} therefore exists which maximizes the SNR. It is given by:

$$\frac{2eB(I_p + I_d)M_{op}^x}{(4KTBF_n/R_L)M_{op}^{-2}} = \frac{2}{x} \qquad (9.24)$$

and therefore

$$M_{op}^{2+x} = \frac{4KTF_n}{xeR_L(I_p + I_d)} \qquad (9.25)$$

The variation in M_{op} for both silicon and germanium APDs is illustrated in Fig. 9.5 [Ref. 5]. This shows a plot of Eq. (9.22) with F_n equal to unity and neglecting the dark current. For good silicon APDs where x is 0.3, the optimum multiplication factor covers a wide range. In the case illustrated in Fig. 9.5 M_{op} commences at about 40 where the possible improvement in SNR above a photodiode without internal gain is in excess of 25 dB. However, for germanium and III–V alloy APDs where x may be equal to unity it can be seen that less SNR improvement is possible (less than 19 dB). Moreover, the maximum is far sharper, occurring at a multiplication factor of about 12. Also it must be noted that Fig. 9.5 demonstrates the variation of M_{op} with x for a specific case, and therefore only represents a general trend. It may be observed from Eq. (9.25) that M_{op} is dependent on a number of other variables apart from x.

Example 9.6

A good silicon APD ($x = 0.3$) has a capacitance of 5 pF, negligible dark current and is operating with a post detection bandwidth of 50 MHz. When the photocurrent before gain is 10^{-7} A and the temperature is 18 °C; determine the maximum SNR improvement between $M = 1$ and $M = M_{op}$ assuming all operating conditions are maintained.

Solution: Determine the maximum value of the load resistor from Eq. (9.20):

$$R_L = \frac{1}{2\pi C_d B} = \frac{1}{2\pi \times 5 \times 10^{-12} \times 50 \times 10^6}$$

$$= 636.5 \ \Omega$$

When $M = 1$, the SNR is given by Eq. (9.22),

$$\frac{S}{N} = \frac{I_p^2}{2eBI_p + \dfrac{4KTB}{R_L}}$$

where $I_d = 0$ and $F_n = 1$

The shot noise is:

$$2eBI_p = 2 \times 1.602 \times 10^{-19} \times 50 \times 10^6 \times 10^{-7}$$

$$= 1.602 \times 10^{-18} \ A^2$$

and the thermal noise is:

$$\frac{4KTB}{R_L} = \frac{4 \times 1.381 \times 10^{-23} \times 291 \times 50 \times 10^6}{636.5}$$

$$= 1.263 \times 10^{-15} \ A^2$$

It may be noted that the thermal noise is dominating.

Therefore

$$\frac{S}{N} = \frac{10^{-14}}{1.602 \times 10^{-18} \times 1.263 \times 10^{-15}} = 7.91$$

and the SNR in dBs is

$$\frac{S}{N} = 10 \log_{10} 7.91 = 8.98 \ dB$$

Thus the SNR when $M = 1$ is 9.0 dB

When $M = M_{op}$ and $x = 0.3$, from Eq. (9.25):

$$M_{op}^{2+x} = \frac{4KT}{xeR_L I_p}$$

where $I_d = 0$ and $F_n = 1$. Hence:

$$M_{op}^{2.3} = \frac{4 \times 1.381 \times 10^{-23} \times 291}{0.3 \times 1.602 \times 10^{-19} \times 636.5 \times 10^{-7}}$$

and

$$M_{op} = (5.255 \times 10^3)^{0.435}$$
$$= 41.54$$

The SNR at M_{op} may be obtained from Eq. (9.22):

$$\frac{S}{N} = \frac{M^2 I_p^2}{2eBI_p M^{2.3} + \dfrac{4KTB}{R_L}}$$

$$= \frac{(41.54)^2 \times 10^{-14}}{\{1.602 \times 10^{-18} \times (41.54)^{2.3}\} + 1.263 \times 10^{-15}}$$

$$= 1.78 \times 10^3$$

and the SNR in dBs is

$$\frac{S}{N} = 10 \log_{10} 1.78 \times 10^3 = 32.50 \, \text{dB}$$

Therefore the SNR when $M = M_{op}$ is 32.5 dB and the SNR improvement over $M = 1$ is 23.5 dB.

Example 9.7

A germanium APD (with $x = 1$) is incorporated into an optical fiber receiver with a 10 kΩ load resistance. When operated at a temperature of 120 K, the minimum photocurrent required to give a SNR of 35 dB at the output of the receiver is found to be a factor of 10 greater than the dark current. If the noise figure of the following amplifier at this temperature is 1 dB and the post detection bandwidth is 10 MHz, determine the optimum avalanche multiplication factor.

Solution: From Eq. (9.22) with $x = 1$ and $M = M_{op}$ (i.e. minimum photocurrent specifies that $M = M_{op}$) the SNR is:

$$\frac{S}{N} = \frac{M_{op}^2 I_p^2}{2eB(I_p + I_d)M_{op}^3 + \dfrac{4KTB}{R_L}}$$

Also from Eq. (9.25)

$$M_{op}^3 = \frac{4KTF_n}{eR_L(I_p + I_d)}$$

Therefore

$$M_{op} = \left\{ \frac{4KTF_n}{eR_L(I_p + I_d)} \right\}^{1/3}$$

Substituting into Eq. (9.22), this gives:

$$\frac{S}{N} = \frac{\left\{ \dfrac{4KTF_n}{eR_L(I_p + I_d)} \right\}^{2/3} I_p^2}{\dfrac{8KTBF_n}{R_L} + \dfrac{4KTBF_n}{R_L}}$$

and as $I_d = 0.1\,I_p$ the SNR is

$$\frac{S}{N} = \frac{\left(\dfrac{4KTF_n}{1.1eR_L} \right)^{2/3} I_p^{4/3}}{\dfrac{12KTBF_n}{R_L}}$$

Therefore the minimum photocurrent I_p:

$$I_p^{4/3} = \left(\frac{S}{N} \right) \frac{\dfrac{12KTBF_n}{R_L}}{\left(\dfrac{4KTF_n}{1.1eR_L} \right)^{2/3}}$$

where the SNR is

$$\frac{S}{N} = 35\,\text{dB} = 3.16 \times 10^3$$

and as $F_n = 1$ dB which is equivalent to 1.26:

$$\frac{12KTBF_n}{R_L} = \frac{12 \times 1.381 \times 10^{-23} \times 120 \times 10^7 \times 1.26}{10^4}$$

$$= 2.51 \times 10^{-17}$$

Also

$$\left(\frac{4KTF_n}{1.1eR_L} \right)^{2/3} = \left(\frac{4 \times 1.381 \times 10^{-23} \times 120 \times 1.26}{1.1 \times 1.602 \times 10^{-19} \times 10^4} \right)^{2/3}$$

$$= 2.82 \times 10^{-4}$$

Therefore

$$I_p = \left(\frac{3.16 \times 10^3 \times 2.51 \times 10^{-17}}{2.82 \times 10^{-4}} \right)^{3/4}$$

$$= 6.87 \times 10^{-8}\,\text{A}$$

To obtain the optimum avalanche multiplication factor we substitute back into Eq.

(9.25), where:

$$M_{op} = \left(\frac{4 \times 1.381 \times 10^{-23} \times 120 \times 1.26}{1.602 \times 10^{-19} \times 10^3 \times 1.1 \times 6.87 \times 10^{-8}} \right)^{1/3}$$

$$= 8.84$$

In example 9.7 the optimum multiplication factor for the germanium APD is found to be approximately 9. It shows the dependence of the optimum multiplication factor on the variables in Eq. (9.25), and although the example does not necessarily represent a practical receiver (some practical germanium APD receivers are cooled to reduce dark current), the optimum multiplication factor is influenced by device and system parameters as well as operating conditions.

9.3.4 Excess Avalanche Noise Factor

The value of the excess avalanche noise factor is dependent upon the detector material, the shape of the electric field profile within the device and whether the avalanche is initiated by holes or electrons. It is often represented as $F(M)$ and we have considered in the previous section one of the approximations for the excess noise factor, where:

$$F(M) = M^x \tag{9.26}$$

and the resulting noise is assumed to be white with a Gaussian distribution. However, a second and more exact relationship is given by [Ref. 6]:

$$F(M) = M \left[1 - (1-k) \left(\frac{M-1}{M} \right)^2 \right] \tag{9.27}$$

where the only carriers are injected electrons and k is the ratio of the ionization coefficients of holes and electrons. If the only carriers are injected holes:

$$F(M) = M \left[1 + \left(\frac{1-k}{k} \right) \left(\frac{M-1}{M} \right)^2 \right] \tag{9.28}$$

The best performance is achieved when k is small, and for silicon APDs k is between 0.02 and 0.10, whereas for germanium and III–V alloy APDs k is between 0.3 and 1.0.

With electron injection in silicon photodiodes, the smaller values of k obtained correspond to a larger ionization rate for the electrons than for the holes. As k departs from unity, only the carrier with the larger ionization rate contributes to the impact ionization and the excess avalanche noise factor is reduced. When the impact ionization is initiated by electrons this corresponds to fewer ionizing collisions involving the hole current which is flowing in the opposite direction (i.e. less feedback). In this case the amplified signal contains less excess noise. The carrier ionization rates in germanium photodiodes are often nearly equal and hence k approaches unity, giving a high level of excess noise.

9.4 RECEIVER STRUCTURES

A full equivalent circuit for the digital optical fiber receiver, in which the optical detector is represented as a current source i_{det}, is shown in Fig. 9.6. The noise sources (i_t, i_{TS} and i_{amp}) and the immediately following amplifier and equalizer are also shown. Equalization [Ref. 7] compensates for distortion of the signal due to the combined transmitter, medium and receiver characteristics. The equalizer is often a frequency shaping filter which has a frequency response that is the inverse of the overall system frequency response. In wideband systems this will normally boost the high frequency components to correct the overall amplitude of the frequency response. To acquire the desired spectral shape for digital systems (e.g. raised cosine, see Fig. 10.37), in order to minimize intersymbol interference, it is important that the phase frequency response of the system is linear. Thus the equalizer may also apply selective phase shifts to particular frequency components.

However, the receiver structure immediately preceding the equalizer is the major concern of this section. In both digital and analog systems it is important to minimize the noise contributions from the sources shown in Fig. 9.6 so as to maximize the receiver sensitivity whilst maintaining a suitable bandwidth. It is therefore useful to discuss various possible receiver structures with regard to these factors.

9.4.1 Low Impedance Front End

Three basic amplifier configurations are frequently used in optical fiber communication receivers. The simplest, and perhaps the most common, is the voltage amplifier with an effective input resistance R_a as shown in Fig. 9.7.

In order to make suitable design choices, it is necessary to consider both bandwidth and noise. The bandwidth considerations in Section 9.3.2 are

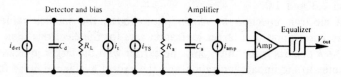

Fig. 9.6 A full equivalent circuit for a digital optical fiber receiver including the various noise sources.

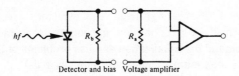

Fig. 9.7   Low impedance front end optical fiber receiver with voltage amplifier.

treated solely with regard to a detector load resistance R_L. However, in most practical receivers the detector is loaded with a bias resistor R_b and an amplifier (see Fig. 9.7). The bandwidth is determined by the passive impedance which appears across the detector terminals which is taken as R_L in the bandwidth relationship given in Eq. (9.20).

However, R_L may be modified to incorporate the parallel resistance of the detector bias resistor R_b and the amplifier input resistance R_a. The modified total load resistance R_{TL} is therefore given by:

$$R_{TL} = \frac{R_b R_a}{R_b + R_a} \qquad (9.29)$$

Considering the expressions given in Eqs. (9.20) and (9.29), to achieve an optimum bandwidth both R_b and R_a must be minimized. This leads to a low impedance front end design for the receiver amplifier. Unfortunately this design allows thermal noise to dominate within the receiver (following Eq. (9.14)), which may severely limit its sensitivity. Therefore this structure demands a trade-off between bandwidth and sensitivity which tends to make it impractical for long-haul, wideband optical fiber communication systems.

9.4.2   High Impedance (Integrating) Front End

The second configuration consists of a high input impedance amplifier together with a large detector bias resistor in order to reduce the effect of thermal noise. However, this structure tends to give a degraded frequency response as the bandwidth relationship given in Eq. (9.20) is not maintained for wideband operation. The detector output is effectively integrated over a large time constant and must be restored by differentiation. This may be performed by the correct equalization at a later stage [Ref. 8] as illustrated in Fig. 9.8. Therefore

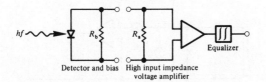

Fig. 9.8   High impedance integrating front end optical fiber receiver with equalized voltage amplifier.

the high impedance (integrating) front end structure gives a significant improvement in sensitivity over the low impedance front end design, but it creates a heavy demand for equalization and has problems of limited dynamic range (the ratio of maximum to minimum input signals).

The limitations on dynamic range result from the attenuation of the low frequency signal components by the equalization process which causes the amplifier to saturate at high signal levels. When the amplifier saturates before equalization has occurred the signal is heavily distorted. Thus the reduction in dynamic range is dependent upon the amount of integration and subsequent equalization employed.

9.4.3 The Transimpedance Front End

This configuration largely overcomes the drawbacks of the high impedance front end by utilizing a low noise, high input impedance amplifier with negative feedback. The device therefore operates as a current mode amplifier where the high input impedance is reduced by negative feedback. An equivalent circuit for an optical fiber receiver incorporating a transimpedance front end structure is shown in Fig. 9.9. In this equivalent circuit the parallel resistances and capacitances are combined into R_{TL} and C_T respectively. The open loop current to voltage transfer function $H_{OL}(\omega)$ for this transimpedance configuration corresponds to the transfer function for the two structures described previously which do not employ feedback (i.e. the low and high impedance front ends). It may be written as:

$$H_{OL}(\omega) = -G\,\frac{V_{in}}{i_{det}} = -G\,\frac{R_{TL}\,\dfrac{1}{j\omega C_T}}{R_{TL} + \dfrac{1}{j\omega C_T}} = \frac{-GR_{TL}}{1 + j\omega R_{TL} C_T}\ \text{V A}^{-1} \quad (9.30)$$

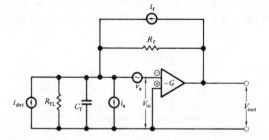

Fig. 9.9 An equivalent circuit for the optical fiber receiver incorporating a transimpedance (current mode) preamplifier.

where G is the open loop voltage gain of the amplifier and ω is the angular frequency of the input signal.

In this case the bandwidth (without equalization) is constrained by the time constant given in Eq. (9.20).*

When the feedback is applied, the closed loop current to voltage transfer function $H_{CL}(\omega)$ for the transimpedance configuration is given by (see Appendix E),

$$H_{CL}(\omega) \simeq \frac{-R_f}{1 + (j\omega R_f C_T/G)} \text{ V A}^{-1} \qquad (9.31)$$

where R_f is the value of the feedback resistor. In this case the permitted electrical bandwidth B (without equalization) may be written as:

$$B \leqslant \frac{G}{2\pi R_f C_T} \qquad (9.32)$$

Hence, comparing Eq. (9.32) with Eq. (9.20) it may be noted that the transimpedance (or feedback) amplifier provides a much greater bandwidth than do the amplifiers without feedback. This is particularly pronounced when G is large.

Moreover, it is interesting to consider the thermal noise generated by the transimpedance front end. Using a referred impedance noise analysis it can be shown [Ref. 12] that to a good approximation the feedback resistance (or impedance) may be referred to the amplifier input in order to establish the noise performance of the configuration. Thus when $R_f \ll R_{TL}$, the major noise contribution is from thermal noise generated in R_f. The noise performance of this configuration is therefore improved when R_f is large, and it approaches the noise performance of the high impedance front end when $R_f = R_{TL}$. Unfortunately, the value of R_f cannot be increased indefinitely due to problems of stability with the closed loop design. Furthermore it may be observed from Eq. (9.32) that increasing R_f reduces the bandwidth of the transimpedance configuration. This problem may be alleviated by making G as large as the stability of the closed loop will allow. Nevertheless it is clear that the noise in the transimpedance amplifier will always exceeed that incurred by the high impedance front end structure.

Example 9.8

A high input impedance amplifier which is employed in an optical fiber receiver has an effective input resistance of 4 MΩ which is matched to a detector bias resistor of the same value. Determine:

(a) The maximum bandwidth that may be obtained without equalization if the total capacitance C_T is 6 pF.

---

* The time constant can be obtained directly from Eq. (9.30) where the maximum bandwidth is defined by $\omega = 2\pi B = 1/R_{TL} C_T$.

(b) The mean square thermal noise current per unit bandwidth generated by this high input impedance amplifier configuration when it is operating at a temperature of 300 K.

(c) Compare the values calculated in (a) and (b) with those obtained when the high input impedance amplifier is replaced by a transimpedance amplifier with a 100 kΩ feedback resistor and an open loop gain of 400. It may be assumed that $R_f \ll R_{TL}$, and that the total capacitance remains 6 pF.

Solution: (a) Using Eq. (9.29), the total effective load resistance:

$$R_{TL} = \frac{(4 \times 10^6)^2}{8 \times 10^6} = 2 \text{ M}\Omega$$

Hence from Eq. (9.20) the maximum bandwidth is given by:

$$B = \frac{1}{2\pi R_{TL} C_T} = \frac{1}{2\pi \times 2 \times 10^6 \times 6 \times 10^{-12}}$$

$$= 1.33 \times 10^4 \text{ Hz}$$

The maximum bandwidth that may be obtained without equalization is 13.3 kHz.

(b) The mean square thermal noise current per unit bandwidth for the high impedance configuration following Eq. (9.14) is:

$$\overline{i_t^2} = \frac{4KT}{R_{TL}} = \frac{4 \times 1.381 \times 10^{-23} \times 300}{2 \times 10^6}$$

$$= 8.29 \times 10^{-27} \text{ A}^2 \text{ Hz}^{-1}$$

(c) The maximum bandwidth (without equalization) for the transimpedance configuration may be obtained using Eq. (9.32), where

$$B = \frac{G}{2\pi R_f C_T} = \frac{400}{2\pi \times 10^5 \times 6 \times 10^{-12}}$$

$$= 1.06 \times 10^8 \text{ Hz}$$

Hence a bandwidth of 106 MHz is permitted by the transimpedance design.

Assuming $R_f \ll R_{TL}$, the mean square thermal noise current per unit bandwidth for the transimpedance configuration is given by:

$$\overline{i_t^2} = \frac{4KT}{R_f} = \frac{4 \times 1.381 \times 10^{-23} \times 300}{10^5}$$

$$= 1.66 \times 10^{-25} \text{ A}^2 \text{ Hz}^{-1}$$

The mean square thermal noise current in the transimpedance configuration is therefore a factor of 20 greater than that obtained with the high input impedance configuration.

The equivalent value in decibels of the ratio of these noise powers is:

$$\frac{\text{noise power in the transimpedance configuration}}{\text{noise power in the high input impedance configuration}} = 10 \log_{10} 20$$

$$= 13 \text{ dB}$$

Thus the transimpedance front end in example 9.8 provides a far greater bandwidth without equalization than the high impedance front end. However, this advantage is somewhat offset by the 13 dB noise penalty incurred with the transimpedance amplifier over that of the high input impedance configuration. Nevertheless it is apparent, even from this simple analysis, that transimpedance amplifiers may be optimized for noise performance, although this is usually obtained at the expense of bandwidth. This topic is pursued further in Ref. 13. However, wideband transimpedance designs generally give a significant improvement in noise performance over the low impedance front end structures using simple voltage amplifiers (see problem 9.18). Finally it must be emphasized that the approach adopted in example 9.8 is by no means rigorous and includes two important simplifications: firstly, that the thermal noise in the high impedance amplifier is assumed to be totally generated by the effective input resistance of the device; and secondly, that the thermal noise in the transimpedance configuration is assumed to be totally generated by the feedback resistor when it is referred to the amplifier input. Both these assumptions are approximations, the accuracy of which is largely dependent on the parameters of the particular amplifier. For example, another factor which tends to reduce the bandwidth of the transimpedance amplifier is the stray capacitance C_f generally associated with the feedback resistor R_f. When C_f is taken into account the closed loop response of Eq. (9.31) becomes:

$$H_{CL}(\Omega) \simeq \frac{-R_f}{1 + j\omega R_f(C_T/G + C_f)} \tag{9.33}$$

However, the effects of C_f may be cancelled by employing a suitable compensating network [Ref. 14].

The other major advantage which the transimpedance configuration has over the high impedance front end is a greater dynamic range. This improvement in dynamic range obtained using the transimpedance amplifier is a result of the different attenuation mechanism for the low frequency components of the signal. The attenuation is accomplished in the transimpedance amplifier through the negative feedback and therefore the low frequency components are amplified by the closed loop rather than the open loop gain of the device. Hence for a particular amplifier the improvement in dynamic range is approximately equal to the ratio of the open loop to the closed loop gains. The transimpedance structure therefore overcomes some of the problems encountered with the other configurations and is often preferred for use in wideband optical fiber communication receivers [Ref. 15].

9.5 FET PREAMPLIFIERS

The lowest noise amplifier device which is widely available is the silicon field effect transistor (FET). Unlike the bipolar transistor, the FET operates by con-

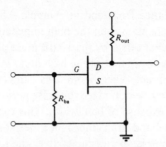

Fig. 9.10   Grounded source FET configuration for the front end of an optical fiber receiver amplifier.

trolling the current flow with an electric field produced by an applied voltage on the gate of the device (see Fig. 9.10) rather than with a base current. Thus the gate draws virtually no current, except for leakage, giving the device an extremely high input impedance (can be greater than 10^{14} ohms). This, coupled with its low noise and capacitance (no greater than a few picofarads), makes the silicon FET appear an ideal choice for the front end of the optical fiber receiver amplifier. However, the superior properties of the FET over the bipolar transistor are limited by its comparatively low transconductance g_m (no better than 5 millisiemens in comparison with at least 40 millisiemens for the bipolar). It can be shown [Ref. 13] that a figure of merit with regard to the noise performance of the FET amplifier is g_m/C_T^2. Hence the advantage of high transconductance together with low total capacitance C_T is apparent. Moreover, as $C_T = C_d + C_a$, it should be noted that the figure of merit is optimized when $C_a = C_d$. This requires FETs to be specifically matched to particular detectors, a procedure which device availability does not generally permit in current optical fiber receiver design. As indicated above, the gain of the FET is restricted. This is especially the case for silicon FETs at frequencies above 25 MHz where the current gain drops to values near unity as the transconductance is fixed with a decreasing input impedance. Therefore at frequencies above 25 MHz, the bipolar transistor is a more useful amplifying device.*

Figure 9.10 shows the grounded source FET configuration which increases the device input impedance especially if the amplifier bias resistor R_{ba} is large. A large bias resistor has the effect of reducing the thermal noise but it will also increase the low frequency impedance of the detector load which tends to integrate the signal (i.e. high impedance integrating front end). Thus compensation through equalization at a later stage is generally required.

---

* The figure of merit in relation to noise performance for the bipolar transistor amplif.er may be shown [Ref. 13] to be $(h_{FE})^{\frac{1}{2}}/C_T$ where h_{FE} is the common emitter current gain of the device. Hence the noise performance of the bipolar amplifier may be optimized in a similar manner to that of the FET amplifier.

9.5.1   Gallium Arsenide MESFETs

Although silicon FETs have a limited useful bandwidth, much effort has been devoted to the development of high performance microwave FETs over the last decade. These FETs are fabricated from gallium arsenide and being Schottky barrier devices [Refs. 16–19] are called GaAs MESFETs. They overcome the major disadvantages of silicon FETs in that they will operate with both low noise and high gain at microwave frequencies (GHz). Thus in optical fiber communication receiver design they present an alternative to bipolar transistors for wideband operation. These devices have therefore been incorporated into high performance receiver designs using both $p–i–n$ and avalanche photodiode detectors [Refs. 21–32]. However, there is, in particular, growing interest in hybrid integrated receiver circuits utilizing $p–i–n$ photodiodes with GaAs MESFET amplifier front ends.

9.5.2   PIN–FET Hybrids

The $p–i–n$/FET, or PIN–FET, hybrid receiver utilizes a high performance $p–i–n$ photodiode followed by a low noise preamplifier often based on a GaAs MESFET, the whole of which is fabricated using thick film integrated circuit technology. This hybrid integration on a thick film substrate reduces the stray capacitance to negligible levels giving a total input capacitance which is very low (e.g. 0.4 pF). The MESFETs employed have a transconductance of approximately 15 millisiemens at the bandwidths required (e.g. 140 Mbits s^{-1}). Early work [Refs. 22 and 23] in the 0.8–0.9 μm wavelength band utilizing a silicon $p–i–n$ detector showed the PIN–FET hybrid receiver to have a sensitivity of –45.8 dBm for a 10^{-9} bit error rate which is only 4 dB worse than current silicon RAPD receivers (see Section 8.9.2).

The work was subsequently extended into the longer wavelength band (1.1–1.6 μm) utilizing III–V alloy $p–i–n$ photodiode detectors. An example of a PIN–FET hybrid high impedance (integrating) front end receiver for operation at a wavelength of 1.3 μm using an InGaAs $p–i–n$ photodiode is shown in Fig. 9.11 [Refs. 24–27]. This design, used by British Telecom, consists of a preamplifier with a GaAs MESFET and microwave bipolar transistor cascode followed by an emitter follower output buffer. The cascode circuit is chosen to ensure sufficient gain is obtained from the first stage to give an overall gain of 18 dB. As the high impedance front end effectively integrates the signal, the following digital equalizer is necessary. The pulse shaping and noise filtering circuits comprise two passive filter sections to ensure that the pulse waveform shape is optimized and the noise is minimized. Equalization for the integration (i.e. differentiation) is performed by monitoring the change in the integrated waveform over one period with a subminiature coaxial delay line followed by a high speed low level comparator. The receiver is designed for use at a transmission rate of 140 Mbit s^{-1} where its performance is found to be comparable

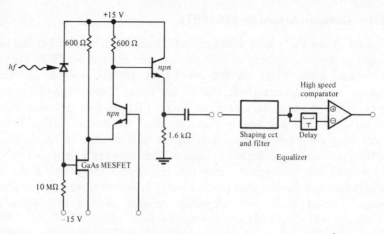

Fig. 9.11 PIN-FET hybrid high impedance integrating front end receiver [Refs. 24–27].

to germanium and III–V alloy APD receivers. For example, the receiver sensitivity at a bit error rate of 10^{-9} is −44.2 dBm.

When compared with the APD receiver the PIN–FET hybrid has both cost and operational advantages especially in the longer wavelength region. The low voltage operation (e.g. +15 and −15 V supply rails) coupled with good sensitivity and ease of fabrication makes the incorporation of this receiver into wideband optical fiber communication systems commercially attractive. A major drawback with the PIN–FET receiver is the possible lack of dynamic range. However, the configuration shown in Fig. 9.11 gave adequate dynamic range via a control circuit which maintained the mean voltage at the gate at

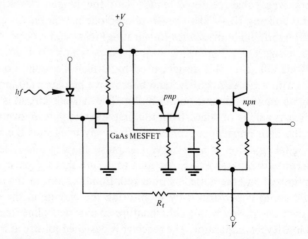

Fig. 9.12 PIN-FET hybrid transimpedance front end receiver [Ref. 29].

0 V by applying a negative voltage proportional to the mean photocurrent to the MESFET bias resistor. With a −15 V supply rail an optical dynamic range of some 20 dB was obtained. This was increased to 27 dB by reducing the value of the MESFET bias resistor from 10 to 2 MΩ which gave a slight noise penalty of 0.5 dB. These figures compare favorably with practical APD receivers.

Transimpedance front end receivers have also been fabricated using the PIN–FET hybrid approach. An example of this type of circuit [Ref. 29] is shown in Fig. 9.12. The amplifier consists of a GaAs MESFET followed by two complementary bipolar microwave transistors. A silicon *p–i–n* photodiode was utilized with the amplifier and the receiver was designed to accept data at a rate of 274 Mbits s^{-1}. In this case the effective input capacitance of the receiver was 4.5 pF giving a sensitivity around −35 dBm for a bit error rate of 10^{-9}.

These figures are somewhat worse than the high impedance front end design discussed previously. However, this design has the distinct advantage of a flat frequency response to a wider bandwidth which requires little, if any, equalization.

PROBLEMS

9.1 Briefly discuss the possible sources of noise in optical fiber receivers. Describe in detail what is meant by quantum noise. Consider this phenomenon with regard to:
(a) digital signalling;
(b) analog transmission,
giving any relevant mathematical formulae.

9.2 A silicon photodiode has a responsivity of 0.5 A W^{-1} at a wavelength of 0.85 μm. Determine the minimum incident optical power required at the photodiode at this wavelength in order to maintain a bit error rate of 10^{-7}, when utilizing ideal binary signalling at a rate of 35 Mbits s^{-1}.

9.3 An analog optical fiber communication system requires an SNR of 40 dB at the detector with a post detection bandwidth of 30 MHz. Calculate the minimum optical power required at the detector if it is operating at a wavelength of 0.9 μm with a quantum efficiency of 70%. State any assumptions made.

9.4 A digital optical fiber link employing ideal binary signalling at a rate of 50 MBits s^{-1} operates at a wavelength of 1.3 μm. The detector is a germanium photodiode which has a quantum efficiency of 45% at this wavelength. An alarm is activated at the receiver when the bit error rate drops below 10^{-5}. Calculate the theoretical minimum optical power required at the photodiode in order to keep the alarm inactivated. Comment briefly on the reasons why in practice the minimum incident optical power would need to be significantly greater than this value.

9.5   Discuss the implications of the load resistance on both thermal noise and post detection bandwidth in optical fiber communication receivers.

9.6   A silicon $p-i-n$ photodiode has a quantum efficiency of 65% at a wavelength of 0.8 μm. Determine:
   (a) the mean photocurrent when the detector is illuminated at a wavelength of 0.8 μm with 5 μW of optical power;
   (b) the rms quantum noise current in a post detection bandwidth of 20 MHz;
   (c) the SNR in dB, when the mean photocurrent is the signal.

9.7   The photodiode in problem 9.6 has a capacitance of 8 pF. Calculate:
   (a) the minimum load resistance corresponding to a post detection bandwidth of 20 MHz;
   (b) the rms thermal noise current in the above resistor at a temperature of 25 °C;
   (c) the SNR in dB resulting from the illumination in problem 9.6 when the dark current in the device is 1 nA.

9.8   The photodiode in problems 9.6 and 9.7 is used in a receiver where it drives an amplifier with a noise figure of 2 dB and an input capacitance of 7 pF. Determine:
   (a) the maximum amplifier input resistance to maintain a post detection bandwidth of 20 MHz without equalization;
   (b) the minimum incident optical power required to give an SNR of 50 dB.

9.9   A germanium photodiode incorporated into an optical fiber receiver working at a wavelength of 1.55 μm has a dark current of 500 nA at the operating temperature. When the incident optical power at this wavelength is 10^{-6} W and the responsivity of the device is 0.6 A W^{-1}, shot noise dominates in the receiver. Determine the SNR in dB at the receiver when the post detection bandwidth is 100 MHz.

9.10   Discuss the expression for the SNR in an APD receiver given by:

$$\frac{S}{N} = \frac{M^2 I_p^2}{2eB(I_p + I_d)M^{2+x} + \dfrac{4KTBF_n}{R_L}}$$

with regard to the various sources of noise present in the receiver. How may this expression be modified to give the optimum avalanche multiplication factor?

9.11   A silicon RAPD has a quantum efficiency of 95% at a wavelength of 0.9 μm, has an excess avalanche noise factor of $M^{0.3}$ and a capacitance of 2 pF. It may be assumed that the post detection bandwidth (without equalization) is 25 MHz, and that the dark current in the device is negligible at the operating temperature of 290 K. Determine the minimum incident optical power which can yield an SNR of 23 dB.

9.12   With the device and conditions given in problem 9.11, calculate:
   (a) the SNR obtained when the avalanche multiplication factor for the RAPD falls to half the optimum value calculated;

      (b) the increased optical power necessary to restore the SNR to 23 dB with $M = 0.5M_{op}$.

9.13 What is meant by the excess avalanche noise factor $F(M)$? Give two possible ways of expressing this factor in analytical terms. Comment briefly on their relative merits.

9.14 A germanium APD (with $x = 1.0$) operates at a wavelength of 1.35 μm where its responsivity is 0.45 A W^{-1}. The dark current is 200 nA at the operating temperature of 250 K and the device capacitance is 3 pF. Determine the maximum possible SNR when the incident optical power is 8×10^{-7} W and the post detection bandwidth without equalization is 560 MHz.

9.15 The photodiode in problem 9.14 drives an amplifier with a noise figure of 3 dB and an input capacitance of 3 pF. Determine the new maximum SNR when they are operated under the same conditions.

9.16 Discuss the three main amplifier configurations currently adopted for optical fiber communications. Comment on their relative merits and drawbacks.

      A high impedance integrating front end amplifier is used in an optical fiber receiver in parallel with a detector bias resistor of 10 MΩ. The effective input resistance of the amplifier is 6 MΩ and the total capacitance (detector and amplifier) is 2 pF.

      It is found that the detector bias resistor may be omitted when a transimpedance front end amplifier design is used with a 270 kΩ feedback resistor and an open loop gain of 100.

      Compare the bandwidth and thermal noise implications of these two cases, assuming an operating temperature of 290 K.

9.17 A $p–i–n$ photodiode operating at a wavelength of 0.83 μm has a quantum efficiency of 50% and a dark current of 0.5 nA at a temperature of 295 K. The device is unbiased but loaded with a current mode amplifier with a 50 kΩ feedback resistor and an open loop gain of 32. The capacitance of the photodiode is 1 pF and the input capacitance of the amplifier is 6 pF.

      Determine the incident optical power required to maintain a SNR of 55 dB when the post detection bandwidth is 10 MHz. Is equalization necessary?

9.18 A voltage amplifier for an optical fiber receiver is designed with an effective input resistance of 200 Ω which is matched to the detector bias resistor of the same value. Determine:

      (a) The maximum bandwidth that may be obtained without equalization if the total capacitance (C_T) is 10 pF.

      (b) The rms thermal noise current generated in this configuration when it is operating over the bandwidth obtained in (a) and at a temperature of 290 K. The thermal noise generated by the voltage amplifier may be assumed to be from the effective input resistance to the device.

      (c) Compare the values calculated in (a) and (b) with those obtained when the voltage amplifier is replaced by a transimpedance amplifier with a 10 kΩ feedback resistor and an open loop gain of 50. It may be assumed that the feedback resistor is also used to bias the detector, and the total capacitance remains 10 pF.

9.19 What is a PIN–FET hybrid receiver? Discuss in detail its merits and possible drawbacks in comparison with the APD receiver.

Answers to Numerical Problems

9.2 −70.4 dBm

9.3 −37.2 dBm

9.4 −70.1 dBm

9.6 (a) 2.01 μA; (b) 3.59 nA;
(c) 55.0 dB

9.7 (a) 994.7 Ω; (b) 18.19 nA;
(c) 39.3 dB

9.8 (a) 1.137 kΩ; (b) 19.58 μW

9.9 40.1 dB

9.11 −50.3 dBm

9.12 (a) 14.2 dB; (b) −49.6 dBm

9.14 23.9 dB

9.15 21.9 dB

9.16 High impedance front end:
21.22 kHz,
4.27×10^{-27} A^2 Hz^{-1};
Transimpedance front end:
29.47 MHz, 5.93×10^{-26}
A^2 Hz^{-1}

9.17 −23.1 dBm, equalization is
unnecessary

9.18 (a) 159.13 MHz; (b) 160 nA;
(c) 79.56 MHz, 11.3 nA, noise
power 23 dB down

REFERENCES

1 (a) M. Schwartz, *Information Transmission, Modulation and Noise* (3rd Edn.), McGraw-Hill, 1980.
(b) F. R. Conner, *Noise*, (2nd Edn.), Edward Arnold, 1982.

2 P. Russer, 'Introduction to optical communications', in M. J Howes and D. V. Morgan (Eds.), *Optical Fibre Communications*, pp. 1–26, John Wiley, 1980.

3 M. Garbuny, *Optical Physics*, Academic Press, 1965.

4 W. M. Hubbard, 'Efficient utilization of optical frequency carriers for low and moderate bit rate channels', *Bell Syst. Tech. J.*, **50**, pp. 713–718, 1973.

5 I. Garrett, 'Receivers for optical fibre communications', *Electron. and Radio Eng.*, **51**(7/8), pp. 349–361, 1981.

6 P. P Webb, R. J. McIntyre and J. Conradi, 'Properties of avalanche photodiodes', *RCA Rev.*, **35**, pp. 234–278, 1974.

7 W. R. Bennett and J. R. Davey, *Data Transmission*, McGraw-Hill, 1965.

8 S. D. Personick, 'Receiver design for digital fiber optic communication systems (Part I and II)', *Bell Syst. Tech. J.*, **52**, pp. 843–886, 1973.

9 T. P. Lee and T. Li, 'Photodetectors', in S. E. Miller and A. G. Chynoweth (Eds.), *Optical Fiber Telecommunications*, pp. 593–623, Academic Press, 1979.

10 S. D. Personick, 'Receiver design', in S. E. Miller and A. G. Chynoweth (Eds.), *Optical Fiber Telecommunications*, pp. 627–651, Academic Press, 1979.

11 J. E. Goell, 'Input amplifiers for optical PCM receivers', *Bell Syst. Tech. J.*, **54**, pp. 1771–1793, 1974.

12 J. L. Hullett and T. V. Muoi, 'Referred impedance noise analysis for feedback amplifiers', *Electron. Lett.*, **13**(13), pp. 387–389, 1977.

13 R. G. Smith and S. D. Personick, 'Receiver design for optical fiber communication systems', in H. Kressel (Ed.), *Semiconductor Devices for Optical Communication*, (2nd edn.), Springer-Verlag, 1982.

14 J. L. Hullett, 'Optical communication receivers', *Proc. IREE Australia*, **40**(4), pp. 127–136, 1979.

15 J. L. Hullett and T. V. Muoi, 'A feedback receiver amplifier for optical transmission systems', *Trans. IEEE*, **COM 24**, pp. 1180–1185, 1976.

16 J. S. Barrera, 'Microwave transistor review, Part 1. GaAs field-effect transistors'. *Microwave J.*, (USA), **19**(2), pp. 28–31, 1976.

17 B. S. Hewitt, H. M. Cox, H. Fukui, J. V. Dilorenzo, W. O. Scholesser and D. E. Iglesias, 'Low noise GaAs MESFETs', *Electron. Lett.*, **12**(12), pp. 309–310, 1976.

18 D. V. Morgan, F. H. Eisen and A. Ezis, 'Prospects for ion bombardment and ion implantation in GaAs and InP device fabrication', *IEE Proc.*, **128**(1–4), pp. 109–129, 1981.

19 J. Mun, J. A. Phillips and B. E. Barry, 'High-yeild process for GaAs enhancement-mode MESFET integrated circuits', *IEE Proc.*, **128**(1–4), pp. 144–147, 1981.

20 S. D. Personick, P. Balaban, J. H. Bobsin and P. R. Kumar, 'A detailed comparison of four approaches to the calculation of the sensitivity of optical fiber system receivers', *IEEE Trans. Commun.*, **COM-25**, pp. 541–549, 1977.

21 S. D. Personick, 'Design of receivers and transmitters for fiber systems', in M. K. Barnoski (Ed.), *Fundamentals of Optical Fiber Communications*, (2nd edn.), Academic Press, 1981.

22 D. R. Smith, R. C. Hooper and I. Garrett, 'Receivers for optical communications: A comparison of avalanche photodiodes with PIN–FET hybrids', *Opt. Quant. Electron.*, **10**, pp. 293–300, 1978.

23 R. C. Hooper and D. R. Smith, 'Hybrid optical receivers using PIN photodiodes', *IEE (London) Colloquium on Broadband High Frequency Amplifiers*, pp. 9/1–9/5, 1979.

24 K. Ahmad and A. W. Mabbitt, 'Ga$_{1-x}$In$_x$As photodetectors for 1.3 micron PIN–FET receiver', *IEEE NY (USA) International Electronic Devices Meeting* (Washington, DC), pp. 646–649, 1978.

25 D. R. Smith, R. C. Hooper and R. P. Webb, 'High performance digital optical receivers with PIN photodiodes', *IEEE (NY) Proceedings of the International Symosim on Circuits and Systems* (Tokyo), pp. 511–514, 1979.

26 D. R. Smith, R. C. Hooper, K. Ahmad, D. Jenkins, A. W. Mabbitt and R. Nicklin, 'p–i–n/FET hybrid optical receiver for longer wavelength optical communication systems', *Electron. Lett.*, **16**(2), pp. 69–71, 1980.

27 R. C. Hopper, D. R. Smith and B. R. White, 'PIN–FET Hybrids for digital optical receivers', *IEEE NY (USA) 30th Electronic Components Conference*, San Francisco, pp. 258–260, 1980.

28 S. Hata, Y. Sugeta, Y. Mizushima, K. Asatani and K. Nawata, 'Silicon p–i–n photodetectors with integrated transistor amplifiers', *IEEE Trans. Electron. Devices*, **ED-26**(6), pp. 989–991, 1979.

29 K. Ogawa and E. L. Chinnock, 'GaAs FET transimpedance front-end design for a wideband optical receiver', *Electron. Lett.*, **15**(20), pp. 650–652, 1979.

30 S. M. Abbott and W. M. Muska, 'Low noise optical detection of a 1.1 Gb/s optical data stream, *Electron. Lett.*, **15**(9), pp. 250–251, 1979.

31 L. A. Godfrey, 'Designing for the fastest response ever — ultra high speed photodetection', *Opt. Spectra (USA)*, **13**(10), pp. 43, 46, 1979.

32 R. I. MacDonald, 'High gain optical detection with GaAs field effect transistors', *Appl. Opt. (USA)*, **20**(4), pp. 591–594, 1981.

10

Optical Fiber Systems

10.1 INTRODUCTION

The transfer of information in the form of light propagating within an optical fiber requires the successful implementation of an optical fiber communication system. This system, in common with all systems, is composed of a number of discrete components which are connected together in a manner that enables them to perform a desired task. Hence, to achieve reliable and secure communication using optical fibers it is essential that all the components within the transmission system are compatible so that their individual performances, as far as possible, enhance rather than degrade the overall system performance.

The principal components of a general optical fiber communication system for either digital or analog transmission are shown in the system block schematic of Fig. 10.1. The transmit terminal equipment consists of an information encoder or signal shaping circuit preceding a modulation or electronic driver stage which operates the optical source. Light emitted from the source is launched into an optical fiber incorporated within a cable which constitutes the transmission medium. The light emerging from the far end of the transmission medium is converted back into an electrical signal by an optical detector positioned at the input of the receive terminal equipment. This electrical signal is then amplified prior to decoding or demodulation in order to obtain the information originally transmitted.

The operation and characteristics of the optical components of this general system have been discussed in some detail within the previous chapters. However, to enable the successful incorporation of these components into an optical fiber communication system it is necessary to consider the interaction of one component with another, and then to evaluate the overall performance of the system. Furthermore, to optimize the system performance for a given application it is often helpful to offset a particular component characteristic by trading it off against the performance of another component, in order to provide a net gain within the overall system. The electronic components play an important role in this context, allowing the system designer further choices which, depending on the optical components utilized, can improve the system performance.

386

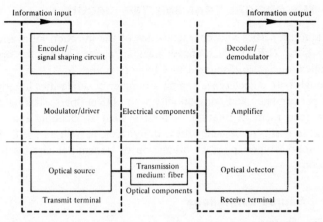

Fig. 10.1 The principal components of an optical fiber communication system.

The purpose of this chapter is to bring together the important performance characteristics of the individual system elements, and to consider their interaction within optical fiber communication systems. It is intended that this will provide guidance in relation to the various possible component configurations which may be utilized for different system applications, whilst also giving an insight into system design and optimization. Hence the optical components and the associated electronic circuits will be discussed prior to consideration of general system design procedures. Although the treatment is by no means exhaustive, it will indicate the various problems involved in system design and provide a description of the basic techniques and practices which may be adopted to enable successful system implementation.

We commence in Section 10.2 with a discussion of the optical transmitter circuit. This includes consideration of the source limitations prior to description of various LED and laser drive circuits for both digital and analog transmission. In Section 10.3 we present a similar discussion for the optical receiver including examples of preamplifier and main amplifier circuits. General system design considerations are then dealt with in Section 10.4. This is followed by a detailed discussion of digital systems commencing with an outline of the operating principles of pulse code modulated (PCM) systems in Section 10.5 before continuing to consider the various aspects of digital optical fiber systems in Section 10.6. Analog optical fiber systems are then dealt with in Section 10.7 where the various possible analog modulation techniques are described and analyzed. Finally, we conclude in Section 10.8 with a brief account of coherent optical fiber systems which are currently attracting much interest.

10.2   THE OPTICAL TRANSMITTER CIRCUIT

The unique properties and characteristics of the injection laser and the light emitting diode (LED) which make them attractive sources for optical fiber communications were discussed in Chapters 6 and 7.

Although both device types exhibit a number of similarities in terms of their general performance and compatibility with optical fibers, striking differences exist between them in relation to both system application and transmitter design. It is useful to consider these differences, as well as the limitations of the two source types, prior to discussion of transmitter circuits for various applications.

10.2.1   Source Limitations

10.2.1.1   *Power*
The electrical power required to operate both injection lasers and LEDs is generally similar with typical current levels of between 20 and 300 mA (certain laser thresholds may be substantially higher than this—of the order of 1–2 A), and voltage drops across the terminals of 1.5–2.5 V. However, the optical output power against current characteristic for the two devices varies considerably, as indicated in Fig. 10.2. The injection laser is a threshold device which must be operated in the region of stimulated emission (i.e. above the threshold) where continuous optical output power levels are typically in the range 1–10 mW.

Much of this light output may be coupled into an optical fiber because the isotropic distribution of the narrow linewidth, coherent radiation is relatively directional. In addition the spatial coherence of the laser emission allows it to be readily focused by appropriate lenses within the numerical aperture of the

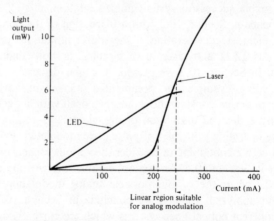

Fig. 10.2   Light output (power) emitted into air as a function of d.c. drive current for a typical high radiance LED and for a typical injection laser. The curves exhibit nonlinearity at high currents due to junction heating.

fiber. Coupling efficiencies near 30% may be obtained by placing a fiber close to a laser mirror, and these can approach 80% with a suitable lens arrangement [Refs. 1 and 2]. Therefore injection lasers are capable of launching between 0.5 and several milliwatts of optical power into a fiber.

LEDs are capable of similar optical output power levels to injection lasers depending on their structure and quantum efficiency as indicated by the typical characteristic for a surface emitter shown in Fig. 10.2. However, the spontaneous emission of radiation over a wide linewidth from the LED generally exhibits a Lambertian intensity distribution which gives poor coupling into optical fibers. Consequently only between 1 and perhaps 10% (using a good edge emitter) of the emitted optical power from an LED may be launched into a multimode fiber, even with appropriate lens coupling (see Section 7.3.4). These considerations translate into optical power levels from a few to several hundred microwatts launched into individual multimode fibers. Thus the optical power coupled into a fiber from an LED can be 10–20 dB below that obtained with a typical injection laser. The power advantage gained with the injection laser is a major factor in the choice of source, especially when considering a long-haul optical fiber link.

10.2.1.2 *Linearity*
Linearity of the optical output power against current characteristic is an important consideration with both the injection laser and LED. It is especially pertinent to the design of analog optical fiber communication systems where source nonlinearities may cause severe distortion of the transmitted signal. At first sight the LED may appear to be ideally suited to analog transmission as its output is approximately proportional to the drive current. However, most LEDs display some degree of nonlinearity in their optical output against current characteristic because of junction heating effects which may either prohibit their use, or necessitate the incorporation of a linearizing circuit within the optical transmitter. Certain LEDs (e.g. etched well surface emitters) do display good linearity, with distortion products (harmonic and intermodulation) between 35 and 45 dB below the signal level [Refs. 3 and 4].

An alternative approach to obtaining a linear source characteristic is to operate an injection laser in the light-generating region above its threshold, as indicated in Fig. 10.2. This may prove more suitable for analog transmission than would the use of certain LEDs. However, gross nonlinearities due to mode instabilities may occur in this region. These are exhibited as kinks in the laser output characteristic (see Section 6.5.3). Therefore many of the multimode injection lasers have a limited use for analog transmission without additional linearizing circuits within the transmitter, although some of the single mode structures have demonstrated linearity suitable for most analog applications. Alternatively, digital transmission, especially utilizing a binary (2 level) format, is far less sensitive to source nonlinearities and is therefore often preferred when using both injection lasers and LEDs.

10.2.1.3 *Thermal*

The thermal behavior of both injection lasers and LEDs can limit their operation within the optical transmitter. However, as indicated in Section 6.9.1 the variation of injection laser threshold current with the device·junction temperature can cause a major operating problem. Threshold currents of typical AlGaAs devices increase by approximately 1% per degree centigrade increase in junction temperature. Hence any significant increase in the junction temperature may cause loss of lasing and a subsequent dramatic reduction in the optical output power. This limitation cannot usually be overcome by simply cooling the device on a heat sink, but must be taken into account within the transmitter design, through the incorporation of optical feedback, in order to obtain a constant optical output power level from the device.

The optical output from an LED is also dependent on the device junction temperature as indicated in Section 7.4.2. Most LEDs exhibit a decrease in optical output power following an increase in junction temperature, which is typically around −1% per degree centigrade. This thermal behavior, however, although significant is not critical to the operation of the device due to its lack of threshold. Nevertheless this temperature dependence can result in à variation in optical output power of several decibels over the temperature range 0–70 °C. It is therefore a factor within system design considerations which, if not tolerated, may be overcome by providing a circuit within the transmitter which adjusts the LED drive current with temperature.

10.2.1.4 *Response*

The speed of response of the two types of optical source is largely dictated by their respective radiative emission mechanisms. Spontaneous emission from the LED is dependent on the effective minority carrier lifetime in the semiconductor material (see Section 7.4.3). In heavily doped ($10^{18}-10^{19}$ per cm^3) gallium arsenide this is typically between 1 and 10 nanoseconds. However, the response of an optical fiber source to a current step input is often specified in terms of the 10–90% rise time, a parameter which is reciprocally related to the device frequency response (see Section 10.6.5). The rise time of the LED is at least twice the effective minority carrier lifetime, and often much longer because of junction and stray capacitance. Hence, the rise times for currently available LEDs lie between 2 and 50 nanoseconds, give 3 dB bandwidths of around 7 to at best 175 MHz. Therefore LEDs are inherently restricted to lower bandwidth applications, although suitable drive circuits can maximize their bandwidth capabilities (i.e. reduce rise times).

Stimulated emission from injection lasers occurs over a much shorter period giving rise times of the order of 0.1–1 ns, thus allowing 3 dB bandwidths above 1 GHz. However, injection laser performance is limited by the device switch-on delay (see Section 6.9.2). To achieve the highest speeds it is therefore necessary to minimize the switch-on delay. Transmitter circuits which prebias the laser to just below or just above threshold in conjunction with high speed drive currents which take the device well above threshold, prove useful in the reduction of this limitation.

10.2.1.5 *Spectral Width*

The finite spectral width of the optical source causes pulse broadening due to material dispersion on an optical fiber communication link. This results in a limitation on the bandwidth–length product which may be obtained using a particular source and fiber. The incoherent emission from an LED usually displays a spectral linewidth of between 20 and 50 nm (full width at half power (FWHP) points) when operating in the 0.8–0.9 μm wavelength range. This limits the bandwidth–length product with a silica fiber to around 100 and 160 MHz km at wavelengths of 0.8 and 0.9 μm respectively. Hence the overall system bandwidth for an optical fiber link over several kilometers may be restricted by material dispersion rather than the response time of the source.

The problem may be alleviated by working at a longer wavelength where the material dispersion in high silica fibers approaches zero (i.e. near 1.3 μm, see Section 3.8.1). In this region the source spectral width is far less critical and bandwidth–length products approaching 1 GHz km are feasible using LEDs.

Alternatively, an optical source with a narrow spectral linewidth may be utilized in place of the LED. The coherent emission from an injection laser generally has a linewidth of 1 nm or less (FWHP). Use of the injection laser greatly reduces the effect of material dispersion within the fiber, giving bandwidth–length products of 1 GHz km at 0.8 μm, and far higher at longer wavelengths. Hence, the requirement for a system operating at a particular bandwidth over a specific distance will influence both the choice of source and operating wavelength.

10.2.2 LED Drive Circuits

Although the LED is somewhat restricted in its range of possible applications in comparison with the more powerful, higher speed injection laser, it is generally far easier to operate. Therefore in this section we consider some of the circuit configurations that may be used to convert the information voltage signal at the transmitter into a modulation current suitable for an LED source. In this context it is useful to discuss circuits for digital and analog transmission independently.

10.2.2.1 *Digital Transmission*

The operation of the LED for binary digital transmission requires the switching on and off of a current in the range of several tens to several hundreds of milliamperes. This must be performed at high speed in response to logic voltage levels at the driving circuit input. A common method of achieving this current switching operation for an LED is shown in Fig. 10.3. The circuit illustrated uses a bipolar transistor switch operated in the common emitter

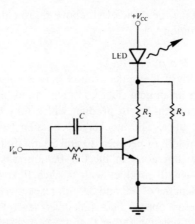

Fig. 10.3   A simple drive circuit for binary digital transmission consisting of a common emitter saturating switch.

mode. This single stage circuit provides current gain as well as giving only a small voltage drop across the switch when the transmitter is in saturation (i.e. when the collector–base junction is forward biassed, the emitter to collector voltage V_{CE} (sat) is around 0.3 V).

The maximum current flow through the LED is limited by the resistor R_2 whilst independent bias to the device may be provided by the incorporation of resistor R_3. However, the switching speed of the common emitter configuration is limited by space charge and diffusion capacitance; thus bandwidth is

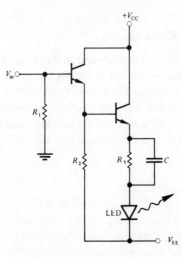

Fig. 10.4   Low impedance drive circuit consisting of an emitter follower with compensating matching network [Ref. 5].

traded for current gain. This may, to a certain extent, be compensated by over-driving (pre-emphasizing) the base current during the switch-on period. In the circuit shown in Fig. 10.3 pre-emphasis is accomplished by use of the speed up capacitor C.

Increased switching speed may be obtained from an LED without a pulse shaping or speed up element by use of a low impedance driving circuit, whereby charging of the space charge and diffusion capacitance occurs as rapidly as possible. This may be achieved with the emitter follower drive circuit shown in Fig. 10.4 [Ref. 5]. The use of this configuration with a compensating matching network $(R_3 C)$ provides fast direct modulation of LEDs with relatively low drive power. A circuit, with optimum values for the matching network, is capable of giving optical rise times of 2.5 ns for LEDs with capacitance of 180 pF, thus allowing 100 Mbits s^{-1} operation [Ref. 6].

Another type of low impedance driver is the shunt configuration shown in Fig. 10.5. The switching transistor in this circuit is placed in parallel with the LED providing a low impedance path for switching off the LED by shunting current around it. The switch-on performance of the circuit is determined by the combination of resistor R and the LED capacitance. Stored space charge may be removed by slightly reverse biassing the LED when the device is switched off. This may be achieved by placing the transistor emitter potential V_{EE} below ground. In this case a Schottky clamp (shown dotted) may be incorporated to limit the extent of the reverse bias without introducing any extra minority carrier stored charge into the circuit.

A frequent requirement for digital transmission is the interfacing of the LED drive circuit with a common logic family as illustrated in the block schematic of Fig. 10.6(a). In this case the logic interface must be considered along with possible drive circuits. Compatibility with TTL may be achieved by use of commercial integrated circuits as shown in Figs. 10.6(b) and (c). The configuration shown in Fig. 10.6(b) uses a Texas Instruments' 74S140 line driver

Fig. 10.5 Low impedance drive circuit consisting of a simple shunt configuration.

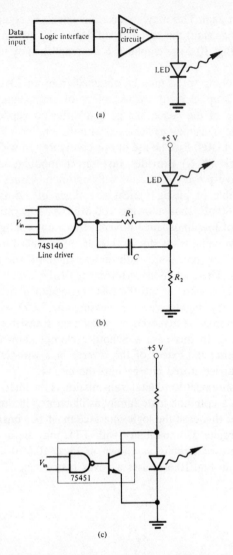

Fig. 10.6   Logic interfacing for digital transmission: (a) block schematic showing the interfacing of the LED drive circuit with logic input levels; (b) a simple TTL compatible LED drive circuit employing a Texas Instruments' 74S140 line driver [Ref. 7]; (c) a TTL compatible shunt drive circuit using a commercially available integrated circuit [Ref. 7].

which provides a drive current of around 60 mA to the LED when R_1 is 50 Ω. Moreover, the package contains two sections which may be connected in parallel in order to obtain a drive current of 120 mA. The incorporation of a suitable speed up capacitor (e.g. $C = 47$ pF) gives optical rise times of around 5 ns when using LEDs with between 150 and 200 pF capacitance [Ref. 7].

Figure 10.6(c) illustrates the shunt configuration using a standard TTL 75451 integrated circuit. The rise time of this shunt circuit may be improved through maintenance of charge on the LED capacitance by placing a resistor between the shunt switch collector and the LED [Ref. 7].

An alternative important drive circuit configuration is the emitter coupled circuit shown in Fig. 10.7 [Ref. 7]. The LED acts as a load in one collector so that the circuit provides current gain and hence a drive current for the device. Thus the circuit resembles a linear differential amplifier, but it is operated outside the linear range and in the switching mode. Fast switching speeds may be obtained due to the configuration's nonsaturating characteristic which avoids switch-off time degradations caused by stored charge accumulation on the transistor base region. The lack of saturation also minimizes the base drive requirements for the transistors thus preserving their small signal current gain. The emitter coupled driver configuration shown in Fig. 10.7 is compatible with commercial emitter coupled logic (ECL). However, to achieve this compatibility the circuit includes two level shifting transistors which give ECL levels (high −0.8 V, low −1.8 V) when the positive terminal of the LED is at earth potential. The response of this circuit is specified [Ref. 7] at up to 50 Mbit s^{-1}, with a possible extension to 300 Mbit s^{-1} when using a faster ECL logic family and high speed transistors. The emitter coupled drive circuit configuration may also be interfaced with other logic families, and a TTL compatible design is discussed in Ref. 8.

10.2.2.2 Analog Transmission

For analog transmission the drive circuit must cause the light output from an LED source to follow accurately a time-varying input voltage waveform in

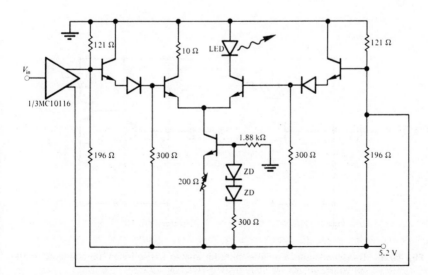

Fig. 10.7 An emitter coupled drive circuit which is compatible with ECL logic [Ref. 7].

both amplitude and phase. Therefore, as indicated previously, it is important that the LED output power responds linearly to the input voltage or current. Unfortunately, this is not always the case because of inherent nonlinearities within LEDs which create distortion products on the signal. Thus the LED itself tends to limit the performance of analog transmission systems unless suitable compensation is incorporated into the drive circuit. However, unless extremely low distortion levels are required, simple transistor drive circuits may be utilized.

Two possible high speed drive circuit configurations are illustrated in Fig. 10.8. Figure 10.8(a) shows a driver consisting of a common emitter transconductance amplifier which converts an input base voltage into a collector current. The circuit is biassed for a class A mode of operation with the quiescent collector current about half the peak value. A similar transconductance configuration which utilizes a Darlington transistor pair in order to reduce the impedance of the source is shown in Fig. 10.8(b). A circuit of this type has been used to drive high radiance LEDs at frequencies of 70 MHz [Ref. 9].

Another simple drive circuit configuration is shown in Fig. 10.9. It consists of a differential amplifier operated over its linear region which directly modulates the LED. The LED operating point is controlled by a reference voltage V_{ref} whilst the current generator provided by the transistor T_3 feeding the differential stage (T_1 and T_2) limits the maximum current through the device. The transimpedance of the driver is reduced through current series feedback provided by the two resistors R_1 and R_2 which are normally assigned equal values. Furthermore, variation between these feedback resistors can be used to compensate for the transfer function of both the drive circuit and the LED.

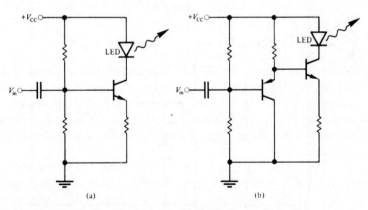

Fig. 10.8 Transconductance drive circuits for analog transmission: (a) common emitter configuration; (b) Darlington transistor pair.

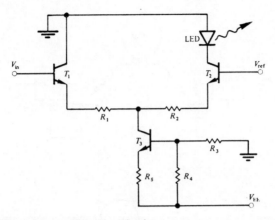

Fig. 10.9 A differential amplifier drive circuit.

Although in many communication applications where a single analog signal is transmitted certain levels of amplitude and phase distortion can be tolerated, this is not the case in frequency multiplexed systems (see Section 10.4.2) where a high degree of linearity is required in order to minimize interference between individual channels caused by the generation of intermodulation products. Also baseband video transmission of TV signals requires the maintenance of extremely low levels of amplitude and phase distortion. For such applications the simple drive circuits described previously are inadequate without some form of linearization to compensate for both LED and drive circuit non-linearities. A number of techniques have been reported [Ref. 10], some of which are illustrated in Fig. 10.10. Figure 10.10(a) shows the complementary distortion technique [Ref. 11] where additional nonlinear devices are included in the system. It may take the form of predistortion compensation (before the source drive circuit) or postdistortion compensation (after the receiver). This approach has been shown [Ref. 12] to reduce harmonic distortion by up to 20 dB over a limited range of modulation amplitudes.

In the negative feedback compensation technique shown in Fig. 10.10(b), the LED is included in the linearization scheme. The optical output is detected and compared with the input waveform, the amount of compensation being dependent on the gain of the feedback loop. Although the technique is straightforward, large bandwidth requirements (i.e. video) can cause problems at high frequencies [Ref. 13].

The technique shown in Fig. 10.10(c) employs phase shift modulation for selective harmonic compensation using a pair of LEDs with similar characteristics [Ref. 14]. The input signal is divided into equal parts which are phase shifted with respect to each other. These signals then modulate the two LEDs giving a cancellation of the second and third harmonic with a 90° and 60°

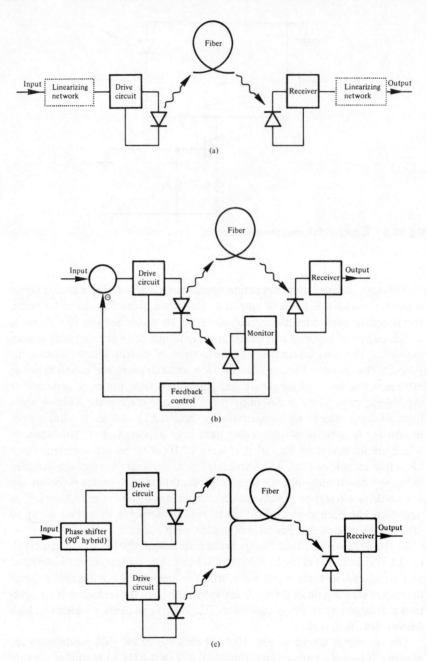

Fig. 10.10   Block schematics of some linearization methods for LED drive circuits: (a) complementary distortion technique; (b) negative feedback compensation technique; (c) selective harmonic compensation technique.

phase shift respectively. However, although there is a high degree of distortion cancellation, both harmonics cannot be reduced simultaneously.

Other linearization techniques include cascade compensation |Ref. 15|, feedforward compensation |Ref. 16| and quasi-feedforward compensation |Refs. 17 and 18|.

10.2.3 Laser Drive Circuits

A number of configurations described for use as LED drive circuits for both digital and analog transmission may be adapted for injection laser applications with only minor changes. The laser, being a threshold device, has somewhat different drive current requirements from the LED. For instance, when digital transmission is considered, the laser is usually given a substantial applied bias, often referred to as prebias, in the off state. Reasons for biassing the laser near but below threshold in the off state are:

(a) it reduces the switch-on delay and minimizes any relaxation oscillations;
(b) it allows easy compensation for changes in ambient temperature and device ageing;
(c) it reduces the junction heating caused by the digital drive current since the on and off currents are not widely different for most lasers.

Although biassing near threshold causes spontaneous emission of light in the off state, this is not normally a problem for digital transmission because the stimulated emission in the on state is generally greater by, at least, a factor of 10.

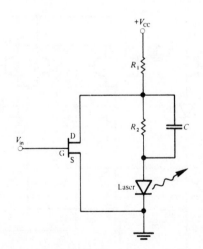

Fig. 10.11   A shunt drive circuit for use with an injection laser.

A simple laser drive circuit for digital transmission is shown in Fig. 10.11. This circuit is a shunt driver utilizing a field effect transistor (FET) to provide high speed laser operation. Sufficient voltage is maintained in series with the laser using the resistor R_2 and the compensating capacitor C such that the FET is biassed into its active or pinch-off region. Hence for a particular input voltage V_{in} (i.e. V_{GS}) a specific amount of the total current flowing through R_1 is diverted around the laser leaving the balance of the current to flow through R_2 and provide the off state for the device. Using suitable gallium arsenide MESFETs (see Section 9.5.1) the circuit shown in Fig. 10.11 has modulated lasers at rates in excess of 1 Gbit s^{-1} [Ref. 19].

An alternative high speed laser drive circuit employing bipolar transistors is shown in Fig. 10.12 [Ref. 20]. This circuit configuration, again for digital transmission, consists of two differential amplifiers connected in parallel. The input stage, which is ECL compatible, exhibits a 50 Ω input impedance by use of an emitter follower T_1 and a 50 Ω resistor in parallel with the input. The transistor T_2 acts as a current source with the zener diode ZD adjusting the signal level for ECL operation. The two differential amplifiers provide sufficient modulation current amplitude for the laser under the control of a d.c. control current I_E through the two emitter resistors R_{E1} and R_{E2}. I_E is provided by an optical feedback control circuit to be discussed shortly. Finally, a prebias current is applied to the laser from a separate current source. This circuit when utilizing microwave transistors was operated with a return to zero digital format (see Section 3.7) at 1 Gbit s^{-1} [Ref. 20].

A major difference between the drive circuits of Figs. 10.11 and 10.12 is the absence and use respectively of feedback control for adjustment of the laser

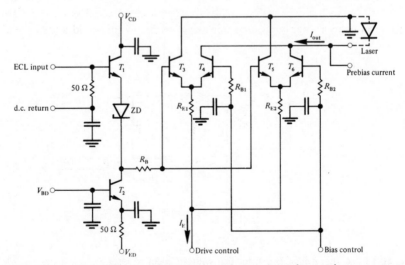

Fig. 10.12   An ECL compatible high speed laser drive circuit [Ref. 20].

output level. For this reason it is unlikely that the shunt drive circuit of Fig. 10.11 would be used for a system application. Some form of feedback control is generally required to ensure continuous laser operation because the device lasing threshold is a sensitive function of temperature. Also the threshold level tends to increase as the laser ages following an increase in internal device losses. Although lasers may be cooled to compensate for temperature variations, ageing is not so easily accommodated by the same process. However, both problems may be overcome through control of the laser bias using a feedback technique. This may be achieved using low speed feedback circuits which adjust the generally static bias current when necessary. For this purpose it is usually found necessary to monitor the light output from the laser in order to keep some aspect constant.

Several strategies of varying complexity are available to provide automatic output level control for the laser. The simplest and perhaps most common form of laser drive circuit incorporating optical feedback is the mean power control circuit shown in Fig. 10.13. Often the monitor detector consists of a cheap, slow photodiode positioned next to the rear face of the laser package as indicated in Fig. 10.13. Alternatively, an optical coupler at the fiber input can be used to direct some of the radiation emitted from the laser into the monitor photodiode. The detected signal is integrated and compared with a reference by an operational amplifier which is used to servo-control the d.c. bias applied to the laser. Thus the mean optical power is maintained constant by varying the threshold current level. This technique is suitable for both digital and analog transmission.

An alternative control method for digital systems which offers accurate threshold tracking and very little device dependence is the switch-on delay technique illustrated in Fig. 10.14 [Ref. 24]. This circuit monitors the switch-on delay of an optical pulse in order to control the laser bias current. The switch-on delay is measured for a zero level set below threshold and the feed-

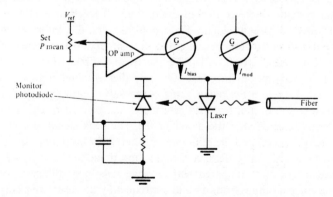

Fig. 10.13   Mean power feedback circuit for control of the laser bias current.

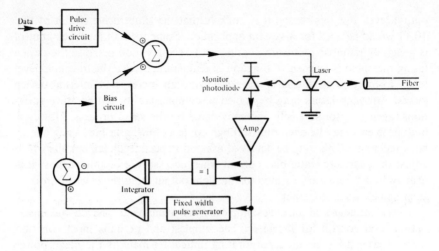

Fig. 10.14   Switch-on delay feedback laser control circuit [Ref. 24].

back is set to a constant fixed delay to control it. Hence, the circuit provides a reference signal proportional to the delay period. This signal is used to control the bias level. The technique requires a fast monitor photodiode as well as a wideband amplifier to allow measurement of the small delay periods. It is also essential that the zero level is set below the lasing threshold because the feedback loop will only stabilize for a finite delay (i.e. the delay falls to zero at the threshold).

A major disadvantage, however, with just controlling the laser bias current is that it does not compensate for variations in the laser slope efficiency. The modulation current for the device is preset and does not take into account any slope changes with temperature and ageing. In order to compensate for such changes, the a.c. and d.c. components of the monitored light output must be processed independently. This is especially important in the case of high bit rate digital systems where control of the on and off levels as well as the light level is required. A circuit which utilizes both a.c. and d.c. information in the laser output to control the device drive current and bias independently is shown in Fig. 10.15 [Ref. 20]. The electrical output from the monitor photodiode is fed into a low drift d.c. amplifier $A1$ and into a wideband amplifier $A2$. Therefore the mean value of the laser output power $P_e(\text{ave})$ is proportional to the output from $A1$ whilst the a.c. content of the monitoring signal is peak detected after the amplifier $A2$. The peak signals correspond to the maximum $P_e(\text{max})$ and the minimum $P_e(\text{min})$ laser output powers within a certain time interval. The difference signal proportional to $(P_e(\text{max}) - P_e(\text{min}))$ is acquired in $A3$ and compared with a drive reference voltage in order to control the current output from $A4$ and consequently the laser drive current. In this way the modulation amplitude of the laser is controlled. Control of the

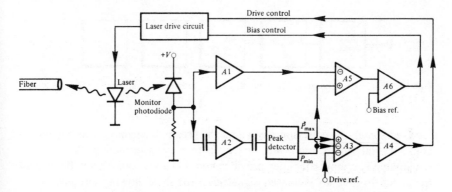

Fig. 10.15 A laser feedback control circuit which uses a.c. and d.c. information in the monitored light output to control the laser drive and bias currents independently [Ref. 20].

laser bias current is achieved from the difference between the output signal of $A1$ (P_e(ave)) and P_e(min) which is acquired in $A5$. The output voltage of $A5$ which is proportional to P_e(min) is compared with a bias reference voltage in $A6$ which supplies a current output to control the laser d.c. bias. This feedback control circuit was designed for use with the laser drive circuit shown in Fig. 10.12 to give digital operation at bit rates in the gigahertz range.

10.3 THE OPTICAL RECEIVER CIRCUIT

The noise performance for optical fiber receivers incorporating both major detector types (the p–i–n and avalanche photodiode) was discussed in Chapter 9. Receiver noise is of great importance within optical fiber communications as it is the factor which limits receiver sensitivity and therefore can dictate the overall system design. It was necessary within the analysis given in Chapter 9 to consider noise generated by electronic amplification (i.e. within the pre-amplifier) of the low level signal as well as the noise sources associated with the optical detector. Also the possible strategies for the configuration of the pre-amplifier were considered (see Section 9.4) as a guide to optimization of the receiver noise performance for a particular application. In this section we extend the discussion to consider different possible circuit arrangements which may be implemented to achieve low noise preamplification as well as further amplification (main amplification) and processing of the detected optical signal.

A block schematic of an optical fiber receiver is shown in Fig. 10.16. Following the linear conversion of the received optical signal into an electrical current at the detector, it is amplified to obtain a suitable signal level. Initial amplification is performed in the preamplifier circuit where it is essential that additional noise is kept to a minimum in order to avoid corruption of the

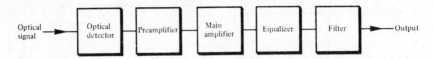

Fig. 10.16 Block schematic showing the major elements of an optical fiber receiver.

received signal. As noise sources within the preamplifier may be dominant, its configuration and design are major factors in determining the receiver sensitivity. The main amplifier provides additional low noise amplification of the signal to give an increased signal level for the following circuits.

Although optical detectors are very linear devices and do not themselves introduce significant distortion onto the signal, other components within the optical fiber communication system may exhibit nonlinear behavior. For instance, the received optical signal may be distorted due to the dispersive mechanisms within the optical fiber. Alternatively the transfer function of the preamplifier–main amplifier combination may be such that the input signal becomes distorted (especially the case with the high impedance front end pre-amplifier). Hence, to compensate for this distortion and to provide a suitable signal shape for the filter, an equalizer is often included in the receiver. It may precede or follow the main amplifier, or may be incorporated in the functions of the amplifier and filter. In Fig. 10.16 the equalizer is shown as a separate element following the amplifier and preceding the filter.

The function of the final element in the receiver, the filter, is to maximize the received signal to noise ratio whilst preserving the essential features of the signal. In digital systems the function of the filter is primarily to reduce intersymbol interference, whereas in analog systems it is generally required to hold the amplitude and phase response of the received signal within certain limits. The filter is also designed to reduce the noise bandwidth as well as in-band noise levels.

Finally, the general receiver consisting of the elements depicted in Fig. 10.16 is often referred to as a linear channel because all operations on the received optical signal may be considered to be mathematically linear.

10.3.1 The Preamplifier

The choice of circuit configuration for the preamplifier is largely dependent upon the system application. Bipolar or field effect transistors (FETs) can be operated in three useful connections. These are the common emitter or source, the common base or gate, and the emitter or source follower for the bipolar and field effect transistors respectively. Each connection has characteristics which will contribute to a particular preamplifier configuration. It is therefore useful to discuss the three basic preamplifier structures (low impedance, high impedance and transimpedance front end) and indicate possible choices of

transistor connection. In this context the discussion is independent of the type
of optical detector utilized. However, it must be noted that there are a number
of significant differences in the performance characteristics between the *p–i–n*
and avalanche photodiode (see Chapter 8) which must be considered within
the overall design of the receiver.

The simplest preamplifier structure is the low input impedance voltage
amplifier. This design is usually implemented using a bipolar transistor configuration because of the high input impedance of FETs. The common emitter
and the grounded emitter (without an emitter resistor) amplifier shown in Fig.
10.17 are favored connections, as they may be designed with reasonably low
input impedance and therefore give operation over a moderate bandwidth
without the need for equalization. However, this is achieved at the expense of
increased thermal noise due to the low effective load resistance presented to the
detector. Nevertheless it is possible to reduce the thermal noise contribution of
this preamplifier by choosing a transistor with characteristics which give a
high current gain at a low emitter current in order to maintain the bandwidth
of the stage. Also an inductance may be inserted at the collector to provide
partial equalization for any integration performed by the stage. The alternative
connection giving very low input impedance is the common base circuit.
Unfortunately this configuration has an input impedance which gives
insufficient power gain when connected to the high impedance of the optical
detector.

The preferred preamplifier configurations for low noise operation use either
a high impedance integrating front end or a transimpedance amplifier (see Sections 9.4.2 and 9.4.3). Careful design employing these circuit structures can
facilitate high gain coupled with low noise performance and therefore

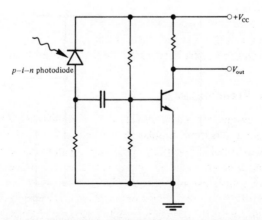

Fig. 10.17   A *p–i–n* photodiode with a grounded emitter, low input impedance voltage
preamplifier.

enhanced receiver sensitivity. Although the bipolar transistor incorporated in the emitter follower circuit may be used to realize a high impedance front end amplifier, the FET is generally employed for this purpose because of its low noise operation. It was indicated in Section 9.5 that the grounded source FET connection was a useful circuit to provide a high impedance front end amplifier. The same configuration with a source resistor (common source connection) shown in Fig. 10.18 provides a similar high input impedance and may also be used (often both configurations are referred to as the common source connection). When operating in this mode the FET power gain and output impedance are both high, which tends to minimize any noise contributions from the following stages. It is especially the case when the voltage gain of the common source stage is minimized in order to reduce the Miller capacitance [Ref. 27] associated with the gate to drain capacitance of the FET. This may be achieved by following the common source stage with a stage having a low input impedance.

Two configurations which provide a low input impedance stage are shown in Fig. 10.19. Figure 10.19(a) shows the grounded source FET followed by a bipolar transistor in the common emitter connection with shunt feedback over the stage. Another favored configuration to reduce Miller capacitance in the first stage FET is shown in Fig. 10.19(b). In this case the second stage consists of a bipolar transistor in the common base configuration which, with the initial grounded source FET, forms the cascode configuration.

The high impedance front end structure provides a very low noise preamplifier design but suffers from two major drawbacks. The first is with regard to equalization which must generally be tailored to the amplifier in order to compensate for distortion introduced onto the signal. Secondly, the high input–impedance approach suffers from a lack of dynamic range which occurs

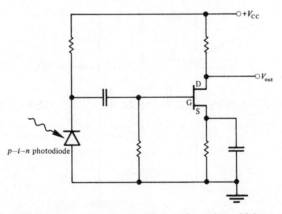

Fig. 10.18   An FET common source preamplifier configuration which provides high input impedance for the *p–i–n* photodiode.

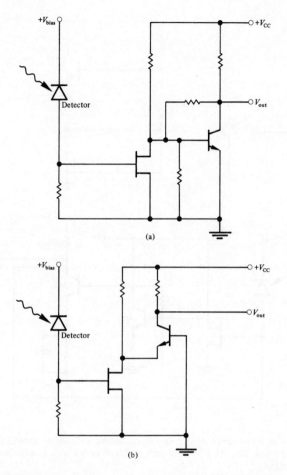

Fig. 10.19 High input impedance preamplifier configurations: (a) grounded source FET followed by common emitter connection with shunt feedback; (b) cascode connection. The separate bias voltage indicates the use of either *p–i–n* or avalanche photodiode.

because the charge on the input capacitance from the low frequency components in the signal builds up over a period of time, causing premature saturation of the amplifier at high input signal levels. Therefore although the circuits shown in Fig. 10.19 are examples of possible high impedance integrating front end amplifier configurations, similar connections may be employed with overall feedback (to the first stage) to obtain a transimpedance preamplifier.

The transimpedance or shunt feedback amplifier finds wide application in preamplifier design for optical fiber communications. This front end structure which acts as a current–voltage converter gives low noise performance without the severe limitations on bandwidth imposed by the high input impedance front

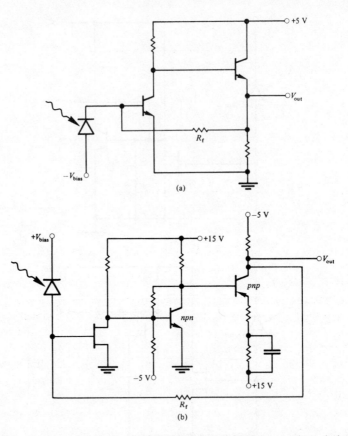

Fig. 10.20   Transimpedance front end configurations: (a) bipolar transistor design [Refs. 28 and 29]; (b) FET front end and bipolar transistor cascade structure [Ref. 32].

end design. It also provides greater dynamic range than the high input impedance structure. However, in practice the noise performance of the transimpedance amplifier is not quite as good as that achieved with the high impedance structure due to the noise contribution from the feedback resistor (see Section 9.4.3). Nevertheless the transimpedance design incorporating a large value of feedback resistor can achieve a noise performance which approaches that of the high impedance front end.

Two examples of transimpedance front end configurations are shown in Fig. 10.20. Figure 10.20(a) illustrates a bipolar transistor structure consisting of a common emitter stage followed by an emitter follower [Refs. 28 and 29] with overall feedback through resistor R_f. The output signal level from this transimpedance pair may be increased by the addition of a second common emitter stage [Ref. 30] after the emitter follower. This stage is not usually

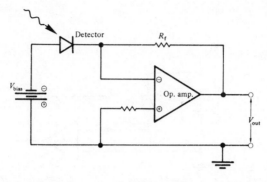

Fig. 10.21   A typical circuit for an operational amplifier transimpedance front end [Ref. 33].

included in the feedback loop. An FET front end transimpedance design is shown in Fig. 10.20(b) [Ref. 32].

The circuit consists of a grounded source configuration followed by a bipolar transistor cascade with feedback over the three stages. In this configuration the bias currents for the bipolar stages and the feedback resistance may be chosen to give good open loop bandwidth whilst making the noise contribution from these stages negligible.

Finally, for lower-bandwidth, shorter-haul applications an FET operational amplifier front end is often adequate [Ref. 33]. Such a transimpedance preamplifier circuit which is generally used with a p–i–n photodiode is shown in Fig. 10.21. The choice of the operational amplifier is dependent on the gain versus bandwidth product for the device. In a simple digital receiver design all that may be required in addition to the circuit shown in Fig. 10.21 is a logic (e.g. TTL) interface stage following the amplifier.

10.3.2   Automatic Gain Control (AGC)

It may be noted from the previous section that the receiver circuit must provide a steady reverse bias voltage for the optical detector. With a p–i–n photodiode this is not critical and a voltage of between 5 and 80 V supplying an extremely low current is sufficient. The avalanche photodiode requires a much larger bias voltage of between 100 and 400 V which defines the multiplication factor for the device. An optimum multiplication factor is usually chosen so that the receiver signal to noise ratio is maximized (see Section 9.3.3). The multiplication factor for the APD varies with the device temperature (see Section 8.9.5) making provision of fine control for the bias voltage necessary in order to maintain the optimum multiplication factor. However, the multiplication factor can be held constant by some form of automatic gain control (AGC). An additional advantage in the use of AGC is

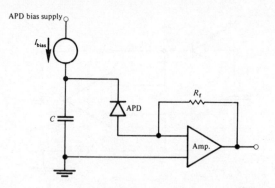

Fig. 10.22 Bias of an APD with a constant current source to provide simple AGC.

that it reduces the dynamic range of the signals applied to the preamplifier giving increased optical dynamic range at the receiver input.

One method of providing AGC is simply to bias the APD with a constant d.c. current source I_{bias} as illustrated in Fig. 10.22. The constant current source is decoupled by a capacitor C at all signal frequencies to prevent gain modulation. When the mean optical input power is known, the mean current to the APD is defined by the bias which gives a constant multiplication factor (gain) at all temperatures. Any variation in the multiplication factor will produce a variation in the charge on C, thus adjusting the biassing of the APD back to the required multiplication factor. Therefore the output current from the photodetector is only defined by the input current from the constant current source, giving full automatic gain control. However, this simple AGC technique is dependent on a constant, mean optical input power level, and takes no account of dark current generated within the detector.

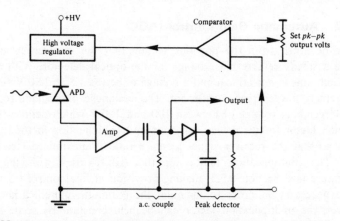

Fig. 10.23 Bias of an APD by peak detection and feedback to provide AGC.

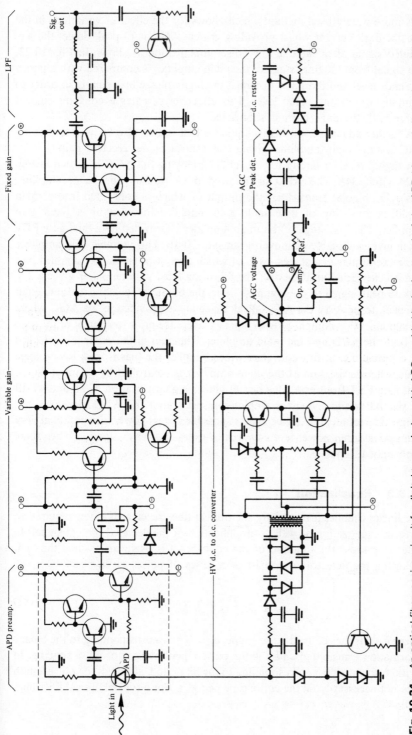

Fig. 10.24 An optical fiber receiver circuit for digital transmission with AGC provided by both control of the APD bias and the main amplifier gain [Ref. 34].

A more widely used method which allows for the effect of variations in the detector dark current whilst providing critical AGC is to peak detect the a.c. coupled signal after suitable low noise amplification as shown in Fig. 10.23. The signal from the final stage of the main amplifier is compared with a preset reference level and fed back to adjust the high voltage bias supply in order to maintain a constant signal level. This effectively creates a constant current source with the dark current subtracted.

A further advantage of this technique is that it may also be used to provide AGC for the main amplifier giving full control of the receiver gain.

A digital receiver circuit for an APD employing full AGC is shown in Fig. 10.24 [Ref. 34]. The APD is followed by a transimpedance preamplifier employing bipolar transistors, the output of which is connected into a main amplifier consisting of a variable gain amplifier followed by a fixed gain amplifier. The first stage of the main amplifier is provided by a dual gate FET which gives a variable gain over a range of 20 dB. This variable gain amplifier also incorporates two stages, each of which consist of an emitter coupled pair with a gain variation of 14 dB. The following fixed gain amplifier gives a 2 V peak to peak signal to the low pass filter, the output of which is maintained at 1 V peak to peak by the AGC. Peak detection is provided in the AGC where the signal level is compared with a preset reference prior to control of the gain for both the APD and the main amplifier. The gain of the APD is controlled via a simple d.c. to d.c. converter which supplies the bias from a low voltage input, whereas the gain of the main amplifier is controlled by an input on the dual gate FET front end. This circuit allows a gain variation of 26 and 47 dB for the APD and the main amplifier respectively. The APD bias circuit is designed to protect the device against possible excess power dissipation at very high optical input power levels as well as excess power dissipation when there is no optical input.

10.3.3 Equalization

The linear channel provided by the optical fiber receiver is often required to perform equalization as well as amplification of the detected optical signal. In order to discuss the function of the equalizer it is useful to assume the light falling on the detector to consist of a series of pulses given by:

$$P_o(t) = \sum_{k=-\infty}^{+\infty} a_k h_p(t - k\tau) \qquad (10.1)$$

where $h_p(t)$ is the received pulse shape, $a_k = 0$ or 1 corresponding to the binary information transmitted and τ is the pulse repetition time or pulse spacing. In digital transmission τ corresponding to the bit period, although the pulse length does not necessarily fill the entire time period τ. For a typical optical fiber link, the received pulse shape is dictated by the transmitted pulse shape $h_t(t)$ and the fiber impulse response $h_f(t)$ following:

$$h_p(t) = h_t(t) * h_f(t) \tag{10.2}$$

where $*$ denotes convolution. Hence determination of the received pulse shape requires knowledge of the fiber impulse response which is generally difficult to characterize. However, it can be shown |Ref. 37| for fiber which exhibits mode coupling, that the impulse response is close to a Gaussian shape in both the time and frequency domain.

It is likely that the pulses given by Eq. (10.1) will overlap due to pulse broadening caused by dispersion on the link giving intersymbol interference (ISI). Following detection and amplification Eq. (10.1) may be written in terms of a voltage $v_A(t)$ as:

$$v_A(t) = \sum_{k=-\infty}^{+\infty} a_k h_A(t - k\tau) \tag{10.3}$$

where the response $h_A(t)$ includes any equalization required to compensate for distortion (e.g. integration) introduced by the amplifier. Therefore, although there is equalization for degradations caused by the amplifier, distortion caused by the channel and the resulting intersymbol interference is still included in $h_A(t - k\tau)$. The pulse overlap causing this intersymbol interference may be reduced through the incorporation of a suitable equalizer with a frequency response $H_{eq}(\omega)$ such that:

$$H_{eq}(\omega) = \frac{\mathcal{F}\{h_{out}(t)\}}{\{h_A(t)\}} = \frac{H_{out}(\omega)}{H_A(\omega)} \tag{10.4}$$

where $h_{out}(t)$ is the desired output pulse shape and $\mathcal{F}$ indicates Fourier transformation. A block diagram indicating the pulse shapes in the time and frequency domains at the various points in an optical fiber system is shown in Fig. 10.25.

An equalizer characterized by Eq. (10.4) will provide high frequency enhancement in the linear channel to compensate for high frequency roll off in the received pulses, thus giving the desired pulse shape. However, in order to construct such an equalizer we require knowledge of $h_A(t)$ and therefore $h_p(t)$. In turn this necessitates information on the fiber impulse response $h_f(t)$ which may not be easily obtained.

Nevertheless the conventional transversal equalizer shown in Fig. 10.26 may be incorporated into the linear channel to keep ISI at tolerable levels, even if it is difficult to design a circuit which gives the optimum system response indicated in Eq. (10.4).

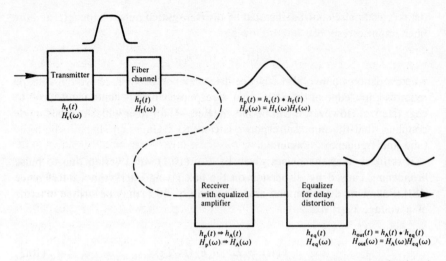

Fig. 10.25 Block schematic of an optical fiber system illustrating the transmitted and received optical pulse shapes together with electrical pulse shape at the linear channel output.

The transversal equalizer consists of a delay line tapped at τ_E second intervals. Each tap is connected through a variable gain device with tap coefficients c_i to a summing amplifier. Intersymbol interference is reduced by filtering the input signal and by the computing values for the tap coefficients which minimize the peak ISI. It is likely that further reduction in ISI will be accomplished using adaptive equalization which has yet to be rigorously applied to optical fiber communications. This topic is discussed further in Ref. 40.

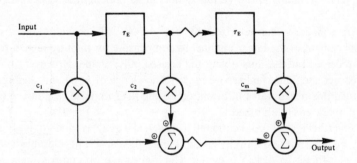

Fig. 10.26 The transversal equalizer employing a tapped delay line.

10.4 SYSTEM DESIGN CONSIDERATIONS

Many of the problems associated with the design of optical fiber communication systems occur as a result of the unique properties of the glass fiber as a transmission medium. However, in common with metallic line transmission systems, the dominant design criteria for a specific application using either digital or analog transmission techniques are the required transmission distance and the rate of information transfer.

Within optical fiber communications these criteria are directly related to the major transmission characteristics of the fiber, namely optical attenuation and dispersion. Unlike metallic conductors where the attenuation (which tends to be the dominant mechanism) can be adjusted by simply changing the conductor size, entirely different factors limit the information transfer capability of optical fibers (see Chapter 3). Nevertheless it is mainly these factors, together with the associated constraints within the terminal equipment, which finally limit the maximum distance that may be tolerated between the optical fiber transmitter and receiver. Where the terminal equipment is more widely spaced than this maximum distance, as in long-haul telecommunication applications, it is necessary to insert repeaters at regular intervals as shown in Fig. 10.27. The repeater incorporates a line receiver in order to convert the optical signal back into the electrical regime where, in the case of analog transmission, it is amplified and equalized (see Section 10.3.3) before it is retransmitted as an optical signal via a line transmitter. When digital transmission techniques are used the repeater also regenerates the original digital signal in the electrical regime (a regenerative repeater which is often simply called a regenerator) before it is retransmitted as a digital optical signal. In this case the repeater may additionally provide alarm, supervision and engineering order wire facilities.

The installation of repeaters substantially increases the cost and complexity of any line communication system. Hence a major design consideration for long-haul telecommunication systems is the maximum distance of unrepeated transmission so that the number of intermediate repeaters may

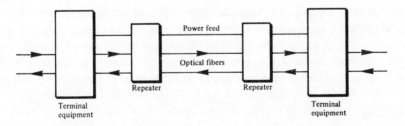

Fig. 10.27 The use of repeaters in a long-haul optical fiber communication system.

be reduced to a minimum. In this respect optical fiber systems display a marked improvement over alternative line transmission systems using metallic conductors. However, this major advantage of optical fiber communications is somewhat reduced due to the present requirement for electrical signal processing at the repeater. This necessitates the supply of electrical power to the intermediate repeaters via metallic conductors as may be observed in Fig. 10.27.

Before any system design procedures can be initiated it is essential that certain basic system requirements are specified. These specifications include:

(a) transmission type: digital or analog;
(b) acceptable system fidelity generally specified in terms of the received BER for digital systems or the received SNR and signal distortion for analog systems;
(c) required transmission bandwidth;
(d) acceptable spacing between the terminal equipment or intermediate repeaters;
(e) cost;
(f) reliability.

However, the exclusive use of the above specifications inherently assumes that system components are available which will allow any system, once specified, to be designed and implemented. Unfortunately this is not always the case, especially when the desired result is a wideband, long-haul system. In this instance it may be necessary to make choices by considering factors such as availability, reliability, cost and ease of installation and operation, before specifications (a)–(d) can be fully determined. A similar approach must be adopted in lower-bandwidth, shorter-haul applications where there is a requirement for the use of specific components which may restrict the system performance. Hence it is likely that the system designer will find it necessary to consider the possible component choices in conjunction with the basic system requirements.

10.4.1 Component Choice

The system designer has many choices when selecting components for an optical fiber communication system. In order to exclude certain components at the outset it is useful if the operating wavelength of the system is established (i.e. shorter wavelength region 0.8–0.9 µm or longer wavelength region 1.1–1.6 µm). This decision will largely be dictated by the overall requirements for the system performance, the ready availability of suitable reliable components, and cost. Hence the major component choices are:

(a) Optical fiber type and parameters. Multimode or single mode; size, refractive index profile, attenuation, dispersion, mode coupling, strength, cabling, jointing, etc.

(b) Source type and characteristics. Laser or LED; optical power launched into the fiber, rise and fall time, stability, etc.

(c) Transmitter configuration. Design for digital or analog transmission; input impedance, supply voltage, dynamic range, optical feedback, etc.

(d) Detector type and characteristics, p–n, p–i–n, or avalanche photodiode; responsivity, response time, active diameter, bias voltage, dark current, etc.

(e) Receiver configuration. Preamplifier design (low impedance, high impedance or transimpedance front end), BER or SNR, dynamic range, etc.

(f) Modulation and coding. Source intensity modulation; using pulse modulation techniques for either digital (e.g. pulse code modulation, adaptive delta modulation) or analog (pulse amplitude modulation, pulse frequency modulation, pulse width modulation, pulse position modulation) transmission. Also encoding schemes for digital transmission such as biphase (Manchester) and delay modulation (Miller) codes [Ref. 7]. Alternatively analog transmission using direct intensity modulation or frequency modulation of the electrical subcarrier (subcarrier FM). In the latter technique the frequency of an electrical subcarrier is modulated rather than the frequency of the optical source as would be the case with direct frequency modulation. The electrical subcarrier, in turn, intensity modulates the optical source (see Section 10.7.5).

Digital and analog modulation techniques which require coherent detection are under investigation but system components which will permit these modulation methods to be utilized are not widely available (see Section 10.8).

Decisions in the above areas are interdependent and may be directly related to the basic system requirements. The potential choices provide a wide variety of economic optical fiber communication systems. However, it is necessary that the choices are made in order to optimize the system performance for a particular application.

10.4.2 Multiplexing

In order to maximize the information transfer over an optical fiber communication link it is usual to multiplex several signals onto a single fiber. It is possible to convey these multichannel signals by multiplexing in the electrical time or frequency domain, as with conventional electrical line or radio communication, prior to intensity modulation of the optical source. Hence, digital pulse modulation schemes may be extended to multichannel operation by time division multiplexing (TDM) narrow pulses from multiple modulators under the control of a common clock. Pulses from the individual channels are interleaved and transmitted sequentially, thus enhancing the bandwidth utilization of a single fiber link.

Alternatively, a number of baseband channels may be combined by frequency division multiplexing (FDM). In FDM the optical channel bandwidth is divided into a number of nonoverlapping frequency bands and each signal is assigned one of these bands of frequencies. The individual signals can be extracted from the combined FDM signal by appropriate electrical filtering at the receive terminal. Hence frequency division multiplexing is generally performed electrically at the transmit terminal prior to intensity modulation of a single optical source. However, it is possible to utilize a number of optical sources each operating at a different wavelength on the single fiber link. In this technique, often referred to as wavelength division multiplexing (WDM), the separation and extraction of the multiplexed signals (i.e. wavelength separation) is performed with optical filters (e.g. interference filters, diffraction grating filters, or prism filters) [Ref. 41].

Finally, a multiplexing technique which does not involve the application of several message signals onto a single fiber is known as space division multiplexing (SDM). In SDM each signal channel is carried on a separate fiber within a fiber bundle or multifiber cable form. The good optical isolation offered by fibers means that cross coupling between channels can be made negligible. However, this technique necessitates an increase in the number of optical components required (e.g. fiber, connectors, sources, detectors) within a particular system and therefore is not widely used.

10.5 DIGITAL SYSTEMS

Most of the future expansion of the telecommunication network is being planned around digital telephone exchanges linked by digital transmission systems. The shift towards digitizing the network followed the introduction of digital circuit techniques and, especially, integrated circuit technology which made the transmission of discrete time signals both advantageous and economic. Digital transmission systems generally give superior performance over their analog counterparts as well as providing an ideal channel for data communications and compatibility with digital computing techniques.

Optical fiber communication is well suited to baseband digital transmission in several important ways. For instance, it offers a tremendous advantage with regard to the acceptable signal to noise ratio (SNR) at the optical fiber receiver over analog transmission by some 20–30 dB (for practical systems) as indicated in the noise considerations of Section 9.2. Also the use of baseband digital signalling reduces problems involved with optical source (and sometimes detector) nonlinearities and temperature dependence which may severely affect analog transmission. Therefore, most high capacity optical fiber communication systems convey digital information in the baseband using intensity modulation (IM) of the optical source.

In common with electrical transmission systems, analog signals (e.g. speech) may be digitized for transmission utilizing pulse code modulation (PCM). Encoding the analog signal into a digital bit pattern is performed by initially sampling the analog signal at a frequency in excess of the Nyquist rate (i.e. greater than twice the maximum signal frequency). Within the European telecommunication network where the 3 dB telephone bandwidth is defined as 3.4 kHz, the sampling rate is 8 kHz. Hence, the amplitude of the constant width sampling pulses varies in proportion to the sample values of the analog signal giving a discrete time signal known as pulse amplitude modulation (PAM) as indicated in Fig. 10.28. The sampled analog signal is then quantized into a number of discrete levels, each of which are designated by a binary code which provides the PCM signal. This process is also illustrated in Fig. 10.28 using a linear quantizer with eight levels (or seven steps) so that each PAM sample is encoded into three binary bits. The analog signal is thus digitized and may be transmitted as a baseband signal or alternatively be modulated by amplitude, frequency or phase shift keying |Ref. 43|. However, in practical PCM systems for speech transmission, nonlinear encoding (A law in Europe and μ law in North America) is generally employed over 256 levels (2^8) giving eight binary bits per sample (seven bits for code levels plus one polarity bit). Hence, the bandwidth requirement for PCM transmission is substantially greater (in this case by a factor of approximately 16) than the corresponding baseband analog transmission. This is not generally a problem with optical fiber communications because of the wideband nature of the optical channel.

Nonlinear encoding may be implemented via a mechanism known as companding where the input signal is compressed before transmission to give a nonlinear encoding characteristic and expanded again at the receive terminal after decoding. A typical nonlinear input–output characteristic giving compression is shown in Fig. 10.29. Companding is used to reduce the quantization error on small amplitude analog signal levels when they are encoded from PAM to PCM. The quantization error (i.e. the rounding off to the nearest

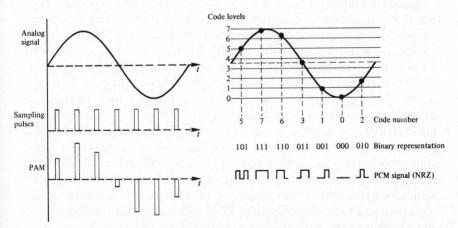

Fig. 10.28 The quantization and encoding of an analog signal into PCM using a linear quantizer with eight levels.

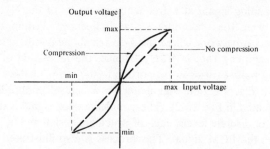

Fig. 10.29 A typical nonlinear input–output characteristic which provides compression.

discrete level) is exhibited as distortion or noise on the signal (often called quantization noise). Companding tapers the step size, thus reducing the distance between levels for small amplitude signals whilst increasing the distance between levels for higher amplitude signals. This substantially reduces the quantization noise on small amplitude signals at the expense of slightly increased quantization noise, in terms of signal amplitude, for the larger signal levels. The corresponding SNR improvement for small amplitude signals significantly reduces the overall signal degradation of the system due to the quantization process.

A block schematic of a simplex (one direction only) baseband PCM system is shown in Fig. 10.30(a). The optical interface is not shown but reference may be made to Fig. 10.1 which illustrates the general optical fiber communication system. It may be noted from Fig. 10.30(a) that the received PCM waveform is decoded back to PAM via the reverse process to encoding, and then simply passed through a low pass filter to recover the original analog signal.

The conversion of a continuous analog waveform into a discrete PCM signal allows a number of analog channels to be time division multiplexed (TDM) for simultaneous transmission down one optical fiber link as illustrated in Fig. 10.30(b). The encoded samples from the different channels are interleaved within the multiplexer to give a single composite signal consisting of all the interleaved pulses. This signal is then transmitted over the optical channel. At the receive terminal the interleaved samples are separated by a synchronous switch or demultiplexer before each analog signal is reconstructed from the appropriate set of samples. Time division multiplexing a number of channels onto a single link can be used with any form of digital transmission and is frequently employed in the transmission of data as well as with the transmission of digitized analog signals. However, the telecommunication network is primarily designed for the transmission of analog speech signals although the compatibility of PCM with data signals has encouraged the adoption of digital transmission systems.

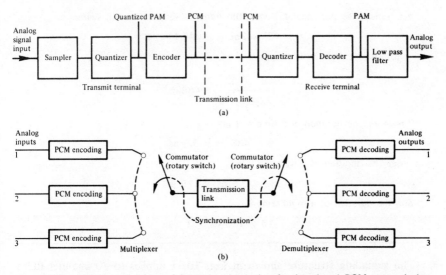

Fig. 10.30 PCM transmission: (a) block schematic of a baseband PCM transmission system for single channel transmission; (b) time division multiplexing of three PCM channels onto a single transmission link and subsequent demultiplexing at the link output.

A current European standard for speech transmission using PCM on metallic conductors (i.e. coaxial line) is the 30 channel system. In this system the PAM samples from each channel are encoded into eight binary bits which are incorporated into a single time slot. Time slots from respective channels are interleaved (multiplexed) into a frame consisting of 32 time slots. The two additional time slots do not carry encoded speech but signalling and synchronization information. Finally, 16 frames are incorporated into a multiframe which is a self-contained timing unit. The timing for this line signalling structure is shown in Fig. 10.31 and calculated in example 10.1.

Example 10.1

The sampling rate for each speech channel on the 30 channel PCM system is 8 kHz and each sample is encoded into eight bits. Determine:

(a) the transmission or bit rate for the system;
(b) the duration of a time slot;
(c) the duration of a frame and multiframe.

Solution: (a) The 30 channel PCM system has 32 time slots each eight bits wide which make up a frame. Therefore,

$$\text{number of bits in a frame} = 32 \times 8 = 256 \text{ bits}$$

This frame must be transmitted within the sampling period and thus 8×10^3 frames are transmitted per second. Hence, the transmission rate for the system is:

$$8 \times 10^3 \times 256 = 2.048 \text{ Mbit s}^{-1}$$

(b) The bit duration is simply:

$$\frac{1}{2.048 \times 10^6} = 488 \text{ ns}$$

Therefore, the duration of a time slot is:

$$8 \times 488 \text{ ns} = 3.9 \text{ μs}$$

(c) The duration of a frame is thus:

$$32 \times 3.9 \text{ μs} = 125 \text{ μs}$$

and the duration of a multiframe is:

$$16 \times 125 \text{ μs} = 2 \text{ ms}$$

The signalling structure shown in Fig. 10.31 applies to 30 channel PCM systems which were originally designed to transmit over metallic conductors

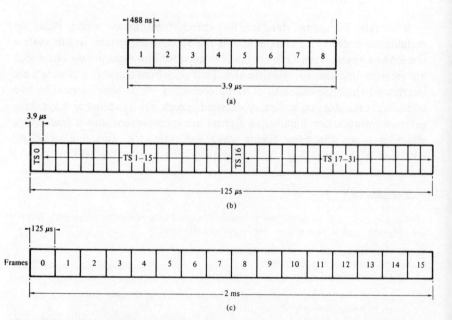

Fig. 10.31   The timing for the line signalling structure of the European standard 30 channel PCM system: (a) bits per time slot; (b) time slots per frame; (c) frames per multiframe.

Table 10.1 Digital bit rates for multichannel PCM transmission transmission in Europe and North America

Europe		North America	
Telephone channels	Bit rates Mbit s^{-1}	Telephone channels	Bit rates Mbit s^{-1}
30	2.048	24	1.544
120	8.448	48	3.152
480	34.368	96	6.312
1920	139.364	672	44.736
7680	565.000	4032	274.176

using a high density bipolar line code (HDB 3). The increased bandwidth with optical fiber communications allows transmission rates far in excess of 2.048 Mbit s^{-1}. Therefore an increased number of telephone channels may be sampled, encoded, multiplexed and transmitted on an optical fiber link. In Europe the increased bit rates were chosen as multiples of the 30 channel system, whereas in North America they tend to be multiples of a 24 channel system. These bit rates and the corresponding number of transmitted telephone channels are specified in Table 10.1.

It must be noted that a bipolar code with a zero mean level (i.e. with positive and negative going pulses in the electrical regime) such as HDB 3 cannot be transmitted directly over an optical fiber link unless the mean level is raised to allow both positive and negative going pulses to be transmitted by the intensity modulated optical source. The resultant ternary (three level) optical transmission is not always suitable for telecommunication applications and therefore binary coding after appropriate scrambling, biphase (Manchester encoding), delay modulation (Miller encoding), etc., is often employed. This involves additional complexity at the transmit and receive terminals as well as necessitating extra redundancy (i.e. bits which do not contain the transmitted information, thus giving a reduction in the information per transmitted symbol) in the line code. This topic is considered in greater detail in Section 10.6.7.

10.6 DIGITAL SYSTEM PLANNING CONSIDERATIONS

The majority of digital optical fiber communication systems for the telecommunication network or local data applications utilize binary intensity modulation of the optical source. Therefore we choose to illustrate the planning considerations for digital transmission based on this modulation technique. Baseband PCM transmission using source intensity modulation is usually designated as PCM-IM.

10.6.1   The Regenerative Repeater

In the case of the long-haul, high-capacity digital systems, the most important overall system performance parameter is the spacing of the regenerative repeaters. It is therefore useful to consider the performance of the digital repeater, especially as it is usually designed with the same optical components as the terminal equipment. Figure 10.32 shows the functional parts of a typical regenerative repeater for optical fiber communications. The attenuated and dispersed optical pulse train is detected and amplified in the receiver unit. This consists of a photodiode followed by a low noise preamplifier. The electrical signal thus acquired is given a further increase in power level in a main amplifier prior to reshaping in order to compensate for the transfer characteristic of the optical fiber (and the amplifier) using an equalizer. Depending on the photodiode utilized, automatic gain control may be provided at this stage for both the photodiode bias current and the main amplifier (see Section 10.3.2).

Accurate timing (clock) information is then obtained from the amplified and equalized waveform using a timing extraction circuit such as a ringing circuit or phase locked loop. This enables precise operation of the following regenerator circuit within the bit intervals of the original pulse train. The function of the regenerator circuit is to reconstitute the originally transmitted pulse train, ideally without error. This can be achieved by setting a threshold above which a binary one is registered, and below which a binary zero is recorded, as indicated in Fig. 10.32. The regenerator circuit makes these decisions at times corresponding to the center of the bit intervals based on the clock information provided by the timing circuit.

Hence the decision times are usually set at the mid-points between the decision level crossings of the pulse train. The pulse train is sampled at a regular frequency equal to the bit rate, and at each sample instant a decision is made of the most probable symbol being transmitted. The symbols are then regenerated in their original form (either a binary one or zero) before

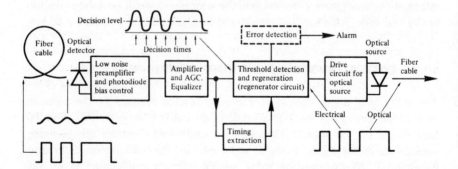

Fig. 10.32   Block schematic showing a typical regenerative repeater for digital optical fiber communications.

retransmission as an optical signal using a source operated by an electronic drive circuit. Hence the possible regeneration of an exact replica of the originally transmitted waveform is a major advantage of digital transmission over corresponding analog systems. Repeaters in analog systems filter, equalize and amplify the received waveform, but are unable to reconstitute the originally transmitted waveform entirely free from distortion and noise. Signal degradation in long-haul analog systems is therefore accumulative being a direct function of the number of repeater stages. In contrast the signal degradation encountered in PCM systems is purely a function of the quantization process and the system bit error rate.

Errors may occur in the regeneration process when:

(a) The signal to noise ratio at the decision instant is insufficient for an accurate decision to be made. For instance, with high noise levels, the binary zero may occur above the threshold and hence be registered as a binary one.

(b) There is intersymbol interference due to dispersion on the optical fiber link. This may be reduced by equalization which forces the transmitted binary one to pass through zero at all neighboring decision times.

(c) There is a variation in the clock rate and phase degradations (jitter) such as distortion of the zero crossings and static decision time misalignment.

A method which is often used to obtain a qualitative indication of the performance of a regenerative repeater or a PCM system is the examination of the received waveform on an oscilloscope using a sweep rate which is a fraction of the bit rate. The display obtained over two bit intervals duration, which is the result of superimposing all possible pulse sequences is called an eye pattern or diagram. An illustration of an eye pattern for a binary system with little distortion and no additive noise is shown in Fig. 10.33(a). It may be observed that the pattern has the shape of a human eye which is open and that the decision time corresponds to the center of the opening. To regenerate the pulse sequence without error the eye must be open thereby indicating a decision area exists, and the decision crosshair (provided by the decision time and the decision threshold) must be within this open area. The effect of practical degradations on the pulses (i.e. intersymbol interference and noise) is to reduce the size of, or close, the eye as shown in Fig. 10.33(b). Hence for reliable transmission it is essential that the eye is kept open, the margin against an error occurring being the minimum distance between the decision crosshair and the edge of the eye.

In practice, a low bit error rate (BER) in the region 10^{-7}–10^{-10} may be tolerated with PCM transmission. However, with data transmission (e.g. computer communications) any error can cause severe problems, and it is necessary to incorporate error detecting and possibly correcting circuits into the regenerator. This invariably requires the insertion of a small amount of redundancy into the transmitted pulse train (see Section 10.6.7).

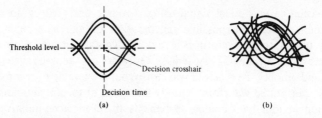

Fig. 10.33   Eye patterns in binary digital transmission: (a) the pattern obtained with a bandwidth limitation but no additive noise (open eye); (b) the pattern obtained with a bandwidth limitation and additive noise (partially closed eye).

Calculation of the possible repeater spacing must take account of the following system component performances:

(a) The average optical power launched into the fiber based on the end of life transmitter performance.
(b) The receiver input power required to achieve an acceptably low BER (e.g. 10^{-9}), taking into account component deterioration during the system's lifetime.
(c) The installed fiber cable loss, including jointing and coupling (to source and detector) losses as well as the effects of ageing and from anticipated environmental changes.
(d) The temporal response of the system including the effects of pulse dispersion on the channel. This becomes an important consideration with high bit rate multimode fiber systems which may be dispersion limited.

These considerations are discussed in detail in the following sections.

10.6.2   The Optical Transmitter

The average optical power launched into the fiber from the transmitter depends upon the type of source used and the required system bit rate as indicated in Section 10.2.1. These factors may be observed in Fig. 10.34 [Ref. 45] which compares the optical power available from an injection laser and an LED for transmission over a multimode fiber with a core diameter of 50 µm and a numerical aperture of 0.2. Typically the laser launches around 1 mW whereas usually the LED is limited to about 100 µW. It may also be noted that both device types emit less optical power at higher bit rates. However, the LED gives reduced output at modulation bandwidths in excess of 50 MHz whereas laser output is unaffected below 200 MHz. Also the fact that generally the optical power which may be launched into a fiber from an LED even at low bit rates is 10–15 dB down on that available from a laser is an important consideration, especially when receiver noise is a limiting factor within the system.

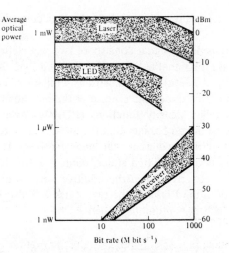

Fig. 10.34 The average power launched into multimode optical fiber from typical injection lasers and LEDs as a function of digital bit rate (upper bands). Also included in the lower band is the received optical power required for binary NRZ pulses transmitted with a BER of 10^{-9}. Reproduced with permission from D. C. Gloge and T. Li, 'Multimode-fiber technology for digital transmission', *Proc. IEEE*, **68**, p. 1269, 1980. Copyright © 1980 IEEE.

10.6.3 The Optical Receiver

The input optical power required at the receiver is a function of the detector combined with the electrical components within the receiver structure. It is strongly dependent upon the noise (i.e. quantum, dark current and thermal) associated with the optical fiber receiver. The theoretical minimum pulse energy or quantum limit required to maintain a given BER was discussed in Section 9.2.4.

It was predicted that approximately 21 incident photons were necessary at an ideal photodetector in order to register a binary one with a BER of 10^{-9}. However, this is a fundamental limit which cannot be achieved in practice and therefore it is essential that estimates of the minimum required optical input power are made in relation to practical devices and components.

Although the statistics of quantum noise follow a Poisson distribution, other important sources of noise within practical receivers (e.g. thermal) are characterized by a Gaussian probability distribution. Hence estimates of the required SNR to maintain particular bit error rates may be obtained using the procedure adopted for error performance of electrical digital systems where the noise distribution is considered to be white Gaussian. This Gaussian approximation [Ref. 46] is sufficiently accurate for design purposes and is far easier to evaluate than the more exact probability distribution within the receiver [Ref. 47]. The receiver sensitivities calculated by using the Gaussian approximation are generally within 1 dB of those calculated by other methods [Ref. 29].

Although the transmitted signal consists of two well-defined light levels, in the presence of noise the signal at the receiver is not as well defined. This situation is shown in Fig. 10.35(a) which illustrates a binary signal in the presence of noise. The signal plus the additive noise at the detector may be defined in terms of the probability density functions (PDFs) shown in Fig. 10.35(b). These PDFs describe the probability that the input current (or output voltage) has a value i (or v) within the incremental range di (or dv). The expected values of the signal in the two transmitted states, namely 0 and 1, are indicated by $p_0(x)$ and $p_1(x)$ respectively. When the additive noise is assumed to have a Gaussian distribution, the PDFs of the two states will also be Gaussian. The Gaussian PDF which is continuous is defined by:

$$p(x) = \frac{1}{\sigma \sqrt{(2\pi)}} \exp -|(x - m)^2 / 2\sigma^2| \qquad (10.5)$$

where m is the mean value and σ the standard deviation of the distribution. When $p(x)$ describes the probability of detecting a noise current or voltage, σ corresponds to the rms value of that current or voltage.

If a decision threshold D is set between the two signal states as indicated in Fig. 10.35, signals greater than D are registered as a one and those less than D as a zero. However, when the noise current (or voltage) is sufficiently large it can either decrease a binary one to a zero or increase a binary zero to a one. These error probabilities are given by the integral of the signal probabilities

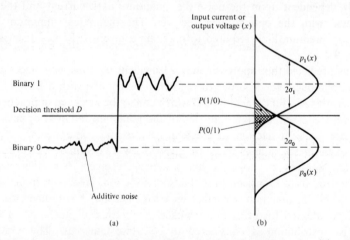

Fig. 10.35   Binary transmission: (a) the binary signal with additive noise; (b) probability density functions for the binary signal showing the decision case. $P(0/1)$ is the probability of falsely identifying a binary one and $P(1/0)$ is the probability of falsely identifying a binary zero.

outside the decision region. Hence the probability that a signal transmitted as a 1 is received as a 0, $P(0/1)$, is proportional to the shaded area indicated in Fig. 10.35(b). The probability that a signal transmitted as a 0 is received as a 1, $P(1/0)$, is similarly proportional to the other shaded area shown in the figure. If $P(1)$ and $P(0)$ are the probabilities of transmission for binary ones and zeros respectively, then the total probability of error $P(e)$ may be defined as:

$$P(e) = P(1)P(0/1) + P(0)P(1/0) \tag{10.6}$$

Now let us consider a signal current i_{sig} together with an additive noise current i_N and a decision threshold set at $D = i_D$. If at any time when a binary 1 is transmitted the noise current is negative such that:

$$i_N < -(i_{sig} - i_D) \tag{10.7}$$

then the resulting current $i_{sig} + i_N$ will be less than i_D and an error will occur. The corresponding probability of the transmitted 1 being received as a 0 may be written as:

$$P(0/1) = \int_{-\infty}^{i_D} p(i, i_{sig}) \, di \tag{10.8}$$

and following Eq. (10.5):

$$p_1(x) = p(i, i_{sig}) = \frac{1}{(\overline{i_N^2})^{\frac{1}{2}} \sqrt{(2\pi)}} \exp - \left[\frac{(i - i_{sig})^2}{2(\overline{i_N^2})} \right] \tag{10.9}$$

$$= \text{Gsn} \, [i, i_{sig}, (\overline{i_N^2})^{\frac{1}{2}}] \tag{10.10}$$

where i is the actual current, i_{sig} is the peak signal current during a binary 1 (this corresponds to the peak photocurrent I_p when only a signal component is present), and $\overline{i_N^2}$ is the mean square noise current. Substituting Eq. (10.10) into Eq. (10.8) gives:

$$P(0/1) = \int_{-\infty}^{i_D} \text{Gsn} \, [i, i_{sig}, (\overline{i_N^2})^{\frac{1}{2}}] \, di \tag{10.11}$$

Similarly, the probability that a binary 1 will be received when a 0 is transmitted is the probability that the received current will be greater than i_D at some time during the zero bit interval. It is given by:

$$P(1/0) = \int_{i_D}^{\infty} p(i, 0) \tag{10.12}$$

Assuming the mean square noise current in the zero state is equal to the mean square noise current in the one state $(\overline{i_N^2})$ (this is an approximation if shot noise is dominant), and that for a zero bit $i_{sig} = 0$, then following Eq. (10.5):

$$p_o(x) = p(i, 0) = \frac{1}{(\overline{i_N^2})^{\frac{1}{2}} \sqrt{(2\pi)}} \exp - \left[\frac{(i - 0)^2}{2(\overline{i_N^2})} \right] \tag{10.13}$$

$$= \text{Gsn} \, |i, 0, (\overline{i_N^2})^{\frac{1}{2}}| \tag{10.14}$$

Hence substituting Eq. (10.14) into Eq. (10.12) gives:

$$P(1/0) = \int_{i_D}^{\infty} \text{Gsn} \, |i, 0, (\overline{i_N^2})^{\frac{1}{2}}| \, di \tag{10.15}$$

The integrals of Eqs. (10.11) and (10.15) are not readily evaluated but may be written in terms of the error function (erf)* where:

$$\text{erf} \, (u) = \frac{2}{\sqrt{\pi}} \int_0^u \exp(-z^2) \, dz \tag{10.16}$$

and the complementary error function is:

$$\text{erfc} \, (u) = 1 - \text{erf} \, (u) = \frac{2}{\sqrt{\pi}} \int_u^{\infty} \exp(-z^2) \, dz \tag{10.17}$$

Hence

$$P(0/1) = \frac{1}{2} \left[1 - \text{erf} \, \left(\frac{|i_{\text{sig}} - i_D|}{(\overline{i_N^2})^{\frac{1}{2}} \sqrt{2}} \right) \right]$$

$$= \tfrac{1}{2} \text{erfc} \, \left(\frac{|i_{\text{sig}} - i_D|}{(\overline{i_N^2})^{\frac{1}{2}} \sqrt{2}} \right) \tag{10.18}$$

and

$$P(1/0) = \tfrac{1}{2} \text{erfc} \, \left(\frac{|0 - i_D|}{(\overline{i_N^2})^{\frac{1}{2}} \sqrt{2}} \right) = \tfrac{1}{2} \text{erfc} \, \left(\frac{|-i_D|}{(\overline{i_N^2})^{\frac{1}{2}} \sqrt{2}} \right) \tag{10.19}$$

If we assume that a binary code is chosen such that the number of transmitted ones and zeros are equal, then $P(0) = P(1) = \frac{1}{2}$, and the net probability of error is one half the sum of the shaded areas in Fig. 10.35(b). Therefore Eq. (10.6) becomes:

$$P(e) = \tfrac{1}{2} |P(0/1) + P(1/0)| \tag{10.20}$$

and substituting for $P(0/1)$ and $P(1/0)$ from Eqs. (10.18) and (10.19) gives:

---

* Another form of the error function denoted by Erf is defined in problem 10.10.

$$P(e) = \frac{1}{2} \left[\frac{1}{2} \operatorname{erfc} \left(\frac{|i_{\text{sig}} - i_D|}{(\overline{i_N^2})^{\frac{1}{2}} \sqrt{2}} \right) + \frac{1}{2} \operatorname{erfc} \left(\frac{|-i_D|}{(\overline{i_N^2})^{\frac{1}{2}} \sqrt{2}} \right) \right] \quad (10.21)$$

Equation (10.21) may be simplified by setting the threshold decision level at the mid-point between zero current and the peak signal current such that $i_D = i_{\text{sig}}/2$. In electrical systems this situation corresponds to an equal minimum probability of error in both states due to the symmetrical nature of the PDFs. It must be noted that for optical fiber systems this is not generally the case since the noise in each signal state contains shot noise contributions proportional to the signal level. Nevertheless assuming a Gaussian distribution for the noise and substituting $i_D = i_{\text{sig}}/2$ into Eq. (10.21) we obtain:

$$P(e) = \frac{1}{2} \left[\frac{1}{2} \operatorname{erfc} \left(\frac{|i_{\text{sig}}/2|}{(\overline{i_N^2})^{\frac{1}{2}} \sqrt{2}} \right) + \frac{1}{2} \operatorname{erfc} \left(\frac{|-i_{\text{sig}}/2|}{(\overline{i_N^2})^{\frac{1}{2}} \sqrt{2}} \right) \right]$$

$$= \frac{1}{2} \operatorname{erfc} \left(\frac{i_{\text{sig}}}{2(\overline{i_N^2})^{\frac{1}{2}} \sqrt{2}} \right) \quad (10.22)$$

The electrical SNR at the detector may be written in terms of the peak signal power to rms noise power (mean square noise current) as:

$$\frac{S}{N} = \frac{i_{\text{sig}}^2}{\overline{i_N^2}} \quad (10.23)$$

Comparison of Eq. (10.23) with Eq. (10.22) allows the probability of error to be expressed in terms of the analog SNR as:

$$P(e) = \frac{1}{2} \operatorname{erfc} \left(\frac{(S/N)^{\frac{1}{2}}}{2\sqrt{2}} \right) \quad (10.24)$$

Estimates of the required SNR to maintain a given error rate may be obtained using the standard table for the complementary error function. A plot of $P(e)$ against $\frac{1}{2} \operatorname{erfc}(u)$ is shown in Fig. 10.36(a). This may be transposed into the characteristic illustrated in Fig. 10.36(b) where the bit error rate which is equivalent to the error probability $P(e)$ is shown as a function of the SNR following Eq. (10.24).

Example 10.2

Using the Gaussian approximation determine the required signal to noise ratios (optical and electrical) to maintain a BER of 10^{-9} on a baseband binary digital optical fiber link. It may be assumed that the decision threshold is set midway between the one and the zero level and that $2 \times 10^{-9} \simeq \operatorname{erfc} 4.24$.

Solution: Under the above conditions, the probability of error is given by

Eq. (10.24) where,

$$P(e) = \tfrac{1}{2} \operatorname{erfc} \left(\frac{(S/N)^{\frac{1}{2}}}{2\sqrt{2}} \right) = 10^{-9}$$

Hence

$$\operatorname{erfc} \left(\frac{(S/N)^{\frac{1}{2}}}{2\sqrt{2}} \right) = 2 \times 10^{-9}$$

and

$$\frac{(S/N)^{\frac{1}{2}}}{2\sqrt{2}} = 4.24$$

giving

$$(S/N)^{\frac{1}{2}} = 4.24 \times 2\sqrt{2} \simeq 12$$

The optical SNR may be defined in terms of the peak signal current and rms noise current as $i_{\text{sig}}/(\overline{i_N^2})^{\frac{1}{2}}$. Therefore using Eq. (10.23):

$$\frac{i_{\text{sig}}}{(\overline{i_N^2})^{\frac{1}{2}}} = \left(\frac{S}{N} \right)^{\frac{1}{2}} = 12 \text{ or } 10.8 \text{ dB}$$

The electrical SNR is defined by Eq. (10.23) as:

$$\frac{i_{\text{sig}}^2}{\overline{i_N^2}} = \frac{S}{N} = 144 \text{ or } 21.6 \text{ dB}$$

These results for the SNRs may be seen to correspond to a bit error rate of 10^{-9} on the curve shown in Fig. 10.36(b).

However, the plot shown in Fig. 10.36(b) does not reflect the best possible results, or those which may be obtained with an optimized receiver design. In this case, if the system is to be designed with a particular BER, the appropriate value of the error function is established prior to adjustment of the parameter values (signal levels, decision threshold level, avalanche gain, component values, etc.) in order to obtain this BER [Ref. 48]. It is therefore necessary to use the generalized forms of Eqs. (10.18) and (10.19) where:

$$P(0/1) = \tfrac{1}{2} \operatorname{erfc} \left(\frac{|i_{\text{sig}\,1} - i_D|}{(i_{N1}^2)^{\frac{1}{2}} \sqrt{2}} \right) \tag{10.25}$$

$$P(1/0) = \tfrac{1}{2} \operatorname{erfc} \left(\frac{|i_D - i_{\text{sig}\,0}|}{(i_{N0}^2)^{\frac{1}{2}} \sqrt{2}} \right) \tag{10.26}$$

where $i_{\text{sig}\,1}$ and $i_{\text{sig}\,0}$ are the signal currents, in the 1 and 0 states respectively, and $\overline{i_{N1}^2}$ and $\overline{i_{N0}^2}$ are the corresponding mean square noise currents which may

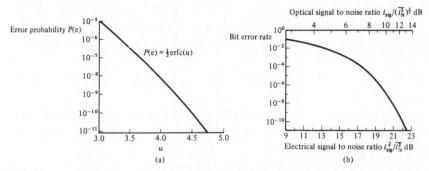

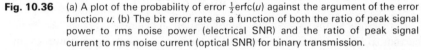

(a)   (b)

Fig. 10.36   (a) A plot of the probability of error $\frac{1}{2}$erfc(u) against the argument of the error function u. (b) The bit error rate as a function of both the ratio of peak signal power to rms noise power (electrical SNR) and the ratio of peak signal current to rms noise current (optical SNR) for binary transmission.

include both shot and thermal noise terms. Equations (10.25) and (10.26) allow a more exact evaluation of the error performance of the digital optical fiber system under the Gaussian approximation |Refs. 48 and 49|. Unfortunately this approach does not give a simple direct relationship between the BER and the analog SNR as the one shown in Eq. (10.24). Thus for estimates of SNR within this text we will make use of the slightly poorer approximation given by Eq. (10.24). Although this approximation does not give the correct decision threshold level or optimum avalanche gain it is reasonably successful at predicting bit error rate as a function of signal power and hence provides realistic estimates of the number of photons required at a practical detector in order to maintain given bit error rates.

For instance, let us consider a good avalanche photodiode receiver which we assume to be quantum noise limited. Hence we ignore the shot noise contribution from the dark current within the APD as well as the thermal noise generated by the electronic amplifier. In practice this assumption holds when the multiplication factor M is chosen to be sufficiently high to ensure that the SNR is determined by photon noise rather than by electronic amplifier noise, and the APD used has a low dark current. To determine the SNR for this ideal APD receiver it is useful to define the quantum noise on the primary photocurrent I_p within the device in terms of shot noise following Eq. (9.8). Therefore, the mean square shot noise current is given by:

$$\overline{i_s^2} = 2eBI_pM^2 \qquad (10.27)$$

where e is the electronic charge and B is the post detection or effective noise bandwidth. It may be observed that the mean square shot noise current $\overline{i_s^2}$ given in Eq. (10.27) is increased by a factor M^2 due to avalanche gain in the APD. However, Eq. (10.27) does not give the total noise current at the output of the APD as there is an additional noise contribution from the random gain mechanism. The excess avalanche noise factor $F(M)$ incurred was discussed in

Section 9.3.4 and defined by Eqs. (9.27) and (9.28). Equation (9.27) may be simplified |Ref. 50| to give an expression for electron injection in the low frequency limit of:

$$F(M) = kM + \left(2 - \frac{1}{M}\right)(1 - k) \qquad (10.28)$$

where k is the ratio of the carrier ionization rates. Hence the excess avalanche noise factor may be combined into Eq. (10.27) to give a total mean square shot noise current $\overline{i_n^2}$ as:

$$\overline{i_n^2} = 2eBI_p M^2 F(M) \qquad (10.29)$$

Furthermore, the avalanche multiplication mechanism raises the signal current to MI_p and therefore the SNR in terms of the peak signal power to rms noise power may be written as:

$$\frac{S}{N} = \frac{(MI_p)^2}{2eBI_p M^2 F(M)} = \frac{I_p}{2eBF(M)} \qquad (10.30)$$

Now, if we let z_{md} correspond to the average number of photons detected in a time period of duration τ, then

$$I_p = \frac{z_{md}e}{\tau} = \frac{z_m e\eta}{\tau} \qquad (10.31)$$

where z_m is the average number of photons incident on the APD and η is the quantum efficiency of the device. Substituting for I_p in Eq. (10.30) we have:

$$\frac{S}{N} = \frac{z_m \eta}{2B\tau F(M)} \qquad (10.32)$$

Rearranging Eq. (10.32) gives an expression for the average number of photons required within the signalling interval τ to detect a binary one in terms of the received SNR for the good APD receiver as:

$$z_m = \frac{2B\tau F(M)}{\eta}\left(\frac{S}{N}\right) \qquad (10.33)$$

A reasonable pulse shape obtained at the receiver in order to reduce intersymbol interference has the raised cosine spectrum shown in Fig. 10.37. The raised cosine spectrum for the received pulse gives a pulse response resulting in a binary pulse train passing through either full or zero amplitude at the centers of the pulse intervals and with transitions passing through half amplitude at points which are midway in time between pulse centers. For raised cosine pulse shaping and full τ signalling $B\tau$ is around 0.6. Hence the

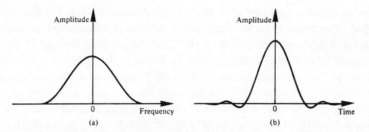

Fig. 10.37 (a) Raised cosine spectrum. (b) Output of a system with a raised cosine output spectrum for a single input pulse.

average number of photons required to detect a binary one using a good APD receiver at a specified BER may be estimated using Eq. (10.33) in conjunction with Eq. (10.24).

Example 10.3

A good APD is used as a detector in an optical fiber PCM receiver designed for baseband binary transmission with a decision threshold set midway between the zero and one signal levels. The APD has a quantum efficiency of 80%, a ratio of carrier ionization rates of 0.02 and is operated with a multiplication factor of 100. Assuming a raised cosine signal spectrum at the receiver, estimate the average number of photons which must be incident on the APD to register a binary one with a BER of 10^{-9}.

Solution: The electrical SNR required to obtain a BER of 10^{-9} at the receiver is given by the curve shown in Fig. 10.36(b), or the solution to example 10.2 as 21.6 dB or 144. Also the excess avalanche noise factor $F(M)$ may be determined using Eq. (10.28) where,

$$F(M) = kM + \left(2 - \frac{1}{M}\right)(1 - k)$$

$$= 2 + (2 - 0.01)(1 - 0.02)$$

$$= 3.95 \simeq 4$$

The average number of photons which must be incident at the receiver in order to maintain the BER can be estimated using Eq. (10.33) (assuming $B\tau = 0.6$ for the raised cosine pulse spectrum) as:

$$z_m = \frac{2B\tau F(M)}{\eta}\left(\frac{S}{N}\right)$$

$$= \frac{2 \times 0.6 \times 4 \times 144}{0.8}$$

$$= 864 \text{ photons}$$

The estimate in example 10.3 gives a more realistic value for the average number of incident photons required at a good APD receiver in order to register a binary one with a BER of 10^{-9} than the quantum limit of 21 photons determined for an ideal photodetector in example 9.1. However, it must be emphasized that the estimate in example 10.3 applies to a good silicon APD receiver (with high sensitivity and low dark current) which is quantum noise limited, and that no account has been taken of the effects of either dark current within the APD or thermal noise generated within the preamplifier. It is therefore likely that at least 1000 incident photons are required at a good APD receiver to register a binary one and provide a BER of 10^{-9} [Ref. 51]. Nevertheless somewhat lower values may be achieved by setting the decision threshold below the half amplitude level because the shot noise on the zero level is lower than the shot noise on the one level.

The optical power required at the receiver P_o is simply the optical energy divided by the time interval over which it is incident. The optical energy E_o may be obtained directly from the average number of photons required at the receiver in order to maintain a particular BER following:

$$E_o = z_m hf \tag{10.34}$$

where hf is the energy associated with a single photon which is given by Eq. (6.1). In order that a binary one is registered at the receiver, the optical energy E_o must be incident over the bit interval τ. For system calculations we can assume a zero disparity code which has an equal density of ones and zeros. In this case the optical power required to register a binary one may be considered to be incident over two bit intervals giving:

$$P_o = \frac{E_o}{2\tau} \tag{10.35}$$

Substituting for E_o from Eq. (10.34) we obtain:

$$P_o = \frac{z_m hf}{2\tau} \tag{10.36}$$

Also as the bit rate B_T for the channel is the reciprocal of the bit interval τ, Eq. (10.36) may be written as:

$$P_o = \frac{z_m hf B_T}{2} \tag{10.37}$$

Equation (10.37) allows estimates of the incident optical power required at a good APD receiver in order to maintain a particular BER, based on the average number of incident photons. In system calculations these optical power levels are usually expressed in dBm. It may also be observed that the required incident optical power is directly proportional to the bit rate B_T which typifies a shot noise limited receiver.

Example 10.4

The receiver of example 10.3 operates at a wavelength of 1 µm. Assuming a zero disparity binary code, estimate the incident optical power required at the receiver to register a binary one with a BER of 10^{-9} at bit rates of 10 Mbit s^{-1} and 140 Mbit s^{-1}.

Solution: Under the above conditions, the required incident optical power may be obtained using Eq. (10.37) where,

$$P_0 = \frac{z_m h f B_T}{2} = \frac{z_m h c B_T}{2\lambda}$$

At 10 Mbit s^{-1},

$$P_0 = \frac{864 \times 6.626 \times 10^{-34} \times 2.998 \times 10^8 \times 10^7}{2 \times 1 \times 10^{-6}}$$

$$= 858.2 \text{ pW}$$

$$= -60.7 \text{ dBm}$$

At 140 Mbit s^{-1},

$$P_0 = \frac{864 \times 6.626 \times 10^{-34} \times 2.998 \times 10^8 \times 14 \times 10^7}{2 \times 1 \times 10^{-14}}$$

$$= 12.015 \text{ nW}$$

$$= -49.2 \text{ dBm}$$

Example 10.4 illustrates the effect of direct proportionality between the optical power required at the receiver and the system bit rate. In the case considered, the required incident optical power at the receiver to give a BER of 10^{-9} must be increased by around 11.5 dB (factor of 14) when the bit rate is increased from 10 to 140 Mbit s^{-1}. Also comparison with example 9.1 where a similar calculation was performed for an ideal photodetector operating at 10 Mbit s^{-1} emphasizes the necessity of performing the estimate for a practical photodiode. The good APD receiver considered in example 10.4 exhibits around 16 dB less sensitivity than the ideal photodetector (i.e. quantum limit).

The assumptions made in the evaluation of examples 10.3 and 10.4 are not generally valid when considering $p-i-n$ photodiode receivers because these devices are seldom quantum noise limited due to the absence of internal gain within the photodetector. In this case thermal noise generated within the electronic amplifier is usually the dominating noise contribution and is typically 1×10^3 to 3×10^3 times larger than the peak response produced by the displacement current of a single electron–hole pair liberated in the detector. Hence, for reliable performance with a BER of 10^{-9}, between 1 and 3×10^4 photons must be detected when a binary one is incident on the receiver [Ref. 53].

This translates into sensitivities which are about 30 dB or more, less than the quantum limit. Finally for a thermal noise limited receiver the input optical power is proportional to the square root of both the post detection or effective noise bandwidth and the SNR (i.e. $P_o \propto |(S/N)B|^{\frac{1}{2}}$). However, this result is best obtained from purely analog SNR considerations and therefore is dealt with in Section 10.7.1.

10.6.4   Channel Losses

Another important factor when estimating the permissible separation between regenerative repeaters or the overall link length is the total loss encountered between the transmitter(s) and receiver(s) within the system. Assuming there are no dispersion penalties on the link, the total channel loss may be obtained by simply summing in decibels the installed fiber cable loss, the fiber–fiber jointing losses and the coupling losses of the optical source and detector. The fiber cable loss in decibels per kilometer α_{fc} is normally specified by the manufacturer, or alternatively it may be obtained by measurement (see Sections 5.2 and 5.7). It must be noted that the cabled fiber loss is likely to be greater than the uncabled fiber loss usually measured in the laboratory due to possible microbending of the fiber within the cabling process (see Section 4.6.2).

Loss due to joints (generally splices) on the link may also, for simplicity, be specified in terms of an equivalent loss in decibels per kilometer α_j. In fact, it is more realistic to regard α_j as a distributed loss since the optical attenuation resulting from the disturbed mode distribution at a joint does not only occur in the vicinity of the joint. Finally the loss contribution attributed to the connectors α_{cr} (in decibels) used for coupling the optical source and detector to the fiber must be included in the overall channel loss. Hence the total channel loss C_L (in decibels) may be written as:

$$C_L = (\alpha_{fc} + \alpha_j)L + \alpha_{cr} \qquad (10.38)$$

where L is the length in kilometers of the fiber cable either between regenerative repeaters or between the transmit and receive terminals for a link without repeaters.

Example 10.5

An optical fiber link of length 4 km comprises a fiber cable with an attenuation of 5 dB km^{-1}. The splice losses for the link are estimated at 2 dB km^{-1}, and the connector losses at the source and detector are 3.5 and 2.5 dB respectively. Ignoring the effects of dispersion on the link determine the total channel loss.

Solution: The total channel loss may be simply obtained using Eq. (10.38) where:

$$C_L = (\alpha_{fc} + \alpha_j)L + \alpha_{cr}$$

$$= (5 + 2)4 + 3.5 + 2.5$$

$$= 34 \text{ dB}$$

10.6.5 Temporal Response

The system design considerations must also take into account the temporal response of the system components. This is especially the case with regard to pulse dispersion on the optical fiber channel. The formula given in Eq. (10.38) allows determination of the overall channel loss in the absence of any pulse broadening due to the dispersion mechanisms within the transmission medium. However, the finite bandwidth of the optical system may result in overlapping of the received pulses or intersymbol interference, giving a reduction in sensitivity at the optical receiver. Therefore either a worse BER must be tolerated, or the ISI must be compensated by equalization within the receiver (see Section 10.3.3). The latter necessitates an increase in optical power at the receiver which may be considered as an additional loss penalty. This additional loss contribution is usually called the dispersion–equalization or ISI penalty. The dispersion–equalization penalty D_L becomes especially significant in high bit rate multimode fiber systems and has been determined analytically for Gaussian shaped pulses [Ref. 48]. In this case it is given by:

$$D_L = 2 \left(\frac{\tau_e}{\tau} \right)^4 \text{ dB} \qquad (10.39)$$

where τ_e is the $1/e$ full width pulse broadening due to dispersion on the link and τ is the bit interval or period. For Gaussian shaped pulses, τ_e may be written in terms of the rms pulse width σ as (see Appendix F):

$$\tau_e = 2\sigma\sqrt{2} \qquad (10.40)$$

Hence, substituting into Eq. (10.39) for τ_e, and writing the bit rate B_T as the reciprocal of the bit interval τ gives:

$$D_L = 2(2\sigma B_T \sqrt{2})^4 \text{ dB} \qquad (10.41)$$

Since the dispersion–equalization penalty as defined by Eq. (10.41) is measured in decibels, it may be included in the formula for the overall channel loss given by Eq. (10.38). Therefore the total channel loss including the dispersion–equalization penalty C_{LD} is given by:

$$C_{LD} = (\alpha_{fc} + \alpha_j)L + \alpha_{cr} + D_L \text{ dB} \qquad (10.42)$$

The dispersion–equalization penalty is usually only significant in wideband multimode fiber systems which exhibit intermodal as well as intramodal dispersion. Single mode fiber systems which are increasingly being utilized for wideband long-haul applications are not generally limited by pulse broadening on the channel because of the absence of intermodal dispersion. However, it is often the case that intermodal dispersion is the dominant mechanism within multimode fibers. In Section 3.9.1 intermodal pulse broadening was considered to be a linear function of the fiber length L. Furthermore it was indicated that the presence of mode coupling within the fiber made the pulse broadening increase at a slower rate proportional to $L^{\frac{1}{2}}$. Hence it is useful to consider the dispersion–equalization penalty in relation to fibers without and with mode coupling operating at various bit rates.

Example 10.6

The rms pulse broadening resulting from intermodal dispersion within a multimode optical fiber is $0.6\,\text{ns km}^{-1}$. Assuming this to be the dominant dispersion mechanism, estimate the dispersion–equalization penalty over an unrepeatered fiber link of length 8 km at bit rates of (a) 25 Mbit s^{-1} and (b) 150 Mbit s^{-1}. In both cases evaluate the penalty without and with mode coupling. The pulses may be assumed to have a Gaussian shape.

Solution: (a) *Without mode coupling.* The total rms pulse broadening over 8 km is given by:

$$\sigma_T = \sigma \times L = 0.6 \times 8 = 4.8 \text{ ns}$$

The dispersion–equalization penalty is given by Eq. (10.41) where:

$$D_L = 2(2\sigma_T B_T \sqrt{2})^{4'} = 2(2 \times 4.8 \times 10^{-9} \times 25 \times 10^6 \sqrt{2})^4$$

$$= 0.03 \text{ dB}$$

With mode coupling. The total rms pulse broadening is:

$$\sigma_T \simeq \sigma\sqrt{L} = 0.6 \times \sqrt{8} = 1.7 \text{ ns}$$

Hence the dispersion–equalization penalty is:

$$D_L = 2(2 \times 1.7 \times 10^{-9} \times 25 \times 10^6 \sqrt{2})^4$$

$$= 4.2 \times 10^{-4} \text{ dB (i.e. negligible)}$$

(b) *Without mode coupling.*

$$\sigma_T = 4.8 \text{ ns}$$

$$D_L = 2(2 \times 4.8 \times 10^{-9} \times 150 \times 10^6 \sqrt{2})^4 = 34.38 \text{ dB}$$

With mode coupling.

$$\sigma_T = 1.7 \text{ ns}$$

$$D_L = 2(2 \times 1.7 = 10^{-9} \times 150 \times 10^6 \sqrt{2})^4 = 0.54 \text{ dB}$$

Example 10.6(a) demonstrates that at low bit rates the dispersion–equalization penalty is very small if not negligible. In this case the slight advantage of the effect of mode coupling on the penalty is generally outweighed by increased attenuation on the link because of the mode coupling, which may be of the order of 1 dB km^{-1}. Example 10.6(b) indicates that at higher bit rates with no mode coupling the dispersion–equalization penalty dominates to the extent that it would be necessary to reduce the repeater spacing to between 4 and 5 km. However, it may be observed that encouragement of mode coupling on the link greatly reduces this penalty and outweighs any additional attenuation incurred through mode coupling within the fiber. In summary, it is clear that the dispersion–equalization penalty need only be applied when considering wideband systems. Moreover it is frequently the case that lower bit rate systems may be up-graded at a later date to a higher capacity without incurring a penalty which might necessitate a reduction in repeater spacing.

An alternative approach involving the calculation of the system rise time can be employed to determine the possible limitation on the system bandwidth resulting from the temporal response of the system components. Therefore, if there is not a pressing need to obtain the maximum possible bit rate over the maximum possible distance, it is sufficient within the system design to establish that the total temporal response of the system is adequate for the desired system bandwidth. Nevertheless this approach does allow for a certain amount of optimization of the system components, but at the exclusion of considerations regarding equalization and the associated penalty.

The total system rise time may be determined from the rise times of the individual system components which include the source (or transmitter), the fiber cable, and the detector (or receiver). These times are defined in terms of a Gaussian response as the 10–90% rise (or fall) times of the individual components. The fiber cable 10–90% rise time may be separated into rise times arising from intermodal T_n and intramodal or chromatic dispersion T_c. The total system rise time is given by [Ref. 56]:

$$T_{\text{syst}} = 1.1(T_S^2 + T_n^2 + T_c^2 + T_D^2)^{\frac{1}{2}} \qquad (10.43)$$

where T_S and T_D are the source and detector 10–90% rise times respectively, and all the rise times are measured in nanoseconds. Comparison of the rise time edge with the overall pulse dispersion results in the weighting factor of 1.1.

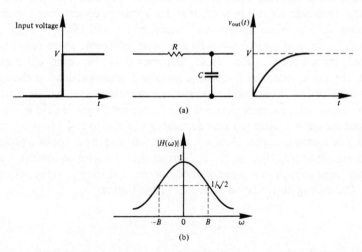

Fig. 10.38 (a) The response of a low pass RC filter circuit to a voltage step input. (b) The transfer function $H(\omega)$ for the circuit in (a).

The maximum system bit rate $B_T(\text{max})$ is usually defined in terms of T_{syst} by consideration of the rise time of the simple RC filter circuit shown in Fig. 10.38(a). For a voltage step input of amplitude V, the output voltage waveform $v_{\text{out}}(t)$ as a function of time t is:

$$v_{\text{out}}(t) = V(1 - e^{-t/RC}) \qquad (10.44)$$

Hence the 10–90% rise time t_r for the circuit is given by:

$$t_r = \frac{2.2}{RC} \qquad (10.45)$$

The transfer function for this circuit is shown in Fig. 10.38(b) and is given by:

$$|H(\omega)| = \frac{1}{(1 + \omega^2 C^2 R^2)^{\frac{1}{2}}} \qquad (10.46)$$

Therefore the 3 dB bandwidth for the circuit is

$$B = \frac{1}{2\pi RC} \qquad (10.47)$$

Combining Eqs. (10.45) and (10.47) gives,

$$t_r = \frac{2.2}{2\pi B} = \frac{0.35}{B} \qquad (10.48)$$

The result for the 10–90% rise time indicated in Eq. (10.48) is of general validity, but a different constant term may be obtained with different filter circuits. However, for rise time calculations involving optical fiber systems the constant 0.35 is often utilized and hence in Eq. (10.48), $t_r = T_{\text{syst}}$. Alternatively, if an ideal (unrealizable) filter with an arbitrarily sharp cutoff is considered, the constant in Eq. (10.48) becomes 0.44. However, although this value for the constant is frequently employed when calculating the optical bandwidth of fiber from pulse dispersion measurements (see Section 5.3.1), the more conservative estimate obtained using a constant term of 0.35 is generally favored for use in system rise time calculations [Refs. 56 and 57]. Also, in both cases it is usually accepted [Ref. 43] that to conserve the shape of a pulse with a reasonable fidelity through the RC circuit then the 3 dB bandwidth must be at least large enough to satisfy the condition $B\tau = 1$, where τ is the pulse duration. Combining this relation with Eq. (10.48) gives:

$$T_{\text{syst}} = t_r = 0.35\tau \qquad (10.49)$$

For an RZ pulse format, the bit rate $B_T = B = 1/\tau$ (see Section 3.7) and hence substituting into Eq. (10.49) gives:

$$B_T(\max) = \frac{0.35}{T_{\text{syst}}} \tag{10.50}$$

Alternatively for an NRZ pulse format $B_T = B/2 = 1/2\tau$ and therefore the maximum bit rate is given by:

$$B_T(\max) = \frac{0.7}{T_{\text{syst}}} \tag{10.51}$$

Thus the upper limit on T_{syst} should be less than 35% of the bit interval for an RZ pulse format and less than 70% of the bit interval for an NRZ pulse format.

The effects of mode coupling are usually neglected in calculations involving system rise time, and hence the pulse dispersion is assumed to be a linear function of the fiber length. This results in a pessimistic estimate for the system rise time and therefore provides a conservative value for the maximum possible bit rate.

Example 10.7

An optical fiber system is to be designed to operate over an 8 km length without repeaters. The rise times of the chosen components are:

Source (LED)	8 ns
Fiber: intermodal	5 ns km^{-1}
(pulse broadening) intramodal	1 ns km^{-1}
Detector (p–i–n photodiode)	6 ns

From system rise time considerations, estimate the maximum bit rate that may be achieved on the link when using an NRZ format.

Solution: The total system rise time is given by Eq. (10.43) as:

$$
\begin{aligned}
T_{\text{syst}} &= 1.1(T_{\text{S}}^2 + T_{\text{n}}^2 + T_{\text{c}}^2 + T_{\text{D}}^2)^{\frac{1}{2}} \\
&= 1.1(8^2 + (8 \times 5)^2 + (8 \times 1)^2 + 6^2)^{\frac{1}{2}} \\
&= 46.2 \text{ ns}
\end{aligned}
$$

Hence the maximum bit rate for the link using an NRZ format is given by Eq. (10.51) where:

$$B_T(\max) = \frac{0.7}{T_{\text{syst}}} = \frac{0.7}{46.2 \times 10^{-9}} \simeq 15.2 \text{ Mbit s}^{-1}$$

The rise time calculations indicate that the link will support a maximum bit rate of 15.2 Mbit s^{-1} which for an NRZ format is equivalent to a 3 dB optical bandwidth of 7.6 MHz (i.e. the NRZ format has two bit intervals per wavelength).

Once it is established that pulse dispersion is not a limiting factor, the major design exercise is the optical power budget for the system.

10.6.6 Optical Power Budgeting

Power budgeting for a digital optical fiber communication system is performed in a similar way to power budgeting within any communication system. When the transmitter characteristics, fiber cable losses and receiver sensitivity are known, the relatively simple process of power budgeting allows the repeater spacing or the maximum transmission distance for the system to be evaluated. However, it is necessary to incorporate a system margin into the optical power budget so that small variations in the system operating parameters do not lead to an unacceptable decrease in system performance. The operating margin is often included in a safety margin M_a which also takes into account possible source and modal noise together with receiver impairments such as equalization error, noise degradations and eye opening impairments. The safety margin depends to a large extent on the system components as well as the system design procedure and is typically in the range 5–10 dB. Systems using an injection laser transmitter generally require a larger safety margin (e.g. 9 dB) than those using an LED source (e.g. 7 dB) because the temperature variation and ageing of the LED are less pronounced.

The optical power budget for a system is given by the following expression:

$$P_i = P_o + C_L + M_a \text{ dB} \tag{10.52}$$

where P_i is the mean input optical power launched into the fiber, P_o is the mean incident optical power required at the receiver and C_L (or C_{LD} when there is a dispersion–equalization penalty) is the total channel loss given by Eq. (10.38) (or Eq. (10.42)). Therefore the expression given in Eq. (10.52) may be written as:

$$P_i = P_o + (\alpha_{fc} + \alpha_j)L + \alpha_{cr} + M_a \text{ dB} \tag{10.53}$$

Alternatively, when a dispersion–equalization penalty is included Eq. (10.52) becomes:

$$P_i = P_o + (\alpha_{fc} + \alpha_j)L + \alpha_{cr} + D_L + M_a \text{ dB} \tag{10.54}$$

Equations (10.53) and (10.54) allow the maximum link length without repeaters to be determined, as demonstrated in example 10.8.

Example 10.8

The following parameters are established for a long-haul single mode optical fiber system operating at a wavelength of 0.85 μm.

Mean power launched from the laser transmitter	−3 dBm
Cabled fiber loss	1.9 dB km⁻¹
Splice loss	0.3 dB km⁻¹
Connector losses at the transmitter and receiver	1 dB each
Mean power required at the APD receiver:	
when operating at 35 Mbit s⁻¹ (BER 10⁻¹)	−55 dBm
when operating at 500 Mbit s⁻¹ (BER 10⁻⁹)	−44 dBm
Required safety margin	9 dB

Estimate:

(a) the maximum possible link length without repeaters when operating at 35 Mbit s^{-1} (BER 10^{-9}). It may be assumed that there is no dispersion–equalization penalty at this bit rate.

(b) the maximum possible link length without repeaters when operating at 500 Mbit s^{-1} (BER 10^{-9}) and assuming no dispersion–equalization penalty.

(c) the reduction in the maximum possible link length without repeaters of (b) when there is a dispersion–equalization penalty of 5 dB. It may be assumed for the purposes of this estimate that the reduced link length has the 5 dB penalty.

Solution: (a) When the system is operating at 35 Mbit s^{-1} an optical power budget may be performed using Eq. (10.53), where

$$P_i - P_o = (\alpha_{fc} + \alpha_j)L + \alpha_{cr} + M_a \text{ dB}$$

$$-3 \text{ dBm} - (-55 \text{ dBm}) = (\alpha_{fc} + \alpha_j)L + \alpha_{cr} + M_a$$

Hence,

$$(\alpha_{fc} + \alpha_j)L = 52 - \alpha_{cr} - M$$

$$2.1L = 52 - 2 - 9$$

$$L = \frac{41}{2.1} = 19.5 \text{ km}$$

(b) Again using Eq. (10.53) when the system is operating at 500 Mbit s^{-1}:

$$-3 \text{ dBm} - (-44 \text{ dBm}) = (\alpha_{fc} + \alpha_j)L + \alpha_{cr} + M$$

$$(\alpha_{fc} + \alpha_j)L = 41 - 2 - 9$$

$$L = \frac{30}{2.1} = 14.3 \text{ km}$$

(c) Performing the optical power budget using Eq. (10.54) gives:

$$P_i - P_o = (\alpha_{fc} + \alpha_j)L + \alpha_{cr} + D_L + M_a$$

Hence,

$$2.1L = 41 - 2 - 5 - 9$$

and

$$L = \frac{25}{2.1} = 11.9 \text{ km}$$

Thus there is a reduction of 2.4 km in the maximum possible link length without repeaters.

Although we have demonstrated in example 10.8 the use of the optical power budget to determine the maximum link length without repeaters, it is also frequently used to aid decisions with relation to the combination of components required for a particular optical fiber communication system. In this case the maximum transmission distance and the required bandwidth may already be known. Therefore the optical power budget is used to provide a basis for optimization in the choice of the system components, whilst also establishing that a particular component configuration meets the system requirements.

Example 10.9

Components are chosen for a digital optical fiber link of overall length 7 km and operating at a 20 Mbit s^{-1} using an RZ code. It is decided that an LED emitting at 0.85 µm with graded index fiber to a *p–i–n* photodiode is a suitable choice for the system components, giving no dispersion–equalization penalty. An LED which is capable of launching an average of 100 µW of optical power (including the connector loss) into a 50 µm core diameter graded index fiber is chosen. The proposed fiber cable has an attenuation of 2.6 dB km^{-1} and requires splicing every kilometer with a loss of 0.5 dB per splice. There is also a connector loss at the receiver of 1.5 dB. The receiver requires mean incident optical power of −41 dBm in order to give the necessary BER of 10^{-10}, and it is predicted that a safety margin of 6 dB will be required.

Write down the optical power budget for the system and hence determine its viability.

Solution:

Mean optical power launched into the fiber from the transmitter (100 µW)	−10 dBm
Receiver sensitivity at 20 Mbit s^{-1} (BER 10^{-10})	−41 dBm
Total system margin	31 dB
Cabled fiber loss (7 × 2.6 dB km^{-1})	18.2 dB
Splice losses (6 × 0.5 dB)	3.0 dB
Connector loss (1 × 1.5 dB)	1.5 dB
Safety margin	6.0 dB
Total system loss	28.7 dB
Excess power margin	2.3 dB

Based on the figures given the system is viable and provides a 2.3 dB excess power margin. This could give an extra safety margin to allow for possible future splices if these were not taken into account within the original safety margin.

10.6.7 Line Coding

The previous discussions of digital system design have assumed that only information bits are transmitted, and that the 0 and 1 symbols are equally likely. However, within digital line transmission there is a requirement for redundancy in the line coding to provide efficient timing recovery and synchronization (frame alignment) as well as possible error detection and correction at the receiver. Line coding also provides suitable shaping of the transmitted signal power spectral density. Hence the choice of line code is an important consideration within digital optical fiber system design.

Binary line codes are generally preferred because of the large bandwidth available in optical fiber communications. In addition these codes are less susceptible to any temperature dependence of optical sources and detectors. Under these conditions two level codes are more suitable than codes which utilize an increased number of levels (multilevel codes). Nevertheless, these factors do not entirely exclude the use of multilevel codes, and it is likely that ternary codes (three levels 0, $\frac{1}{2}$, 1) which give increased information transmission per symbol over binary codes will be considered for some system applications. The corresponding symbol transmission rate (i.e. bit rate) for a ternary code may be reduced by a factor of 1.58 ($\log_2 3$) whilst still providing the same information transmission rate as a similar system using a binary code. It must be noted that this gain in information capacity for a particular bit rate is obtained at the expense of the dynamic range between adjacent levels as there are three levels inserted in place of two. This is exhibited as a 3 dB SNR penalty at the receiver when compared with a binary system at a given BER. Therefore ternary codes (and higher multilevel codes) are not attractive for long-haul systems.

For the reasons described above most digital optical fiber communication systems currently in use employ binary codes. In practice, binary codes are designed which insert extra symbols into the information data stream on a regular and logical basis to minimize the number of consecutive identical received symbols, and to facilitate efficient timing extraction at the receiver by producing a high density of decision level crossings. The reduction in consecutive identical symbols also helps to minimize the variation in the mean signal level which provides a reduction in the low frequency response requirement of the receiver. This shapes the transmitted signal spectrum by reducing the d.c. component. However, this factor is less important for optical fiber systems where a.c. coupling is performed with capacitors unlike metallic cable systems where transformers are often used, and the avoidance of d.c. components is critical. A further advantage is apparent within the optical receiver with a line code which is free from long identical symbol sequences, and where the continuous presence of 0 and 1 levels aids decision level control and avoids gain instability effects.

Two level block codes of the $nBmB$ type fulfil the above requirements through the addition of a limited amount of redundancy. These codes convert blocks of n bits into blocks of m bits where $m > n$ so that the difference between the number of transmitted ones and zeros is on average zero. A simple code of this type is the $1B2B$ code in which a 0 may be transmitted as 01, and a 1 as 10. This encoding format is shown in Fig. 10.39(b) and is commonly referred to as biphase or Manchester encoding. It may be observed that with this code there are never more than two consecutive identical symbols, and that two symbols must be transmitted for one information bit, giving 50% redundancy. Thus twice the transmission bandwidth is required for the $1B2B$ code which restricts its use to systems where pulse dispersion is not a limiting factor. Another example of a $1B2B$ code which is illustrated in Fig. 10.39(c) is the coded mark inversion (CMI) code. In this code a digit 0 is transmitted as 01 and the digit 1 alternately as 00 or 11.

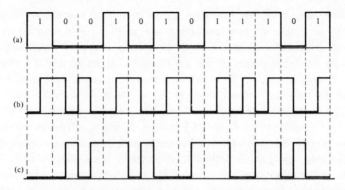

Fig. 10.39 Examples of binary 1*B*2*B* codes used in optical fiber communications: (a) unencoded NRZ data; (b) biphase or Manchester encoding; (c) coded mark inversion (CMI) encoding.

Timing information is obtained from the frequent positive to negative transitions, but once again the code is highly redundant requiring twice as many transmitted bits as input information bits.

More efficient codes of this type requiring less redundancy exist such as the 3*B*4*B*, 5*B*6*B* and the 7*B*8*B* codes. There is a trade-off within this class of code between the complexity of balancing the number of zeros and ones, and the added redundancy. The increase in line symbol rate (bit rate) and the corresponding power penalty over uncoded binary transmission is given by the ratio $m:n$. Hence, considering the widely favored 5*B*6*B* code, the symbol rate is increased by a factor of 1.2 whilst the power penalty is also equal to 1.2 or about 0.8 dB. It is therefore necessary to take into account the increased bandwidth requirement and the power penalty resulting from coding within the optical fiber system design.

Simple error monitoring may be provided with block codes, at the expense of a small amount of additional redundancy, by parity checking. Each block of *N* bits can be made to have an even (even parity) or odd (odd parity) number of ones so that any single error in a block can be identified. More extensive error detection and error correction may be provided with increased redundancy and equipment complexity. This is generally not considered worthwhile unless it is essential that the digital transmission system is totally secure (e.g. data transmission applications). Alternatively, error monitoring when using block codes may be performed by measuring the variation in disparity between the numbers of ones and zeros within the received bit pattern. Any variation in the accumulated disparity above an upper limit or below a lower limit allowed by a particular code is indicated as an error. Further discussion of error correction with relation to disparity may be found in Ref. 65.

10.7 ANALOG SYSTEMS

In Section 10.5 we indicated that the vast majority of optical fiber communication systems are designed to convey digital information (e.g. analog speech encoded as PCM). However, in certain areas of the telecommunication network or for particular applications, information transfer in analog form is still likely to remain for some time to come, or be advantageous. Therefore, analog optical fiber transmission will undoubtedly have a part to play in future communication networks, especially in situations where the optical fiber link is part of a larger analog network (e.g. microwave relay network). Use of analog transmission in these areas avoids the cost and complexity of digital terminal equipment, as well as degradation due to quantization noise. This is especially the case with the transmission of video signals over short distances where the cost of high speed analog to digital (A–D) and D–A converters is not generally justified. Hence, there are many applications such as direct cable television and common antenna television (CATV) where analog optical fiber systems may be utilized.

There are limitations, however, inherent to analog optical fiber transmission, some of which have been mentioned previously. For instance, the unique requirements of analog transmission over digital are for high signal to noise ratios at the receiver output which necessitates high optical input power (see Section 9.2.5), and high end to end linearity to avoid distortion and prevent cross talk between different channels of a multiplexed signal (see Section 10.4.2). Furthermore, it is instructive to compare the SNR constraints for typical analog optical fiber and coaxial cable systems.

In a coaxial cable system the fundamental limiting noise is $4KTB$, where K is Boltzmann's constant, T is the absolute temperature, and B is the effective noise bandwidth for the channel. If we assume for simplicity that the coaxial cable loss is constant and independent of frequency, the SNR for a coaxial system is

$$\left(\frac{S}{N}\right)_{coax} = \frac{V^2 \exp(-\alpha_N)}{Z_0 4KTB} \tag{10.55}$$

where α_N is the attenuation in nepers between the transmitter and receiver, V is the peak output voltage, and Z_0 is the impedance of the coaxial cable.

The SNR for an analog optical fiber system may be obtained by referring to Eq. (9.11) where

$$\left(\frac{S}{N}\right)_{fiber} = \frac{\eta P_0}{2hfB} \tag{10.56}$$

The expression given in Eq. (10.56) includes the fundamental limiting noise for optical fiber systems which is $2hfB$. Although Eq. (10.56) is sufficiently accurate for the purpose of comparison it applies to an unmodulated optical carrier. A more accurate expression would take into account the depth of modulation for the analog optical fiber system which cannot be unity [Ref. 53].* The average received optical power P_0 may be expressed in terms of the average input (transmitted) optical power P_i as

$$P_0 = P_i \exp(-\alpha_N) \tag{10.57}$$

Substituting for P_0 into Eq. (10.56) gives

$$\left(\frac{S}{N}\right)_{fiber} = \frac{P_i \exp(-\alpha_N)}{2hfB} \tag{10.58}$$

Equations (10.55) and (10.58) allow a simple comparison to be made of available SNR (or CNR) between analog coaxial and optical fiber systems as demonstrated in example 10.10.

Example 10.10

A coaxial cable system operating at a temperature of 17 °C has a transmitter peak output voltage of 5 V with a cable impedance of 100 Ω. An analog optical fiber system uses an injection laser source emitting at 0.85 μm and launches an average of 1 mW of optical power into the fiber cable. The optical receiver comprises a photodiode with a quantum efficiency of 70%. Assuming the effective noise bandwidth and the attenuation between the transmitter and receiver for the two systems is identical, estimate in decibels the ratio of the SNR of the coaxial system to the SNR of the fiber system.

Solution: Using Eqs. (10.55) and (10.58) for the SNRs of the coaxial and fiber systems respectively:

$$\text{Ratio} = \frac{\left(\dfrac{S}{N}\right)_{coax}}{\left(\dfrac{S}{N}\right)_{fiber}} = \frac{\dfrac{V^2 \exp(-\alpha_N)}{Z_0 4KTB}}{\dfrac{\eta P_i \exp(-\alpha_N)}{2hfB}} = \frac{V^2 hf}{2KTZ_0 \eta P_i}$$

$$= \frac{V^2 hc}{2KTZ_0 \eta P_i \lambda}$$

Hence,

$$\text{Ratio} = \frac{25 \times 6.626 \times 10^{-34} \times 2.998 \times 10^8}{2 \times 1.385 \times 10^{-23} \times 290 \times 100 \times 0.7 \times 1 \times 10^{-3} \times 0.85 \times 10^{-6}}$$

$$= 1.04 \times 10^4 \simeq 40 \text{ dB}$$

---

* Strictly speaking Eq. (10.56) depicts the optical carrier to noise ratio (CNR).

The optical fiber channel in example 10.10 has around 40 dB less SNR available than the alternative coaxial channel exhibiting similar channel losses. This results both from $2hfB$ being larger than $4KTB$ and from the far smaller transmitted power with the optical system. Furthermore it must be noted that the comparison was made using an injection laser transmitter. If an LED transmitter with 10–20 dB less optical output power was compared, the coaxial system would display an advantage in the region 50–60 dB. For this reason it is difficult to match with fiber systems the SNR requirements of some analog coaxial links, even though the fiber cable attenuation may be substantially lower than that of the coaxial cable.

The analog signal can be transmitted within an optical fiber communication system using one of several modulation techniques. The simplest form of analog modulation for optical fiber communications is direct intensity modulation (D–IM) of the optical source. In this technique the optical output from the source is modulated simply by varying the current flowing in the device around a suitable bias or mean level in proportion to the message. Hence the information signal is transmitted directly in the baseband.

Alternatively the baseband signal can be translated onto an electrical subcarrier by means of amplitude, phase or frequency modulation using standard techniques, prior to intensity modulation of the optical source. Pulse analog techniques where a sequence of pulses are used for the carrier may also be utilized. In this case a suitable parameter such as the pulse amplitude, pulse width, pulse position or pulse frequency is electrically modulated by the baseband signal. Again the modulated electrical carrier is transmitted optically by intensity modulation of the optical source.

Direct modulation of the optical source in frequency, phase or polarization rather than by intensity requires these parameters to be well defined throughout the optical fiber system. Although there is much interest in this area (see Section 10.8) present optical component technology does not as yet completely provide for practical system implementation.

10.7.1 Direct Intensity Modulation (D–IM)

A block schematic for an analog optical fiber system which uses direct modulation of the optical source intensity with the baseband signal is shown in Fig. 10.40(a). Obviously no electrical modulation or demodulation is required with this technique, making it both inexpensive and easy to implement.

The transmitted optical power waveform as a function of time $P_{opt}(t)$, an example of which is illustrated in Fig. 10.40(b), may written as:

$$P_{opt}(t) = P_i(1 + m(t)) \qquad (10.59)$$

where P_i is the average transmitted optical power (i.e. the unmodulated carrier power) and $m(t)$ is the intensity modulating signal which is proportional to the source message $a(t)$. For a cosinusoidal modulating signal:

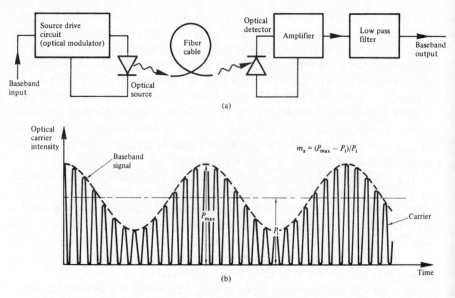

Fig. 10.40 (a) Analog optical fiber system employing direct intensity modulation. (b) Time domain representation showing direct intensity modulation of the optical carrier with a baseband analog signal.

$$m(t) = m_a \cos \omega_m t \qquad (10.60)$$

where m_a is the modulation index or the ratio of the peak excursion from the average to the average power as shown in Fig. 10.40(b), and ω_m is the angular frequency of the modulating signal. Combining Eqs. (10.59) and (10.60) we get:

$$P_{opt}(t) = P_i(1 + m_a \cos \omega_m t) \qquad (10.61)$$

Furthermore, assuming the transmission medium has zero dispersion, the received optical power will be of the same form as Eq. (10.61), but with an average received optical power P_o. Hence the secondary photocurrent $I(t)$ generated at an APD receiver with a multiplication factor M is given by:

$$I(t) = I_p M(1 + m_a \cos \omega_m t) \qquad (10.62)$$

where the primary photocurrent obtained with an unmodulated carrier I_p is given by Eq. (8.8) as,

$$I_p = \frac{\eta e}{hf} P_o \qquad (10.63)$$

The mean square signal current $\overline{i_{sig}^2}$ which is obtained from Eq. (10.62) is given by:

$$\overline{i_{\text{sig}}^2} = \tfrac{1}{2}(m_a M I_p)^2 \qquad (10.64)$$

The total average noise in the system is composed of quantum, dark current, and thermal (circuit) noise components. The noise contribution from quantum effects and detector dark current may be expressed as the mean square total shot noise current for the APD receiver $\overline{i_{\text{SA}}^2}$ given by Eq. (9.21) where the excess avalanche noise factor is written following Eq. (9.26) as $F(M)$ such that:

$$\overline{i_{\text{SA}}^2} = 2eB(I_p + I_d)M^2 F(M) \qquad (10.65)$$

where B is the effective noise or post detection bandwidth.

Thermal noise generated by the load resistance R_L and the electronic amplifier noise can be expressed in terms of the amplifier noise figure F_n referred to R_L as given by Eq. (9.17). Thus the total mean square noise current $\overline{i_N^2}$ may be written as:

$$\overline{i_N^2} = 2eB(I_p + I_d)M^2 F(M) + \frac{4KTBF_n}{R_L} \qquad (10.66)$$

The SNR defined in terms of the ratio of the mean square signal current to the mean square noise current (rms signal power to rms noise power) for the APD receiver is therefore given by:

$$\left(\frac{S}{N}\right)_{\text{rms}} = \frac{\overline{i_{\text{sig}}^2}}{\overline{i_N^2}} = \frac{\tfrac{1}{2}(m_a M I_p)^2}{2eB(I_p + I_d)M^2 F(M) + (4KTBF_n/R_L)} \quad \text{(APD)} \qquad (10.67)$$

It must be emphasized that the SNR given in Eq. (10.67) is defined in terms of rms signal power rather than peak signal power used previously. When a unity gain photodetector is utilized in the receiver (i.e. p–i–n photodiode) Eq. (10.67) reduces to:

$$\left(\frac{S}{N}\right)_{\text{rms}} = \frac{\tfrac{1}{2}(m_a I_p)^2}{2eB(I_p + I_d) + (4KTBF_n/R_L)} \quad (p\text{–}i\text{–}n) \qquad (10.68)$$

Moreover, the SNR for video transmission is often defined in terms of the peak to peak picture signal power to the rms noise power and may include the ratio of luminance to composite video b. Using this definition in the case of the unity gain detector gives:

$$\left(\frac{S}{N}\right)_{\text{p-p}} = \frac{(2m_a I_p b)^2}{2eB(I_p + I_d) + (4KTBF_n/R_L)} (p\text{–}i\text{–}n) \qquad (10.69)$$

It may be observed that, excluding b, the SNR defined in terms of the peak to peak signal power given in Eq. (10.69) is a factor of 8 (or 9 dB) greater than that defined in Eq. (10.68).

Example 10.11

A single TV channel is transmitted over an analog optical fiber link using direct intensity modulation. The video signal which has a bandwidth of 5 MHz and a ratio of luminance to composite video of 0.7 is transmitted with a modulation index of 0.8. The receiver contains a p–i–n photodiode with a responsivity of 0.5 A W^{-1} and a preamplifier with an effective input impedance of 1 MΩ together with a noise figure of 1.5 dB. Assuming the receiver is operating at a temperature of 20 °C and neglecting the dark current in the photodiode, determine the average incident optical power required at the receiver (i.e. receiver sensitivity) in order to maintain a peak to peak signal power to rms noise power ratio of 55 dB.

Solution: Neglecting the photodiode dark current, the peak to peak signal power to rms noise power ratio is given following Eq. (10.69) as:

$$\left(\frac{S}{N}\right)_{p-p} = \frac{(2 m_a I_p b)^2}{2 e B I_p + (4 K T B F_n / R_L)}$$

The photocurrent I_p may be expressed in terms of the average incident optical power at the receiver P_o using Eq. (8.4) as:

$$I_p = R P_o$$

where R is the responsivity of the photodiode. Hence,

$$\left(\frac{S}{N}\right)_{p-p} = \frac{(2 m_a R P_o b)^2}{2 e B R P_o + (4 K T B F_n / R_L)}$$

and

$$\left(\frac{S}{N}\right)_{p-p} \left(2 e B R P_o + \frac{4 K T B F_n}{R_L}\right) = (2 m_a R P_o b)^2$$

Rearranging,

$$(2 m_a R b)^2 P_o^2 - \left(\frac{S}{N}\right)_{p-p} 2 e B R P_o - \left(\frac{S}{N}\right)_{p-p} \frac{4 K T B F_n}{R_L} = 0$$

where

$$(2 m_a R b)^2 = 4 \times 0.64 \times 0.25 \times 0.49$$

$$= 0.314$$

$$\left(\frac{S}{N}\right)_{p-p} 2 e B R = 3.162 \times 10^5 \times 2 \times 1.602 \times 10^{-19} \times 5 \times 10^6 \times 0.5$$

$$= 2.533 \times 10^{-7}$$

$$\left(\frac{S}{N}\right)_{p-p} \frac{4 K T B F_n}{R_L} = \frac{3.162 \times 10^5 \times 4 \times 1.381 \times 10^{-23} \times 293 \times 5 \times 10^6 \times 1.413}{10^6}$$

$$= 3.616 \times 10^{-14}$$

Therefore,

$$0.314 P_o^2 - 2.533 \times 10^{-7} P_o - 3.616 \times 10^{-14} = 0$$

and

$$P_o = \frac{2.533 \times 10^{-7} \pm \sqrt{[(2.533 \times 10^{-7})^2 - (-4 \times 0.314 \times 3.616 \times 10^{-14})]}}{0.628}$$

$$= 0.93 \ \mu W$$

$$= -30.3 \ dBm$$

It must be noted that the low noise preamplification depicted in example 10.11 may not always be obtained, and that higher thermal noise levels will adversely affect the receiver sensitivity for a given SNR. This is especially the case with lower SNRs as illustrated in the peak to peak signal power to rms noise power ratio against average received optical power characteristics for a video system shown in Fig. 10.41. The performance of the system for various values of mean square thermal noise current $\overline{i_t^2} = 4KTBF_n/R_L$, where $\overline{i_t^2}$ is expressed as a spectral density in $A^2 \ Hz^{-1}$, is indicated. The value for the receiver sensitivity obtained in example 10.11 is approaching the quantum limit, also illustrated in Fig. 10.41, which is the best that could possibly be achieved with a noiseless amplifier.

The quantum or shot noise (when ignoring the photodetector dark current) limit occurs with large values of signal current (i.e. primary photocurrent) at the receiver. Considering a p–i–n photodiode receiver, this limiting case which

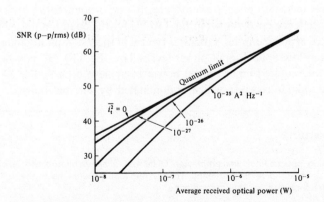

Fig. 10.41   Peak to peak signal power to rms noise power ratio against the average received optical power for a direct intensity modulated video system and various levels of thermal noise given by $\overline{i_t^2}$. Reproduced with permission from G. G. Windus, *Marconi Rev.*, **XLIV**, p. 77, 1981.

corresponds to large SNR is given by Eq. (10.68) when neglecting the device dark current as:

$$\left(\frac{S}{N}\right)_{rms} \simeq \frac{m_a^2 I_p}{4eB} \quad \text{(quantum noise limit)} \tag{10.70}$$

Using the relationship between the average received optical power P_o and the primary photocurrent given in Eq. (10.63) allows Eq. (10.70) to be expressed as:

$$P_o \simeq \frac{4hf}{m_a^2 \eta} \left(\frac{S}{N}\right)_{rms} B \tag{10.71}$$

Equation (10.71) indicates that for a quantum noise limited analog receiver, the optical input power is directly proportional to the effective noise or post detection bandwidth B. A similar result was obtained in Eq. (10.37) for the digital receiver.

Alternatively at low SNRs thermal noise is dominant, and the thermal noise limit when I_p is small, which may also be obtained from Eq. (10.68), is given by:

$$\left(\frac{S}{N}\right)_{rms} \simeq \frac{(m_a I_p)^2 R_L}{8KTBF_n} \quad \text{(thermal noise limit)} \tag{10.72}$$

Again substituting for I_p from Eq. (10.63) gives:

$$P_o \simeq \frac{hf}{e\eta m_a^2} \left(\frac{8KTF_n}{R_L}\right)^{\frac{1}{2}} \left(\frac{S}{N}\right)_{rms}^{\frac{1}{2}} B^{\frac{1}{2}} \tag{10.73}$$

Therefore it may be observed from Eq. (10.73) that in the thermal noise limit the average incident optical power is directly proportional to $B^{\frac{1}{2}}$ instead of the direct dependence on B shown in Eq. (10.71) for the quantum noise limit. The dependence expressed in Eq. (10.73) is typical of the p–i–n photodiode receiver operating at low optical input power levels. Thus Eq. (10.73) may be used to estimate the required input optical power to achieve a particular SNR for a p–i–n photodiode receiver which is dominated by thermal noise.

Example 10.12

An analog optical fiber link employing D–IM has a p–i–n photodiode receiver in which thermal noise is dominant. The system components have the following characteristics and operating conditions.

p–i–n photodiode quantum efficiency	60%
effective load impedance for the photodiode	50 kΩ
preamplifier noise figure	6 dB

operating wavelength	1 μm
operating temperature	300 K
receiver post detection bandwidth	10 MHz
modulation index	0.5

Estimate the required average incident optical power at the receiver in order to maintain an SNR, defined in terms of the mean square signal current to mean square noise current, of 45 dB.

Solution: The average incident optical power for a thermal noise limited p–i–n photodiode receiver may be estimated using Eq. (10.73) where:

$$P_o \simeq \frac{hf}{e\eta m_a^2} \left(\frac{8KTF_n}{R_L} \right)^{\frac{1}{2}} \left(\frac{S}{N} \right)^{\frac{1}{2}}_{rms} B^{\frac{1}{2}}$$

and

$$\frac{hf}{e\eta m_a^2} = \frac{hc}{e\eta m_a^2 \lambda} = \frac{6.626 \times 10^{-34} \times 2.998 \times 10^8}{1.602 \times 10^{-19} \times 0.6 \times 0.25 \times 1 \times 10^{-6}}$$

$$= 8.267$$

$$\left(\frac{8KTF_n}{R_L} \right)^{\frac{1}{2}} = \left(\frac{8 \times 1.381 \times 10^{-23} \times 300 \times 4}{50 \times 10^3} \right)^{\frac{1}{2}}$$

$$= 1.628 \times 10^{-12}$$

$$\left(\frac{S}{N} \right)^{\frac{1}{2}}_{rms} B^{\frac{1}{2}} = (3.162 \times 10^4 \times 10^7)^{\frac{1}{2}}$$

$$= 5.623 \times 10^5$$

Hence,

$$P_o \simeq 8.267 \times 1.628 \times 10^{-12} \times 5.623 \times 10^5$$

$$= 7.57 \ \mu W$$

$$= -21.2 \ dBm$$

Therefore, as anticipated, the receiver sensitivity in the thermal noise limit is low.

10.7.2 System Planning

Many of the general planning considerations for optical fiber systems outlined in Section 10.4 may be applied to analog transmission. However, extra care must be taken to ensure that the optical source and, to a lesser extent, the detector have linear input–output characteristics, in order to avoid distortion of the transmitted optical signal. Furthermore, careful optical power budgeting is often necessary with analog systems because of the generally high SNRs required at the optical receiver (40–60 dB) in comparison with digital systems (20–25 dB), to obtain a similar fidelity. Therefore, although analog system optical power budgeting may be carried out in a similar manner to digital systems (see Section 10.6.6), it is common for the system margin, or the difference between the optical power launched into the fiber and the required optical power at the receiver, for analog systems to be quite small (perhaps only 10–20 dB when using an LED source to $p-i-n$ photodiode receiver). Consequently analog systems employing direct intensity modulation of the optical source tend to have a limited transmission distance without repeaters which generally prohibits their use for long-haul applications.

Example 10.13

A D–IM analog optical fiber link of length 2 km employs an LED which launches mean optical power of −10 dBm into a multimode optical fiber. The fiber cable exhibits a loss of 3.5 dB km^{-1} with splice losses calculated at 0.7 dB km^{-1}. In addition there is a connector loss at the receiver of 1.6 dB. The $p-i-n$ photodiode receiver has a sensitivity of −25 dBm for an SNR (i_{sig}^2/i_N^2) of 50 dB and with a modulation index of 0.5. It is estimated that a safety margin of 4 dB is required. Assuming there is no dispersion–equalization penalty:

(a) Perform an optical power budget for the system operating under the above conditions and ascertain its viability.
(b) Estimate any possible increase in link length which may be achieved using an injection laser source which launches mean optical power of 0 dBm into the fiber cable. In this case the safety margin must be increased to 7 dB.

Solution: (a) Optical power budget:

Mean power launched into the fiber cable from the LED transmitter	−10 dBm
Mean optical power required at the $p-i-n$ photodiode receiver for SNR of 50 dB and a modulation index of 0.5	−25 dBm
Total system margin	**15 dB**
Fiber cable loss (2 × 3.5)	7.0 dB
Splice losses (2 × 0.7)	1.4 dB
Connector loss at the receiver	1.6 dB
Safety margin	4.0 dB
Total system loss	**14.0 dB**
Excess power margin	**1.0 dB**

Hence the system is viable, providing a small excess power margin.

(b) In order to calculate any possible increase in link length when using the injection laser source we refer to Eq. (10.53), where

$$P_i - P_0 = (\alpha_{fc} + \alpha_j)L + \alpha_{cr} + M_a \text{ dB}$$

Therefore,

$$0 \text{ dBm} - (-25 \text{ dBm}) = (3.5 + 0.7)L + 1.6 + 7.0$$

and

$$4.2L = 25 - 8.6 = 16.4 \text{ dB}$$

giving

$$L = \frac{16.4}{4.2} = 3.9 \text{ km}$$

Hence the use of the injection laser gives a possible increase in the link length of 1.9 km or almost a factor of 2. It must be noted that in this case the excess power margin has been reduced to zero.

The transmission distance without repeaters for the analog link of example 10.13 could be extended further by utilizing an APD receiver which has increased sensitivity. This could facilitate an increase in the maximum link length to around 7 km, assuming no additional power penalties or excess power margin. Although this is quite a reasonable transmission distance, it must be noted that a comparable digital system could give in the region of 13 km transmission without repeaters.

The temporal response of analog systems may be determined from system rise time calculations in a similar manner to digital systems (see Section 10.6.5). The maximum permitted 3 dB optical bandwidth for analog systems in order to avoid dispersion penalties follows from Eq. (10.49) and is given by:

$$B_{opt}(max) = \frac{0.35}{T_{syst}} \qquad (10.74)$$

Hence calculation of the total system 10–90% rise time T_{syst} allows the maximum system bandwidth to be estimated. Often this calculation is performed in order to establish that the desired system bandwidth may be achieved using a particular combination of system components.

Example 10.14

The 10–90% rise times for possible components to be used in a D–IM analog optical fiber link are specified below:

Source (LED)	10 ns
Fiber cable: intermodal	9 ns km^{-1}
intramodal	2 ns km^{-1}
Detector (APD)	3 ns

The desired link length without repeaters is 5 km and the required optical bandwidth is 6 MHz. Determine whether the above combination of components give an adequate temporal response.

Solution: Equation (10.74) may be used to calculate the maximum permitted system rise time which gives the desired bandwidth where:

$$T_{syst}(max) = \frac{0.35}{B_{opt}} = \frac{0.35}{6 \times 10^6} = 58.3 \text{ ns}$$

The total system rise time using the specified components can be estimated using Eq. (10.43) as:

$$T_{syst} = 1.1(T_S^2 + T_n^2 + T_c^2 + T_D^2)^{\frac{1}{2}}$$

$$= 1.1(10^2 + (9 \times 5)^2 + (2 \times 5)^2 + 3^2)^{\frac{1}{2}}$$

$$\simeq 52 \text{ ns}$$

Therefore the specified components give a system rise time which is adequate for the bandwidth and distance requirements of the optical fiber link. However, there is little leeway for upgrading the system in terms of bandwidth or distance without replacing one or more of the system components.

10.7.3 Subcarrier Intensity Modulation

Direct intensity modulation of the optical source is suitable for the transmission of a baseband analog signal. However, if the wideband nature of the optical fiber medium is to be fully utilized it is essential that a number of baseband channels are multiplexed onto a single fiber link. This may be achieved with analog transmission through frequency division multiplexing of the individual baseband channels. Initially, the baseband channels must be translated onto carriers of different frequency by amplitude modulation (AM), frequency modulation (FM) or phase modulation (PM) prior to being simultaneously transmitted as an FDM signal. The frequency translation may be performed in the electrical regime where the baseband analog signals modulate electrical subcarriers and are then frequency division multiplexed to form a composite electrical signal prior to intensity modulation of the optical source.

A block schematic of an analog system employing this technique which is known as subcarrier intensity modulation is shown in Fig. 10.42. The baseband signals are modulated onto radiofrequency (RF) subcarriers by either AM, FM or PM and multiplexed before being applied to the optical source drive circuit. Hence an intensity modulated (IM) optical signal is obtained which may be either AM–IM, FM–IM or PM–IM. In practice, however, system output SNR considerations dictate that generally only the

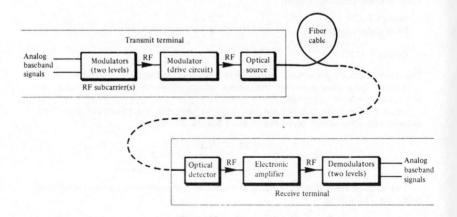

Fig. 10.42   Subcarrier intensity modulation system for analog optical fiber transmission.

latter two modulation formats are used. Nevertheless systems may incorporate two levels of electrical modulation whereby the baseband channels are initially amplitude modulated prior to frequency or phase modulation [Ref. 66]. The FM or PM signal thus obtained is then used to intensity modulate the optical source. At the receive terminal the transmitted optical signal is detected prior to electrical demodulation and demultiplexing (filtering) to obtain the originally transmitted baseband signals.

A further major advantage of subcarrier intensity modulation is the possible improvement in SNR that may be obtained during subcarrier demodulation. In order to investigate this process it is necessary to obtain a general expression for the SNR of the intensity modulated optical carrier which may then be applied to the subcarrier intensity modulation formats. Therefore, as with D–IM, considered in the previous section, an electrical signal $m(t)$ modulates the source intensity. The transmitted optical power waveform is of the same form as Eq. (10.59), where:

$$P_{\text{opt}}(t) = P_{\text{i}}(1 + m(t)) \tag{10.75}$$

Also the secondary photocurrent $I(t)$ generated at an APD receiver following Eq. (10.62) is given by:

$$I(t) = I_{\text{p}}M(1 + m(t)) \tag{10.76}$$

The mean square signal current $\overline{i_{\text{sig}}^2}$ may be written as [Ref. 65]:

$$\overline{i_{\text{sig}}^2} = (I_{\text{p}}M)^2 P_m \tag{10.77}$$

where P_m is the total power of $m(t)$, which can be defined in terms of the spectral density $S_m(\omega)$ of $m(t)$ occupying a one-sided bandwidth B_m Hz as:

$$P_m = \frac{1}{2\pi} \int_{-2\pi B_m}^{2\pi B_m} S_m(\omega)\, d\omega \tag{10.78}$$

Hence the SNR defined in terms of the mean square signal current to mean square noise current (i.e. rms signal power to rms noise power) using Eqs. (10.77) and (10.66) can now be written as:

$$\left(\frac{S}{N}\right)_{\text{rms}} = \frac{\overline{i_{\text{sig}}^2}}{\overline{i_{\text{N}}^2}} = \frac{(I_{\text{p}}M)^2 P_m}{2eB_m(I_{\text{p}} + I_{\text{d}})M^2 F(M) + (4KTBF_{\text{n}}/R_{\text{L}})}$$

$$= \frac{I_{\text{p}}^2 P_m}{2B_m e(I_{\text{p}} + I_{\text{d}})F(M) + (4KTBF_{\text{n}}/M^2 R_{\text{L}})}$$

$$= \frac{(RP_{\text{o}})^2 P_m}{2B_m N_{\text{o}}} \quad \text{(D–IM)} \tag{10.79}$$

where we substitute for I_p from Eq. (8.4) and for notational simplicity write:

$$N_o = e(I_p + I_d)F(M) + \frac{4KTBF_n}{M^2 R_L} \tag{10.80}$$

The result obtained in Eq. (10.79) gives the SNR for a direct intensity modulated optical source where the total modulating signal power is P_m. In this context Eq. (10.79) is simply a more general form of Eq. (10.67). However, we are now in a position to examine the signal to noise performance of various subcarrier intensity modulation formats.

10.7.4 Subcarrier Double Sideband Modulation (DSB–IM)

A simple way to translate the spectrum of the baseband message signal $a(t)$ is by direct multiplication with the subcarrier waveform $A_c \cos \omega_c t$ giving the modulated waveform $m(t)$ as:

$$m(t) = A_c a(t) \cos \omega_c t \tag{10.81}$$

where A_c is the amplitude, and ω_c the angular frequency of the subcarrier waveform. For a cosinusoidal modulating signal ($\cos \omega_m t$) the subcarrier electric field $E_m(t)$ becomes:

$$E_m(t) = \frac{A_c}{2} \cos (\omega_c + \omega_m)t + \cos (\omega_c - \omega_m)t \tag{10.82}$$

giving the upper and lower sidebands. The time and frequency domain representations of the modulated waveform are shown in Fig. 10.43. It may be

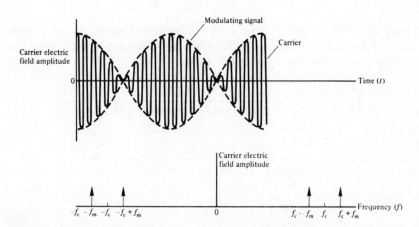

Fig. 10.43 Time and frequency domain representations of double sideband modulation.

observed from the frequency domain representation that only the two sideband components are present as indicated in Eq. (10.82). This modulation technique is known as double sideband modulation (DSB) or double sideband suppressed carrier (DSBSC) amplitude modulation. It provides a more efficient method of translating the spectrum of the baseband message signal than conventional full amplitude modulation where a large carrier component is also present in the modulated waveform.

The DSB signal shown in Fig. 10.43 intensity-modulates the optical source. Therefore the transmitted optical power waveform is obtained by combining Eqs. (10.75) and (10.81) where for simplicity we set the carrier amplitude A_c to unity, giving:

$$P_{opt}(t) = P_i(1 + a(t) \cos \omega_c t) \tag{10.83}$$

Furthermore, in order to prevent overmodulation, the value of the message signal is normalized such that $|a(t) \leqslant 1|$ with power $P_a \leqslant 1$. The DSB modulated electrical subcarrier occupies a bandwidth $B_m = 2B_a$, and with a carrier amplitude of unity, $P_m = P_a/2$. Hence, the ratio of rms signal power to rms noise power obtained within the subcarrier bandwidth at the input to the double sideband demodulator is given by Eq. (10.79), where:

$$\left(\frac{S}{N}\right)_{rms} \text{input DSB} = \frac{(RP_o)^2 P_a/2}{2 \times 2B_a N_o} = \frac{(RP_o)^2 P_a}{8B_a N_o} \tag{10.84}$$

However, an ideal DSB demodulator gives a detection gain of 2 or 3 dB improvement in SNR [Ref. 67]. This yields an output SNR of:

$$\left(\frac{S}{N}\right)_{rms} \text{output DSB} = 2 \left(\frac{S}{N}\right)_{rms} \text{input DSB} = \frac{(RP_o)^2 P_a}{4B_a N_o} \tag{10.85}$$

Comparison of the result obtained in Eq. (10.85) to that using direct intensity modulation of the baseband signal given by Eq. (10.79) shows a 3 dB degradation in SNR when employing DSB–IM under the same conditions of bandwidth (i.e. $B_m = B_a$), modulating signal power (i.e. $P_m = P_a$), detector photocurrent and noise. For this reason DSB–IM systems (and also AM–IM systems in general) are usually not considered efficient for optical fiber communications. Therefore far more attention is devoted to both FM–IM and PM–IM systems.

10.7.5 Subcarrier Frequency Modulation (FM–IM)

In this modulation format, the subcarrier is frequency modulated by the message signal. The conventional form for representing the baseband signal which intensity modulates the optical source is [Ref. 67]:

$$m(t) = A_c \cos \left[\omega_c t + k_f \int_0^t a(\tau) \, d\tau \right] \tag{10.86}$$

where k_f is the angular frequency deviation in radians per second per unit of $a(t)$. To prevent intensity over modulation, the carrier amplitude $A_c \leqslant 1$. The generally accepted expression for the bandwidth which is referred to as Carson's rule is given by:

$$B_m \simeq 2(D_f + 1)B_a \tag{10.87}$$

where D_f is the frequency deviation ratio defined by:

$$D_f = \frac{\text{peak frequency deviation}}{\text{bandwidth of } a(t)} = \frac{f_d}{B_a} \tag{10.88}$$

The peak frequency deviation in the subcarrier FM signal f_d is given by:

$$f_d = k_f \max |a(t)| \tag{10.89}$$

Hence the SNR at the input to the subcarrier FM demodulator is:

$$\left(\frac{S}{N} \right)_{\text{rms}} \text{input FM} = \frac{(RP_o)^2 (A_c^2/2)}{2B_m N_o} \tag{10.90}$$

The subcarrier demodulator operating above threshold yields an output SNR [Ref. 65]:

$$\left(\frac{S}{N} \right)_{\text{rms}} \text{output FM} = 6D_f^2(D_f^2 + 1) \frac{P_a(RP_o)^2(A_c^2/2)}{2B_m N_o} \tag{10.91}$$

Substituting for B_m from Eq. (10.87) gives:

$$\left(\frac{S}{N} \right)_{\text{rms}} \text{output FM} = \frac{3D_f^2 P_a(RP_o)^2(A_c^2/2)}{2B_a N_o} \tag{10.92}$$

The result obtained in Eq. (10.92) indicates that a significant improvement in the postdetection SNR may be achieved by using wideband FM–IM as demonstrated in the following example.

Example 10.15

(a) A D–IM and an FM–IM optical fiber communication system are operated under the same conditions of modulating signal power and bandwidth, detector photocurrent and noise. Furthermore, in order to maximize the SNR in the FM–IM system, the amplitude of the subcarrier is set to unity. Derive an expression for the improvement in post detection SNR of the FM–IM system over the D–IM system. It may be assumed that the SNR is defined in terms of the rms signal power to rms noise power.

(b) The FM–IM system described in (a) has an 80 MHz subcarrier which is modulated by a baseband signal with a bandwidth of 4 kHz such that the peak frequency deviation is 400 kHz. Use the result obtained in (a) to determine the improvement in post detection SNR (in decibels) over the D–IM system operating under the same conditions. Also estimate the bandwidth of the FM signal.

Solution: (a) The output SNR for the D–IM system is given by Eq. (10.79) where we can write $P_m = P_a$ and $B_m = B_a$. Hence:

$$\left(\frac{S}{N}\right)_{rms} \text{output D–IM} = \frac{(RP_0)^2 P_a}{2B_a N_0}$$

The corresponding output SNR for the FM–IM system is given by Eq. (10.92) where setting A_c to unity gives:

$$\left(\frac{S}{N}\right)_{rms} \text{output FM} = \frac{3D_f^2 P_a (RP_0)^2}{4B_a N_0}$$

Therefore the improvement in SNR of the FM–IM system over the D–IM system is given by:

$$\text{SNR improvement} = \frac{[3D_f^2 P_a (RP_0)^2]/(4B_a N_0)}{[(RP_0)^2 P_a]/(2B_a N_0)}$$

$$= \frac{3D_f^2}{2}$$

and,

$$\text{SNR improvement in decibels} = 10 \log_{10} \frac{3}{2} D_f^2$$

$$= 1.76 + 20 \log_{10} D_f$$

(b) The frequency deviation ratio is given by Eq. (10.88) where:

$$D_f = \frac{f_d}{B_a} = \frac{400 \times 10^3}{4 \times 10^3} = 100$$

Therefore the SNR improvement is:

$$\text{SNR improvement} = 1.76 + 20 \log_{10} 100$$

$$= 41.76 \text{ dB}$$

The bandwidth of the FM–IM signal may be estimated using Eq. (10.87) where:

$$B_m \simeq 2(D_f + 1)B_a = 2(100 + 1)4 \times 10^3$$

$$= 808 \text{ kHz}$$

This result indicates that the system is operating as a wideband FM–IM system.

Example 10.15 illustrates that a substantial improvement in the post detection SNR over D–IM may be obtained using FM–IM. However, it must be noted that this is at the expense of a tremendous increase in the bandwidth required (808 kHz) for transmission of the 4 kHz baseband channel.

10.7.6 Subcarrier Phase Modulation (PM–IM)

With this modulation technique the instantaneous phase of the subcarrier is set proportional to the modulating signal. Hence in a PM–IM system the modulating signal $m(t)$ may be written as [Ref. 67]:

$$m(t) = A_c \cos(\omega_c t + k_p a(t)) \tag{10.93}$$

where k_p is the phase deviation constant in radians per unit of $a(t)$. Again the carrier amplitude $A_c \leqslant 1$ to prevent intensity overmodulation. Moreover, the bandwidth of the PM–IM signal is given by Carson's rule as:

$$B_m \simeq 2(D_p + 1)B_a \tag{10.94}$$

where D_p is the frequency deviation ratio for the PM–IM system. In common with subcarrier frequency modulation the frequency deviation ratio is defined as:

$$D_p = \frac{f_d}{B_a} \tag{10.95}$$

where f_d is the peak frequency deviation of the subcarrier PM signal, which is given by:

$$f_d = k_p \max \left| \frac{da(t)}{dt} \right| \tag{10.96}$$

The SNR at the input to the subcarrier PM demodulator is:

$$\left(\frac{S}{N} \right)_{rms} \text{input PM} = \frac{(RP_o)^2 A_c^2 / 2}{2 B_m N_o} \tag{10.97}$$

The output SNR from an ideal subcarrier PM demodulator operating above threshold is [Ref. 65]:

$$\left(\frac{S}{N} \right)_{rms} \text{output PM} = \frac{D_p^2 P_a (RP_o)^2 A_c^2 / 2}{2 B_a N_o} \tag{10.98}$$

The result given in Eq. (10.98) suggests that an improvement in SNR over D–IM may be obtained using PM–IM, especially when the SNR is maximized with $A_c = 1$. However, comparison of PM–IM with FM–IM indicates that the latter modulation format gives the greatest improvement.

Example 10.16

A PM–IM and an FM–IM optical fiber communication system are operated under the same conditions of bandwidth, baseband signal power, subcarrier amplitude, frequency deviation, detector photocurrent and noise. Assuming the demodulators for both systems are ideal, determine the ratio (in decibels) of the output SNR from the FM–IM system to the output SNR from the PM–IM system.

Solution: The output SNR from the FM–IM system is given by Eq. (10.92) where:

$$\left(\frac{S}{N}\right)_{rms} \text{output FM} = \frac{3D_f^2 P_a (RP_o)^2 A_c^2/2}{2B_a N_o}$$

Substituting for D_f from Eq. (10.88) gives:

$$\left(\frac{S}{N}\right)_{rms} \text{output FM} = \frac{3f_d^2 P_a (RP_o)^2 A_c^2/2}{2B_a^3 N_o}$$

The output SNR for the PM–IM system is given by Eq. (10.98) where:

$$\left(\frac{S}{N}\right)_{rms} \text{output PM} = \frac{D_p^2 P_a (RP_o)^2 A_c^2/2}{2B_a N_o}$$

Substituting for D_p from Eq. (10.95) gives:

$$\left(\frac{S}{N}\right)_{rms} \text{output PM} = \frac{f_d^2 P_a (RP_o)^2 A_c^2/2}{2B_a^3 N_o}$$

The ratio of the output SNRs from the FM–IM and the PM–IM system is:

$$\text{Ratio} = \frac{[3f_d^2 P_a (RP_o)^2 A_c^2/2]/(2B_a^3 N_o)}{[f_d^2 P_a (RP_o)^2 A_c^2/2]/(2B_a^3 N_o)}$$

$$= 3$$

$$= 4.77 \text{ dB}$$

Example 10.16 shows that the FM–IM system has a superior output SNR by some 4.77 dB over the corresponding PM–IM system. Nevertheless, this does not prohibit the use of PM–IM systems for analog optical fiber communications as they still exhibit a substantial improvement in output SNR over D–IM systems, as well as allowing frequency division multiplexing. It should be noted, however, that a similar bandwidth penalty to FM–IM is incurred using this modulation format.

10.7.7 Pulse Analog Techniques

Pulse modulation techniques for analog transmission, rather than encoding the analog waveform into PCM, were mentioned within the system design considerations of Section 10.4. The most common techniques are pulse amplitude modulation (PAM), pulse width modulation (PWM), pulse position modulation (PPM) and pulse frequency modulation (PFM). All the pulse analog techniques employ pulse modulation in the electrical regime prior to intensity modulation of the optical source. However, PAM–IM is affected by source nonlinearities and is less efficient than D–IM, and therefore is usually discounted. PWM–IM is also inefficient since a large part of the transmitted energy conveys no information as only variations of the pulse width about a nominal value are of interest. Alternatively, PPM–IM and PFM–IM offer distinct advantages since the modulation affects the timing of the pulses, thus allowing the transmission of very narrow pulses. Hence, PPM–IM and PFM–IM provide similar signal to noise performance to subcarrier phase and frequency modulation whilst avoiding problems involved with source linearity. These techniques therefore prove advantageous for longer-haul analog fiber links. Although PPM–IM is slightly more efficient it provides less SNR improvement over D–IM, than that gained with PFM–IM, where wideband FM gain may be obtained. Furthermore, the terminal equipment required for PFM–IM is less complex and therefore it is generally the preferred pulse analog technique [Refs. 68 and 72–76]. For these reasons the system aspects of pulse analog transmission will be considered in relation to PFM–IM.

A block schematic of a PFM–IM optical fiber system is shown in Fig. 10.44. Pulse frequency modulation in which the pulse repetition rate is varied in sympathy with the modulating signal is performed in the PFM modulator which consists of a voltage controlled oscillator (VCO). This in turn operates the optical source by means of either a fixed pulse width or a fixed duty cycle (e.g. 50%). Demodulation in the system shown in Fig. 10.44 is by regenerative baseband recovery, whereby the individual pulses are detected in a wideband receiver before they are regenerated with a limiter and monostable. This provides the desired modulating signal as a baseband component which is recovered through a low pass filter.

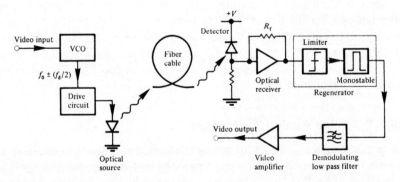

Fig. 10.44   A PFM-IM optical fiber system employing regenerative baseband recovery.

Regenerative baseband recovery gives the best SNR at the system output. A simpler PFM demodulation technique for fixed width pulse transmission is direct baseband recovery. In this case, because a baseband component is generated at the transmit terminal, detection may be performed with a low bandwidth receiver and the modulating signal obtained directly from a low pass filter. However, this technique gives a reduced SNR for a given optical power and therefore does not find wide application.

The SNR in terms of the peak to peak signal power to rms noise power of a PFM–IM system using regenerative baseband recovery is given by [Ref. 68]:

$$\left(\frac{S}{N}\right)_{p-p} = \frac{3(T_0 f_D M R P_{po})^2}{(2\pi T_R B)^2 \overline{i_N^2}} \tag{10.99}$$

where T_0 is the nominal pulse period which is equivalent to the reciprocal of the pulse rate f_0, f_D is the peak to peak frequency deviation, R is the photodiode responsivity, M is the photodiode multiplication factor, P_{po} is the peak received optical power, T_R is the pulse rise time at the regenerator circuit input, B is the post detection or effective baseband noise bandwidth and $\overline{i_N^2}$ is the receiver mean square noise current. It may be noted that improved SNRs are obtained with short rise time detected pulses. Moreover, the pulse rise time at the regenerator circuit input is dictated by the overall 10–90% system rise time T_{syst}, so there is no advantage in using a wideband receiver with a better pulse rise time than this. In fact such a receiver would degrade the system performance by passing increased front end noise. Therefore in an optimized PFM regenerative receiver design, $T_R = T_{syst}$ and following Eq. (10.43):

$$T_R \simeq 1.1 \, (T_S^2 + T_n^2 + T_c^2 + T_D^2)^{\frac{1}{2}} \tag{10.100}$$

where T_S, T_n, T_c and T_D are the rise times of the source (or transmitter), the fiber (intermodal and intramodal) and the detector (or receiver) respectively.

Example 10.17

An optical fiber PFM–IM system for video transmission employs regenerative baseband recovery. The system uses graded index fiber and an APD detector and has the following operational parameters:

Nominal pulse rate	20 MHz
Peak to peak frequency deviation	5 MHz
APD responsivity	0.7
APD multiplication factor	60
Total system 10–90% rise time	12 ns
Baseband noise bandwidth	6 MHz
Receiver mean square noise current	1×10^{-17} A^2

Calculate: (a) the optimum receiver bandwidth; (b) the peak to peak signal power to rms noise power ratio obtained when the peak input optical power to the receiver is −40 dBm.

Solution: (a) For an optimized design the pulse rise time at the regenerator circuit is equal to the total system rise time, hence $T_R = 12$ ns. The optimum receiver bandwidth is simply obtained by taking the reciprocal of T_R.

Solution: (a) For an optimized design the pulse rise time at the regenerator circuit is equal to the total system rise time, hence $T_R = 12$ ns. The optimum receiver bandwidth is simply obtained by taking the reciprocal of T_R giving 83.3 MHz.

(b) The nominal pulse period $T_0 (= 1/f_0)$ is 5×10^{-8} s and the peak optical power at the receiver is 1×10^{-7} W. Therefore, the peak to peak signal to rms noise ratio may be obtained using Eq. (10.99), where:

$$
\left(\frac{S}{N}\right)_{p-p} = \frac{3(T_0 \, f_D \, MRP_{po})^2}{(2\pi T_R B)^2 \overline{i_N^2}}
$$

$$
= \frac{3(5 \times 10^{-8} \times 5 \times 10^6 \times 60 \times 0.7 \times 10^{-7})^2}{(2\pi \times 12 \times 10^{-9} \times 6 \times 10^6)^2 \times 10^{-17}}
$$

$$
= 1.62 \times 10^6
$$

$$
= 62.1 \text{ dB}
$$

The result of example 10.17(b) illustrates the possibility of acquiring high SNRs at the output to a PFM–IM system using a regenerative receiver with achievable receiver noise levels and with moderate input optical signal power to the receiver.

10.8 COHERENT SYSTEMS

The detection of an intensity modulated optical carrier is basically a photon counting process where each detected photon is converted into an electron–hole pair (or in the case of the APD a number of pairs due to avalanche gain). It was indicated in Section 7.5 that this process which ignores the phase and polarization of the electromagnetic carrier may be readily implemented with currently available optical components. Thus all the previous discussion in this chapter involving both digital and analog systems has been applied to an intensity modulated optical carrier. However, receivers designed to detect the intensity modulated lightwave are often limited by noise generated in the electronic amplifier, except at very high SNRs. This reduces receiver sensitivity below the quantum noise limited conditions by at least 10–20 dB. It has been demonstrated that improved sensitivity may be obtained using well-known coherent detection techniques for the optical signal (i.e. heterodyne or homodyne detection). These methods have been successfully applied in free space optical communication systems using gas lasers [Ref. 77].

Coherent optical fiber transmission requires the direct modulation of the amplitude (direct AM), frequency (direct FM) or phase (direct PM) of a coherent optical carrier prior to demodulation using coherent detection. In the case of digital transmission this implies amplitude, frequency or phase shift keying (ASK, FSK or PSK) modulation techniques [Ref. 67]. Coherent detection by heterodyning involves the mixing of the incoming signal with a local optical oscillator to produce an intermediate frequency (IF), which is a difference signal (or difference frequency) carrying the modulating information signal. This signal may then be processed using standard techniques [Ref. 67]. Alternatively, with homodyne detection a local oscillator is set at the same frequency and phase as the coherent carrier, prior to optical mixing, which then translates the required information signal into the baseband.

A principal advantage of coherent detection for optical systems is the fact that the local oscillator power may be set such that the receiver noise is dominated by the shot noise contribution from the local oscillator. Furthermore, the signal power is proportional to the local oscillator power giving, for large local oscillator powers, an SNR which is independent of electronic amplifier noise. This allows a significant improvement in SNR at the receiver. In free space optical communications the SNR at the receiver output when using heterodyne detection may be shown [Ref. 78] to be at least a factor of 8 (or 9 dB) higher than that for subcarrier intensity modulation; the SNR improvement using homodyne detection in a similar system is at least a factor of 4 (or 6 dB). Conversely, investigation of digital optical modulation formats for coherent transmission [Ref. 79] indicates that PSK homodyne detection gives the best sensitivity by some 16–22 dB over ASK baseband direct detection. PSK heterodyne detection also proves to be the most sensitive heterodyn-

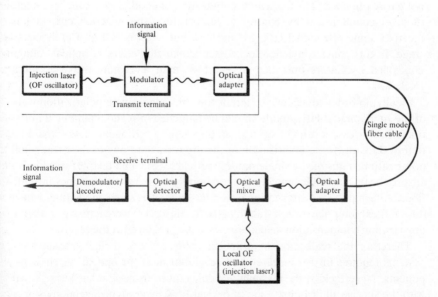

Fig. 10.45 A coherent optical fiber system employing heterodyne detection [Ref. 84].

ing technique with a minimum detectable power only 3 dB higher than that of PSK homodyne detection. Hence this shot noise limited detection offers to provide a distinct increase in the potential repeater spacing for optical fiber links. Moreover, coherent transmission will allow high capacity information transfer using frequency division multiplexing with fine frequency separation. This in turn will allow full exploitation of the low loss transmission windows in the longer wavelength region (around 1.3 μm and 1.55 μm).

However, there are problems associated with coherent optical fiber transmission which can be observed by considering the block schematic of a possible system employing heterodyne detection shown in Fig. 10.45 |Ref. 80|. The transmitter comprises an injection laser source which acts as an optical frequency oscillator, the output of which is modulated by the information signal using a suitable optical modulator. The directly modulated optical carrier is then launched into a single mode fiber through an optical adapter. The received optical signal is mixed with the output from the local optical oscillator, which is also provided by a suitable injection laser. To permit satisfactory heterodyne detection, the optical mixer must combine the polarized optical signal field with the similarly polarized local optical field in the most efficient manner. Ideally this device would be realized using integrated optical techniques (see Section 11.7). Finally, the signal is demodulated to obtain the originally transmitted information. For such a system to be practical both the source and local oscillator must be spectrally pure and frequency stabilized with respect to each other. Although this has proved possible using gas lasers, suitable semiconductor injection lasers are not easily obtained. The frequency stabilization of good, pure frequency, single mode injection lasers by control of temperature and injection current has received some attention [Refs. 81 and 82], but work is still at a preliminary stage. In this context injection locked semiconductor lasers or optical voltage controlled oscillators operating a very low input level may prove essential [Ref. 83].

Ideally, in order to achieve coherent transmission a single polarization state of the fundamental HE_{11} mode should be launched into and maintained in the fiber. In this case a suitable optical adapter (e.g. birefringent plate) would be required to adjust the polarization of the input wave to one of the orthogonal polorization states of the single mode fiber cable. Such an adapter would also ensure the best possible matching to maximize the launching efficiency into the fiber. Alternatively, if conventional circular symmetric single mode fiber which does not exhibit polarization stability (see Section 3.12) were employed, then a polarization compensation mechanism would be required at the receiver.

Therefore the realization of practical coherent optical fiber transmission systems requires further research and development of the optical system components. Nevertheless, the feasibility of coherent transmission has been shown with the successful demonstration of optical FSK heterodyne detection using a semiconductor laser source and local oscillator [Ref. 84]. Furthermore the potential benefits of these systems, especially in the area of long-haul, wideband optical fiber communications ensures that interesting developments may be anticipated in the near future.

PROBLEMS

10.1 Discuss the major considerations in the design of digital drive circuits for:
(a) an LED source;
(b) an injection laser source.
Illustrate your answer with an example of a drive circuit for each source.

10.2 Outline, with the aid of suitable diagrams, possible techniques for:
(a) the linearization of LED transmitters;
(b) the maintenance of constant optical output power from an injection laser transmitter.

10.3 Discuss, with the aid of a block diagram, the function of the major elements of an optical fiber receiver. In addition, describe possible techniques for automatic gain control in APD receivers.

10.4 Equalization within an optical receiver may be provided using the simple frequency 'rollup' circuit shown in Fig. 10.46(a). The normalized frequency response for this circuit is illustrated in Fig. 10.46(b).
The amplifier indicated in Fig. 10.46(a) presents a load of 5 kΩ to the photodetector and together with the photodetector gives a total capacitance of 5 pF. However, the desired response from the amplifier–equalizer configuration has an upper 3 dB point or corner frequency at 30 MHz. Assuming R_2 is fixed at 100 Ω determine the required values for C_1 and R_1 in order to obtain such a response.

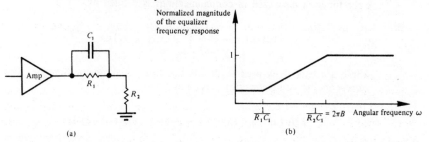

Fig. 10.46 The equalizer of problem 10.4: (a) the frequency 'rollup' circuit; (b) the spectral transfer characteristic for the circuit.

10.5 Describe the conversion of an analog signal into a pulse code modulated waveform for transmission on a digital optical fiber link. Furthermore indicate how several signals may be multiplexed onto a single fiber link.
A speech signal is sampled at 8 kHz and encoded using a 256 level binary code. What is the minimum transmission rate for this single pulse code modulated speech signal? Comment on the result.

10.6 A 1.5 MHz information signal with a dynamic range of 64 mV is sampled, quantized and encoded using a direct binary code. The quantization is linear with 512 levels. Determine:
    (a) the maximum possible bit duration;
    (b) the amplitude of one quantization level.

10.7 Describe, with the aid of a suitable block diagram, the operation of an optical fiber regenerative repeater. Indicate reasons for the occurrence of bit errors in the regeneration process and outline a technique for establishing the quality of the channel.

10.8 Twenty-four 4 kHz speech channels are sampled, quantized, encoded and then time division multiplexed for transmission as binary PCM on a digital optical fiber link. The quantizer is linear with 0.5 mV steps over a dynamic range of 2.048 V.
    Calculate:
    (a) the frame length of the PCM transmission, assuming an additional channel time slot is used for signalling and synchronization;
    (b) the required channel bandwidth assuming NRZ pulses.

10.9 Develop a relationship between the error probability and the received SNR (peak signal power to rms noise power ratio) for a baseband binary optical fiber system. It may be assumed that the number of ones and zeros are equiprobable and that the decision threshold is set midway between the one and zero level.
    The electrical SNR (defined as above) at the digital optical receiver is 20.4 dB. Determine:
    (a) the optical SNR;
    (b) the BER.
It may be assumed that erfc $(3.71) \simeq 1.7 \times 10^{-7}$.

10.10 The error function (erf) is defined in the text by Eq. (10.16). However an error function also used in communications is defined as:

$$\text{Erf}\,(u) = \frac{1}{\sqrt{(2\pi)}} \int_{-\infty}^{u} \exp\,(-x^2/2)\,\mathrm{d}x$$

where a capital E is used to denote this form of the error function. The corresponding complementary error function is:

$$\text{Erfc}\,(u) = 1 - \text{Erf}\,(u) = \frac{1}{\sqrt{(2\pi)}} \int_{u}^{\infty} \exp\,(-x^2/2)\,\mathrm{d}x$$

This complementary error function is also designated as $Q(u)$ in certain texts. Use of Erfc (u) or $Q(u)$ is sometimes considered more convenient within communication systems.
    Develop a relationship for erfc (u) in terms of Erfc (u). Hence obtain an expression for the error probability $P(e)$ as a function of the Erfc for a binary digital optical fiber system where the decision threshold is set midway between the one and zero levels and the number of transmitted ones and zeros are equiprobable. In addition given that Erfc $(4.75) \simeq 1 \times 10^{-6}$ estimate the required peak signal power to rms noise power ratios (both

optical and electrical) at the receiver of such a system in order to maintain a BER of 10^{-6}.

10.11 Show that Eq. (9.27) reduces to Eq. (10.28). Hence determine $F(M)$ when $k = 0.3$ and $M = 20$.

10.12 A silicon APD detector is utilized in a baseband binary PCM receiver where the decision threshold is set midway between the one and zero signal level. The device has a quantum efficiency of 70% and a ratio of carrier ionization rates of 0.05. In operation the APD has a multiplication factor of 65. Assuming a raised cosine signal spectrum and a zero disparity code, and given that erfc $(4.47) \simeq 2 \times 10^{-10}$:

    (a) estimate the number of photons required at the receiver to register a binary one with a BER of 10^{-10};

    (b) calculate the required incident optical power at the receiver when the system is operating at a wavelength of 0.9 μm and a transmission rate of 34 Mbit s^{-1};

    (c) indicate how the value obtained in (b) should be modified to compensate for a $3B4B$ line code.

10.13 A p–i–n photodiode receiver requires 2×10^4 incident photons in order to register a binary one with a BER of 10^{-9}. The device has a quantum efficiency of 65%. Estimate in decibels the additional signal level required in excess of the quantum limit for this photodiode to maintain a BER of 10^{-9}.

10.14 An optical fiber system employs an LED transmitter which launches an average of 300 μW of optical power at a wavelength of 0.8 μm into the optical fiber cable. The cable has an overall attenuation (including joints) of 4 dB km^{-1}. The APD receiver requires 1200 incident photons in order to register a binary one with a BER of 10^{-10}. Determine the maximum transmission distance (without repeaters) provided by the system when the transmission rate is 1 Mbit s^{-1} and 1 Gbit s^{-1} such that a BER of 10^{-10} is maintained.

    Hence sketch a graph showing the attenuation limit on transmission distance against the transmission rate for the system.

10.15 An optical fiber system uses fiber cable which exhibits a loss of 7 dB km^{-1}. Average splice losses for the system are 1.5 dB km^{-1}, and connector losses at the source and detector are 4 dB each. After safety margins have been allowed, the total permitted channel loss is 37 dB. Assuming the link to be attenuation limited, determine the maximum possible transmission distance without a repeater.

10.16 Assuming a linear increase in pulse broadening with fiber length, show that the transmission rate $B_T(\text{DL})$ at which a digital optical fiber system becomes dispersion limited is given by:

$$B_T(\text{DL}) \simeq \frac{(\alpha_{dB} + \alpha_j)}{5\sigma_T(\text{km})} \frac{1}{10 \log_{10}(P_i/P_o)}$$

where $\sigma_T(\text{km})$ is the total rms pulse broadening per kilometer on the link (hint: refer to Eqs. (3.3) and (3.10)).

(a) A digital optical fiber system using an injection laser source displays rms pulse broadening of 1 ns km^{-1}. The fiber cable has an attenuation of 3.5 dB km^{-1} and joint losses average out to 1 dB km^{-1}. Estimate the transmission rate at the dispersion limit when the difference in optical power levels between the input and output is 40 dB.

(b) Calculate the dispersion limited transmission distance for the system described in (b) when the transmission rates are 1 Mbit s^{-1} and 1 Gbit s^{-1}. Hence sketch a graph showing the dispersion limit on transmission distance against the transmission rate for the system.

10.17 The digital optical fiber system described in problem 10.16(a) has a transmission rate of 50 Mbit s^{-1} and operates over a distance of 12 km without repeaters. Assuming Gaussian shaped pulses, calculate the dispersion–equalization penalty exhibited by the system for the cases when:
(a) there is no mode coupling; and
(b) there is mode coupling.

10.18 A digital optical fiber system uses an RZ pulse format. Show that the maximum bit rate for the system $B_T(max)$ may be estimated using the expression:

$$B_T(max) = \frac{0.35}{T_{syst}}$$

where T_{syst} is the total system rise time. Comment on the possible use of the factor 0.44 in place of 0.35 in the above relationship.

An optical fiber link is required to operate over a distance of 10 km without repeaters. The fiber available exhibits a rise time due to intermodal dispersion of 0.7 ns km^{-1}, and a rise time due to intramodal dispersion of 0.2 ns km^{-1}. In addition the APD detector has a rise time of 1 ns. Estimate the maximum rise time allowable for the source in order for the link to be successfully operated at a transmission rate of 40 Mbit s^{-1} using an RZ pulse format.

10.19 A digital single mode optical fiber system is designed for operation at a wavelength of 1.5 μm and a transmission rate of 560 Mbit s^{-1} over a distance of 50 km without repeaters. The single mode injection laser is capable of launching a mean optical power of −13 dBm into the fiber cable which exhibits a loss of 0.35 dB km^{-1}. In addition, average splice losses are 0.1 dB at 1 km intervals. The connector loss at the receiver is 0.5 dB and the receiver sensitivity is −39 dBm. Finally an extinction ratio penalty of 1 dB is predicted for the system. Perform an optical power budget for the system and determine the safety margin.

10.20 Briefly discuss the reasons for the use of block codes in digital optical fiber transmission. Indicate the advantages and drawbacks when a 5B6B code is employed.

10.21 A D–IM analog optical fiber system utilizes a p–i–n photodiode receiver. Derive an expression for the rms signal power to rms noise power ratio in the quantum limit for this system.

The p–i–n photodiode in the above system has a responsivity of 0.5 at the operating wavelength of 0.85 μm. Furthermore the system has a modulation index of 0.4 and transmits over a bandwidth of 5 MHz. Sketch a graph of the quantum limited receiver sensitivity against the received SNR (rms signal power to rms noise power) for the system over the range 30–60 dB. It may be assumed that the photodiode dark current is negligible.

10.22   In practice the analog optical fiber receiver of problem 10.21 is found to be thermal noise limited. The mean square thermal noise current for the receiver is 2×10^{-23} A^2 Hz^{-1}. Determine the peak to peak signal power to rms noise power ratio at the receiver when the average incident optical power is -17.5 dBm.

10.23   An analog optical fiber system has a modulation bandwidth of 40 MHz and a modulation index of 0.6. The system utilizes an APD receiver with a responsivity of 0.7 and is quantum noise limited. An SNR (rms signal power to rms noise power) of 35 dB is obtained when the incident optical power at the receiver is -30 dBm. Assuming the detector dark current may be neglected, determine the excess avalanche noise factor at the receiver.

10.24   A simple analog optical fiber link operates over a distance of 15 m. The transmitter comprises an LED source which emits an average of 1 mW of optical power into air when the drive current is 40 mA. Plastic fiber cable with an attenuation of 500 dB km^{-1} at the transmission wavelength is utilized. The minimum optical power level required at the receiver for satisfactory operation of the system is 5 µW. The coupling losses at the transmitter and receiver are 8 and 2 dB respectively. In addition a safety margin of 4 dB is necessary. Calculate the minimum LED drive current required to maintain satisfactory system operation.

10.25   An analog optical fiber system employs an LED which emits 3 dBm mean optical power into air. However, a coupling loss of 17.5 dB is encountered when launching into a fiber cable. The fiber cable which extends for 6 km without repeaters exhibits a loss of 5 dB km^{-1}. It is spliced every 1.5 km with an average loss of 1.1 dB per splice. In addition there is a connector loss at the receiver of 0.8 dB. The PIN–FET receiver has a sensitivity of -54 dBm at the operating bandwidth of the system. Assuming there is no dispersion–equalization penalty, perform an optical power budget for the system and establish a safety margin.

10.26   Indicate the techniques which may be used for analog optical fiber transmission where an electrical subcarrier is employed. Illustrate your answer with a system block diagram showing the multiplexing of several signals onto a single analog optical fiber link.

10.27   A narrowband FM–IM optical signal has a maximum frequency deviation of 120 kHz when the frequency deviation ratio is 0.2. Compare the post detection SNR of this signal with that of a DSB–IM optical signal having the same modulating signal power, bandwidth, detector photocurrent and noise. Also estimate the bandwidth of the FM–IM signal. Comment on both results.

10.28   A frequency division multiplexed optical fiber system uses FM–IM. It has 50 equal amplitude voice channels each bandlimited to 3.5 kHz. A 1 kHz guard band is provided between the channels and below the first channel. The peak frequency deviation for the system is 1.35 MHz. Determine the transmission bandwidth for this FDM system.

10.29 An FM–IM system utilizes pre-emphasis and de-emphasis to enhance its performance in noise [Ref. 65]. The de-emphasis filter is a first order RC low pass filter placed at the demodulator to reduce the total noise power. This filter may be assumed to have an amplitude response $H_{de}(\omega)$ given by:

$$|H_{de}(\omega)| = \frac{1}{1 + (\omega/\omega_c)^2}$$

where $\omega_c = 2\pi f_c = 1/RC$.

The SNR improvement over FM–IM without pre-emphasis and de-emphasis is given by:

$$\text{SNR}_{de} \text{ improvement} = \frac{1}{3} \left(\frac{B_a}{f_c}\right)^2$$

where B_a is the bandwidth of the baseband signal and $f_c \ll B_a$.

(a) Write down an expression for the amplitude response of the pre-emphasis filter so that there is no overall signal distortion.

(b) Deduce an expression for the post detection SNR improvement in decibels for the FM–IM system with pre-emphasis and de-emphasis over a D–IM system operating under the same conditions of modulating signal power and bandwidth, photocurrent and noise. It may be assumed that $f_c \ll B_a$.

(c) A baseband signal with a bandwidth of 300 kHz is transmitted using the FM–IM system with pre-emphasis and de-emphasis. The maximum frequency deviation for the system is 4 MHz. In addition the de-emphasis filter comprises a 500 Ω resistor and a 0.1 μF capacitor. Determine the post detection SNR improvement for this system over a D–IM system operating under the same conditions of modulating signal power and bandwidth, photocurrent and noise.

10.30 A PM–IM optical fiber system operating above threshold has a frequency deviation ratio of 15 and a transmission bandwidth of 640 kHz.

(a) Estimate the bandwidth of the baseband message signal.

(b) Compute the post detection SNR improvement for the system over a D–IM system operating with the same modulating signal power and bandwidth, detector photocurrent and noise.

10.31 Discuss the advantages and drawbacks of the various pulse analog techniques for optical fiber transmission. Describe the operation of a PFM–IM optical fiber system employing regenerative baseband recovery.

10.32 An optical fiber PFM–IM system uses regenerative baseband recovery. The optical receiver which incorporates a p–i–n photodiode has an optimized bandwidth of 125 MHz. The other system parameters are:

Nominal pulse rate	35 MHz
Peak to peak frequency deviation	8 MHz
p–i–n photodiode responsivity	0.6 A W^{-1}
Baseband noise bandwidth	10 MHz
Receiver mean square noise current	3×10^{-25} A^2 Hz^{-1}

(a) Calculate the peak level of incident optical power necessary at the receiver to maintain a peak to peak signal power to rms noise power ratio of 60 dB.

(b) The source and detector have rise times of 3.5 and 5.0 ns respectively. Estimate the maximum permissible total rise time for the fiber cable utilized in the system such that satisfactory operation is maintained. Comment on the value obtained.

10.33 Discuss, with the aid of a suitable system block diagram, coherent optical fiber transmission. Indicate the system component requirements necessary to achieve successful coherent operation.

Answers to Numerical Problems

10.4	53.0 pF, 472 Ω	**10.19**	2.1 dB
10.5	64 kbit s^{-1}	**10.21**	40 nW, 40 µW
10.6	(a) 37 ns; (b) 125 µV	**10.22**	49.0 dB
10.8	(a) 300 bits; (b) 1.2 MHz	**10.23**	3.1
10.9	(a) 10.2 dB; (b) 3.4×10^{-7}	**10.24**	35.5 mA
10.10	$P(e) = \text{Erfc} \,[(S/N)^{\frac{1}{2}}/2]$, 9.8 dB, 19.6 dB	**10.25**	5.4 dB
10.11	7.4	**10.27**	Ratio of output SNRs (rms signal power to rms noise power) for FM–IM to DSB–IM is −9.21 dB, 1200 kHz
10.12	(a) 1400; (b) −52.8 dBm; (c) −51.8 dBm		
10.13	27.9 dB	**10.28**	3.15 MHz
10.14	15.76 km, 8.26 km	**10.29**	(a) $1 + (\omega/\omega_c)^2$;
10.15	3.41 km		(b) $-3 + 20 \log_{10} (D_f B_a / f_c)$;
10.16	(a) 22.5 Mbit s^{-1}; (b) 200 km, 0.2 km		(c) 59.0 dB
		10.30	(a) 20 kHz; (b) 20.5 dB
10.17	(a) 16.6 dB; (b) 0.1 dB	**10.32**	(a) −24.4 dBm; (b) 4.6 ns
10.18	3.04 ns		

REFERENCES

1　C. C. Timmermann, 'Highly efficient light coupling from GaAlAs lasers into optical fibers', *Appl. Opt.*, **15**(10), pp. 2432–2433, 1976.

2　M. Maeda, I. Ikushima, K. Nagano, M. Tanaka, H. Naskshima, R. Itoh, 'Hybrid laser to fiber coupler with a cylindrical lens', *Appl. Opt.*, **16**(7), pp. 1966–1970, 1977.

3　R. W. Dawson, 'Frequency and bias dependence of video distortion in Burrus-type homostructure and heterostructure LED's', *IEEE Trans. Electron. Devices*, **ED-25**(5), pp. 550–553, 1978.

4　J. Strauss, 'The nonlinearity of high-radiance light-emitting diodes', *IEEE J. Quantum Electron.*, **QE-14**(11), pp. 813–819, 1978.

5　K. Asatani and T. Kimura, 'Non-linear phase distortion and its compensation in LED direct modulation', *Electron. Lett.*, **13**(6), pp. 162–163, 1977.

6　G. White and C. A. Burrus, 'Efficient 100 Mb/s driver for electroluminescent diodes', *Int. J. Electron.*, **35**(6), pp. 751–754, 1973.

7　P. W. Shumate Jr and M. DiDomenico Jr, 'Lightwave transmitter', in H. Kressel (Ed.), *Semiconductor Devices for Optical Communications, Topics in Applied Physics*, Volume 39, pp. 161–200, Springer-Verlag, 1982.

8   L. Foltzer, 'Low-cost transmitters, receivers serve well in fibre-optic links', *EDN*, pp. 141–146, 20 October 1980.

9   A. Albanese and H. F. Lenzing, 'Video transmission tests, performed on intermediate-frequency light wave entrance links', *J. SMPTE (USA)*, **87**(12), pp. 821–824, 1978.

10 J. Strauss, 'Linearized transmitters for analog fiber links', *Laser Focus (USA)*, **14**(10), pp. 54–61, 1978.

11 A. Prochazka, P. Lancaster, R. Neumann, 'Amplifier linearization by complementary pre or post distortion', *IEEE Trans. Cable Telev.*, **CATV-1**(1), pp. 31–39, 1976.

12 K. Asatani and T. Kimura, 'Nonlinear distortions and their compensations in light emitting diodes', *Proceedings of International Conference on Integrated Optics and Optical Fiber Communications*, p. 105, 1977.

13 K. Asatani and T. Kimura, 'Linearization of LED nonlinearity by predistortions', *IEEE J. Solid State Circuits*, **SC-13**(1), pp. 133–138, 1978.

14 J. Strauss, A. J. Springthorpe and O. I. Szentesi, 'Phase shift modulation technique for the linearisation of analogue optical transmitters', *Electron. Lett.*, **13**(5), pp. 149–151, 1977.

15 J. Strauss and D. Frank, 'Linearisation of a cascaded system of analogue optical links', *Electron. Lett.*, **14**(14), 436–437, 1978.

16 H. S. Black, US Patent 1686792, issued Oct 9, 1929.

17 B. S. Kawasaki ands K. O. Hill, 'Low-loss access coupler for multimode optical fiber distribution network', *Appl. Opt.*, **16**(7) p. 1794, 1977.

18 J. Strauss and O. I. Szentesi, 'Linearisation of optical transmitters by a quasifeedforward compensation technique', *Electron. Lett.*, **13**(6), pp. 158–159, 1977.

19 S. M. Abbott, W. M. Muska, T. P. Lee, A. G. Dentai ands C. A. Burrus, '1.1 Gb/s pseudorandsom pulse-code modulation of 1.27 μm wavelength CW InGaAsP/InP DH lasers', *Electron. Lett.*, **14**(11), pp. 349–350, 1978.

20 J. Gruber, P. Marten, R. Petschacher and P. Russer, 'Electronic circuits for high bit rate digital fiber optic communication systems', *IEEE Trans. Commun.*, **COM-26**(7), pp. 1088–1098, 1978.

21 P. K. Runge, 'An experimental 50 Mb/s fiber optic PCM repeater', *IEEE Trans. Commun.*, **COM-24**(4), pp. 413–418, 1976.

22 U. Wellens, 'High-bit-rate pulse regeneration and injection laser modulation using a diode circuit', *Electron. Lett.*, **13**(18), pp. 529–530, 1977.

23 A. Chappell (Ed.), *Optoelectronics: Theory and Practice*, McGraw-Hill, 1978.

24 S. R. Salter, D. R. Smith, B. R. White and R. P. Webb, 'Laser automatic level control for optical communications systems', Third European Conf. on Optical Communications, Munich, September 1977, VDE-Verlag GmbH, Berlin, 1977.

25 A. Fausone, 'Circuit considerations', *Optical Fibre Communication*, by Tech. Staff of CSELT, pp. 777–800, McGraw-Hill, 1981.

26 A. Moncalvo and R. Pietroiusti, 'Transmission systems using optical fibres', *Telecommunication J. (Switzerland)*, **49**, pp. 84–92, 1982.

27 S. D. Personick. 'Design of receivers and transmitters for fiber systems', in M. K. Barnoski (Ed.), *Fundamentals of Optical Fiber Communications* (2nd Ed.), pp. 295–328. Academic Press, 1981.

28 T. L. Maione and D. D. Sell, 'Experimental fiber-optic transmission system for interoffice trunks', *IEEE Trans. Commun.*, **COM-25**(5), pp. 517–522, 1977.

29 R. G. Smith and S. D. Personick, 'Receiver design for optical fibre communication systems', in H. Kressel (Ed.), *Semiconductor Devices for Optical Communications*, Topics in Advanced Physics, Vol. 39, pp. 88–160, Springer-Verlag, 1982.

30 R. G. Smith, C. A. Brackett and H. W. Reinbold, 'Atlanta fiber system experiment, optical detector package', *Bell Syst. Tech. J.*, **57**(6), pp. 1809–1822, 1978.

31 J. L. Hullett and T. V. Muoi, 'A feedback amplifier for optical transmission systems', *IEEE Trans. Commun.*, **COM-24**, pp. 1180–1185, 1976.

32 J. L. Hullett, 'Optical communication receivers', *Proc. IREE Australia*, pp. 127–134, September 1979.

33 N. J. Bradley, 'Fibre optic systems design', *Electronic Eng.*, pp. 98–101. mid April 1980.

34 T. L. Maione, D. D. Sell and D. H. Wolaver, 'Practical 45 Mb/s regenerator for lightwave transmission', *Bell Syst. Tech. J.*, **57**(6), pp. 1837–1856, 1978.

35 S. D. Personick, 'Receiver design for optical systems', *Proc. IEEE*, **65**(12), pp. 1670–1678, 1977.

36 J. E. Goell, 'Input amplifiers for optical PCM receivers', *Bell Syst. Tech. J.*, **53**(9), pp. 1771–1793, 1974.

37 S. D. Personick, 'Time dispersion in dielectric waveguides', *Bell Syst. Tech. J.*, **50**(3), pp. 843–859, 1971.

38 I. Garrett, 'Receivers for optical fibre communications', *Radio Electron. Eng. J. IERE*, **51**(7/8), pp. 349–361, 1981.

39 S. D. Personick, 'Receiver design', in S. E. Miller and A. G. Chynoweth (Eds.), *Optical Fiber Telecommunications*, pp. 627–651, Academic Press Inc., 1979.

40 A. Moncalvo and L. Sacchi, 'System considerations', *Optical Fibre Communication*, by Tech. Staff of CSELT, pp. 723–776, McGraw-Hill, 1981.

41 M. Rocks and R. Kerstein, 'Increase in fiber bandwidth for digital systems by means of multiplexing', ICC 80 1980 International Conf. on Commun., Seattle, WA, USA, Part 28 5/1–5, June 1980.

42 W. Koester and F. Mohr, 'Bidirectional optical link', *Electrical Commun.*, **55**(4), pp. 342–349, 1980.

43 H. Taub and D. L. Schilling, *Principles of Communication System*, McGraw-Hill, 1971.

44 P. Hensel and R. C. Hooper, 'The development of high performance optical fibre data links', *IERE Conference Proceedings Fibre Optics*, 1–2 March 1982 (London), pp. 91–98, 1982.

45 D. C. Gloge and T. Li, 'Multimode-fiber technology for digital transmission', *Proc. IEEE*, **68**(10), pp. 1269–1275, 1980.

46 G. E. Stillman, 'Design considerations for fibre optic detectors', *Proc. Soc. Photo-opt. Instrum. Eng.*, **239**, pp. 42–52, 1980.

47 P. P. Webb, R. J. McIntyre and J. Conradi, 'Properties of avalanche photodiodes', *RCA Rev.*, **35**, pp. 234–278, 1974.

48 J. E. Midwinter, *Optical Fibers for Transmission*, John Wiley, 1979.

49 I. Garrett and J. E. Midwinter, 'Optical communication systems', in M. J. Howes and D. V. Morgan (Eds.), *Optical Fibre Communications: Devices, Circuits, and Systems*, pp. 251–300, John Wiley, 1980.

50 R. J. McIntyre and J. Conradi, 'The distribution of gains in uniformly multiplying avalanche photodiodes', *IEEE Trans. Electron. Devices*, **ED-19**, pp. 713–718, 1972.

51 K. Mouthaan, 'Telecommunications via glass–fibre cables', *Philips Telecommun. Rev.*, **37**(4), pp. 201–214, 1979.

52 H. F. Wolf, 'System aspects', in H. F. Wolf (Ed.), *Handbook of Fiber Optics: Theory and Applications*, pp. 377–427, Granada, 1979.

53 S. D. Personick, N. L. Rhodes, D. C. Hanson and K. H. Chan, 'Contrasting fiber-optic-component-design requirements in telecommunications, analog, and local data communications applications', *Proc. IEEE*, **68**(10), pp. 1254–1262, 1980.

54   S. E. Miller, 'Transmission system design', in S. E. Miller and A. G. Chynoweth (Eds.), *Optical Fiber Telecommunications*, pp. 653–683, Academic Press, 1979.

55   C. K. Koa, *Optical Fiber Systems: Technology, Design and Applications*, McGraw-Hill, 1982.

56   C. Kleekamp and B. Metcalf, *Designer's Guide to Fiber Optics*, Cahners Publishing Company, 1978.

57   G. R. Elion and H. A. Elion, *Fiber Optics in Communications Systems*, Marcel Dekker, 1978.

58   S. Shimada, 'Systems engineering for long-haul optical-fiber transmission', *Proc. IEEE*, **68**(10), pp. 1304–1309, 1980.

59   J. H. C. Van Heuven, 'Techniques for optical transmission', in *Proceedings of 11th European Microwave Conference*, Amsterdam, Netherlands, pp. 3–10, 1981.

60   R. Tell and S. T. Eng, 'Optical fiber communication at 5 Gbit/sec', *Appl. Opt.*, **20**(22), pp. 3853–3858, 1981.

61   P. Wells, 'Optical-fibre systems for telecommunications', *GEC J. Sci. Tech.*, **46**(2), pp. 51–60, 1980.

62   M. Chown and K. C. Koa, 'Some broadband fiber system design considerations', in *Proceedings of IEEE International Conference on Communications*, Philadelphia PA, 1972, pp. 12/1–5, IEEE, 1972.

63   I. Garrett and C. J. Todd, 'Optical fiber transmission systems at 1.3 and 1.5 μm wavelength', in *Proceedings of IEEE 1981 International Conference on Communications*, New York, Vol. 1, Pt. 16.2/1–5, IEEE, 1981.

64   J. L. Hullett and T. V. Muoi, 'Optical fiber systems analysis', *Proc. IREE Australia*, **38**(1–2), pp. 390–397, 1977.

65   A. Luvison, 'Topics in optical fibre communication theory', *Optical Fibre Communications*, by Technical Staff of CSELT, pp. 647–721, McGraw-Hill, 1981.

66   M. Chown, A. W. Davis, R. E. Epworth and J. G. Farrington, 'System design', in C. P. Sandbank (Ed.), *Optical Fibre Communication Systems*, pp. 206–283, John Wiley, 1980.

67   K. Sam Shanmugan, *Digital and Analog Communication Systems*, John Wiley, 1979.

68   G. G. Windus, 'Fibre optic systems for analogue transmission', *Marconi Rev.*, **XLIV**(221), pp. 78–100, 1981.

69   W. Horak, 'Analog TV signal transmission over multimode optical waveguides', *Siemens Research and Development Reports*, **5**(4), pp. 192–202, 1976.

70   K. Sato and K. Asatani, 'Analogue baseband TV transmission experiments using semiconductor laser diodes', *Electron. Lett.*, **15**(24), pp. 794–795, 1979.

71   R. M. Gagliadi and S. Karp, *Optical Communications*, John Wiley, 1976.

72   C. C. Timmerman, 'Signal to noise ratio of a video signal transmitted by a fiber-optic system using pulse-frequency modulation', *IEEE Trans. Broadcasting*, **BC-23**(1), pp. 12–16, 1976.

73   C. C. Timmerman, 'A fiber optical system using pulse frequency modulation', *NTZ*, **30**(6), pp. 507–508, 1977.

74   D. J. Brace and D. J. Heatley, 'The application of pulse modulation schemes for wideband distribution to customers (integrated optical fibre systems)', in *Sixth European Conference on Optical Communication*, York, UK, 16–19 Sept. 1980, pp. 446–449, 1980.

75   E. Yoneda, T. Kanada and K. Hakoda, 'Design and performance of optical fibre transmission systems for color television signals', *Rev. Elect. Commun. Lab.*, **29**(11–12), pp. 1107–1117, 1981.

76 T. Kanada, K. Hakoda and E. Yoneda, 'SNR fluctuation and nonlinear distortion in PFM optical NTSC video transmission systems', *IEEE Trans. Commun.*, **COM-30**(8), pp. 1868–1875, 1982.

77 M. C. Teich, 'Homodyne detection of infrared radiation from a moving diffuse target', *Proc. IEEE*, **57**(5), pp. 789–792, 1969.

78 W. K. Pratt, *Laser Communication Systems*, John Wiley, 1969.

79 Y. Yamamoto, 'Receiver performance evaluation of various digital optical modulation–demodulation systems in the 0.5–10 μm wavelength region', *IEEE J. Quantum Electron.*, **QE-16**(11), pp. 1251–1259, 1980.

80 F. Favre, L. Jeunhomme, I. Joindot, M. Monerie and J. C. Simon, 'Progress towards heterodyne-type single-mode fibre communication systems', *IEEE J. Quantum Electron.*, **QE-17**(6), pp. 897–905, 1981.

81 T. Okoshi and K. Kikuchi, 'Frequency stabilisation of semiconductor lasers for heterodyne type optical communication systems', *Electron. Lett.*, **16**(5), pp. 179–181, 1980.

82 F. Favre and D. Le Guen, 'High frequency stability of laser diode for heterodyne communication systems', *Electron. Lett.*, **16**, pp. 709–710, 1980.

83 R. Lang, 'Injection locking properties of a semiconductor laser', *IEEE J. Quantum Electron.*, **QE-18**(6), pp. 976–983, 1982.

84 S. Saito, Y. Yamamoto and T. Kimura, 'Optical FSK heterodyne detection experiments using semiconductor laser transmitter and local oscillator', *IEEE J. Quantum Electron.*, **QE-17**(6), pp. 935–941, 1981.

85 Y. Yamamoto and T. Kimura, Coherent optical fiber transmission systems', *IEEE J. Quantum Electron.*, **QE-17**(6), pp. 919–934, 1982.

86 B. E. A. Saleh and M. I. Irshid, 'Coherence and intersymbol interference in digital fiber optic communication systems', *IEEE J. Quantum Electron.*, **QE-18**(6), pp. 944–951, 1982.

87 T. Kimura and Y. Yamamoto, 'Progress of coherent optical fibre communication systems', *Opt. Quantum Electron.*, **15**, pp. 1–39, 1983.

11

Applications and Future Developments

11.1 INTRODUCTION

In order to appreciate the many areas in which the application of lightwave transmission via optical fibers may be beneficial, it is useful to review the advantages and special features provided by this method of communication. The primary advantages ·obtained using optical fibers for line transmission were discussed in Section 1.3 and may be summarized as follows:

(a) enormous potential bandwidth;
(b) small size and weight;
(c) electrical isolation;
(d) immunity to interference and crosstalk;
(e) signal security;
(f) low transmission loss;
(g) ruggedness and flexibility;
(h) system reliability and ease of maintenance;
(i) potential low cost.

Although this list is very impressive, it is not exhaustive and several other attributes associated with optical fiber communications have become apparent as the technology has developed. Perhaps the most significant are the reduced power consumption exhibited by optical fiber systems in comparison with their metallic cable counterparts and their ability to provide for an expansion in the system capability often without fundamental and costly changes to the system configuration. For instance, a system may be upgraded by simply changing from an LED to an injection laser source, by replacing a $p-i-n$ photodiode with an APD detector, or alternatively by operating at a longer wavelength without replacing the fiber cable.

The use of fibers for optical communication does have some drawbacks in practice. Hence to provide a balanced picture these disadvantages must be considered. They are:

(a) the fragility of the bare fibers;

(b) the small size of the fibers and cables which creates some difficulties with splicing and forming connectors;
(c) some problems involved with forming low loss T-couplers;
(d) some doubt in relation to the long-term reliability of optical fibers in the presence of moisture (effects of stress corrosion—see Section 4.6);
(e) an independent electrical power feed is required for any repeaters;
(f) new equipment and field practices are required;
(g) testing procedures tend to be more complex.

A number of these disadvantages are not just inherent to optical fiber systems but are always present at the introduction of a new technology. Furthermore, both continuing developments and experience with optical fiber systems are generally reducing the other problems.

The combination of the numerous attributes and surmountable problems makes optical fiber transmission a very attractive proposition for use within national and international telecommunication networks (PTT applications). To date applications for optical fiber systems in this area have proved the major impetus for technological developments in the field. The technology has progressed from what may be termed first generation systems using multimode step index fiber and operating in the shorter wavelength region (0.8–0.9 µm), to second generation systems utilizing multimode graded index fiber operating in both the shorter and longer wavelength regions (0.8–1.6 µm). Furthermore, fully engineered third generation systems incorporating single mode fiber predominantly for operation in the longer wavelength region (1.1–1.6 µm) have been accepted for commercial operation in the public telecommunications network. In addition many alternative fiber systems applications have become apparent in other areas of communications where often first and second generation systems provide an ideal solution. Also the growing utilization of optical fiber systems has stimulated tremendous research efforts towards enhanced fiber design. This has resulted in improvement of the associated optoelectronics as well as investigation of 'passive' optics which are likely to provide an advance in the current 'state of the art' of optical fiber communications together with an expansion in its areas of use. Hence, what may be termed fourth generation systems are already close to realization being concerned with both coherent transmission (see Section 10.8) and integrated optics (see Section 11.7). Even fifth generation systems could be on the horizon judging from the results of preliminary investigations [Ref. 1 and 2]. These seek to utilize nonlinear pulse propagation in optical fibers to provide greatly increased channel capacity whilst exhibiting compatibility with integrated optical systems which function in a nonlinear environment.

In this chapter we consider current and potential applications of optical fiber communication systems together with some likely future developments in the general area of optical transmission and associated components. The discussion is primarily centered around application areas including the public network, military, civil and consumer, industrial and computer systems which are dealt with in Sections 11.2 to 11.6. However, this discussion is extended in Sections 11.7 and 11.8 with a brief review of integrated optical techniques and devices so that the reader may obtain an insight into the technological developments which it is generally believed will instigate further generations of optical fiber communication systems.

11.2 PUBLIC NETWORK APPLICATIONS

The public telecommunications network provides a variety of applications for optical fiber communication systems. It was in this general area that the suitability of optical fibers for line transmission first made an impact. The current plans of the major PTT administrations around the world feature the installation of increasing numbers of optical fiber links as an alternative to coaxial and high frequency pair cable systems. In addition it is indicated [Ref. 3] that administrations appear to be abandoning plans for millimetric waveguide transmission (see Section 1.1) in favor of optical fiber communications.

11.2.1 Trunk Network

The trunk or toll network is used for carrying telephone traffic between major conurbations. Hence there is generally a requirement for the use of transmission systems which have a high capacity in order to minimize costs per circuit. The transmission distance for trunk systems can vary enormously from under 20 km to over 300 km, and occasionally to as much as 1000 km. Therefore transmission systems which exhibit low attenuation and hence give a maximum distance of unrepeatered operation are the most economically viable. In this context optical fiber systems with their increased bandwidth and repeater spacings offer a distinct advantage. This may be observed from Fig. 11.1 [Ref. 4] which shows a cost comparison of different high capacity line transmission media. It may be observed that optical fiber systems show a significant cost advantage over coaxial cable systems and compete favorably with millimetric waveguide systems at all but the highest capacities. It may also be noted that only digital systems are compared. This is due to the advent of the fully integrated digital public network which invariably means that the majority of trunk routes will employ digital transmission systems.

The speed of operation of most digital trunk optical fiber systems are based on the principal digital hierarchies for Europe and North America which were shown in Table 10.1. Proprietary systems (where one contractor supplies the complete system in order to minimize interface problems) operating at 34 Mbit s^{-1} and 140 M bit s^{-1} have been installed in the trunk network in the

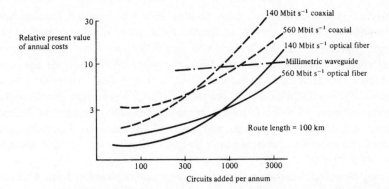

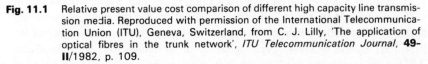

Fig. 11.1 Relative present value cost comparison of different high capacity line transmission media. Reproduced with permission of the International Telecommunication Union (ITU), Geneva, Switzerland, from C. J. Lilly, 'The application of optical fibres in the trunk network', *ITU Telecommunication Journal*, **49**-II/1982, p. 109.

UK on low and high growth rate trunk routes respectively. In the main these systems operate in the 0.85–0.9 μm wavelength region using injection laser sources via graded index fiber to silicon APD detectors with repeater spacings of between 8 and 10 km. A typical system power budget for a 140 Mbit s^{-1} system operating over 8 km of multimode graded index fiber at a wavelength of 0.85 μm is shown in Table 11.1 [Ref. 5]. The mean power launched from the laser into the fiber may be improved by over 3 dB using lens coupling rather than the butt launch indicated.

High radiance LED sources emitting at 1.3 μm are also being used with multimode graded index fiber in proprietary trunk systems operating at both

Table 11.1 A typical optical power budget for a 140 Mbit s^{-1} trunk system operating over 8 km of multimode graded index fiber at a wavelength of 0.85 μm

Mean power launched from the laser transmitter (butt coupling)	−4.5 dBm
APD receiver sensitivity at 140 Mbit s^{-1} (BER 10^{-9})	−48.0 dBm
Total system margin	43.5 dB
Cabled fiber loss (8 × 3 dB km^{-1})	24.0 dB
Splice losses (9 × 0.3 dB each)	2.7 dB
Connector loss (2 × 1 dB each)	2.0 dB
Dispersion–equalization penalty	6.0 dB
Safety margin	7.0 dB
Total system loss	41.7 dB
Excess power margin	1.8 dB

34 Mbit s^{-1} and 140 Mbit s^{-1} most notably in a link between London and Birmingham which is 205 km in length. Field trials of single mode fiber systems operating in the longer wavelength region have demonstrated repeaterless transmission at 565 Mbit s^{-1} over 62 km at 1.3 µm and 140 Mbit s^{-1} over 91 km at 1.5 µm |Ref. 6|. These field trials have been followed by the installation of proprietary long wavelength single mode systems utilizing PIN–FET hybrid receivers between Luton and Milton Keynes, a distance of 27.3 km, and over 52 km between Liverpool and Preston. The provisional optical power budget for the former system is given in Table 11.2 [Ref. 7].

It may be noted that the single mode fiber system depicted in Table 11.2 has a lower total system margin than the multimode fiber system operating at 0.85 µm outlined in Table 11.1. This is caused by the difficulty of launching light into the smaller single mode fiber as well as the reduced sensitivity of the PIN–FET hybrid receiver operating at 1.3 µm in comparison with the silicon APD receiver operating at 0.85 µm. Nevertheless the low loss (0.6 dB km^{-1}) of the single mode fiber when operating in the longer wavelength region together with the absence of any dispersion–equalization penalties more than compensates for these drawbacks; in this case allowing a transmission distance of 27.3 km without repeaters. In the UK public network a 30 km unrepeatered transmission distance (or repeater spacing) is quite sufficient since it is the maximum spacing between existing surface stations and hence power feed points. This removes any requirement for the installation of a metallic conductor for power feed within the system as well as allowing any repeaters to be installed above ground in a protected internal environment. Benefits gained include significantly reduced system costs along with additional reliability and ease of maintenance.

Table 11.2   Optical power budget for 140 Mbit s^{-1} single mode fiber trunk system operating over 27.3 km at a wavelength of 1.3 µm

Mean power launched from the laser transmitter	−6 dBm
PIN–FET hybrid receiver sensitivity at 140 Mbit s^{-1} at 1.3 µm wavelength (BER 10^{-9})	−36 dBm
Total system margin	30 dB
Cabled fiber loss (27.3 × 0.6 dB km^{-1})	16.4 dB
Splice losses (27 × 0.15 dB average)	4.1 dB
Connector loss (2 × 1.5 dB)	3.0 dB
Dispersion–equalization penalty	0 dB
Safety margin	6.0 dB
Total system loss	29.5 dB
Excess power margin	0.5 dB

The preferred transmission rate for optical fiber trunk systems based on the 1.5 Mbit s^{-1} digital hierarchy (i.e. North America) is at present 45 Mbit s^{-1}. This is largely due to the fact that much higher growth rates are required for the high speed systems operating at 274 Mbit s^{-1} and above. It is indicated |Ref. 8| that these high speed systems are more appropriate to very long haul trunk routes (up to 6400 km) where repeater spacings in excess of 25 km are required. Hence the incorporation of systems with transmission rates in excess of 45 Mbit s^{-1} into the trunk network awaits the advent of commercial long wavelength single mode fiber systems capable of operating at 274 Mbit s^{-1} over long repeater spans. However, this position is unlikely to remain static for very long as Bell Laboratories have already demonstrated 100 km unrepeatered transmission at this rate |Ref. 6|. Furthermore an experimental single mode trunk system operating at a wavelength of i.3 µm and a transmission rate of 400 Mbit s^{-1} has been in operation in Japan since 1981 |Ref. 9|. This system which utilizes an APD receiver allows repeater spacings of up to 20 km.

11.2.2 Junction Network

The junction or interoffice network usually consists of routes within major conurbations over distances of typically 5–20 km. However, the distribution of distances between switching centers (telephone exchanges) or offices in the junction network of large urban areas varies considerably for various countries as indicated in Fig. 11.2 |Ref. 10|. It may be observed from Fig. 11.2 that the benefits of long unrepeated transmission distances offered by optical fiber

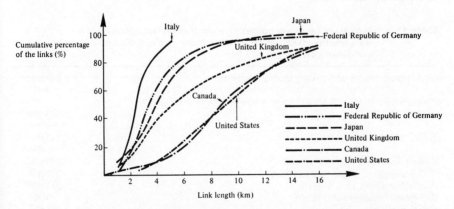

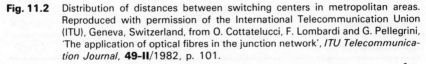

Fig. 11.2 Distribution of distances between switching centers in metropolitan areas. Reproduced with permission of the International Telecommunication Union (ITU), Geneva, Switzerland, from O. Cottatelucci, F. Lombardi and G. Pellegrini, 'The application of optical fibres in the junction network', *ITU Telecommunication Journal*, **49-II**/1982, p. 101.

systems are not as apparent in the junction network due to the generally shorter link lengths. Nevertheless optical fiber junction systems are often able to operate using no intermediate repeaters whilst alleviating duct congestion in urban areas.

In Europe optical fiber systems with transmission rates of 8 Mbit s^{-1} and for busy routes 34 Mbit s^{-1} are finding favor in the junction network. A number of proprietary systems predominantly operating at 8 Mbit s^{-1} using both injection laser and LED sources via multimode graded index fiber to APD detectors are in operation in the UK with repeater spacings between 7.5 and 12 km. A typical optical power budget for such a system operating at a wavelength of 0.88 µm over 12 km is shown in Table 11.3 |Ref. 5|. It may be noted that the mean power launched from the laser is reduced below the level obtained with similar dimensioned multimode graded index fiber in the optical power budget shown in Table 11.1. This is due to the lower duty factor when using a 2B3B code on the 8 Mbit s^{-1} system in comparison with a 7B8B code used on the 140 Mbit s^{-1} system (see Section 10.5).

In North America, 6 Mbit s^{-1} systems offer flexibility whereas 45 Mbit s^{-1} systems prove suitable for junction traffic requirements of crowded areas. However, economic studies for the USA have indicated that 45 Mbit s^{-1} systems are the most economic choice for the initial service |Ref. 10|. Hence a significant number of commercial 45 Mbit s^{-1} junction systems have been installed. These operate in the shorter wavelength region utilizing injection laser sources, multimode graded index fiber and APD detectors with repeater spacings up to 7.5 km. In addition several experimental 32 Mbit s^{-1} junction systems have been in operation in Japan since 1980. These systems, which utilize both injection laser and LED sources to APD detectors, have repeater spacings up to 21 km.

Table 11.3 Typical optical power budget for a junction system operating at a wavelength of 0.88 µm and a transmission rate of 8 Mbit s^{-1} over an unrepeatered distance of 12 km

Mean power launched from the laser transmitter	−6.0 dBm
Receiver sensitivity at 8 Mbit s^{-1} and a wavelength of 0.88 µm (BER 10^{-9})	−63.0 dBm
Total system margin	57.0 dB
Cabled fiber loss (12 × 3.5 dB km^{-1})	42.0 dB
Splice losses (13 × 0.3 dB)	3.9 dB
Connector losses (2 × 1 dB)	2.0 dB
Dispersion–equalization penalty	0 dB
Safety margin	7.0 dB
Total system loss	54.9 dB
Excess power margin	2.1 dB

11.2.3   Local and Rural Networks

The local and rural network or subscriber loop connects telephone subscribers to the local switching center or office. Possible network configurations are shown in Fig. 11.3 and include a ring, tree and star topology from the local switching center. In a ring network (Fig. 11.3(a)) any information fed into the network by a subscriber passes through all the network nodes and hence a number of transmission channels must be provided between all nodes. This may be supplied by a time division multiplex system utilizing a broadband transmission medium. In this case only information addressed to a particular subscriber is taken from the network at that subscriber node. The tree network, which consists of several branches as indicated in Fig. 11.3(b), must also

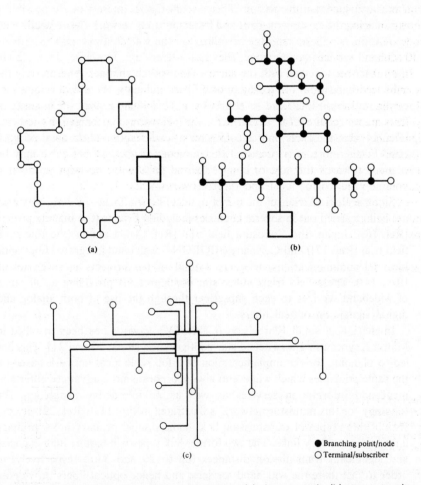

● Branching point/node
○ Terminal/subscriber

Fig. 11.3   Local and rural network configurations: (a) ring network; (b) tree network; (c) star network.

provide a number of transmission channels on its common links. However, in comparison with the ring network it has the advantage of greater flexibility in relation to topological enlargement. Nevertheless in common with the ring network, the number of subscribers is limited by the transmission capacity of the links used.

In contrast, the star network (Fig. 11.3(c)) provides a separate link for every subscriber to the local switching center. Hence the amount of cable required is considerably increased over the ring or tree network, but is offset by enhanced reliability and availability for the subscribers. In addition simple subscriber equipment is adequate (i.e. no TDM) and network expansion is straight-forward. Thus virtually all local and rural telephone networks utilize a star configuration based on copper conductors (twisted pair) for full duplex (bothway) speech transmission. There is substantial interest in the possibility of replacing the existing narrowband local and rural network twisted pairs with optical fibers. These can also be utilized in the star configuration to provide wideband services (videophone, television, stereo hi-fi, facsimile, data, etc.) to the subscriber together with the narrowband speech channel. Alternatively the enhanced bandwidth offered by optical fibers will allow the use of ring or tree configurations in local and rural networks. This would reduce the quantity of fiber cable required for subscriber loops. However, investigations indicate [Ref. 14] that the cable only accounts for a small fraction of the total network cost. Furthermore, it is predicted that the cost of optical fiber cable may be reduced towards the cost of copper twisted pairs with the large production volume required for local and rural networks.

Small scale field trials of the use of optical fibers in local and rural networks are being carried out in several countries including France (the Biarritz project [Ref. 16]), Japan (the Yokosuta field trial [Ref. 14]), Canada (the Elie rural field trial [Ref. 17]) and Germany (BIGFON—wideband integrated fiber optic local telecommunications network; a total of ten projects in seven towns [Ref. 14]). These field trials utilize star configurations providing a full range of wideband services to each subscriber through the use of both analog and digital signals on optical fibers.

In the UK a small Fibrevision (Cable TV) network has been installed in Milton Keynes using a switched star configuration [Refs. 20 and 21]. This has led to planning for the implementation of a full scale local network based on the same principles which will eventually incorporate more advanced entertain-ment and information services as well as an interactive capability. The topology for this multistar network is illustrated in Fig. 11.4 [Ref. 22]. It con-sists of three types of transmission link, namely super-primary links, primary links and secondary links. The two former link types will require high capacity and significant transmission distances (up to 20 and 5 km respectively) in order to distribute the wideband services, and hence optical fibers provide an ideal solution. The primary links service wideband switch points (WSPs) from hub sites giving access to all program material and services. In the larger

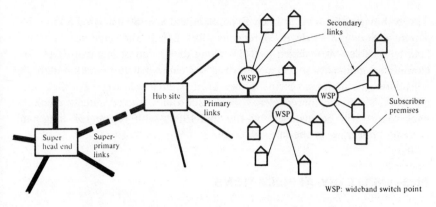

Fig. 11.4   Multistar wideband local network configuration [Ref. 22].

schemes these hub sites will in turn be fed from a super head end where most of the program material and services will originate. Short secondary links (up to 1 km) will fan out from the WSPs to the subscriber premises. The use of wideband switching reduces the capacity required on these secondary links. Economic considerations, especially when the cost of the optoelectronic interface equipment is included, suggest that at present coaxial cable provides the best solution, on cost grounds, for these secondary links. However, it is indicated [Ref. 22] that a large scale integration approach for the optical equipment, to give a low unit cost, may allow optical fibers to be utilized for these secondary links in the future.

11.2.4   Submerged Systems

Undersea cable systems are an integral part of the international telecommunications network. They find application on shorter routes especially in Europe. On longer routes, such as across the Atlantic, they provide route diversity in conjunction with satellite links. The number of submerged cable routes and their capacities are steadily increasing and hence there is a desire to minimize the costs per channel. In this context digital optical fiber communication systems appear to offer substantial advantages over current analog FDM and digital PCM coaxial cable systems. High capacity coaxial cable systems require high quality, large diameter cable to overcome attenuation, and still only allow repeater spacings of around 5 km. By comparison, it is predicted [Ref. 23] that single mode optical fiber systems operating at 1.3 or 1.55 μm will provide repeater spacings of 25–50 km and eventually even longer.

   Research and development of single mode fiber submerged cable systems is progressing in a number of countries including the UK, France, the USA and Japan. A successful field trial of a 140 Mbit s^{-1} system was carried out by

STC Submarine Systems in Loch Fyne, Scotland in 1980 using a 9.5 km cable length, including a single PCM repeater |Ref. 24|. In the same year a 10 km field trial cable was installed by NTT along the Izu coast in Japan |Ref. 23|. Component reliability together with deep sea cable structure and strength are considered the major problems. However, it is envisaged |Ref. 23| that commercial systems for nonrepeated short-haul routes, and repeatered medium-haul routes will be installed by the mid-1980s with repeatered long-haul systems appearing in the late 1980s.

11.3 MILITARY APPLICATIONS

In these applications, although economics are important, there are usually other, possibly overriding, considerations such as size, weight, deployability, survivability (in both conventional and nuclear attack |Ref. 26|) and security. The special attributes of optical fiber communication systems therefore often lend themselves to military use.

11.3.1 Mobiles

One of the most promising areas of military application for optical fiber communications is within military mobiles such as aircraft, ships and tanks. The small size and weight of optical fibers provide an attractive solution to space problems in these mobiles which are increasingly equipped with sophisticated electronics. Also the wideband nature of optical fiber transmission will allow the multiplexing of a number of signals onto a common bus. Furthermore, the immunity of optical transmission to electromagnetic interference (EMI) in the often noisy environment of military mobiles is a tremendous advantage. This also applies to the immunity of optical fibers to lightning and electromagnetic pulses (EMP) especially within avionics. The electrical isolation, and therefore safety, aspect of optical fiber communications also proves invaluable in these applications, allowing routing through both fuel tanks and magazines.

The above advantages were demonstrated with preliminary investigations involving fiber bundles [Ref. 3] and design approaches now include multi-terminal data systems [Ref. 27] using single fibers, and use of an optical data bus |Ref. 28|. In the former case, the time division multiplex system allows ring or star configurations to be realized, or mixtures of both to create bus networks. The multiple access data highway allows an optical signal injected at any access point to appear at all other other access points. An example is shown in Fig. 11.5 [Ref. 5] which illustrates the interconnection of six terminals using two four-way transmissive star couplers. These devices give typically 10 dB attenuation between any pair of ports.

An experimental optical data bus has been installed in the Mirage 4000 aircraft |Ref. 28|. However, significant problems were encountered with optical

Terminal

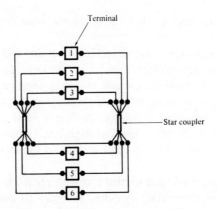

Star coupler

Fig. 11.5 Multiple access bus showing the interconnection of six terminals using two four-way transmissive star couplers [Ref. 5].

connection, fiber fragility and low light levels from the LED source. Nevertheless it was concluded that these drawbacks could be reduced by the use of spliced connections, star couplers rather than T-couplers and smaller diameter fibers (100–150 μm), which would make it possible to produce cables with smaller radii of curvature, and in which the fiber would be freer.

Studies are also underway into the feasibility of using a 1 Mbit s^{-1} optical data bus for flight control, avionic weapons and internal data systems within a helicopter. It is intended that a demonstration system will be installed in an operational Lynx helicopter by 1987 [Ref. 29]. Also an optical fiber data highway is to be installed in the Harrier GR5 aircraft, for operational use. It will be incorporated between the communications, navigation and identification data converter located towards the rear of the aircraft and the amplifier control situated beneath the cockpit. The original system specification calls for large core diameter plastic fiber for operation around 2.4 kbit s^{-1} [Ref. 29].

11.3.2 Communication Links

The other major area for the application of optical fiber communications in the military sphere includes both short and long distance communication links. Short distance optical fiber systems may be utilized to connect closely spaced items of electronic equipment in such areas as operations rooms and computer installations. A large number of these systems have already been installed in military installations in the UK. These operate distances from several centimeters to a few hundred meters at transmission rates between 50 bauds and 4.8 kbit s^{-1} [Ref. 29]. In addition a small number of 7 MHz video links operating over distances of up to 100 m are in operation. There is also a requirement for long distance communication between military installations which could benefit from the use of optical fibers. In both these cases advantages may be gained in terms of bandwidth, security and immunity to electrical interference and earth loop problems over conventional copper systems.

Other long distance applications include torpedo and missile guidance, information links between military vessels and maritime, towed sensor arrays. In these areas the available bandwidth and long unrepeatered transmission distances of optical fiber systems provide a solution which is not generally available with conventional technology. A fiber guided weapons system is illustrated in Fig. 11.6 whereby a low loss, high tensile strength fiber is used to relay a video signal back to a control station to facilitate targeting by an operator.

Investigations have also been carried out with regard to the use of optical fibers in tactical communication systems. In order to control sophisticated weapons systems in conjunction with dispersed military units there is a requirement for tactical command and control communications (often termed C^3). These communication systems must be rapidly deployable, highly mobile, reliable and have the ability to survive in military environments. Existing multi-channel communication cable links employing coaxial cable or wire pairs do not always meet these requirements [Ref. 30]. They tend to be bulky, difficult to install (requiring long installation times), and are susceptible to damage. In contrast optical fiber cables offer special features which may overcome these operational deficiencies. These include small size, light weight, increased flexibility, enhanced bandwidth, low attenuation removing the need for inter-mediate repeaters, and immunity to both EMI and EMP. Furthermore, optical fiber cables generally demonstrate greater ruggedness than conventional deployable cables, making them appear ideally suited for this application.

Optical fiber cables have been installed and tested within the Ptarmigan tactical communication system developed for the British Army [Ref. 31]. They may be utilized as a direct replacement for the HF quad cable system pre-viously employed for the intranodal multichannel cable links within the system [Ref. 5]. The optical fiber element of the system comprises an LED source emitting at a wavelength of 0.9 μm, graded index fiber and an APD detector. It

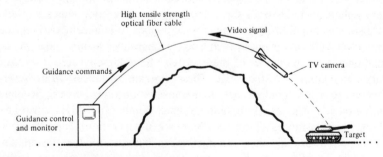

Fig. 11.6   A fiber guided weapons system.

is designed to operate over a range of up to 2 km at data rates of 256, 512 and 2048 kbit s^{-1} without the use of intermediate repeaters. The optical fiber cable assemblies are about half the weight of the HF quad cable, and are quick and easy to deploy in the field. Furthermore, special ruggedized expanded beam optical connectors (see Section 4.12) have been shown to be eminently suitable for use in conditions involving dust, dirt, rough handling and extreme climates. Successful integration of an optical fiber system into a more complex tactical communication system for use in the military environment has demonstrated its substantial operational and technical advantages over HF metallic cable systems.

In summary, it appears that confidence is being established in this new technology such that its widescale use in military applications in the future is ensured.

11.4 CIVIL AND CONSUMER APPLICATIONS

11.4.1 Civil

The introduction of optical fiber communication systems into the public network has stimulated investigation and application of these transmission techniques by public utility organizations which provide their own communication facilities over moderately long distances. For example these transmission techniques may be utilized on the railways and along pipe and electrical power lines. In these applications, although high capacity transmission is not usually required, optical fibers may provide a relatively low cost solution; also giving enhanced protection in harsh environments, especially in relation to EMI and EMP. Experimental optical fiber communication systems are under investigation with a number of organizations in Europe, North America and Japan. For instance, British Rail have successfully demonstrated a 2 Mbit s^{-1} system suspended between the electrical power line gantries over a 6 km route in Cheshire |Ref. 3|. Also the major electric power companies are showing a great deal of interest with regard to the incorporation of optical fibers within the metallic earth of overhead electric power lines |Ref. 34|.

It was indicated in Section 11.2.3 that optical fibers are eminently suitable for video transmission. Thus optical fiber systems are starting to find use in commercial television transmission. These applications include short distance links between studio and outside broadcast vans, links between studios and broadcast or receiving aerials, and close circuit television (CCTV) links for security and traffic surveillance. In addition, the implementation of larger networks for cable and common antenna television (CATV) has demonstrated the successful use of optical fiber communications in this area where it provides significant advantages, in terms of bandwidth and unrepeatered transmission distance, over conventional video links.

One of the first commercial optical fiber video systems was installed in Hastings, UK, in 1976 by Rediffusion Limited for the transmission of television signals over a 1.4 km link for distribution to 34,000 customers. Another early optical fiber CATV field trial was the Hi-OVIS project carried out in Japan [Ref. 36]. The project involved the installation of an interactive video system, plus FM audio and digital data to 160 home subscribers and 8 local studio terminals in various public premises. The system operated over a 6 km distribution cable consisting of 36 fibers plus additional branches to the various destination points: no repeaters were used in this entire network.

Various techniques have been utilized for video transmission including baseband intensity modulation, subcarrier intensity modulation (e.g. FM–IM), pulse analog techniques (e.g. PFM–IM) and digital pulse code modulation (PCM–IM). Generally digital transmission is preferred on larger CATV networks as it allows time division multiplexing as well as greater unrepeatered transmission distance [Ref. 38]. It also avoids problems associated with the nonlinearities of optical sources. An example of commercial digital video transmission is a 7.8 km optical fiber trunk system operating at 322 Mbit s^{-1} in London, Ontario, Canada [Ref. 39]. This system carries 12 video channels and 12 FM stereo channels along 8 fibers installed in a 13 mm cable. A similar digital trunk system has been installed in a CATV network in Denmark [Ref. 40]. This link, using 12 fibers again operating over a distance of 7.8 km, has a capacity of 8 video channels and 12 FM stereo channels.

However, digital transmission of video signals is not always economic, owing to the cost and complexity of the terminal equipment. Hence, optical fiber systems using direct intensity modulation often provide an adequate performance for a relatively small system cost. For example, a block schematic of a long distance analog baseband video link for monitoring railway line appearances such as road crossings, tunnels and snowfall areas is shown in Fig. 11.7 [Ref. 41]. Video signals from TV cameras installed at monitoring points C, are gathered to the concentrating equipment through local transmission lines. These signals are then multiplexed in time, frequency or wavelength on to the main transmission line to the monitoring center. An

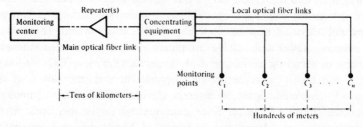

Fig. 11.7 Block schematic of an optical fiber baseband video system for railway line monitoring [Ref. 41].

experimental optical fiber system operating at a wavelength of 1.32 μm using multimode graded index fiber and baseband intensity modulation was installed along a main line of the Japanese national railway |Ref. 41|. Successful video transmission over 16.5 km without intermediate repeaters was achieved, demonstrating the use of fiber systems for this application.

A similar CCTV monitoring system has been implemented for the Kobe Mass Transit System, also in Japan [Ref. 42]. This system, which operates over shorter distances of between 300 m and 5 km, also uses analog intensity modulation together with wavelength division multiplexing of four TV channels at wavelengths of 730, 780, 837 and 879 nm onto multimode step index fiber.

In common with the military applications, other potential civil uses for optical fiber systems include short distance communications within buildings (e.g. broadcast and recording studios) and within mobiles such as aircraft and ships. However, perhaps the largest market for optical fiber systems may eventually be within consumer applications.

11.4.2 Consumer

A major consumer application for optical fiber systems is within automotive electronics. Work is progressing within the automobile industry towards this end together with the use of microcomputers for engine and transmission control as well as control of convenience features such as power windows and seat controls. Optical fiber communication links in this area provide advantages of reduced size and weight together with the elimination of EMI. Furthermore, it is likely they will reduce costs by allowing for an increased number of control signals in the confined space presented by the steering column and internal transmission paths within the vehicle through multiplexing of signals onto a common optical highway.

Such techniques have been under investigation by General Motors for a number of years and a prototype system was reported |Ref. 44| to have demonstrated the feasibility in 1980. This system utilized a bundle of 48 high loss plastic fibers with a simple LED emitting in the visible spectrum. Further developments in the USA and elsewhere suggest that large core diameter (1 mm) single plastic fibers will be utilized in automobile multiplex systems within the passenger compartment, whereas glass fibers will be required to stand the high temperatures (120 °C) encountered in the engine compartment.

Other consumer applications are likely to include home appliances where together with microprocessor technology, optical fibers may be able to make an impact by the late 1980s. However, as with all consumer equipment, progress is very dependent on the instigation of volume production and hence low cost. This is a factor which is likely to delay wider application of optical fiber systems in this area.

11.5 INDUSTRIAL APPLICATIONS

Industrial uses for optical fiber communications cover a variety of generally on-premise applications within a single operational site. Hence the majority of industrial applications tend to fall within the following design criteria [Ref. 47]:

(a) digital transmission at rates from d.c. to 20 Mbit s^{-1}, synchronous or asynchronous, having compatibility with a common logic family (i.e. TTL or CMOS), being independent of the data format and with bit error rates less than 10^{-9};
(b) analog transmission from d.c. to 10 MHz, exhibiting good linearity and low noise;
(c) transmission distances from several meters up to a maximum of kilometers, although generally 1 km will prove sufficient;
(d) a range of environments from benign to harsh, and often exhibiting severe electromagnetic interference from industrial machinery.

Optical fiber systems with performances to meet the above criteria are readily available at a reasonable cost. These systems offer reliable telemetry and control communications for industrial environments where EMI and EMP cause problems for metallic cable links. Furthermore, optical fiber systems provide a far safer solution than conventional electrical monitoring in situations where explosive or corrosive gases are abundant (e.g. chemical processing and petroleum refining plants). Hence the increasing automation of process control, which is making safe, reliable communication in problematical environments essential, is providing an excellent area for the application of optical fiber communication systems.

For example, optical fiber systems have been successfully employed in nuclear testing applications in the USA by the Department of Energy. Two plasma diagnostic experiments developed by the Los Alamos Scientific Laboratory [Ref. 48] were carried out at the Nevada Test Site in Mercury, Nevada. These experiments utilized the unique properties of optical fibers to provide diagnostic capabilities which are not possible with coaxial cable systems. In the first experiment a wideband fiber system (1 GHz bandwidth) was used to record the wideband data from gamma ray sources. The second experiment, a neutron imaging system, provided a time and space resolution for a neutron source on a nanosecond time scale. The neutron source was attenuated and imaged through a pinhole onto an array of scintillator filaments, each of which was aligned to a single PCS fiber for transmission via a graded index fiber to a photomultiplier. A pulsed dye laser was used for system calibration. Both amplitude and overall timing calibration were achieved with an optical time domain reflectometer (see Section 5.7) being used regularly to record the fiber attenuation. It was estimated [Ref. 48] that this system provided a bandwidth advantage of at least a factor of 10 over coaxial cable, at approximately half the cost, and around one-fiftieth of the weight.

11.5.1 Sensor Systems

It has been indicated that optical fiber transmission may be advantageously employed for monitoring and telemetry in industrial environments. The application of optical fiber communications to such sensor systems has stimulated much interest, especially for use in electrically hazardous environments where conventional monitoring is difficult and expensive. There is a requirement for the accurate measurement of parameters such as liquid level, flow rate, position, temperature and pressure in these environments which may be facilitated by optical fiber systems. Early work in this area featured electrical or electro-optical transducers along with optical fiber telemetry systems. A novel approach of this type involved a piezoelectric transducer which was used to apply local deformations to a single fiber highway causing phase modulation of the transmitted signal |Ref. 50|. The unmodulated signal from the same optical source was transmitted via a parallel reference fiber to enable demodulation of the signals from various piezoelectric transducers located on the highway. This technique proved particularly useful when a number of monitoring signals were required at a central control point.

Electro-optical transducers together with optical fiber telemetry systems offer significant benefits over purely electrical systems in terms of immunity to EMI and EMP as well as intrinsic safety in the transmission to and from the transducer. However, they still utilize electrical power at the site of the transducer which is also often in an electrically problematical environment. Therefore much effort is currently being expended in the investigation and development of entirely optical sensor systems. These employ passive optical transducer mechanisms which directly modulate the light for the optical fiber telemetry link. A number of simple optical techniques which enable direct measurement are illustrated in Fig. 11.8. For instance, a fluid level may be detected by the sensor shown in Fig. 11.8(a). When the fluid, which has a refractive index greater than the glass forming the optical dipstick, reaches the chamfered end, total internal reflection ceases and the light is transmitted into the fluid. Hence an indication of the fluid level is obtained at the optical receiver. Although this system is somewhat crude and will not give a continuous measurement of fluid level, it is simple and safe for use with flammable liquids.

Optical sensor mechanisms which provide measurement of displacement are shown in Figs. 11.8(b) and (c). The former is a reflective (often called fotonic) sensor whereby light is transmitted via a fiber(s) to illuminate a target. Light reflected from the target is received by a return fiber(s) and is a function of the distance between the fiber ends and the target d. Hence the position of the target or displacement may be registered at an optical receiver. Figure 11.8(c) illustrates the measurement of displacement using a Moiré fringe modulator. In this case the opaque lined gratings produce dark Moiré fringes. Transverse movement of one grating with respect to the other causes the fringes to move up or down. Therefore a count of the fringes as the gratings are displaced

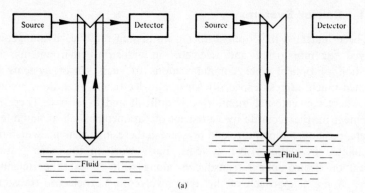

(a)

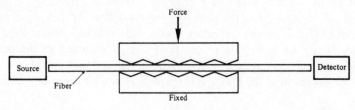

(b)

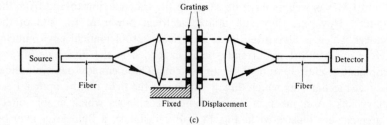

(c)

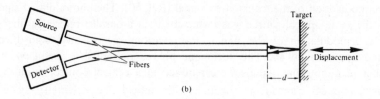

(d)

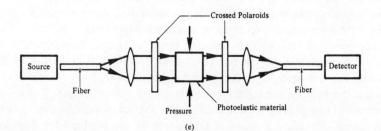

(e)

provides a measurement of the displacement. Unlike the previous techniques the Moiré fringe modulator gives a digital measurement (fringe counting) of displacement which is independent of any drift in the characteristics of the optical source. However, mechanical vibrations may severely affect the measurement accuracy and prove difficult to eradicate. Also there are problems involved with the loss of count if, for any reason, optical power to the sensor is interrupted.

The sensors shown in Figs. 11.8(d) and (e) primarily give a measurement of strain or stress. However, this may easily be converted too from a displacement, temperature or pressure. Figure 11.8(d) illustrates a microbending sensor in which the fiber is bent sharply when a force is applied to metal teeth or an array of pins. Light transmitted down the fiber is lost into the cladding due to the microbending (see Section 4.6.2) providing a measurement of the applied force. Hence changes in the applied force (e.g. strain, displacement, temperature, pressure) cause a change in light intensity of the optical receiver which may be recorded.

With the photoelastic sensor shown in Fig. 11.8(e) the light transmitted is a function of stress within the material. This phenomenon, known as bi-refringence, occurs with the application of mechanical stress to transparent isotropic materials (e.g. polyurethane) whereby they become optically anisotropic. An advantage of this technique is that the stress may be induced directly by pressure without the need for an intermediate mechanism (i.e. pressure to displacement). A drawback, however, is that the birefringence exhibited by photoelastic materials is very temperature-dependent making measurement of a single parameter difficult.

Figure 11.9 shows a possible optical fiber flow meter. A multimode optical fiber is inserted across a pipe such that the liquid flows past the transversely stretched fiber. The turbulence resulting from the fiber's presence causes it to oscillate at a frequency roughly proportional to the flow rate. This results in a corresponding oscillation in the mode power distribution within the fiber giving a similarly modulated intensity profile at the optical receiver. The technique has been used to measure flow rates from 0.3 to 3 m s^{-1} [Ref. 56]. However, it cannot measure flow rates below those at which turbulence occurs.

The most sensitive passive optical sensors to date employ an interferometric approach as illustrated in Fig. 11.10. These devices interfere coherent monochromatic light propagating in a strained or temperature varying fiber with light either directly from the laser source, or (as shown in Fig. 11.10) guided by a reference fiber isolated from the external influence. The effects of

---

Fig. 11.8 Examples of simple optical fiber measurement techniques: (a) optical fluid level detector; (b) reflective or fotonic displacement sensor; (c) Moiré fringe modulation sensor; (d) fiber transmission modulation sensor—microbending transducer; (e) photoelastic sensor.

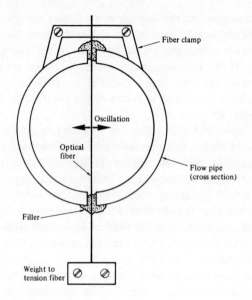

Fig. 11.9 An optical fiber flow meter [Ref. 56].

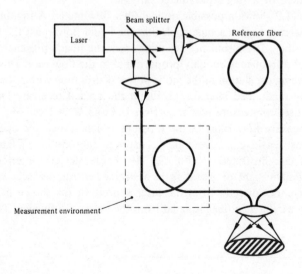

Fig. 11.10 A two-arm interferometric sensor.

strain, pressure or temperature change, give rise to differential optical paths by changing the fiber length, core diameter or refractive index with respect to the reference fiber. This provides a phase difference between the light emitted from the two fibers giving interference patterns, as shown in Fig. 11.10. Very accurate measurements of pressure or temperature may be obtained from these patterns. For example, using fused silica in such a two-arm fiber interferometer, it can be shown that the temperature sensitivity is about 107 rad °C^{-1} m^{-1} [Ref. 57]. Other applications for optical fiber interfero-metric sensors, which are attracting considerable attention, are the optical fiber gyroscope [Ref. 62] and hydrophone [Ref. 63]. The former device is based on the classical Sagnac 'ring' interferometer and provides a measure-ment of rotation, whereas the latter effectively measures acoustic pressure.

As a final example, the polarization sensor shown in Fig. 11.11 may be utilized to measure current along a metallic conductor [Ref. 64]. This device consists of a single polarization-maintaining single mode fiber which passes up from earth to loop around the current-carrying conductor before passing back to earth. A He–Ne laser beam is linearly polarized and launched into the fiber which is then stripped of any cladding modes. The direction of polarization of the light in the fiber core is rotated by the longitudinal magnetic field around the loop, via the action of the Faraday magneto-optic effect [Ref. 65]. A Wollaston prism is used to sense the resulting rotation and resolves the emerg-ing light into two orthogonal components. These components are separately detected with a photodiode prior to generation of a sum and difference signal of the two intensities (I_1 and I_2). The difference signal normalized to the sum

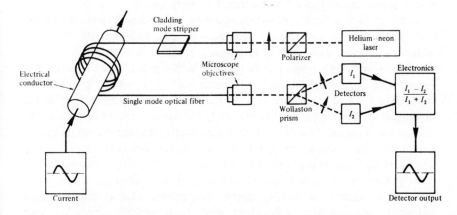

Fig. 11.11 Single mode optical fiber sensor for current measurement [Ref. 64].

gives a parameter which is proportional to the polarization rotation ρ, following [Ref. 64]:

$$K\rho = \frac{I_1 - I_2}{I_1 + I_2} \qquad (11.1)$$

where K is a constant which is dependent on the properties of the fiber. Hence a current measurement (either d.c. or a.c.) may be obtained which is independent of the received light power.

The methods outlined above provide a basis for measurement of the majority of physical parameters using passive optical techniques, and may usually be incorporated into remote optical fiber telemetry systems. At present sensors based on these techniques are generally not commercially available, being very much at the research and development or prototype stages. Although in most cases there are still significant problems to be overcome before practical sensors can be realized, it is likely that devices utilizing these or similar techniques will be on the market in the near future.

11.6 COMPUTER APPLICATIONS

Modern computer systems consist of a large number of interconnections. These range from lengths of a few micrometers (when considering on chip very large scale integration (VLSI) connections) to perhaps thousands of kilometers for terrestrial links in computer networks. The transmission rates over these interconnections also cover a wide range from around 100 bit s^{-1} for some teletype terminals to several hundred Mbit s^{-1} for the on-chip connections. Optical fibers are starting to find application in this connection hierarchy where secure, interference-free transmission is required.

Although in its infancy, integrated optics has stimulated interest in connections within equipment, between integrated circuits, and even within hybrid integrated circuits, using optical techniques. Much of this work is still at the research stage and therefore will be pursued further in Sections 11.7 and 11.8. Nevertheless it is likely that optical transmission techniques and optical fibers themselves will find application within data processing equipment. In addition, investigations have already taken place into the use of optical fibers for mains isolators and digital data buses within both digital telephone exchanges and computers. Their small size, low loss, low radiation properties and freedom from ground loops provide obvious advantages in these applications.

At present, however, a primary potential application for optical fiber communications occurs in interequipment connections. These provide noise immunity, security and removal of earth loop problems, together with increased bandwidth and reduced cable size in comparison with conventional

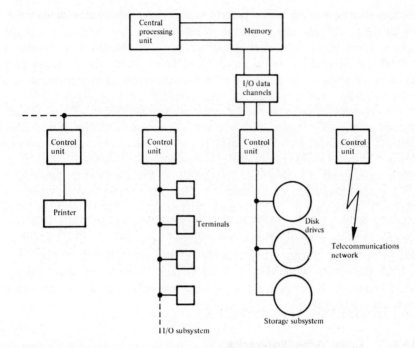

Fig. 11.12   Block schematic of a typical mainframe computer system.

coaxial cable computer system interconnections. The interequipment connection topology for a typical mainframe computer system (host computer) is illustrated in Fig. 11.12. The input/output (I/O) to the host computer is generally handled by a processor, often called a data channel or simply channel, which is attached to the main storage of the host computer. It services all the I/O requirements for the system allowing concurrent instruction processing by the central processing unit (CPU). Each data channel contains an interface to a number of I/O control units. These, in turn, control the I/O devices (e.g. teletypes, visual display units, magnetic disk access mechanisms, magnetic tape drives and printers). When metallic cables are used, the interface between the data channel and the control units comprises a large number (often at least 48) of parallel coaxial lines incorporated into cables. An attractive use of optical fiber interconnection is to serialize this channel interface [Ref. 66] using a multiplex system. This significantly reduces cable and connector bulk and improves connection reliability.

Optical fiber links of this type were demonstrated in 1978 by Sperry Univac [Ref. 67], and subsequently Fujitsu developed a product to perform the same function. However, neither of these systems offered enhanced channel performance as measured by the product of bit rate and link length. Developments are therefore continuing with regard to high performance channel links utilizing new protocols for data exchange. A recently reported |Ref. 66| prototype optical fiber serial subsystem designed by IBM Research operates at 200 Mbit s^{-1} over distances of up to 1 km. This system utilizes a laser chip mounted on a silicon substrate with the fiber encapsulated in monolithic dual in line package, and a single chip, high sensitivity, silicon $p-i-n$ receiver.

The other interconnection requirement for the mainframe computer system is between the I/O control units and the I/O terminals. Again optical fiber systems can provide high speed, multiplexed, secure communication links to replace the multitude of coaxial cables normally required for these interconnections. An example of such a fiber system utilizes a multiplexing system onto a single optical fiber cable for connecting an IBM 3274 controller to its terminals [Ref. 68]. In this case up to 32 terminals and printers can be linked to the controller in either a point to point or multidrop* configuration employing a star coupler or beam splitters. This interconnection requirement is often extended due to the trend of connecting numbers of processors together in order to balance the system work load, increase system reliability and share storage and I/O devices. Hence optical fiber systems are under investigation for use in local area networks.

11.6.1 Local Area Networks

A local area network (LAN) is generally defined as an interconnection topology entirely confined within a geographical area of a few square kilometers. It is therefore usually confined to either a single building or a group of buildings contained within a site or establishment (industrial, military, educational, etc.). Hence, the data processing and peripheral equipment together with any communication links are usually under the control of the owning body rather than a common carrier.† As mentioned previously such a network may support terminals connected to a host computer, or provide communication between multiple processors and terminals or work stations. In the latter application the LAN may also provide an interface to the local telecommunications network. Furthermore within a manufacturing facility it may provide interconnection of a host computer to remote process controllers allowing computer aided manufacture (CAM).

In common with local and rural networks (see Section 11.2.3) LANs may

---

* At present the multidrop bus configuration will not allow interconnection of as many as 32 terminals due to the insertion losses obtained at the beam splitters or T-couplers.

† Another possible definition of a LAN, based on speed and range of operation, is that a LAN typically operates at a transmission rate of between 100 kBit s^{-1} and 100 Mbit s^{-1} over distances of 500 m to 10 km. Hence a LAN is intermediate between a short range, multiprocessor network (usually data bus) and a wide area network which provides relatively low speed data transmission (up to 100 kbit s^{-1}) over very long distances using conventional communications technology. However, it must be noted that there are always exceptions to these general definitions which will doubtless increase in number as the technology advances.

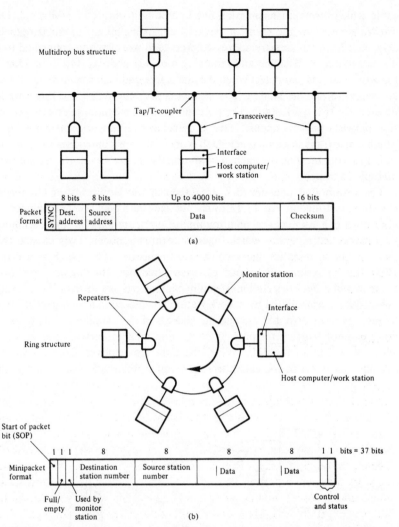

Fig. 11.13 Local area networks: (a) the Ethernet network topology and packet format; (b) the Cambridge ring topology and packet format.

be designed in three major configurations: the star, ring and bus. Work has been in progress for a number of years on various network architectures and protocols, in the main utilizing metallic communication links. To date a standard configuration has not been universally adopted. However, two basic techniques for the implementation of local area computer networks have obtained partial acceptance. These two network topologies are illustrated in Fig. 11.13 and are known as the Ethernet and the Cambridge ring. The Ethernet network, developed by Xerox [Ref. 69], consists of a multidrop bus configuration whereby host computers or work stations are attached to a coaxial cable which forms a transmission line operating at a rate of 3 Mbit s^{-1}. Data are transmitted in the form of packets consisting of up to 4000 data bits (Fig. 11.13(a)). Packets are addressed, and each work station connected to the line is capable of detecting and removing a packet addressed to it. In addition, the work stations can detect when the line is free and hence transmit a message addressed to another station. The system is peer controlled and therefore it is possible for two work stations to transmit at the same time. In this case a collision between packets occurs. This is detected by the work stations involved which cease transmission. When each work station is contending for transmission time on the link in such a manner, the result is known as statistical multiplexing.

The Cambridge ring network developed at the University of Cambridge, UK and illustrated in Fig. 11.13(b) also utilizes data packets. The ring consists of a monitor station as well as a number of work stations (or host computers) with associated repeaters which together form ring nodes. Twin twisted cable pairs are often used for the loop, and transmission takes place at a rate of 10 Mbit s^{-1}. A set number of packets are contained by the ring of a size known to the monitor and work stations (usually minipackets as shown in Fig. 11.13(b)). When data is transmitted by a work station it is placed in a mini-packet format together with its own address and the address of its destination. The work stations monitor the ring for packets containing their respective addresses from which they extract the data. When the data are removed from a packet, a response bit is set to indicate correct receipt of the data. The packet is then

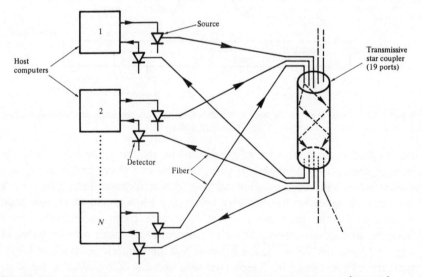

Fig. 11.14   The passive transmissive star network adopted for Fibernet [Ref. 70].

returned to the source which checks it before sending it on empty. Throughout this process the monitor station is continually checking the ring structure by the use of test packets transmitted around the ring.

Although LAN configurations and protocols are still largely under development, optical fibers have been successfully utilized to provide such computer interconnection. An early experiment using multimode step index fiber in a substantially modified Ethernet configuration called Fibernet was undertaken by Xerox [Ref. 70]. Optical fibers, however, displayed drawbacks when used in the multidrop bus network. These resulted from the high insertion losses encountered at optical beam splitters (or T-couplers) which only allowed a small number of work stations to be connected (generally less than ten). Consequently a passive transmissive star network (Fig. 11.14) was adopted for Fibernet. A 19 port transmissive star coupler was utilized which gave an insertion loss of 10 dB between any 2 ports. Using the Ethernet packet switching and protocol, data were successfully transmitted at 150 Mbit s^{-1} and 100 Mbit s^{-1} over distances of 0.5 and 1.1 km respectively, with zero errors.

Perhaps the largest scale applications at present for optical fiber systems within local area networks are with regard to single channel and multiplexed star networks. These network configurations tend to match the communication system design of large mainframes and minicomputers. A typical network is shown in Fig. 11.15. The high bandwidth provided by optical fibers often allows both asynchronous and synchronous terminals to be driven at full rate

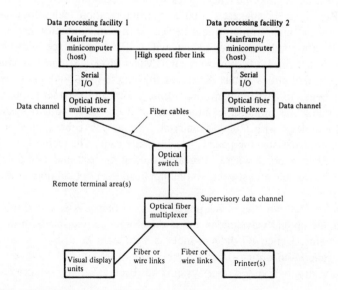

Fig. 11.15 An example of an optical fiber local area network.

without the need for the statistical multiplexing required in the case of Fiber-net. Recent installations of this type include airport data communication networks, traffic monitoring and control schemes, and company on-site or local intersite networks.

Optical fiber systems are also being utilized in the Cambridge ring configuration. They have fuelled the development of a higher performance ring operating at 40 Mbit s^{-1} which promises to find significant application. Furthermore, improvements in optical fiber connector technology and in the optical output power provided by light sources may lead to the extensive use of bus networks in the future.

11.7 INTEGRATED OPTICS

The multitude of potential application areas for optical fiber communications coupled with the tremendous advances in the field have over the last few years stimulated a resurgence of interest in the area of integrated optics (IO). The concept of IO involves the realization of optical and electro-optical elements which may be integrated in large numbers on to a single substrate. Hence, IO seeks to provide an alternative to the conversion of an optical signal back into the electrical regime prior to signal processing by allowing such processing to be performed on the optical signal. Thin transparent dielectric layers on planar substrates which act as optical waveguides are used in IO to produce miniature optical components and circuits.

The birth of IO may be traced back to basic ideas outlined by Anderson in 1965 [Ref. 74]. He suggested that a microfabrication technology could be developed for single mode optical devices with semiconductor and dielectric materials in a similar manner to that which had taken place with electronic circuits. It was in 1969, however, after Miller [Ref. 75] had introduced the term 'integrated optics' whilst discussing the long term outlook in the area, that research began to gain momentum. Although some of the wider implications of IO, including the monolithic integration of a complete optical system in a single technology with a single material, have as yet to be applied in optical fiber communications, the pace is steadily increasing. The technology has now progressed to a point where integrated optical devices and components are becoming available and starting to find application in optical fiber communication systems.

A major factor in the development of IO is that it is essentially based on single mode optical waveguides and is therefore incompatible with multimode fiber systems. Hence IO did not make a significant contribution to first and second generation optical fiber systems. The development, however, of third generation or single mode fiber systems has been aided* by integrated optical

---

* This is especially the case in relation to the fabrication of single mode injection lasers (see Section 6.6).

techniques. Furthermore, it is within single mode fiber systems where IO will allow optical signal processing to replace electronic signal processing through the creation of a family of thin film optical and electro-optical components which may be located on a single substrate. The devices of interest in IO are often the counterparts of microwave or bulk optical devices. These include junctions and directional couplers, switches and modulators, filters and wavelength multiplexers, lasers and amplifiers, detectors and bistable elements. It is envisaged that developments in this technology will provide the basis for the fourth generation systems mentioned in Section 11.1 where full monolithic integration may be achieved.

11.7.1 Planar Waveguides

The use of circular dielectric waveguide structures for confining light is universally utilized within optical fiber communications. IO involves an extension of this guided wave optical technology through the use of planar optical waveguides to confine and guide the light in guided wave devices and circuits. The mechanism of optical confinement in symmetrical planar waveguides was discussed in Section 2.3 prior to investigation of circular structures. In fact the simplest dielectric waveguide structure is the planar slab guide shown in Fig. 11.16. It comprises a planar film of refractive index n_1 sandwiched between a substrate of refractive index n_2 and a cover layer of refractive index n_3 where $n_1 > n_2 \geqslant n_3$. Often the cover layer consists of air where $n_3 = n_0 = 1$, and it exhibits a substantially lower refractive index than the other two layers. In this case the film has layers of different refractive index above and below the guiding layer and hence performs as an asymmetric waveguide.

In the discussions of optical waveguides given in Chapter 2 we were solely concerned with symmetrical structures. When the dimensions of the guide are reduced so are the number of propagating modes. Eventually the waveguide

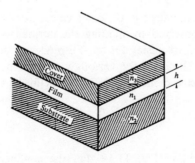

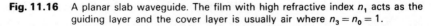

Fig. 11.16 A planar slab waveguide. The film with high refractive index n_1 acts as the guiding layer and the cover layer is usually air where $n_3 = n_0 = 1$.

dimensions are such that only a single mode propagates, and if the dimensions are reduced further this single mode still continues to propagate. Hence there is no cutoff for the fundamental mode in a symmetric guide. This is not the case for an asymmetric guide where the dimensions may be reduced until the structure cannot support any modes and even the fundamental is cutoff. If the thickness or height of the guide layer of a planar asymmetric guide is h (see Fig. 11.16), then the guide can support a mode of order m with a wavelength λ, when [Ref. 76]:

$$h \geqslant \frac{(m + \frac{1}{2})\lambda}{2(n_1^2 - n_2^2)^{\frac{1}{2}}} \tag{11.2}$$

Equation (11.2) which assumes $n_2 > n_3$ defines the limits of the single mode region for h between values when $m = 0$ and $m = 1$. Hence for a typical thin film glass guide with $n_1 = 1.6$ and $n_2 = 1.5$, single mode operation is maintained only when the guide has a thickness in the range $0.45\lambda \leqslant h \leqslant 1.35\lambda$.

An additional consideration of equal importance is the degree of confinement of the light to the guiding layer. The light is not exclusively confined to the guiding region and evanescent fields penetrate into the substrate and cover. An effective guide layer thickness h_{eff} may be expressed as:

$$h_{\text{eff}} = h + x_2 + x_3 \tag{11.3}$$

where x_2 and x_3 are the evanescent field penetration depths for the substrate and cover regions respectively. Furthermore, we can define a normalized effective thickness H for an asymmetric slab guide as:

$$H = kh_{\text{eff}}(n_1^2 - n_2^2)^{\frac{1}{2}} \tag{11.4}$$

where k is the free space propagation constant equal to $2\pi/\lambda$. The normalized frequency (sometimes called the normalized film thickness) for the planar slab guide following Eq. (2.68) is given by:

$$V = kh(n_1^2 - n_2^2)^{\frac{1}{2}} \tag{11.5}$$

An indication of the degree of confinement for the asymmetric slab waveguide may be observed by plotting the normalized effective thickness against the normalized frequency for the TE modes. A series of such plots is shown in Fig. 11.17 [Ref. 77] for various values of the parameter a which indicates the asymmetry of the guide, and is defined as:

$$a = \frac{n_2^2 - n_3^2}{n_1^2 - n_2^2} \tag{11.6}$$

It may be observed in Fig. 11.19 that the confinement improves with decreasing film thickness only up to a point where $V \simeq 2.5$. For example, the

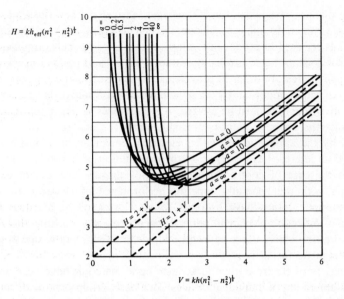

$$H = kh_{eff}(n_1^2 - n_2^2)^{\frac{1}{2}}$$

$$V = kh(n_1^2 - n_2^2)^{\frac{1}{2}}$$

Fig. 11.17 The normalized effective thickness H as a function of the normalized frequency V for a slab waveguide with various degrees of asymmetry. Reproduced with permission from H. Kogelnik and V. Ramaswamy, *Appl. Opt.*, **13**, p. 1857, 1974.

minimum effective thickness for a highly asymmetric guide ($a = \infty$) occurs when $H_{min} = 4.4$ at $V = 2.55$. Using Eq. (11.4) this gives a minimum effective thickness of:

$$(h_{eff})_{min} = \frac{4.4}{k} (n_1^2 - n_2^2)^{-\frac{1}{2}} \tag{11.7}$$

$$= 0.7 \lambda (n_1^2 - n_1^2)^{-\frac{1}{2}}$$

Therefore considering a typical glass waveguide ($n_1 = 1.6$ and $n_2 = 1.5$), we obtain a minimum effective thickness of:

$$(h_{eff})_{min} = 1.26\lambda \tag{11.8}$$

Assuming a minimum operating wavelength to be 0.8 μm limits the effective thickness of the guide, and hence the confinement to around 1 μm. Therefore it appears there is a limit to possible microfabrication with IO which is not present in other technologies* [Ref. 79]. At present there is still ample scope but confinement must be considered along with packing density and the avoidance of crosstalk.

---

* The 1 μm barrier to confinement applies with all suitable waveguide materials. However, metal clad waveguides are not so limited but are plagued by high losses [Ref. 78].

The planar waveguides for IO may be fabricated from glasses and other isotropic materials such as silicon dioxide and polymers. Although these materials are used to produce the simplest integrated optical components, their properties cannot be controlled by external energy sources and hence they are of limited interest. In order to provide external control of the entrapped light to cause deflection, focusing, switching and modulation, active devices employing alternative materials must be utilized. A requirement for these materials is that they have the correct crystal symmetry to allow the local refractive index to be varied by the application of either electrical, magnetic or acoustic energy.*

To date interest has centered on the exploitation of the electro-optic effect due to the ease of controlling electric fields through the use of electrodes together with the generally superior performance of electro-optic devices. Acousto-optic devices have, however, found a lesser role, primarily in the area of beam deflection. Magneto-optic devices [Ref. 80] utilizing the Faraday effect are not widely used, as in general, electric fields are easier to generate than magnetic fields.

A variety of electro-optic and acousto-optic materials have been employed in the fabrication of individual devices. Two basic groups can be distinguished by their refractive indices. These are materials with a refractive index near 2 ($LiNbO_3$, $LiTaO_3$, NbO_5, ZnS and ZnO) and materials with a refractive index greater than 3 (GaAs, InP and compounds of Ga and In with elements of Al, As and Sb).

Planar waveguide structures are produced using several different techniques which have in large part been derived from the microelectronics industry. For example, passive devices may be fabricated by radiofrequency sputtering to deposit thin films of glass onto glass substrates. Alternatively active devices are often produced by titanium (Ti) diffusion into lithium niobate ($LiNbO_3$) or by ion implantation into gallium arsenide [Ref. 81].

The planar slab waveguide shown in Fig. 11.16 confines light in only one direction, allowing it to spread across the guiding layer. In many instances it is useful to confine the light in two dimensions to a particular path on the surface of the substrate. This is achieved by defining the high index guiding region as a thin strip (strip guide) where total internal reflection will prevent the spread of the light beam across the substrate. In addition the strips can be curved or branched as required. Examples of such strip waveguide structures are shown in Fig. 11.18. They may be formed as either a ridge on the surface of the substrate or by diffusion to provide a region of higher refractive index below the substrate, or as a rib of increased thickness within a thin planar slab. Techniques employed to obtain the strip pattern include electron and laser beam lithography as well as photolithography. The rectangular waveguide configurations illustrated in Fig. 11.18 prove very suitable for use with electro-optic

---

* Using the electro-optic, magneto-optic or acousto-optic effects [Ref. 65].

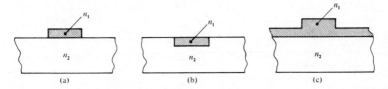

Fig. 11.18 Cross section of some strip waveguide structures: (a) ridge guide; (b) diffused channel (embedded strip) guide; (c) rib guide.

deflectors and modulators giving a reduction in the voltage required to achieve a particular field strength. In addition they allow a number of optical paths to be provided on a given substrate.

Losses exhibited by both slab and strip waveguides are generally much greater than those obtained in optical fibers. Typical losses which are both wavelength and material dependent are in the range 0.5–5 dB cm^{-1}. Furthermore, although the effects of interest in IO are usually exhibited over short distances of the order of a wavelength, efficient devices require relatively long interaction lengths, the effects being cumulative. Thus typical device lengths range from 0.5 to 10 mm.

11.8 INTEGRATED OPTICAL DEVICES

In this section some examples of various types of integrated optical devices together with their salient features are considered. However, the numerous developments in this field exclude any attempt to provide other than general examples in the major areas of investigation which are pertinent to optical fiber communications. The requirement for multichannel communication within the various systems considered in Chapter 10 demands the combination of information from separate channels, transmission of the combined signals over a single optical fiber link, and separation of the individual channels at the receiver prior to routing to their individual destinations. Hence the application of IO in this area is to provide optical methods for multiplexing, modulation and routing. These various functions may be performed with a combination of optical beam splitters, switches, modulators, filters, sources and detectors.

11.8.1 Beam Splitters and Switches

Beam splitters are a basic element of many optical fiber communication systems often providing a Y-junction by which signals from separate sources can be combined, or the received power divided between two or more channels. A passive Y-junction beam splitter fabricated from LiNbO$_3$ is shown in Fig. 11.19. Unfortunately the power transmission through such a splitter

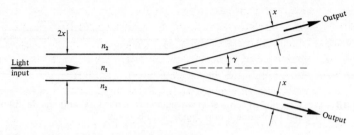

Fig. 11.19   A passive Y-junction beam splitter.

decreases sharply with increasing half angle γ, the power being radiated into the substrate. Hence the total power transmission depends critically upon γ which, for the example chosen, must not exceed 0.5° if an acceptable insertion loss is to be achieved [Ref. 82]. In order to provide effective separation of the output arms so that access to each is possible, the junction must be many times the width of the guide. For example, around 3000 wavelengths are required to give a separation of about 30 µm between the output arms. Therefore, for practical reasons, the device is relatively long.

The passive Y-junction beam splitter finds application where equal power division of the incident beam is required. However, the Y-junction is of wider interest when it is fabricated from an electro-optic material, in which case it may be used as a switch. Such materials exhibit a change in refractive index δn which is directly proportional to an applied electric field* E following,

$$\delta n = \pm \tfrac{1}{2} n_1^3 rE \tag{11.9}$$

where n_1 is the original refractive index, and r is the electro-optic coefficient. Hence an active Y-junction may be fabricated from a single crystal electro-optic material as illustrated in Fig. 11.20. Lithium niobate is often utilized as it combines relatively low loss with large values of electro-optic coefficients† (as high as 30.8×10^{-12} m V^{-1}). Metal electrodes are attached so that when biassing is applied, one side of the waveguide structure exhibits an increased refractive index whilst the value of refractive index on the other side is reduced. The light beam is therefore deflected towards the region of higher refractive index causing it to follow the corresponding output arm. Furthermore, the field is maintained in the electrodes which extend beyond the junction ensuring continuation of the process. With switching voltages around 30 V, these devices prove to be quite efficient allowing for larger junction angles to be tolerated than those of the passive Y-junction beam splitter. However, a physical length of several hundred wavelengths is still required for the switch. These devices therefore serve the function of optical signal routing. In addition, high speed

---

* The linear variation of refractive index with the electric field is known as the Pockels effect [Ref. 65].

† The change in refractive index is related by the applied field via the linear and quadratic electro-optic coefficients [Ref. 80].

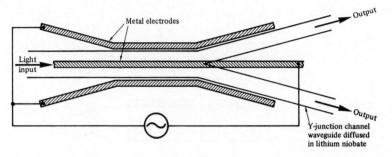

Fig. 11.20 An electro-optic Y-junction switch.

switches can be used to provide time division multiplexing of several lower bit rate channels onto a single mode fiber link.

Switches may also be fabricated by placing two parallel strip waveguides in close proximity to each other as illustrated in Fig. 11.21. The evanescent fields generated outside the guiding region allow transverse coupling between the guides. When the two waveguide modes have equal propagation constants β with amplitudes A and B (Fig. 11.21), then the coupled mode equations may be written as [Ref. 84]:

$$\frac{\mathrm{d}A}{\mathrm{d}z} = -j\beta A + jCB$$

$$\frac{\mathrm{d}B}{\mathrm{d}z} = -j\beta B + jCA$$

(11.10)

where C is the coupling coefficient per unit length. In this case, assuming no losses, all the energy from waveguide X will be transferred to waveguide Y over a coupling length l_0. Furthermore it can be shown [Ref. 85] that for this complete energy transfer l_0 is given by $\pi/2C$. If the waveguide modes have different propagation constants, however, only part of the energy from guide X

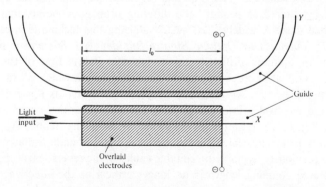

Fig. 11.21 Electro-optically switched directional coupler. The COBRA configuration using two electrodes [Ref. 86].

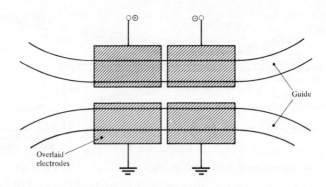

Fig. 11.22   The stepped $\Delta\beta$ reversal coupler switch.

will be coupled into guide Y, and this energy will be subsequently recoupled back into X.

It is also noted that when the propagation constants differ the coupling length l is reduced from the matched value l_0 and although less energy is transferred, the exchange occurs more rapidly. This property may be utilized to good effect in the formation of an optical switch. The mismatch in propagation constants can be adjusted such that the coupling length l is reduced to $l_0/2$. In this case, energy coupled from one guide into the other over a distance $l_0/2$ will be recoupled into the original guide over a similar distance. Hence two distinct cases exist for a switch of length l_0, namely the matched case whereby all the energy is transferred from one guide to the other and the mismatched case when $l = l_0/2$ where over a distance l_0 the energy is recoupled into the original guide.

Optical switches of the above type use electrodes placed on the top of each matched waveguide (Fig. 11.21) so that the refractive indices of the guides are differentially altered to produce the differing propagation constants for the mismatched case. A widely used switch utilizing this technique is called the COBRA (*Commutateur Optique Binaire Rapide*) [Ref. 86] and is normally formed from titanium diffused lithium niobate. Fabrication of the device, however, is critical in order to provide a coupling length which is exactly l_0 or an odd multiple of l_0. An electrode structure which avoids this problem by dividing the electrodes into halves with opposite polarities on each half is shown in Fig. 11.22. With this device, which is called the stepped $\Delta\beta$ reversal coupler, it is always possible to obtain both the matched and mismatched cases described previously by applying suitable values of the reversed voltage. Hence the fabricated coupling length is no longer critical as the effective coupling length of the device may be adjusted electrically to achieve l_0.

11.8.2 Modulators

The limitations imposed by direct current modulation of semiconductor injection lasers currently restricts the maximum achievable modulation frequencies to a few gigahertz. Furthermore, with most injection lasers high speed current modulation also creates undesirable wavelength modulation which imposes problems for systems employing wavelength division multiplexing |Ref. 87|. Thus to extend the bandwidth capability of single mode fiber systems there is a requirement for high speed modulation which can be provided by integrated optical waveguide intensity modulators. Simple on/off modulators may be based on the techniques utilized for the active beam splitters and switches described in Section 11.8.1. In addition a large variety of predominantly electro-optic modulators have been reported |Ref. 88| which exhibit good characteristics. For example, an important waveguide modulator is based upon a Y-branch interferometer which employs optical phase shifting produced by the electro-optic effect.

The change in refractive index exhibited by an electro-optic material with the application of an electric field given by Eq. (11.9) also provides a phase change for light propagating in the material. This phase change $\delta\phi$ is accumulative over a distance L within the material and is given by [Ref. 89]:

$$\delta\phi = \frac{2\pi}{\lambda} \delta n L \qquad (11.11)$$

When the electric field is applied transversely to the direction of optical propagation we may substitute for δn from Eq. (11.9) giving:

$$\delta\phi = \frac{\pi}{\lambda} n_1^3 rE \qquad (11.12)$$

Furthermore taking E equal to VL/d, where V is the applied voltage and d is the distance between electrodes gives:

$$\delta\phi = \frac{\pi}{\lambda} n_1^3 r \frac{VL}{d} \qquad (11.13)$$

It may be noted from Eq. (11.13) that in order to reduce the applied voltage V required to provide a particular phase change, the ratio L/d must be made as large as possible.

A simple phase modulator may therefore be realized on a strip waveguide in which the ratio L/d is large as shown in Fig. 11.23. These devices when, for example, fabricated by diffusion of Nb into $LiTaO_3$ with an L/d ratio of 1000 (i.e. 2 cm/20 μm) provide a phase change of π radians with an applied voltage around 1 V |Ref. 89|. However, as mentioned previously, this property can be

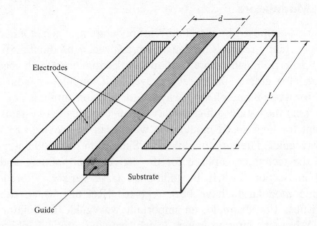

Fig. 11.23 A simple strip waveguide phase modulator.

employed in an interferometric intensity modulator. Such a Mach–Zehnder type interferometer is shown in Fig. 11.24. The device comprises two Y-junctions which give an equal division of the input optical power. With no potential applied to the electrodes, the input optical power is split into the two arms at the first Y-junction and arrives at the second Y-junction in phase giving an intensity maximum at the waveguide output. This condition corresponds to the 'on' state. Alternatively when a potential is applied to the electrodes, which operate in a push–pull mode on the two arms of the interferometer, a differential phase change is created between the signals in the two arms. The subsequent recombination of the signals gives rise to constructive or destructive interference in the output waveguide. Hence the process has the effect of converting the phase modulation into intensity modulation. A phase shift of π between the two arms gives the 'off' state for the device.

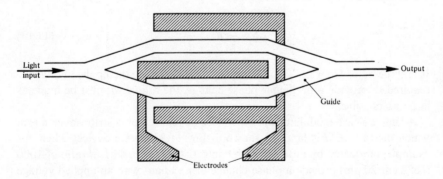

Fig. 11.24 A Y-junction interferometric modulator based on the Mach–Zehnder interferometer.

High speed interferometric modulators have been demonstrated with titanium doped lithium niobate waveguides. A 1.1 GHz modulation bandwidth has been reported [Ref. 90] for a 6 mm interferometer employing a 3.8 V on/off voltage across a 0.9 μm gap. Similar devices incorporating electrodes on one arm only may be utilized as switches and are generally referred to as balanced bridge interferometric switches [Ref. 88].

Useful modulators may also be obtained employing the acousto-optic effect. These devices which deflect a light beam are based on the diffraction of light produced by an acoustic wave travelling through a transparent medium. The acoustic wave produces a periodic variation in density (i.e. mechanical strain) along its path which, in turn, gives rise to corresponding changes in refractive index within the medium due to the photoelastic effect. Therefore, a moving optical phase-diffraction grating is produced in the medium. Any light beam passing through the medium and crossing the path of the acoustic wave is diffracted by this phase grating from the zero order into higher order modes.

Two regimes of operation are of interest: the Bragg regime and the Raman–Nath regime. The interaction, however, is of greatest magnitude in the Bragg regime where the zero order mode is partially deflected into only one higher order (i.e. first order) mode, rather than the multiplicity of higher order modes obtained in the Raman–Nath regime. Hence most acousto-optic modulators operate in the Bragg regime providing the highest modulation depth for a given acoustic power.

The Bragg regime is obtained by effecting a suitably long interaction length for the device so that it performs as a 'thick' diffraction grating. An IO acousto-optic Bragg deflection modulator is shown in Fig. 11.25. It consists of

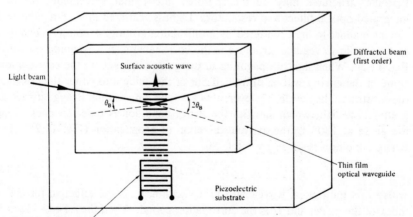

Fig. 11.25   An acousto-optic waveguide modulator. The device gives deflection of a light beam due to Bragg diffraction by surface acoustic waves.

a piezoelectric substrate (e.g. lithium niobate) onto the surface of which a thin film optical waveguide is formed by, for example, titanium indiffusion or lithium outdiffusion. An acoustic wave is launched parallel to the surface of the waveguide forming a surface acoustic wave (SAW) in which most of the wave energy is concentrated within a depth of one acoustic wavelength. The wave is generated from an interdigital electrode system comprising parallel electrodes deposited on the substrate. A light beam guided by the thin film waveguide interacts with the SAW giving beam deflection since both the light and the acoustic energy are confined to the same surface layer. The conditions for Bragg diffraction between the zero and first order mode are met when [Ref. 81]:

$$\sin \theta_B = \frac{\lambda_1}{2\Lambda} \qquad (11.14)$$

where θ_B is the angle between the light beam and the acoustic beam wavefronts, λ_1 is the wavelength of light in the thin film waveguide and Λ is the acoustic wavelength. In this case the light is deflected by $2\theta_B$ from its original path as illustrated in Fig. 11.25.

The fraction of the light beam deflected depends upon the generation efficiency and the width of the SAW, the latter also defining the interaction length for the device. Although diffraction efficiencies are usually low (no more than 20%), the diffracted on/off ratio can be very high. Hence these devices provide effective switches as well as amplitude or frequency modulators.

11.8.3 Periodic Structures for Filters and Injection Lasers

Periodic structures may be incorporated into planar waveguides to form integrated optical filters and resonators. Light is scattered in such a guide in a similar manner to light scattered by a diffraction grating. A common example of a periodic waveguide structure is the corrugated slab waveguide shown in Fig. 11.26. When light propagating in the guide impinges on the corrugation, some of the energy will be diffracted out of the guide into either the cover or the substrate. The device, however, acts as a one-dimensional Bragg diffraction grating, and light which satisfies the Bragg condition is reflected back along the guide at 180° to the original direction of propagation (Fig. 11.26). The Bragg condition is given by [Ref. 80]:

$$\lambda = 2n_e D \qquad (11.15)$$

where λ is the optical wavelength in a vacuum, n_e is the effective refractive index of the guide, and D is the corrugation period. When the reflected light is incident at an angle θ (Fig. 11.26) then:

$$n_e = n_1 \sin 2\theta \qquad (11.16)$$

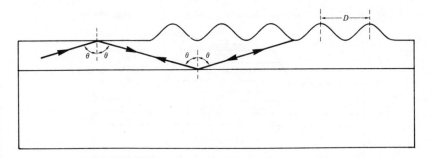

Fig. 11.26 A slab waveguide with surface corrugation giving reflection back along the guide when the Bragg condition is met. Hence the structure performs as a one dimensional Bragg diffraction grating.

where n_1 is the refractive index of the guide. Hence depending on the corrugation period of the structure all the incident power at a particular wavelength will be reflected. Such devices therefore behave as frequency selective rejection filters or mirrors. Narrow bandwidth filters with half power points separated by as little as 1 or 2Å have been realized. These devices may find use in applications such as wavelength demultiplexing. Alternatively, wide bandwidth filters may be obtained by forming gratings which exhibit a gradual change in the corrugation period.*

Integrated optical techniques are used in the fabrication of sources for optical fiber communications. They assisted in the development of the heterojunction and stripe geometry devices described in Chapter 6. The source, however, which is directly compatible with the planar waveguide structure is the single mode injection laser. In common with multimode lasers, these devices are fabricated from the group III–V semiconductor compounds (see Section 6.6). Furthermore, two single mode laser structures incorporate the corrugated gratings discussed above. In this application corrugated gratings are utilized to form a resonator within the device. The gratings are located at the end of the normal active layer of the laser to replace the cleaved end mirrors used in the Fabry–Perot type optical resonator. In this configuration they form the distributed Bragg reflector (DBR) structure illustrated in Fig. 11.27. This device displays the advantage of separating the perturbed regions from the active region but proves to be somewhat lossy due to optical absorption in the unpumped distributed reflectors. In order to avoid this problem, a grating may be applied over the whole active length of the laser where it provides what is known as distributed feedback. This configuration

---

* These gratings are said to have a chirped structure [Ref. 91].

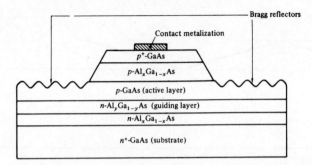

Fig. 11.27 Schematic cross section of a GaAs/AlGaAs DBR laser employing Bragg reflectors rather than cleaved end mirrors to provide optical feedback.

again dispenses with the use of cleaved mirrors. The distributed feedback (DFB) laser structure is shown in Fig. 6.30.

Both the DBR and DFB structures exhibit good longitudinal mode discrimination giving low frequency sensitivity to variations in drive current and hence temperature. Furthermore, these grating resonator lasers are particularly adapted to the construction of integrated optical assemblies in which the device and other components are fabricated on a single semi-conductor chip.

11.8.4 Bistable Optical Devices

Bistable optical devices have been under investigation for a number of years to provide a series of optical processing functions. These include optical logic and memory elements, power limiters and pulse shapers, and A–D converters. An optical device may be made bistable when the optical transmission within it is nonlinear and there is feedback of the optical output to control this transmission. The transfer characteristic for a typical bistable optical device (BOD) is illustrated in Fig. 11.28 where hysteresis may be observed. Bistable optical devices may be separated into two basic classes: all optical devices which utilize a nonlinear optical medium,* and hybrid devices in which an artificial nonlinearity such as an electro-optic medium is combined with an electronic feedback loop. Initial developments [Refs. 94 and 95] of the former devices

---

* The device which has attracted the greatest interest in this category is the nonlinear Fabry–Perot which consists of a medium with a nonlinear refractive index (e.g. sodium vapor, ruby crystal) inside a Fabry–Perot cavity. In this device, the value of the refractive index in the cavity, which is a function of the output light intensity, dictates the optical transmission giving high optical output on resonance and low optical output off resonance.

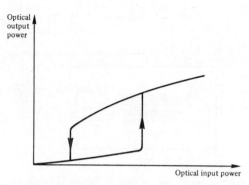
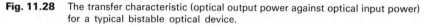

Fig. 11.28 The transfer characteristic (optical output power against optical input power) for a typical bistable optical device.

required extremely high optical power densities and therefore have limited application within IO.

Hybrid devices using electro-optically induced nonlinearities have been fabricated in integrated optical form. The configuration of such a device is shown in Fig. 11.29 [Ref. 96]. It consists of a titanium diffused optical waveguide on a lithium niobate substrate with cleaved and silvered end faces which form a Fabry–Perot resonator. The light emitted from the cavity is detected by an avalanche photodiode. The electrical signal is then fed back to the electrodes deposited on either side of the cavity. Thus the length of the cavity is controlled using the electro-optic effect. In such an interferometric structure, the ratio of input optical power to output light intensity is an

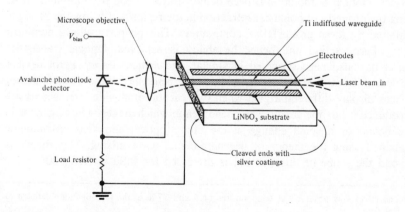

Fig. 11.29 A hybrid integrated bistable optical device [Ref. 96].

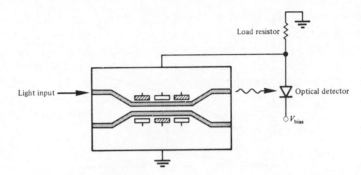

Fig. 11.30   A hybrid bistable optical device using an electro-optically switched directional coupler [Ref. 97].

oscillatory function of the effective cavity length. The device therefore exhibits hysteresis (Fig. 11.28) and bistability.

A hybrid BOD can also be achieved using an electro-optically switched directional coupler as illustrated in Fig. 11.30 [Ref. 97]. The device shown is fabricated from lithium niobate with a titanium indiffusion. Multisection electrodes deposited on the waveguides are connected so that there is a reversal of the electric field at each section in order to obtain low crosstalk switching. One of the output ports is connected to an optical detector which controls the drive voltage to the electrodes. Hence bistability is obtained without using a Fabry–Perot resonator. Furthermore remote optical switching of channels may be performed with this four-port device which could prove useful in optical fiber communication applications.

Recent investigations [Ref. 99] of optical bistability in semiconductors, may eventually produce bistable optical devices which will supersede the hybrid devices. Optical bistability has been demonstrated in cooled* InSb and GaAs using the property of nonlinear refraction in the region just below the bandgap exhibited by these group III–V compounds. This property allows nonlinear Fabry–Perot action and hence bistability is achieved. Optical transistor† action has also been observed in similar one-element Fabry–Perot devices fabricated from InSb. Thus in addition to bistability these semiconductor devices display differential gain when two carbon monoxide laser beams are introduced inside the cavity. The phenomenon which creates a large change in one beam for a small change in the other (optical amplification) may be explained using the theory of degenerate four-wave mixing. This theory is beyond the scope of this text but is discussed for InSb in Ref. 100.

---

* The effect was observed at temperatures of 5 and 77 K in the investigations detailed in Ref. 99.

† Called the transphasor by the authors of Ref. 99.

Unfortunately, the switching and amplifying effects in InSb and GaAs disappear at room temperature which severely limits the usefulness of these materials. Success at room temperature, however, has been reported by Bell Laboratories, USA, and Heriot Watt University, UK, using a derivative of gallium arsenide and thermal nonlinearities in zinc selenide (ZnSe) respectively [Ref. 101]. Work is therefore continuing towards the construction of fundamental logic gates which will provide the basis for integration of a number of elements into a single structure.

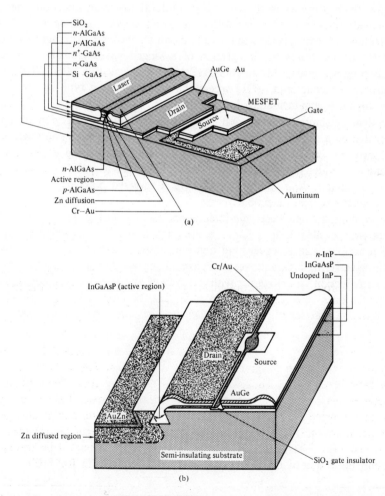

Fig. 11.31 Monolithic integrated transmitter circuits: (a) GaAs/AlGaAs injection laser fabricated with a MESFET on a GaAs substrate; (b) InGaAsP/InP injection laser fabricated with a MISFET on a semi-insulating InP substrate.

11.8.5   Optoelectronic Integration

The integration of interconnected optical and electronic devices is an important area of investigation for applications within optical fiber systems. Monolithic optoelectronic integrated circuits incorporating both optical sources and detectors have been successfully realized over the last few years. Monolithic integration for optical sources is exclusively confined to the use of group III–V semiconductor compounds. These materials prove useful as they possess both optical and electronic properties which can be exploited to produce high performance devices. Circuits are often fabricated from GaAs/AlGaAs for operation in the shorter wavelength region between 0.8 and 0.9 μm. Such a circuit is shown in Fig. 11.31(a) where an injection laser is fabricated on a GaAs substrate with a MESFET (metal-semiconductor FET, see Section 9.5) which is used to bias and modulate the laser. Alternatively Fig. 11.31(b) demonstrates the integration of a longer wavelength (1.1–1.6 μm) injection laser fabricated from InGaAsP/InP together with a MISFET (metal integrated-semiconductor FET) where the conventional n type substrate is replaced by a semi-insulating InP substrate.

Optical detectors for operation in the shorter wavelength region may be integrated on a silicon substrate. An early design of this type of photodetector is shown in Fig. 11.32 [Ref. 104]. Light is coupled into the device via a grating and guided into the silicon substrate at a point where the silicon dioxide layer terminates. It is in this area where a reversed biassed $p–n$ junction performs the optical detection.

More complex optoelectronic integration is shown in Fig. 11.33 [Ref. 105] where two possible designs for monolithic integrated circuits which serve as receive terminals in a wavelength division multiplex system are illustrated.

These wavelength demultiplexers utilize micrograting filters (either transmission or reflection type) together with an array of Schottky barrier photodiodes fabricated on a silicon substrate. In each case the filters pick out individual transmission wavelengths directing them to the appropriate photodiode for

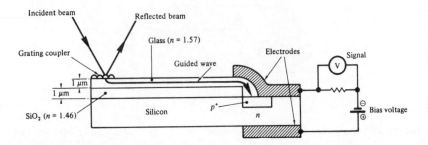

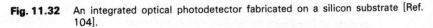

Fig. 11.32   An integrated optical photodetector fabricated on a silicon substrate [Ref. 104].

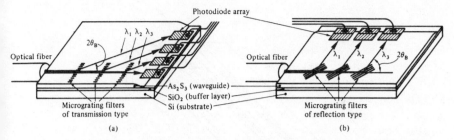

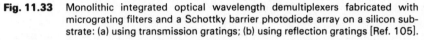

Fig. 11.33 Monolithic integrated optical wavelength demultiplexers fabricated with micrograting filters and a Schottky barrier photodiode array on a silicon substrate: (a) using transmission gratings; (b) using reflection gratings [Ref. 105].

detection. It is likely that integrated optical circuits of this type will find application in WDM systems in the near future.

Monolithic integration of both sources and detectors on the same substrate can be achieved using the group III–V semiconductor compounds. An

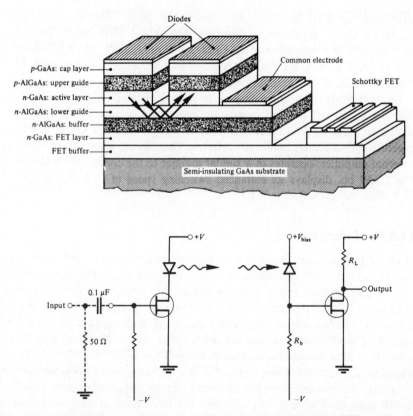

Fig. 11.34 Monolithic integration of an LED, *p–i–n* photodiode and an FET amplifier to provide an integrated transmitter, PIN–FET receiver or repeater [Ref. 106].

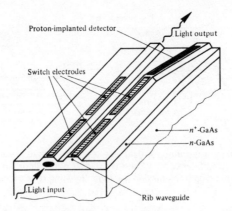

Fig. 11.35   Monolithic integration of a waveguide directional coupler and a detector to provide a hybrid bistable optical device [Ref. 107].

integrated optoelectronic test chip fabricated from AlGaAs on a semi-insulating GaAs substrate is shown in Fig. 11.34 [Ref. 106]. This chip demonstrates the integration of an LED, p–i–n photodiode, and FET together with resistive devices. It may be configured as a linearized source, a PIN–FET receiver, or as an integrated repeater.

Improvements in the performance of hybrid bistable optical devices have also been achieved through monolithic integration. Figure 11.35 [Ref. 107] shows an electro-optic directional coupler integrated on a GaAs substrate together with a proton implanted detector. This circuit, when utilized with an external amplifier, operates in a similar manner to the configuration shown in Fig. 11.29 but displays an enhanced switching speed of 1 µs and an optical switching energy less than 1 nJ.

11.8.6   Summary

Many of the functions provided by integrated optical devices and circuits for use in optical fiber communication systems cannot be fulfilled by other technologies without recourse to the electrical regime. Furthermore, hybrid integration of discrete optical circuit components fabricated on diverse substrates is already reasonably well established and capable of providing integrated optical devices for use with single mode fiber systems. It is, however, within monolithic integration using III–V semiconductor substrates, where sources and detectors together with all other types of both optical and electronic component may be combined, which holds most attraction for fourth generation optical fiber systems. The fabrication of integrated optical transmitters, receivers and repeaters onto a single chip which has already been demonstrated will doubtless soon find application within practical systems. Such integrated optical circuits may appear quite large in comparison to purely electronic integrated circuits due to limits on confinement and interaction length. Their speed of operation, however, displays no such limitation* and it is envisaged that monolithic integrated optical circuits will allow the enormous potential bandwidth of single mode optical fibers to be fully exploited.

In more general terms, developments in the area of optical bistability and hence integrated optical logic devices coupled with advances in monolithic integration have already instigated research towards the optical computer in both the UK and the USA [Ref. 101]. It is predicted [Ref. 109] that a prototype system could be in operation before 1993. The main drive for this development stems from the enhanced switching speeds possible using integrated optical technology. A subsidiary advantage is the ease of interfacing such a device with single mode optical fiber systems. Furthermore, there is likely to be a requirement for such high speed computers at the heart of the digital switching centers of what may become a predominantly optical fiber public telecommunications network.

REFERENCES

1 A. Hasegawa and Y. Kodama, 'Signal transmission by optical solitons in monomode fiber', *Proc. IEEE*, **69**(4), pp. 1145–1150, 1981.

2 A. D. Boardman and G. S. Cooper, 'Simulation of nonlinear pulse propagation in optical fibres', *Proc. SPIE Int. Soc. Opt. Eng. (USA)*, **374**, pp. 25–32, 1983.

3 C. P. Sandbank (Ed.), *Optical Fibre Communication Systems*, Chapter X, John Wiley, 1980.

4 C. J. Lilly, 'The application of optical fibres in the trunk network', *Telecomm. J. (Eng. edn.), Switzerland*, **49**(2), pp. 109–117, 1982.

5 P. E. Radley, 'Systems applications of optical fibre transmission', *Radio Electron. Eng. (IERE J.)*, **51**(7/8), pp. 377–384, 1981.

6 D. R. Smith, 'Advances in optical fibre communications', *Physics Bulletin*, **33**, pp. 401–403, 1982.

7 A. R. Beard, 'High capacity optical systems for trunk networks', *Proc. SPIE Int. Soc. Opt. Eng. (USA)*, **374**, pp. 102–110, 1983.

8 A. Javed, F. McAllum and G. Nault, 'Fibre optic transmission systems: the rationale and application', *Telesis 1981 Two (Canada)*, pp. 2–7, 1981.

9 P. Matthijsse, 'Essential data on optical fibre systems installed in various countries', *Telecomm. J. (Eng. edn.), Switzerland*, **49**(2), pp. 124–130, 1982.

10 O. Cottatellucci, F. Lombardi and G. Pellegrini, 'The application of optical fibres in the junction network', *Telecomm. J. (Eng. edn.), Switzerland*, **49**(2), pp. 101–108, 1982.

---

* The possible speed of operation of integrated optical devices surpasses the highest speeds achieved with current electronic integrated circuit technology by a factor of up to 1000, giving possible switching speeds of around 1 ps [Ref. 79].

11 J. E. Midwinter, 'Optical fiber communications system development in the UK', *IEEE Communications Magazine*, pp. 6–11, January, 1982.

12 Y. Tomaru, 'Optical fibre communication in Japan', *Proc. IREE Australia*, **40**(3), pp. 72–81, 1979.

13 C. K. Koa, 'Optical fibre communication in USA', *Proc. IREE Australia*, **40**(3), pp. 54–56, 1979.

14 G. Schweiger and H. Middel, 'The application of optical fibres in the local and rural networks', *Telecomm. J. (Eng. edn.), Switzerland*, **49**(2), pp. 93–101, 1982.

15 J. E. Midwinter, 'Potential broad-band services', *Proc. IEEE*, **68**(10), pp. 1321–1327, 1980.

16 G. Lentiz, 'The fibering of Biarritz', *Laser Focus (USA)*, **17**(11), pp. 124–128, 1981.

17 G. A. Trough, K. B. Harris, K. Y. Chang, C. I. Nisbet and J. F. Chalmers, 'An integrated fiber optics distribution field trial in Elie, Manitoba', *IEEE International Conference on Communication*, ICC 82, Philadelphia, PA, USA, 13–17 June 1982, p. 4D4/1–5 Vol. 2, IEEE, 1982.

18 W. E. Herold and H. Ohnsorge, 'Optical-fiber system with distributed access', *Proc. IEEE*, **68**(10), pp. 1309–1315, 1980.

19 G. Gara and R. Mills, 'The future: subscriber loop applications', *Telesis 1981 Two (Canada)*, pp. 36–38, 1981.

20 J. R. Fox, D. I. Fordham, R. Wood and D. J. Ahern, 'Initial experience with the Milton Keynes optical fiber cable TV trial', *IEEE Trans. Comm.*, **COM-30**(9), pp. 2155–2162, 1982.

21 W. K. Ritchie, 'Multi-service cable-television distribution systems', *British Telecomm. Eng.*, **1**(4), pp. 205–210, 1983.

22 J. R. Fox, 'Fibre optics in a multi-star wideband local network', *Proc. SPIE Int. Soc. Opt. Eng. (USA)*, **374**, pp. 84–90, 1983.

23 I. Yamashita, Y. Negishi, M. Nunokawa and H. Wakabayashi, 'The application of optical fibres in submarine cable systems', *Telecom. J. (Eng. edn.), Switzerland*, **49**(2), pp. 118–124, 1982.

24 P. Worthington, 'Design and manufacture of an optical fibre cable for submarine telecommunication systems', *Proceedings of Sixth European Conference on Optical Communication*, York, UK, pp. 347–356, 1980.

25 J. Irven, G. J. Cannell, K. C. Byron, A. P. Harrison, R. Worthington and J. G. Lamb, 'Single-mode fibers for submarine cable systems', *IEEE J. Quantum Electron.*, **QE-17**(6), pp. 907–910, 1981.

26 C. S. Grace and R. H. West, 'Nuclear radiation effects on fibre optics', *Proc. No 53, International Conference on Fibre Optics*, London, 1–2 March 1982, pp. 121–128, IERE, 1982.

27 J. G. Farrington and M. Chown, 'An optical fibre multiterminal data system for aircraft', *Fibre and Integrated Optics*, **2**(2), pp. 173–193, 1979.

28 J. J. Mayoux, 'Experimental "Gina" optical data bus in the Mirage 4000 aircraft', *Proc. No 53, International Conference on Fibre Optics*, London, 1–2 March 1982, pp. 147–156, IERE, 1982.

29 D. L. Williams, 'Military applications of fiber optics', *Proc. SPIE Int. Soc. Opt. Eng. (USA)*, **374**, pp. 138–142, 1983.

30 M. K. Barnoski, 'Fiber systems for the military environment', *Proc. IEEE*, **68**(10), pp. 1315–1320, 1980.

31 P. H. Bourne and D. P. M. Chown, 'The Ptarmigan optical fibre subsystem', *Proc. No 53, International Conference on Fibre Optics*, London, 1–2 March 1982, pp. 129–146, IERE, 1982.

32 J. M. Corvino and B. D. DeMarinis, 'Application of fiber optics to tactical communication systems', *Signal*, **34**(5), pp. 55–62, 1980.

33 J. D. Montgomery, 'Systems applications', in H. F. Wolf (Ed.), *Handbook of Fiber Optics: Theory and Applications*, pp. 367–376, Granada, 1979.

34 J. Gladenbeck, K. H. Nolting and G. Olejak, 'Optical fiber cable for overhead line systems', *Proceedings of Sixth European Conference on Optical Communication*, York, UK, pp. 359–362, 1980.

35 S. E. Miller, 'Potential applications', in S. E. Miller and A. G. Chynoweth (Eds.), *Optical Fiber Telecommunications*, pp. 675–683, Academic Press, 1979.

36 T. Nakahara, H. Kumanaru and S. Takeuchi, 'An optical fiber video system', *IEEE Trans. Commun.*, **COM-26**(7), pp. 955–961, 1978.

37 A. C. Deichmiller, 'Progress in fiber optics transmission systems for cable television', *IEEE Trans. Cable Television*, **CATV-5**(2), pp. 50–59, 1980.

38 S. R. Cole, 'Fiber optics for CATV applications', *International Conference on Communications*, pp. 38.3.1–4, IEEE, 1981.

39 E. A. Lacey, *Fiber Optics*, Chapter 1, Prentice-Hall, 1982.

40 P. Tolstrup Nielsen, B. Scharøe Petersen and H. Steffensen, 'A trunk system for CATV using optical transmission at 140 MB/S', *Proceedings of Sixth European Conference on Optical Communication*, York, UK, pp. 406–409, 1980.

41 M. Sekita, T. Kawamura, K. Ito, S. Fujita, M. Ishii and Y. Miyake, 'TV video transmission by analog, baseband modulation of a 1.3 μm-band laser diode', *Proceedings of Sixth European Conference on Optical Communication*, York, UK, pp. 394–397, 1980.

42 M. Kajino, K. Nishimura, F. Hayashida, T. Otsuka, K. Ito, H. Shiono and T. Yamada, 'A 4-channel WDM baseband video optical fiber system for monitoring of an automated guideway transit', *Proceedings of Sixth European Conference on Optical Communication*, York, UK, pp. 442–445, 1980.

43 A. C. Deichmiller and A. H. Kent, 'Fiber optics for CATV', *Optical Spectra*, **15**(9), pp. 42–51, 1981.

44 'GM plans 400-Hz fiber link for cars to cut costs, weight and interference', *Laser Focus*, pp. 60–62, January 1980.

45 J. D. Montgomery, 'Fiber optic applications and markets', *IEEE Trans. Commun.*, **COM-26**(7), pp. 1099–1102, 1978.

46 W. W. Brown, D. C. Hanson, T. Hornak and S. E. Garvey, 'System and circuit considerations for integrated industrial fiber optic data links', *IEEE Trans. Commun.*, **COM-26**(7), pp. 976–982, 1978.

47 D. A. A. Roworth, 'Fibre optics for industrial applications', *Optics and Laser Technology*, **12**(5), pp. 255–259, 1981.

48 P. B. Lyons, E. D. Hodson, L. D. Looney, G. Gow, L. P. Hocker, S. Lutz, R. Malone, J. Manning, M. A. Nelson, R. Selk and D. Simmons, 'Fiber optic application in nuclear testing', Electro Optics/Laser '79 Conf. Exposition, Anahiem, CA, 23–25 October, 1979.

49 W. F. Trover, 'Fiber optics for data acquisition and control communications: case histories', *Wire Technology*, **9**(2), pp. 79–87, 1981.

50 D. E. N. Davies and S. A. Kingsley, 'A novel optical fibre telemetry highway', *Proceedings of First European Conference on Optical Communication*, London, UK, 16–18 September 1975, pp. 165–167, IEE, 1975.

51 T. G. Giallorenzi, 'Fibre optic sensors', *Opt. Laser Technol. (GB)*, **13**(2), pp. 73–78, 1981.

52 D. H. McMahon, A. R. Nelson and W. B. Spillman Jr, 'Fibre-optic transducers', *IEEE Spectrum*, pp. 24–29, December 1981.

53 M. Gottlieb, G. B. Brandt and J. Butler, 'Measurement of temperature with optical fibres', *ISA Transactions*, **19**(4), pp. 55–63, 1980.

54 J. P. Dakin, 'Optical fibre sensors—principles and applications', *Proc. No 53, International Conference on Fibre Optics*, London, 1–2 March 1982, pp. 39–52, IERE, 1982.

55 B. E. Jones and R. C. Spooncer, 'Simple analogue and digital optical sensors', *Tech. Dig. IEE Colloquium on Optical Fibre Sensors*, pp. 7/1–5, May 1982.

56 A. Rogers, 'Measurement using fibre optics', *New Electronics (GB)*, pp. 29–36, 27 October 1981.

57 G. B. Hocker, 'Fiber-optic sensing of pressure and temperature', *Appl. Opt*, **18**(9), pp. 1445–1448, 1979.

58 T. G. Giallorenzi *et al.*, 'Optical fiber sensor technology', *IEEE J. Quantum Electron.*, **18**(4), pp. 626–666, 1982.

59 R. T. Murray, 'A family of grating sensors', *First International Conference on Optical Fibre Sensors*, London, 26–28 April 1983, pp. 114–116, IEE, 1983.

60 W. B. Spillman Jr and D. H. McMahon, 'Multimode fiber optic sensors', *First International Conference on Optical Fibre Sensors*, London, 26–28 April 1983, pp. 160–163, IEE, 1983.

61 I. P. Giles, B. Culshaw and D. E. N. Davies, 'Heterodyne optical fibre gyroscope', *First International Conference on Optical Fibre Sensors*, London, 26–28 April 1983, pp. 151–154, IEE, 1983.

62 K. Böhm, P. Marten, K. Petermann and E. Weidel, 'Fibre gyro for sensitive measurement of rotation', *Proc. No 53, International Conference on Fibre Optics*, London, 1–2 March 1982, pp. 31–38, IERE, 1982.

63 R. F. Cahill and E. Udd, 'Solid state phase-nulling optical gyro', *Appl. Opt.*, **19**(18), pp. 3054–3056, 1980.

64 A. J. Rogers, 'Optical fibre current measurement', *Proc. SPIE Int. Soc. Opt. Eng (USA)*, **374**, pp. 196–201, 1983.

65 J. Wilson and J. F. B. Hawkes, *Optoelectronics: An Introduction*, Chapter 3, Prentice-Hall, 1983.

66 J. D. Crow and M. W. Sachs, 'Optical Fibers for Computer Systems', *Proc. IEEE*, **68**(10), pp. 1275–1280, 1980.

67 J. A. Eibner, *Fiber Optics for Computer Applications*, Connector Symp. Proc., Vol 11, Cherry Hill, NJ, 1978.

68 C. P. Wyles, 'Fibre-optic multiplexing system for the IBM 3270 series', *Proc. SPIE. Int. Soc. Opt. Eng. (USA)*, **374**, pp. 78–83, 1983.

69 R. M. Metcalf and D. R. Boggs, 'Ethernet: distributed packet switching for local computer networks', *Commun. ACM*, **19**(7), p. 395, 1976.

70 E. G. Rawson and R. M. Metcalfe, 'Fibernet: multimode optical fibers for local computer networks', *IEEE Trans-Commun.*, **COM-26**(7), pp. 983–990, 1978.

71 S. Evans and J. Herman, 'Fiber optics successfully links microcomputers', *Digital Design (USA)*, **10**(4), pp. 36–37, 1980.

72 S. J. Healy, S. A. Kahn, R. L. Stewart and S. G. Tolchin, 'A fibre-optic local area communications network', *Johns Hopkins APL Tech. Dig (USA)*, **2**(2), pp. 84–86, 1981.

73 J. Dunphy, 'Local fibre optics', *Systems International*, pp. 21, 22, 24, 26, March 1982.

74 D. B. Anderson, *Optical and Electrooptical Information Processing*, pp. 221–234, MII Press, 1965.

75 S. E. Miller, 'Integrated optics: an introduction', *Bell Syst. Tech. J.*, **48**(7), pp. 2059–2069, 1969.

76 L. Levi, *Applied Optics*, Vol 2, Chapter 13, John Wiley, 1980.

77 H. Kogelnik and V. Ramaswamy, 'Scaling rules for thin-film optical waveguides', *Appl. Opt.*, **13**(8), pp. 1857–1862, 1974.

78 A. Reisinger, 'Attenuation properties of optical waveguides with a metal boundary', *Appl. Phys. Lett.*, **23**(5), pp. 237–239, 1973.

79 H. Kogelnik, 'Limits in integrated optics', *Proc. IEEE*, **69**(2), pp. 232–238, 1981.

80 T. Tamir (Ed.), *Integrated Optics* (2nd Edn.), Springer-Verlag, New York, 1979.

81 P. J. R. Laybourne and J. Lamb, 'Integrated optics: a tutorial review', *Radio Electron. Eng.* (*IERE J.*), **51**(7/8), pp. 397–413, 1981.

82 H. Sasaki and I. Anderson, 'Theoretical and experimental studies on active Y-junctions in optical waveguides', *IEEE J. Quantum Electron*, **QE-14**, pp. 883–892, 1978.

83 A. R. Billings, 'Integrated optics in later generation optical communication systems', *Proc. IREE Australia*, pp. 137–144, September 1979.

84 D. Marcuse, 'The coupling of degenerate modes in two parallel dielectric waveguides', *Bell Syst. Tech. J.*, **50**(6), pp. 1791–1816, 1971.

85 A. Yariv, 'Coupled mode theory for guided wave optics', *IEEE J. Quantum Electron.*, **QE-9**, pp. 919–933, 1973.

86 M. Papuchon, Y. Combemale, X. Mathieu, D. B. Ostrowsky, L. Reiber, A. M. Roy, B. Sejourne and M. Werner, 'Electrically switched optical directional coupler: COBRA', *Appl. Phys. Lett.*, **27**(5), pp. 289–291, 1975.

87 T. Kimura, 'Single-mode systems and components for longer wavelengths', *IEEE Trans. Circuits Syst.*, **CAS-26**, pp. 987–1010, 1979.

88 R. C. Alferness, 'Guided-wave devices for optical communication', *IEEE J. Quantum Electron.*, **QE-17**(6), pp. 946–959, 1981.

89 D. B. Ostrowsky, 'Optical waveguide components' in M. J. Howes and D. V. Morgan (Eds.), *Optical Fibre Communications*, pp. 165–188, John Wiley, 1980.

90 F. Auracher and R. Keil, 'Method for measuring the rf modulation characteristics of Mach-Zehnder-type modulators', *Appl. Phys. Lett.*, **36**, pp. 626–628, 1980.

91 A. Katzir, A. C. Livanos, J. B. Shellan and A. Yariv, 'Chirped gratings in integrated optics', *IEEE J. Quantum Electron.*, **QE-13**(4), pp. 296–304, 1977.

92 I. P. Kaminow and T. Li, 'Modulation techniques', in S. E. Miller and A. G. Chynoweth (Eds.), *Optical Fiber Telecommunications*, pp. 557–591, Academic Press, 1979.

93 A. Scudellari, 'Integrated optics devices', in Technical Staff of CSELT, *Optical Fibre Communication*, pp. 829–845, McGraw-Hill, 1981.

94 H. M. Givvs, S. L. McCall and T. N. C. Venkatesan, 'Differential gain and bistability using sodium filled Fabry-Perot interferometer', *Phys. Rev. Lett.*, **36**, pp. 1135–1138, 1976.

95 T. N. C. Venkatesan and S. L. McCall, 'Optical bistability and differential gain between 85 and 296°K in a Fabry–Perot containing ruby', *Appl. Phys. Lett.*, **30**, pp. 282–284, 1977.

96 P. W. Smith, I. P. Kaminow, P. J. Maloney and L. W. Stulz, 'Integrated bistable optical devices', *Appl. Phys. Lett.*, **33**(1), pp. 24–26, 1978.

97 P. S. Cross, R. V. Schmidt, R. L. Thornton and P. W. Smith, 'Optically controlled two channel integrated optical switch', *IEEE J. Quantum Electron.*, **QE-14**(8), pp. 577–580, 1978.

98 E. Garmire, 'Progress in bistable optical devices', *Proc. Soc. Photo-opto. Instrumen. Eng.*, **176**, pp. 12–16, April 1979.

99 D. A. B. Miller, S. D. Smith and C. T. Seaton, 'Optical bistability in semiconductors', *IEEE J. Quantum Electron.*, **QE-17**(3), pp. 312–317, 1981.

100 D. A. B. Miller, R. G. Harrison, A. M. Johnston, C. T. Seaton and S. D. Smith, 'Degenerate four-wave mixing in InSb at 5 K', *Opt. Commun.*, **32**, pp. 478–480, 1980.

101 'New optical computers are moving fast', *Technology*, **7**(13), p. 27, 1983.

102 H. Matsueda, S. Sasaki and M. Nakamura, 'GaAs optoelectronic integrated light sources', *J. Lightwave Technology*, **LT-1**(1), pp. 261–269, 1983.

103 'Monolithic integrated optoelectronics', *Electronic Product Design*, p. 9, January 1983.

104 D. B. Ostrowsky, R. Poirer, L. M. Reiber and C. Deverdun, 'Integrated optical photodetector', *Appl. Phys. Lett.*, **22**(9), pp. 463–464, 1973.

105 T. Suhara, Y. Hunda, H. Nishihara and J. Koyama, 'Monolithic integrated micrograting and photodiodes for wavelength demultiplexing', *Appl. Phys. Lett.*, **40**(2), pp. 120–122, 1982.

106 N. Forbes, A. C. Carter and R. C. Goodfellow, 'Monolithic integration of active and passive components in GaAlAs/GaAs multilayers', *Proceedings of First European Conference on Integrated Optics*, London, 14–15 Sept. 1981, pp. 83–84, IEE, 1981.

107 A. Carenco and L. Menigaux, 'Optical bistability using a directional coupler and detector monolithically integrated in GaAs', *Appl. Phys. Lett.*, **37**(10), pp. 880–882, 1980.

108 J. C. Campbell, A. G. Dentai, J. A. Copeland and W. S. Holden, 'Optical AND gate', *IEEE J. Quantum Electron.*, **QE-18**(6), pp. 992–995, 1982.

109 'It's no optical illusion, it's a computer', *Guardian* (London), 15 February, 1983.

Appendices

A. THE FIELD RELATIONS IN A PLANAR GUIDE

Let us consider an electromagnetic wave having an angular frequency ω propagating in the z direction with propagation vector (phase constant) β. Then as indicated in Section 2.3.2 the electric and magnetic fields can be expressed as:

$$\mathbf{E} = \text{Re}\{\mathbf{E}_0(x, y) \exp j(\omega t - \beta z)\} \tag{A1}$$

$$\mathbf{H} = \text{Re}\{\mathbf{H}_0(x, y) \exp j(\omega t - \beta z)\} \tag{A2}$$

For the planar guide the Cartesian components of $\mathbf{E}_0$ and $\mathbf{H}_0$ become:

$$\frac{\partial E_z}{\partial y} + j\beta E_y = -j\mu_r \mu_0 \omega H_x \tag{A3}$$

$$j\beta E_x + \frac{\partial E_z}{\partial x} = j\mu_r \mu_0 \omega H_y \tag{A4}$$

$$\frac{\partial E_y}{\partial x} - \frac{\partial E_x}{\partial y} = -j\mu_r \mu_0 \omega H_z \tag{A5}$$

$$\frac{\partial H_z}{\partial y} + j\beta H_y = j\omega \varepsilon_r \varepsilon_0 E_x \tag{A6}$$

$$-j\beta H_x - \frac{\partial H_z}{\partial x} = j\omega \varepsilon_r \varepsilon_0 E_y \tag{A7}$$

$$\frac{\partial H_y}{\partial x} - \frac{\partial H_x}{\partial y} = j\omega \varepsilon_r \varepsilon_0 E_z \tag{A8}$$

If we assume that the planar structure is an infinite film in the y–z plane, then for an infinite plane wave travelling in the z direction the partial derivative with respect to y is zero ($\partial/\partial y = 0$). Employing this assumption we can simplify the above equations to demonstrate fundamental relationships between the fields in such a structure. These are:

$$j\beta E_y = -j\mu_r \mu_0 \omega H_x \quad \text{(TE mode)} \tag{A9}$$

$$j\beta E_x + \frac{\partial E_z}{\partial x} = j\mu_r \mu_0 \omega H_y \quad \text{(TM mode)} \tag{A10}$$

$$\frac{\partial E_y}{\partial x} = -j\mu_r \mu_0 \omega H_z \quad \text{(TE mode)} \tag{A11}$$

$$j\beta H_y = j\omega\varepsilon_r \varepsilon_0 E_x \quad \text{(TM mode)} \tag{A12}$$

$$-j\beta H_x - \frac{\partial H_z}{\partial x} = j\omega\varepsilon_r \varepsilon_0 E_y \quad \text{(TE mode)} \tag{A13}$$

$$\frac{\partial H_y}{\partial x} = j\omega\varepsilon_r \varepsilon_0 E_z \quad \text{(TM mode)} \tag{A14}$$

It may be noted that the fields separate into TE and TM modes corresponding to coupling between E_y, H_x, H_z, $(E_z = 0)$ and H_y, E_x, E_z $(H_z = 0)$ respectively.

B. VARIANCE OF A RANDOM VARIABLE

The statistical mean (or average) value of a discrete random variable X is the numerical average of the values which X can assume weighted by their probabilities of occurrence. For example, if we consider the possible numerical values of X to be x_1, $x_2, \ldots x_i$, with probabilities of occurrence $P(x_1)$, $P(x_2) \ldots P(x_i)$, then as the number of measurements N of X goes to infinity, it would be expected that the outcome $X = x_1$ would occur $NP(x_1)$ times, the outcome $X = x_2$ would occur $NP(x_2)$ times and so on. In this case the arithmetic sum of all N measurements is:

$$x_1 P(x_1)N + x_2 P(x_2)N + \ldots x_i P(x_i)N = N\sum_i x_i P(x_i) \tag{B1}$$

The mean or average value of all these measurements which is equivalent to the mean value of the random variable may be calculated by dividing the sum in Eq. (B1) by the number of measurements N. Furthermore, the mean value for the random variable X which can be denoted as $\overline{X}$ (or m) is also called the expected value of X and may be represented by $E(X)$. Hence:

$$\overline{X} = m = E(X) = \sum_{i=1}^{N} x_i P(x_i) \tag{B2}$$

Moreover Eq. (B2) also defines the first moment of X which we denote as M_1. In a similar manner the second moment M_2 is equal to the expected value of X^2 such that:

$$M_2 = \sum_{i=1}^{N} x_i^2 P(x_i) \tag{B3}$$

M_2 is also called the mean square value of X which may be denoted as $\overline{X^2}$.

For a continuous random variable, the summation of Eq. (B2) approaches an integration over the whole range of X so that the expected value of X:

$$M_1 = E(X) = \int_{-\infty}^{\infty} x p_X(x)\, dx \tag{B4}$$

where $p_X(x)$ is the probability density function of the continuous random variable X. Similarly, the expected value of X^2 is given by:

$$M_2 = E(X^2) = \int_{-\infty}^{\infty} x^2 p_X(x)\, dx \tag{B5}$$

It is often convenient to subtract the first moment $M_1 = m$ prior to computation of the second moment. This is analogous to moments in mechanics which are referred to the center of gravity rather than the origin of the coordinate system. Such a moment is generally referred to as a central moment. The second central moment represented by the symbol σ^2 is therefore defined as:

$$\sigma^2 = E[(X - m)^2] = \int_{-\infty}^{\infty} (x - m)^2 p_X(x)\, dx \tag{B6}$$

where σ^2 is called the variance of the random variable X. Moreover the quantity σ which is known as the standard deviation is the root mean square (rms) value of $(X - m)$.

Expanding the squared term in Eq. (B6) and integrating term by term we find:

$$\begin{aligned}
\sigma^2 &= E[X^2 - 2mX + m^2] \\
&= E(X^2) - 2mE(X) + E(m^2) \\
&= E(X^2) - 2m^2 - m^2 \\
&= E(X^2) - m^2
\end{aligned} \tag{B7}$$

As $E(X^2) = M_2$ and $m = M_1$, the variance may be written as:

$$\sigma^2 = M_2 - (M_1)^2$$

C. VARIANCE OF THE SUM OF INDEPENDENT RANDOM VARIABLES

If a random variable $W = g(X, Y)$ is a function of two random variables X and Y, then extending the definition in Eq. (B4) for expected values gives the expected value of W as:

$$E(W) = \int_{-\infty}^{\infty} \int_{-\infty}^{\infty} g(x, y) p_{XY}(x, y)\, dx\, dy \tag{C1}$$

where $p_{XY}(x, y)$ is the joint probability function. Furthermore the two random variables X and Y are statistically independent when:

$$p_{XY}(x, y) = p_X(x) p_Y(y) \tag{C2}$$

Now let X and Y be two statistically independent random variables with variances σ_X^2 and σ_Y^2 respectively. In addition we assume the sum of these random variables to be another random variable denoted by Z such that $Z = X + Y$, where Z has a variance σ_Z^2. If the mean values of X and Y are zero, employing the definition of variance given in Eq. (B6) together with the expected value for a function of two random variables (Eq. (C1)) we can write:

$$\sigma_Z^2 = \int_{-\infty}^{\infty} \int_{-\infty}^{\infty} (x + y)^2 p_{XY}(x, y) \, dx \, dy \tag{C3}$$

As X and Y are statistically independent we can utilize Eq. (C2) to obtain:

$$\sigma_Z^2 = \int_{-\infty}^{\infty} \int_{-\infty}^{\infty} (x + y)^2 p_X(x) p_Y(y) \, dx \, dy$$

$$= \int_{-\infty}^{\infty} x^2 p_X(x) \, dx + \int_{-\infty}^{\infty} y^2 p_Y(y) \, dy + 2 \int_{-\infty}^{\infty} x p_X(x) \, dx \int_{-\infty}^{\infty} y p_Y(y) \, dx \tag{C4}$$

The two factors in the last term of Eq. (C4) are equal to the mean values of the random variables (X and Y) and hence are zero. Thus:

$$\sigma_Z^2 = \sigma_X^2 + \sigma_Y^2$$

D. SPEED OF RESPONSE OF A PHOTODIODE

Three main factors limit the speed of response of a photodiode. These are:

(a) *Drift time of carriers through the depletion region*
   The speed of response of a photodiode is fundamentally limited by the time it takes photogenerated carriers to drift across the depletion region. When the field in the depletion region exceeds a saturation value then the carriers may be assumed to travel at a constant (maximum) drift velocity v_d. The longest transit time, t_{drift}, is for carriers which must traverse the full depletion layer width w and is given by

$$t_{drift} = \frac{w}{v_d} \tag{D1}$$

A field strength above $2 \times 10^4 \, V \, cm^{-1}$ in silicon gives maximum (saturated) carrier velocities of approximately $10^7 \, cm \, s^{-1}$. Thus the transit time through a depletion layer width of 10 μm is around 0.1 ns.

(b) *Diffusion time of carriers generated outside the depletion region*
   Carrier diffusion is a comparatively slow process where the time taken, t_{diff}, for carriers to diffuse a distance d may be written as

$$t_{diff} = \frac{d^2}{2D_c} \tag{D2}$$

where D_c is the minority carrier diffusion coefficient. For example, the hole diffusion time through 10 μm of silicon is 40 ns whereas the electron diffusion time over a similar distance is around 8 ns.

(c) *Time constant incurred by the capacitance of the photodiode with its load*
A reversed biassed photodiode exhibits a voltage dependent capacitance caused by the variation in the stored charge at the junction. The junction capacitance C_j is given by

$$C_j = \frac{\varepsilon_s A}{w} \tag{D3}$$

where ε_s is the dielectric constant of the semiconductor material and A is the diode junction area. Hence, a small depletion layer width w increases the junction capacitance. The capacitance of the photodiode C_d is that of the junction together with the capacitance of the leads and packaging. This capacitance must be minimized in order to reduce the RC time constant which also limits the detector response time (see Section 9.3.2).

E. CLOSED LOOP TRANSFER FUNCTION FOR THE TRANSIMPEDANCE AMPLIFIER

The close loop transfer function $H_{CL}(\omega)$ for the transimpedance amplifier shown in Fig. 9.9 may be derived by summing the currents at the amplifier input, remembering that the amplifier input resistance is included in R_{TL}. Hence,

$$i_{det} + \frac{V_{out} - V_{in}}{R_f} = V_{in} \left(\frac{1}{R_{TL}} + j\omega C_T \right) \tag{E1}$$

As $V_{in} = -V_{out}/G$, then

$$i_{det} = -V_{out} \left(\frac{1}{R_f} + \frac{1}{GR_f} + \frac{1}{GR_{TL}} + \frac{j\omega C_T}{G} \right) \tag{E2}$$

Therefore,

$$H_{CL}(\omega) = \frac{V_{out}}{i_{det}} = \frac{-R_f}{1 + (1/G) + (R_f/GR_{TL}) + (j\omega C_T R_f/G)}$$

$$= \frac{-R_f/(1 + 1/G + R_f/GR_{TL})}{[1 + j\omega C_T R_f/(1 + R_f/R_{TL} + G)]} \tag{E3}$$

Since,

$$G \gg \left(1 + \frac{R_f}{R_{TL}} \right) \tag{E4}$$

then Eq. (E3) becomes,

$$H_{CL}(\omega) \simeq \frac{-R_f}{1 + (j\omega R_f C_T/G)} \quad \text{VA}^{-1}$$

F. GAUSSIAN PULSE RESPONSE

Many optical fibers, and in particular jointed fiber links, exhibit pulse outputs with a temporal variation that is closely approximated by a Gaussian distribution. Hence the variation in the optical output power with time may be described as:

$$P_o(t) = \frac{1}{\sqrt{(2\pi)}} \exp - \left(\frac{t^2}{2\sigma^2} \right) \tag{F1}$$

where σ and σ^2 are the standard deviation and the variance of the distribution respectively. If t_e represents the time at which $P_o(t_e)/P_o(0) = 1/e$ (i.e. $1/e$ pulse width), then from Eq. (F1) it follows that:

$$t_e = \sigma\sqrt{2}$$

Moreover, if the full width of the pulse at the $1/e$ points is denoted by τ_e then:

$$\tau_e = 2t_e = 2\sigma\sqrt{2}$$

In the case of the Gaussian response given by Eq. (F1) the standard deviation σ is equivalent to the rms pulse width.

The Fourier transform of Eq. (F1) is given by:

$$P(\omega) = \frac{1}{\sqrt{(2\pi)}} \exp - \left(\frac{\omega^2 \sigma^2}{2} \right) \tag{F2}$$

The 3 dB optical bandwidth B_{opt} is defined in Section 7.4.3 as the modulation frequency at which the received optical power has fallen to one half of its constant value. Thus using Eq. (F2):

$$\frac{|\omega(3 \text{ dB opt})|^2}{2} \sigma^2 = 0.693$$

and

$$\omega(3 \text{ dB opt}) = 2\pi B_{opt} = \frac{\sqrt{2} \times 0.8326}{\sigma}$$

Hence

$$B_{opt} = \frac{\sqrt{2} \times 0.8326}{2\pi\sigma} = \frac{0.530}{\tau_e} = \frac{0.187}{\sigma} \quad \text{Hz}$$

When employing return to zero pulse where the maximum bit rate $B_T(\text{max}) = B_{opt}$, then:

$$B_T(\text{max}) \simeq \frac{0.2}{\sigma} \text{ bit s}^{-1}$$

Alternatively, the 3 dB electrical bandwidth B occurs when the received optical power has dropped to $1/\sqrt{2}$ of the constant value (see Section 7.4.3) giving:

$$B = \frac{0.530}{\tau_e \sqrt{2}} = \frac{0.375}{\tau_e} = \frac{0.133}{\sigma} \quad \text{Hz}$$

Index